Lecture Notes in Physics

Bisher erschienen/Already published

Vol. 1: J. C. Erdmann. Wärmeleitung in Kristallen, theoretische Grundlagen und fortgeschrittene experimentelle Methoden. II, 283 Seiten. 1969.

Vol. 2: K. Hepp, Théorie de la renormalisation. III, 215 pages. 1969.

Vol. 3: A. Martin, Scattering Theory: Unitarity, Analyticity and Crossing. IV, 125 pages. 1969.

Vol. 4: G. Ludwig, Deutung des Begriffs „physikalische Theorie" und axiomatische Grundlegung der Hilbertraumstruktur der Quantenmechanik durch Hauptsätze des Messens. 1970. Vergriffen.

Vol. 5: Schaaf, The Reduction of the Product of Two Irreducible Unitary Representations of the Proper Orthochronous Quantummechanical Poincare Group. IV, 120 pages. 1970.

Vol. 6: Group Representations in Mathematics and Physics. Edited by V. Bargmann. V, 340 pages. 1970.

Vol. 7: R. Balescu, J. L. Lebowitz, I. Prigogine, P. Résibois, Z. W. Salsburg, Lectures in Statistical Physics. V, 181 pages. 1971.

Vol. 8: Proceedings of the Second International Conference on Numerical Methods in Fluid Dynamics. Edited by M. Holt. 1971. Out of print.

Vol. 9: D. W. Robinson, The Thermodynamic Pressure in Quantum Statistical Mechanics. V, 115 pages. 1971.

Vol. 10: J. M. Stewart, Non-Equilibrium-Relativistic Kinetic Theory. III, 113 pages. 1971.

Vol. 11: O. Steinmann, Pertubation Expansions in Axiomatic Field Theory. III, 126 pages. 1976.

Vol. 12: Statistical Models and Turbulence. Edited by C. Van Atta and M. Rosenblatt. Reprint of the First Edition. VIII, 492 pages. 1975.

Vol. 13: M. Ryan, Hamiltonian Cosmology. VII, 169 pages. 1972.

Vol. 14: Methods of Local and Global Differential Geometry in General Relativity. Edited by D. Farnsworth, J. Fink, J. Porter, and A. Thompson. V, 188 pages.

Vol. 15: M. Fierz, Vorlesungen zur Entwicklungsgeschichte der Mechanik. V, 97 Seiten. 1972.

Vol. 16: H.-O. Georgii, Phasenübergang 1. Art bei Gittergasmodellen. IX, 167 Seiten. 1972.

Vol. 17: Strong Interaction Physics. Edited by W. Rühl and A. Vancura. V, 405 pages. 1973.

Vol. 18: Proceedings of the Third International Conference on Numerical Methods in Fluid Mechanics, Vol. I. Edited by H. Cabannes and R. Temam. VII, 186 pages. 1973.

Vol. 19: Proceedings of the Third International Conference on Numerical Methods in Fluid Mechanics, Vol. II. Edited by H. Cabannes and R. Temam. VII, 275 pages. 1973.

Vol. 20: Statistical Mechanics and Mathematical Problems. Edited by A. Lenard. VIII, 247 pages. 1973.

Vol. 21: Optimization and Stability Problems in Continuum Mechanics. Edited by P. K. C. Wang. V, 94 pages. 1973.

Vol. 22: Proceedings of the Europhysics Study Conference on Intermediate Processes in Nuclear Reactions. Edited by N. Cindro, P. Kulišic and Th. Mayer-Kuckuk. XIV, 329 pages. 1973.

Vol. 23: Nuclear Structure Physics. Proceedings 1973. Edited by U. Smilansky, I. Talmi, and H. A. Weidenmüller. XII, 296 pages. 1973.

Vol. 24: R. F. Snipes, Statistical Mechanical Theory of the Electrolytic Transport of Non-electrolytes. V, 210 pages. 1973.

Vol. 25: Constructive Quantum Field Theory. The 1973 "Ettore Majorana" International School of Mathematical Physics. Edited by G. Velo and A. Wightman. III, 331 pages. 1973.

Vol. 26: A. Hubert, Theorie der Domänenwände in geordneten Medien. XII, 377 Seiten. 1974.

Vol. 27: R. K. Zeytounian, Notes sur les Ecoulements Rotationnels de Fluides Parfaits. XIII, 407 pages. 1974.

Vol. 28: Lectures in Statistical Physics. Edited by W. C. Schieve and J. S. Turner. V, 342 pages. 1974.

Vol. 29: Foundations of Quantum Mechanics and Ordered Linear Spaces. Advanced Study Institute, Marburg 1973. Edited by A. Hartkämper and H. Neumann. VI, 355 pages. 1974.

Vol. 30: Polarization Nuclear Physics. Proceedings 1973. Edited by D. Fick. IX, 292 pages. 1974.

Vol. 31: Transport Phenomena. Sitges International Schools of Statistical Mechanics, June 1974. Edited by G. Kirczenow and J. Marro. XIV, 517 pages. 1974.

Lecture Notes in Physics

Edited by J. Ehlers, München, K. Hepp, Zürich,
H. A. Weidenmüller, Heidelberg, and J. Zittartz, Köln
Managing Editor: W. Beiglböck, Heidelberg

50

Group Theoretical Methods in Physics

Fourth International Colloquium,
Nijmegen 1975

Edited by A. Janner, T. Janssen, and M. Boon

Springer-Verlag
Berlin Heidelberg GmbH 1976

Editors
Prof. A. Janner
Dr. T. Janssen
Dr. M. Boon
Instituut voor Theoretische Fysica
Katholieke Universiteit Nijmegen
Toernooiveld
Nijmegen/The Netherlands

Library of Congress Cataloging in Publication Data
Main entry under title:

Group theoretical methods in physics.

(Lecture notes in physics ; 50)
Proceedings of the 4th of a series of meetings; held at the Faculty of Science of the University of Nijmegen in 1975.
Includes index.
1. Groups, Theory of--Congresses. 2. Mathematical physics--Congresses. I. Janner, A., 1928- II. Janssen, T. III. Boon, Michael. IV. Series.
QC20.7.G76G76 530.1'5'222 76-18990

ISBN 978-3-540-07789-3 ISBN 978-3-540-38252-2 (eBook)
DOI 10.1007/978-3-540-38252-2

Originally published by Springer-Verlag Berlin Heidelberg New York in 1976

Introduction

The series of colloquia devoted to Group Theoretical Methods in Physics of which the present one is the fourth is based on the idea of H. Bacry of bringing together physicists and mathematicians working in quite different fields but using basically the same group theoretical methods.

The first and third of these colloquia were organized by him and by A. Grossmann at the Centre de Physique Théorique CNRS in Marseilles in 1972 and in 1974, respectively. The second one (in 1973) and the fourth one (in 1975) have been held at the Faculty of Science of the University of Nijmegen. The fifth colloquium will take place in Montreal in 1976 and the sixth one is planned for 1977 in Tübingen.

The proceedings of the first colloquium were published as a joint report of the University of Provence, the University of Aix-Marseilles and of the CNRS. Those of the second and third ones were printed at the Faculty of Science of the University of Nijmegen.

The present proceedings contain invited and contributed papers presented at the fourth colloquium, in June 1975 in Nijmegen. Two elements emerged from the organization of the meeting: The important role played by the advisory committee and the use of poster sessions. The last indeed turned out to be an excellent way of achieving scientific communication between the participants without having to impose any severe limitations on the number of papers presented.

The need for an advisory committee is based on the terrific range of subjects where group theory can be usefully applied, the fact of which the reader of these proceedings can easily convince himself. So one finds papers on elementary particles, on nuclear, molecular, atomic and solid state physics; on classical and on quantum mechanics; on field theory, on statistical physics and, of course, on mathematical physics. It is precisely because of this extreme variety that recurrent colloquia of this type can easily generate a feeling of monotony. In order to reduce this danger without sacrificing one of the basic goals of the colloquium (that of bringing together people working in different branches of physics) the advisory committee proposed a "two dimensional approach" according to which, besides the "dimension" represented by the various fields of application, there is also the one occupied by special topics. These last, chosen in order to focus attention on important developments, should at the same time give to the meeting its specific flavour. Accordingly, in most of the plenary sessions people invited by the various members of the advisory committee lectured on one of the following topics chosen for this fourth colloquium: gauge groups, geometric quantization, broken symmetries, coherent states and recent developments in atomic and nuclear physics.

Gauge groups play an important role in the unification process that is being attempted in the description of all known fundamental interactions: the electromagnetic, the weak, the strong and possibly the gravitational interactions. Thus, if magnetic monopoles exist, their gauge group properties would also be relevant and important implications would follow for this unified approach. Independently, and closely connected with gauge invariant Lagrangian formulations, are attempts to unite particles obeying different statistics (like bosons and fermions) by introducing the concept of super-symmetry group where they appear together in the same irreducible multiplets.

The conceptual and group theoretical importance of geometric quantization goes far beyond the implication of the mere title. Indeed geometric quantization is not only relevant for analyzing the relations between corresponding classical and quantum mechanical systems, but also for the theory of irreducible representations of Lie groups (and quite generally for the algebraic approach in representation theory); for the study of dynamical groups, and of coherent states associated to a Lie group; for the twistor theory used in general relativity; for the study of partial differential equations; in graded Lie algebras; and for the characterization of physical systems which are elementary with respect to a given symmetry group. A selection of contributions which show that this is indeed the case, can be found in these proceedings.

Broken symmetry is a concept which also appears in various domains of physics ranging from the classification of elementary particles to the description of phase transitions in solid state physics.

The definition of coherent states has become, in the progress of time, such a general one (loosely speaking, coherent states are all those generated from a given one by the operations of a Lie group within an irreducible representation) that one feels the need for constraints ensuring some characteristic common physical property of these states. It may be, however, that the above generality is required for ensuring developments on the mathematical level.

The progress in Nuclear Physics and Atomic Physics reported in the plenary sessions concern the dynamical problem of many particle systems approached by means of a group theoretical analysis of the relevant spectra and of the matrix elements of certain many-particle operators defining the Hamiltonian. The review talk devoted to atomic physics was also relevant for solid state physics.

To try to summarize in any way adequately the papers printed here would be a very difficult task, and the editors apologize for not attempting this. They therefore conclude this introduction by thanking the members of the Advisory Committee, who helped the organization committee in such an efficient way: Professor H. Bacry and professor A. Grossmann, Marseilles; Prof. S. Feneuille, Orsay; Prof. P. Kramer, Tübingen; Prof. L. Michel, Bures-sur-Yvette; Prof. L. O'Raifeartaigh and prof. D. Simms, Dublin; prof. J. de Swart, dr. M. Boon and dr. T. Janssen, Nijmegen. The multiple tasks of the secretariat were accomplished by Marianne Alstadt in an excellent manner. Thanks are also expressed to all the members of the central service of the Faculty of Science, who gave their help before, during and after this colloquium, in particular mr. P.J.M. Toll LLM.

Gratitude is expressed to dr. C.J.M. Aarts, Director of the Faculty of Science: without his continuous interest and support, the organization of this colloquium in Nijmegen would not have been possible.

The financial support provided by the "Ministerie van Onderwijs en Wetenschappen", by the "Nijmeegs Universiteitsfonds" and by the "Faculteit der Wiskunde en Natuurwetenschappen" is gratefully acknowledged.

Thanks to the efforts of professor P. Kramer as intermediary and to the Springer Publishing Company, these proceedings are made commercially available, and so concretizing the wishes of the Advisory Committee.

The Editors

A. Janner T. Janssen and M. Boon

Nijmegen, March 1976

The progress in Nuclear Physics and Atomic Physics have led to the [illegible] questions about the dynamical behavior of many-particle systems approached by means of a common theoretical analysis of the relevant aspects [illegible] of the materials [illegible] in many-particle systems [illegible] the [illegible]. The [illegible] was devoted to [illegible] [illegible] was also [illegible] for [illegible]

[illegible]

To try to [illegible] in an [illegible] the [illegible] would be a very [illegible] task, and the [illegible] [illegible] [illegible]. They [illegible] [illegible] by [illegible] the [illegible] of the [illegible] Committee, who helped the organization committee in such an excellent way: Professor [illegible] [illegible], [illegible] [illegible], [illegible] [illegible], Prof. [illegible], [illegible], Prof. A. [illegible], [illegible], Prof. [illegible] [illegible], Prof. [illegible] [illegible], Prof. [illegible] and [illegible] [illegible]. The [illegible] of the [illegible] were [illegible] by [illegible] [illegible] [illegible] [illegible] and [illegible] [illegible] in all the [illegible] of the [illegible] [illegible], [illegible] and gave their help before, during and after the colloquium, in particular [illegible].

Gratitude is expressed to Mr. [illegible], [illegible] of the [illegible] of Science [illegible] [illegible] [illegible] [illegible] [illegible] [illegible] would not have been possible.

The financial support provided by the [illegible] was [illegible] [illegible] by the [illegible], and by the [illegible] [illegible] [illegible] [illegible].

Thanks [illegible] Professor [illegible], as [illegible] and to the [illegible] [illegible] [illegible] [illegible] and [illegible] [illegible] [illegible].

The Editors

A. [illegible], [illegible] [illegible], [illegible] [illegible], [illegible]

[illegible], [illegible] [illegible]

C O N T E N T S

GAUGE GROUPS. ELEMENTARY PARTICLES.

GEOMETRIC QUANTIZATION

---

* plenary sessions

CLASSICAL MECHANICS, QUANTUM MECHANICS, FIELD THEORY, STATISTICAL MECHANICS

EDUCATIONAL

LIST OF PARTICIPANTS

Backhouse, N.B.	University of Liverpool, England
Bacry, H.	C.N.R.S. Marseille, France
Balachandran, A.	Syracuse University, U.S.A.
Beckers, J.J.	Université de Liège, Belgium
Bleuler, K.	University of Bonn, Germany
Bloore, F.J.	University of Liverpool, England
Bohm, A.R.	University of Texas, U.S.A.
Boon, M.	University of Nijmegen, the Netherlands
Bose, A.K.	Université de Montréal, Canada
Boyer, C.P.	University of Mexico, Mexico
Božović, I.B.	University of Belgrade, Yugoslavia
Broek, P. v.d.	University of Nijmegen, the Netherlands
Brunet, M.	University of Tübingen, Germany
Burdet, G.M.	Université de Dijon, France
Butzal, H.D.	University of Regensburg, Germany
Cattaneo, U.G.	University of Kaiserslautern, Germany
Ciccariello, S.	University of Padua, Italy
Combe, Ph.	C.N.R.S. Marseille, France
Cracknell, A.P.	University of Dundee, Scotland
Cronström, C.	University of Helsinki, Finland
Dabboussi, O.B.	University of Petroleum & Minerals, Saudi Arabia
Dalgaard, E.D.	University of Aarhus, Danmark
Dam, H. v.	University of North Carolina, U.S.A.
Dembinski, S.T.	University of British Columbia, Canada
Derome, J.	Canada
Dirl, R.	Technische Hochschule Wien, Austria
Dowker, J.S.	University of Manchester, England
Drechsler, W.	Max-Planck Institute München, Germany
Dreizler, R.M.	Johann Wolfgang Goethe-Universität, Frankfurt/M. Germ.
Drühl, K.	Max-Planck Institute Starnberg, Germany
Dumont-Lepage, M.C.	Facultés Universitaires de Namur, Belgium
Feneuille, S.	Laboratoire Aimé Cotton, Orsay, France
Fritzer, H.	Technische Hochschule Graz, Austria
Garber, W.	University of Göttingen, Germany
Gard, P.	University of Cambridge, England
Gazeau, J.P.	Université de Paris VI, France
Georgelin, Y.	Orsay, France
Giovannini, N.	University of Nijmegen, the Netherlands
Greub, W.	University of Toronto, Canada
Grossmann, A.	C.N.R.S. Marseille, France
Gürsey, F.	I.H.E.S. Bures sur Yvette, France
Hintermann, A.	Ecole Polytechnique Lausanne, Switzerland
Hoek, J.	University of Amsterdam, the Netherlands
Hongoh, M.	Université de Montreal, Canada
Hooft, G. 't	University of Utrecht, the Netherlands
Hoogland, H.	University of Nijmegen, the Netherlands
Hurt, N.E.	Cleveland State University, U.S.A.
Iliopoulos, J.	Ecole Normale Superieure Paris, France
Janner, A.	University of Nijmegen, the Netherlands

Janssen, T.	University of Nijmegen, the Netherlands
Jaspers, M.	Université de Liège, Belgium
Jasselette, P.	Université de Liège, Belgium
John, G.	University of Tübingen, Germany
Joseph, T.	I.H.E.S. Bures sur Yvette, France
Judd, B.R.	John Hopkins University, Baltimore, U.S.A.
Kasperkovitz, P.	Technische Hochschule Wien, Austria
Kéhil-Quesne, C.	Université Libre de Bruxelles, Belgium
Kerner, R.	Université de Paris VI, France
King, R.C.	Southampton University, England
Klein, D.J.	University of Texas, U.S.A.
Kostant, B.	M.I.T. Cambridge, U.S.A.
Kramer, P.	University of Tübingen, Germany
Künzle, H.P.	University of Alberta, Canada
Lévy-Leblond, J.M.	Université de Paris VII, France
Lohe, M.A.	Duke University, Durham, U.S.A.
Louck, J.D.	University of California, Los Alamos, U.S.A.
Madore, J.A.	Institut Henri Poincaré Paris, France
Maksymowicz, A.	Western Washington State College, U.S.A.
Marmo, G.	University of Naples, Italy
McCarthy, P.J.	University of Cambridge, England
McGlynn,	
Michel, L.	I.H.E.S. Bures sur Yvette, France
Michelson, A.M.	Technion, Haifa, Israel
Mooren, L.C.	University of Technology Eindhoven, the Netherlands
Moshinsky, M.	University of Mexico, Mexico
Mozrzymas, J.	University of Wroclaw, Poland
Mundzeck, G.J.	University of Hamburg, Germany
Niederer, U.H.	University of Zürich, Switzerland
Nottrot, R.	University of Technology, Twente, the Netherlands
Onofri, E.O.	University of Parma, Italy
Opechowski, W.	University of British Columbia, Canada
O'Raifeartaigh, L.	Dublin Institute for Advanced Studies, Ireland
Parravicini, G.	Dublin Institute for Advanced Studies, Ireland
Pascolini, A.	University of Padua, Italy
Patera, J.	Université de Montreal, Canada
Perrin, M.J.	Université de Dijon, France
Perroud, M.A.	Université de Montreal, Canada
Pommaret, J.F.	Ecole Polytechnique Boulogne, France
Rawnsley, J.H.	University of Bonn, Germany
Rieckers, A.	University of Tübingen, Germany
Rocca, F.	Université de Nice, France
Rockland, C.	Brandeis University, Waltham, Mass. U.S.A.
Roël, R.W.	University of Amsterdam, the Netherlands
Ronveaux, A.	Facultés Universitaires de Namur, Belgium
Rosen, J.	Tel-Aviv University, Israel
Ruch, E.	Freie Universität Berlin, Germany
Scheerer, K.	University of Tübingen, Germany
Scheurer, P.	University of Nijmegen, the Netherlands
Schranner, R.	University of Tübingen, Germany
Seligman, T.	University of Köln, Germany
Sharp, R.	Université de Montreal, Canada
Shaw, R.	University of Hull, England
Sherry, T.N.	Dublin Institute for Advanced Studies, Ireland

Simms, D.	University of Dublin, Ireland
Simoni, A.	University of Naples, Italy
Śniatycki, J.	University of Calgary, Canada
Solomon, A.	Open University Milton Keynes, England
Sorba, P.S.	C.N.R.S. Marseille, France
Souriau, J.M.	Université de Provence, Aix-en-Provence, France
Strocchi, F.	Scuola Normale Pisa, Italy
Ström, S.	University of Göteborg, Sweden
Swart, J.J. de	University of Nijmegen, the Netherlands
Tip, A.	FOM institute for Atomic & Mol.Phys., Amsterdam, NL
Tripathy, K.C.	Delhi University, India
Tschudi, H.R.	ETH - Zürich, Switzerland
Underhill, J.	University of Liverpool, England
Verbaarschot, J.	University of Utrecht, the Netherlands
Vitale, B.	University of Naples, Italy
Vuillermot, P.	Université de Neuchâtel, Switzerland
Winternitz, P.	Université de Montreal, Canada
Wolf, J.A.	University of California, U.S.A.
Wolfe, J.C.	University of Leuven, Belgium
Woodhouse, N.M.	University of London, King's College, England
Zak, J.	Technion, Haifa, Israel
Zassenhaus, H.J.	California Institute of Technology, Pasadena, U.S.A.

[illegible]	University of Dublin, Ireland
[illegible]	Università di [illegible], Italy
[illegible]	University of Calgary, Canada
[illegible]	Open University, Milton Keynes, England
[illegible]	[illegible] Università, Italia
[illegible]	Université de [illegible], France
[illegible]	Scuola Normale, Italy
[illegible]	University of [illegible], Canada
[illegible]	University of Nijmegen, The Netherlands
[illegible]	Mathematisch Centrum, Amsterdam, Netherlands
[illegible]	Delhi University, India
[illegible]	ETH Zürich, Switzerland
[illegible]	University of Liverpool, England
[illegible]	University of [illegible], The Netherlands
[illegible]	Università di [illegible], Italy
[illegible]	Université de Neuchâtel, Switzerland
[illegible]	University of Montreal, Canada
[illegible]	University of California, U.S.A.
[illegible]	University of Leuven, Belgium
[illegible]	University of London, King's College, England
[illegible]	Technion, Haifa, Israel
[illegible]	California Institute of Technology, Pasadena, U.S.A.

GAUGE GROUPS, ELEMENTARY PARTICLES

MAGNETIC MONOPOLES AND NON-ABELIAN GAUGE GROUPS

G. 't HOOFT

Institute for Theoretical Physics,
University of Utrecht
Utrecht, The Netherlands

See: Nucl. Phys. (1974) B79, 276

It is shown that all those gauge theories in which the electromagnetic group U(1) is taken to be a subgroup of a larger group with a compact covering group, like SU(2) or SU(3), genuine magnetic monopoles can be created as regular solutions of the field equations. Their mass is calculable and of order $137M_W$, where M_W is a typical vector boson mass.

GAUGE THEORIES

J. Iliopoulos, Lab. de Physique Théorique, Ecole Normale
Supérieure, 24 rue Lhomond, 75231 Paris Cedex 05, Frankrijk

I. Introduction: The idea of unifying the weak and electromagnetic interactions is very old and goes back to the classical work of Fermi. On the phenomenological level the two forces present some common features, but also several important differences.

(i) They can both be described as interactions among vector currents. We know that the vector character of the electromagnetic current is due to the fact that the quantum which mediates the electromagnetic interactions (i.e. the photon) has spin equal to one. It is natural to assume that the same is true for the weak interactions (Intermediate Vector Boson hypothesis)

(ii) The e.m. interactions have a long range - The photon is massless. The weak interactions give rise to short range forces. - The Intermediate Vector Boson (I.V.B.), if it exists at all, must be very massive ($m_W \gtrsim$ 10-15GeV)

(iii) The photon is neutral. The I.V.B. must be charged in order to account for the observed weak decays. We often say that the e.m. current is neutral while the weak currents are charged.

(iv) The e.m. current is pure vector - The e.m. interactions conserve parity. The weak currents have both vector and axial parts. - The weak interactions violate parity.

(v) The e.m. current is conserved. The weak currents are not.

These differences have plagued the theoretical study of weak interactions for years. The e.m. interactions were described by a renormalizable field theory which agreed with experiment with a phenomenal accuracy. No consistent field theory existed for weak interactions. The natural framework to look for such a theory was that of non-abelian gauge theories, called Yang-Mills theories.

II. Yang-Mills theories: Let $\mathcal{L}(\varphi^i(x), \partial_\mu \varphi^i(x))$ $i = 1,\ldots,n$ be the Lagrangian density describing the dynamics of a physical system. An "internal" symmetry of the system is normally thought as an invariance of $\mathcal{L}$ under a Lie-group G of transformations acting on the fields $\varphi^i(x)$

$$\varphi^i(x) \longrightarrow \varphi^i(x) + \varepsilon^\alpha (T_\alpha)^i_j \varphi^j(x), \quad \alpha = 1,\ldots,N; \quad i,j = 1,\ldots,n \qquad (1)$$

where:

N: dimension of the Lie algebra of G

ε^α: N c-number, infinitesimal, x-independent parameters

T_α: the matrices of the representation (possibly reducible) to which $\varphi^i(x)$ belong.

We can argue on physical grounds that such "global" transformations, i.e. with x-ind. parameters, are unsatisfactory, because they imply one and the same choice of coordinates in the internal symmetry space over the entire universe. One should instead replace (1) by a set of "local" transformations for which the infinitesimal parameters are functions of x. I will simply call such transformations "gauge transformations". Clearly $\mathcal{L}$ is no more invariant because $\partial_\mu \varphi^i(x)$ does not transform any more like $\varphi^i(x)$.
The standard way to restore invariance is to introduce the afine connections and write:

$$\mathcal{L}\left(\varphi^i(x), \partial_\mu \varphi^i(x)\right) \Rightarrow \mathcal{L}\left(\varphi^i, D_\mu \varphi^i(x)\right) - \frac{1}{4} G^\alpha_{\mu\nu}(x)\, G^{\mu\nu}_\alpha(x) \qquad (2)$$

with

$$D_\mu \varphi^i(x) \equiv \partial_\mu \varphi^i(x) - g\,(T_\alpha)^i_j\, W^\alpha_\mu(x)\, \varphi^j(x) \qquad (3)$$

The "gauge fields" $W^\alpha_\mu(x)$ transform like

$$W^\alpha_\mu(x) \to W^\alpha_\mu(x) + f^\alpha_{cb} W_\mu{}^b(x)\, \varepsilon^c(x) + \frac{1}{g}\partial_\mu \varepsilon^\alpha(x) \qquad (4)$$

where g: arbitrary constant (coupling constant)

f^α_{bc}: structure constants of G.

$$G^\alpha_{\mu\nu}(x) \equiv \partial_\mu W_\nu{}^\alpha(x) - \partial_\nu W_\mu{}^\alpha(x) - g f^\alpha_{bc} W_\mu{}^b(x)\, W_\nu{}^c(x) \qquad (5)$$

Well-known example: Quantum electrodynamics (G = U(1))

Important remark: The gauge fields describe massless spin-one particles, like the photon. We do not know in nature any other such particles (massless spin-one). Conclusion (wrong!): It seems that non-abelian gauge theories have nothing to do with physics in general and the weak interactions in particular.

III. Spontaneously broken symmetries: There exist numerous examples, both in classical and quantum physics, in which a symmetry of the Lagrangian is not reflected in the solutions.

Examples: The problem of the bent rod, or the appearance of a spontaneous magnetization in a Heisenberg ferromagnet.

These are called "spontaneously broken symmetries". Their characteristic features are:

(i) The symmetric solution becomes unstable.

(ii) The ground state of the system is degenerate.

Goldstone theorem: In a relativistically invariant field theory, to every generator of a spontaneously broken symmetry corresponds a massless particle, the so-called Goldstone particle.

The two theoretic ideas we have described so far, namely Yang-Mills theories and spontaneously broken symmetries, each one taken separately, look irrelevant for elementary particle physics, both being hopelessly inflicted with zero-mass particles. And here comes the surprise:

Spontaneously broken gauge symmetries: When the spontaneously broken symmetry is a gauge symmetry, the massless gauge vector bosons acquire a mass and the would-be massless Goldstone particles decouple and disappear! (Higgs mechanism)

IV. Fundamental theorem: A Yang-Mills field theory, broken spontaneously via the Higgs mechanism, remains renormalizable.

The importance of Higgs mechanism: The way to give masses to the gauge bosons (thus avoiding conflict with experiment), still keeping gauge invariance and renormalizability.

The proof of this remarkable theorem opened the way into an avalanche of theoretical papers. Some main points:

V. Models: The principles of model-building are:

(i) Choose a gauge group G.

(ii) Choose the fields and their representations. Include enough scalar mesons some of which will decouple eventually through the Higgs mechanism.

(iii) Write the most general renormalizable Lagrangian invariant under G. At this stage all gauge vector bosons are still massless and all currents conserved.

(iv) Arrange for spontaneous symmetry breaking of the generators of all currents except the e.m. one. This can always be achieved by a suitable choice of the parameters that determine the potential energy of the scalar mesons. As a consequence all vector bosons, but the photon, acquire a mass. The corresponding would-be scalar Goldstone bosons decouple.

As we see, non abelian gauge theories provide only the framework, not a unique model. Detailed comparison with experimental data will determine, hopefully, the right one.-

VI. Experimental consequences:

(i) Existence of the intermediate vector bosons of the weak interactions. They are generally predicted to be heavy, $m_W \sim$ 50-100GeV, out of reach even of F.N.A.L. The reason is that in unified theories the fundamental coupling constant is of the order of e, the electric charge. The Fermi coupling constant is given by $G/\sqrt{2} \sim \alpha/m_W^2$.

Experimentally $G/\sqrt{2} \sim 10^{-5} m^{-2}_{proton}$ and $\alpha \sim \frac{1}{137}$.

Thus, the apparent weakness of weak interactions is a low energy phenomenon due to the large masses of the intermediate vector bosons. At high energies ($E \sim m_W$) weak interactions will become as strong as the electromagnetic ones.

(ii) One can show that a consistent model must contain weak neutral currents and/or heavy leptons. The neutral currents have already been observed experimentally and this gave the first confirmation that with gauge theories we are on the right track.
An experimental study of their properties is essential to an understanding of the detailed form of weak interactions. Heavy leptons, although not excluded by the theory, are no more required.
(iii) Maybe the most important prediction concerns the spectrum of hadrons. If SU(3) is the symmetry group of strong interactions, the charged weak currents have a well-known form, given by Cabibbo. In a gauge theory the weak neutral current is related to the commutator of the two charged ones. We thus predict its properties, in particular we predict that it contributes to $\Delta S = 1$ transitions. This induces decays like $K^0_L \to \mu^+\mu^-$ with appreciable rates. They are absolutely excluded by experiment.
We conclude that the traditional SU(3) scheme for strong interactions is incompatible with the gauge theories we are discussing. We must enlarge the symmetry, thus predicting the existence of new hadronic states carrying new quantum numbers. I call these numbers collectively "charm" and the states which carry them "charmed". The remarkable thing is that their masses are predicted low, of the order of a few GeV. It is easy to understand the general enthousiasm caused by the recent announcement of the discovery of new particles at Brookhaven and SLAC. I strongly believe that they are the manifestations of charm. It is exciting to think that abstract theoretical considerations, based essentially on aesthetic arguments, may have led to the discovery of a new area in particle physics.

VII. Strong interactions: It has been known since several years that nucleons do not interact with the electromagnetic field like point charges but they possess instead a kind of internal structure. Recent experiments, however, at Stanford gave to this vague idea a much more precise and unexpected meaning. They studied the scattering of electrons on nucleons at high energies and large momentum transfers, where, among the final particles, only the outgoing electron was observed (deep inelastic scattering). The astonishing result was that the outcome of these experiments could be explained if one assumed that the target nucleon was made out of an assembly of "elementary" constituents which interacted with the electromagnetic field of the electrons as free, point-like charges. These constituents were given the name"partons". The theoretical question now was: how can the partons be so tightly bound in order to form a nucleon and still act like free particles in deep-inelastic experiments? No renormalizable field theory, in any finite order of pertubation, could reproduce such a result.

The answer to this question turned out to bring once more the non-abelian gauge theories on stage. One can show, using the formalism of the renormalization group, that the effective coupling strength of an interaction described by a renormalizable field theory, is not constant, but depends on the kinematical region one is considering. If the effective coupling of a theory tends to zero for large values of the external momenta, the theory is called "asymptotically free". The remarkable result of the renormalization group is that this property can be discovered by studying the low orders of perturbation theory.

Theorem: Out of all renormalizable field theories, only the non-abelian gauge theories are asymptotically free.

The implications of this theorem are clear: The only way to understand the behaviour of partons in the deep inelastic region, using a field-theoretic framework, is to assume that strong interactions are described by a Yang-Mills theory. The simplest way to realize such a scheme is to introduce a set of new quantum numbers and hence a new symmetry group, called "colour group". This is a new SU(3) symmetry, "colour SU(3)", which is completely different from ordinary SU(3) of hadron physics. All observed hadrons are assumed to be colour singlets. Colour SU(3) is assumed to be realized locally as a Yang-Mills symmetry, thus introducing an octet of gauge vector bosons, called "colour gluons". The symmetry is exact and the eight gauge bosons are massless. The fact that they are not produced in ordinary experiments, as well as the physical absence of colour non-singlet states, are attributed to the singular on-mass-shell structure of unbroken Yang-Mills theories.

VIII. Conclusion : We tried to combine all available experimental results from all processes involving currents, at low energies as well as in the deep inelastic region, and we saw that a consistent picture arises if we postulate that all interactions among elementary particles are described by non-abelian Yang-Mills theories.

The arguments in favor of this postulate are:

(i) Strong interactions: Asymptotic freedom

(ii) Weak + em. interactions: Renormalizability.

(iii) Gravitational interactions: General relativity.

We are thus free to speculate on possible ways to unify all known interactions. And these speculations are no-more in the domain of science-fiction, but in that of serious scientific investigation.

The prices we had to pay are:

(i) Existence of intermediate vector bosons of weak interactions. Their experimental discovery must await for the new generation of accelerators.

(ii) Existence of weak neutral currents and/or heavy leptons. The first have already been observed, thus providing a strong encouragement for gauge theories. The second are no more necessary.

(iii)Existence of the colour group.

Its presence can be detected by very accurate experiments in the deep-inelastic region.

(iv) Existence of charm and charmed hadrons.

If the interpretation of the newly found resonances in terms of charm is confirmed, it will provide, once more, a splendid demonstration of the belief that, the search for symmetry and aesthetic beauty always leads to a more profound understanding of the physical world.

This is only a brief summary of the talk I gave in Nijmegen. The following references contain more detailed accounts on this and related subjects. I quote only some review articles. References to the original papers can be found in them.

1. B.W. Lee, in "Proceedings of the XVI International Conference on High Energy Physics" ed. by J.D. Jackson and A. Roberts (N.A.L., Batavia Ill. 1972). Vol. IV p. 249
2. M. Veltman, in "Proceedings of the 6th International symposium on electron and photon interactions at high energies" Bonn, Aug. 27-31, 1973, ed. by H. Rollnik and W. Pfeil, North Holland publishing Co.1974 p. 429
3. C.H. Llewellyn-Smith, ibid p. 449

4. S. Weinberg, in "Proceedings of the IIe Conference Internationale sur les Particules Elementaires", Aix-en-Provence 1973, Sup. au Journal de Physique Vol. 34 Fasc. 11-12, C1 - 1973, p. 45.
See also: Revs. Mod. Phys. 46 255 (1974).

5. J. Iliopoulos, in "Proceedings of the XVII International Conference on High Energy Physics", London 1974, ed. by J.R. Smith, published by the Science Research Council, Rutherford Laboratory, Chilton, Didcot, Oxon, OX11 OQX, U.K. p. III-89.

6. G. 't Hooft and M. Veltman: "Diagrammar", CERN yellow report, CERN 73-9.

7. E.S. Abers and B.W. Lee: "Gauge theories", Phys. Reports 9C Nb. 1.

8. S. Coleman: "Secret symmetry: An introduction to spontaneous symmetry breakdown and gauge fields".
Lectures given at the 1973 Erice Summer School.

9. M.A.B. Beg and A. Sirlin: Ann.Rev.Nucl.Sci. 24 379 (1974).

10. J. Bernstein: Revs. Mod. Phys. 46 7 (1974).

11. (On spontaneously broken symmetries): L. Michel, talk presented in this Conference.

12. (On the new particles): Proceedings of the August 1975 SLAC Conference.

PRESENT STATUS OF SUPERSYMMETRY

L. O'Raifeartaigh

School of Theoretical Physics

Dublin Institute for Advanced Studies

Dublin, Ireland.

The principles and historical development of the recently developed theory of supersymmetry are reviewed. The present status of the theory (its remarkably elegant properties, experimental problems and outlook for the future) are then discussed in a qualitative manner.

The last two or three years have seen the emergence of a remarkable new symmetry called supersymmetry. This symmetry has the property of allowing fields of different intrinsic spin (and, in particular, fermions and bosons) to appear in the same irreducible multiplets. For this reason a more accurate name might be spin-mixing symmetry.

The basic idea of supersymmetry was first put forward by Ramond[1] in 1971 in the context of the string theory of the dual model for strong interactions. The idea was then developed, and superfields introduced, in the context of the string theory[2] by Iwasaki and Kikkawa, Gervais and Sahita and Neveu and Schwartz all independently. It was then realized by Wess and Zumino[3] in 1973 that the algebra of supersymmetry could be taken from the string theory context, where it operated in 1 + 1 dimensions, and set in the context of conventional field theory in 3 + 1 dimensions. Finally in 1974 the Wess-Zumino theory was formulated in terms of superfields in 3 + 1 dimensions by Salam and Strathdee.

Although supersymmetry appeared in this way only in recent years, it could, in principle, have appeared at any time since 1928. This is because the basic idea of supersymmetry consists in carrying Dirac's factorization of the d'Alembertian operator one step further, and factorizing the Dirac operator itself. Thus, where Dirac succeeded in constructing operators $\gamma^\mu p_\mu = \not{p}$ which satisfied the algebra

$$\{\not{p}, \not{p}\} = 2p^2 , \qquad [p^2, \not{p}] = 0 , \tag{1}$$

supersymmetry succeeds in constructing operators G_α which satisfy the algebra

$$\{G_\alpha, G_\beta\} = (C\not{p})_{\alpha\beta} , \qquad [P_\mu, G_\alpha] = 0 , \tag{2}$$

where $\alpha, \beta = 1 \ldots 4$ are Dirac indices. More precisely, supersymmetry shows that there exist representations of the algebra (2). Furthermore, one finds that for at least two of the representations, renormalizable supersymmetric invariant Lagrangians (Lagrangians invariant with respect to the algebra (2)) can be constructed, and these Lagrangians have very remarkable properties.

Representations of Supersymmetry.

The representations of the algebra (2) are most conveniently defined in terms of superfields. The simplest kind of superfield, corresponding to one of the two types of representation for which a renormalizable supersymmetric Lagrangian has been constructed, is a field $\Phi(x\,\theta)$, which depends not only on the space-time coordinates x, but also on a very special kind of internal variable θ. This variable transforms with respect to the Lorentz group like a real Dirac (Majorana) 4-spinor, and satisfies the anti-commutation relations

$$\{\theta_\alpha, \theta_\beta\} = 0 \tag{3}$$

Note the similarity between (3) and (2). However, the right hand side of (3) is zero, so that the variables θ are independent of P_μ, and are nilpotent. As a consequence of the nilpotency, the expansion of $\Phi(x\,\theta)$ in powers of θ terminates and one obtains

$$\Phi(x\theta) = A(x) + \bar\theta\psi(x) + \tfrac{1}{4}\bar\theta\left[F + \gamma_5 G + i\gamma_\mu\gamma_5 V_\mu\right]\theta + \tfrac{1}{4}(\bar\theta\theta)\,\bar\theta\lambda(x) + \tfrac{1}{32}(\bar\theta\theta)^2 D(x)\,, \tag{4}$$

where $A(x)$, $\psi(x)$ etc. are conventional fields. For simplicity we shall assume that $\Phi(x\,\theta)$ is real and is a Lorentz scalar

$$\Phi^*(x\theta) = \Phi(x\theta)\,, \qquad \left(U(\Lambda)\Phi\right)(x\theta) = \Phi(\Lambda^{-1}x, S^{-1}\theta)\,, \tag{5}$$

where S is the real Dirac (Majorana) four-dimensional representation of $SL(2,C)$. The conventional fields then have the appropriate Lorentz transformation properties; $A(x), F(x), D(x)$ scalars, $G(x)$ pseudo-scalar, $\psi(x)$, $\lambda(x)$ Majorana spinors and $V_\mu(x)$ vector.

To obtain the representation of the algebra (2) on $\Phi(x\theta)$ one now defines the supersymmetric transformations

$$\left(U(\varepsilon)\Phi\right)(x_\mu, \theta) = \Phi\left(x_\mu + \tfrac{i}{2}\bar\varepsilon\gamma_\mu\theta,\ \theta + \varepsilon\right) \tag{6}$$

where the ε are variables similar to the θ and anti-commuting with them. Then if we let G_α denote the infinitesimal generators of the transformations (6) we find that

$$G_\alpha = \frac{\partial}{\partial\theta_\alpha} + \frac{i}{2}(\bar{\theta}\gamma^\mu\partial_\mu)_\alpha \tag{7}$$

and it is trivial to verify that these G_α satisfy the algebra (2). Note that the supersymmetric transformations (6) are direct translations in θ -space, and induce the formal translations $x \to x + \frac{i}{2}\bar{\varepsilon}\gamma\theta$ in x -space. The quantities $x + \frac{i}{2}\bar{\varepsilon}\gamma\theta$ are to be understood in the sense of Taylor expansions of the conventional fields

$$A(x + \tfrac{i}{2}\bar{\varepsilon}\gamma\theta) = A(x) + \tfrac{i}{2}\bar{\varepsilon}\gamma\theta\, A(x)_{,\mu} + \dots \tag{8}$$

where the expansion terminates on account of the nilpotency of θ and ε. From (7), (5) and (8) one can obtain the direct action of the G_α on the conventional fields and it is

$$\begin{aligned}
\delta A &= \psi\cdot\delta\theta \\
\delta\psi &= \tfrac{1}{2}(F + \gamma_5 G + i\gamma_5 \not{V})\delta\theta && - \tfrac{i}{2}(\not\partial A)\delta\theta \\
\delta F &= \tfrac{1}{2}\bar{\chi}\cdot\delta\theta && - \tfrac{i}{2}\delta\bar{\theta}\not\partial\psi \\
\delta G &= \tfrac{1}{2}\bar{\chi}\gamma_5\delta\theta && - \tfrac{i}{2}\delta\bar{\theta}\gamma_5\not\partial\psi \\
\delta V_\mu &= \tfrac{1}{2}\bar{\chi}\gamma_\mu\gamma_5\delta\theta && - \tfrac{i}{2}\delta\bar{\theta}\, i\not\partial\gamma_5\gamma_\mu\psi \\
\delta\chi &= \tfrac{1}{2}D\,\delta\theta && - \tfrac{i}{2}(\not\partial F - \gamma_5\not\partial G + i\gamma_\nu\gamma_5 A_\nu) \\
\delta D &= && - i\delta\bar{\theta}\not\partial\chi
\end{aligned} \tag{9}$$

where

$$\delta X \equiv (G_\alpha X)\delta\theta_\alpha \quad \text{and} \quad \not\partial = \gamma^\mu\partial_\mu, \quad \bar{\psi} = \psi^\dagger\gamma_0 = \tilde{\psi}\gamma_0.$$

Note that the $\frac{\partial}{\partial\theta_\alpha}$ part of G_α acts as a raising operator and the $(\bar{\theta}\not{\partial})_\alpha$ part as a lowering operator, and that the lowering operation is always accompanied by a divergence.

The real scalar superfield just described is one of the representations of the supersymmetric algebra (2) for which a Lagrangian has been constructed. The second kind of superfield for which a Lagrangian has been constructed is obtained by making a "chiral-reduction" of $\Phi(x\theta)$. To make a chiral reduction one notes that

$$\{G_\alpha, \tilde{G}_\beta\} = 0, \qquad \tilde{G}_\alpha = \frac{\partial}{\partial\theta_\alpha} - \frac{i}{2}(\bar{\theta}\not{\partial})_\alpha \tag{10}$$

that is, that the quantities $\tilde{G}_\alpha$, which are similar to the G_α but have a minus sign between the lowering and raising parts, commute with the G_α . This property of the $\tilde{G}_\alpha$ guarantees that the fields $\Psi_\pm(x\theta)$ which satisfy the conditions

$$\left[\left(\frac{1\pm\gamma_5}{2}\right)\tilde{G}\right]_\alpha \Psi_\pm(x\theta) = 0, \tag{11}$$

are again superfields. (The chiral projections $(1\pm\gamma_5)/2$ are used because the condition $\tilde{G}\Psi = 0$ would be too strong and would kill the superfield.) Fields $\Psi_\pm(x\theta)$ satisfying (11) are called chiral scalar superfields. The parity-invariant combination $\Psi(x\theta) = \Psi_+(x\theta) + \Psi_-(x\theta)$ is then a smaller field than $\Phi(x\theta)$ and is thus a reduction of it. It has the same number of formal components as $\Phi(x\theta)$, but the fields are inter-related. For example, for $\Phi(x\theta)$ the spinor fields $\psi(x)$ and $\lambda(x)$ are independent, but for $\Psi(x\theta)$, $\lambda(x) = \not{\partial}\psi(x)$.

Before going on to discuss the supersymmetric Lagrangians for the scalar and chiral-scalar superfields, there is one point that should be mentioned. That is, in the above we have assumed that θ is independent of x , and one might ask what happens if we let θ be x -dependent. The answer is that if we let θ be linear in x , $\theta_\alpha(x) = \mathring{\theta}_\alpha + \theta_\alpha^\mu x_\mu$, then instead of obtaining the generators of the translation group on the right hand side of (2) one obtains the whole conformal group. For arbitrary x -dependence, one obtains the infinite-dimensional Einstein group. The conformal case was actually the case first considered by Wess and Zumino[3]. An interesting feature in the conformal case is that the unrestricted linear transformations generated by $\varepsilon_\alpha(x) = \mathring{\varepsilon}_\alpha + \varepsilon_\alpha^\mu x_\mu$ in θ -space, generate the restricted bilinear conformal transformations $x_\mu \rightarrow x_\mu + \frac{i}{2}\bar{\varepsilon}(x)\gamma_\mu\theta(x)$ in x -space. Thus θ -space appears to be a more basic space for the conformal group.

Supersymmetric Lagrangians.

The construction of renormalizable supersymmetric invariant Lagrangians for is based on two observations:

(1) The product of two scalar superfields is again a scalar superfield;

(2) The supersymmetric variation of the coefficient of $(\bar{\theta}\theta)^2$ in a superfield is a pure divergence.

The first observation follows from the nilpotency of θ, which makes the expansion terminate at $(\theta\bar{\theta})^2$ no matter how many products of superfields are taken, and the fact that the Lorentz and supersymmetric transformations (5) and (6) are carried by the arguments of the fields and hence are the same for products of superfields. The second observation follows from the fact that the lowering operation $((\theta\not{\partial})_\alpha$ part of $G_\alpha)$ is always a pure divergence, while for the coefficient of $(\theta\bar{\theta})^2$ the raising operation $((\partial/\partial\theta)_\alpha$ part of $G_\alpha)$ vanishes.

The procedure for forming Lagrangians is then simple in principle (in practice it can become quite complicated): To construct a kinetic or mass-term for a Lagrangian one takes a product of two superfields, expands it in powers of θ and takes the coefficient of $(\bar{\theta}\theta)^2$ as the Lagrangian density. To construct an interaction one takes a product of three superfields and repeats the process. (A product of more than three will also lead to a Lagrangian but it will not be renormalizable.) Since the supersymmetric variation of the Lagrangian densities constructed in the above way will then be a pure divergence, the Lagrangian itself will be supersymmetric invariant. Similar procedures hold for chiral scalar superfields.

Let us now simply describe what emerges from the procedures just described for the superfields $\Phi(x\theta)$ and $\Psi_{\pm}(x\theta)$. Initially $\Psi_{\pm}(x\theta)$ has the expansion

$$\Psi_{\pm}(x\theta) = A_{\pm}(x) + \bar{\theta}\cdot\psi_{\pm}(x) + \tfrac{i}{4}\bar{\theta}\gamma_{\mu}\gamma_5\theta A_{\pm}(x)_{,\mu} + \ldots \qquad (12)$$

where the terms omitted are, like the third term, dependent on $A_{\pm}$ and $\psi_{\pm}$, where A_{+} is a complex scalar and ψ_{+} is the positive chiral projection of a Majorana spinor. The $\Psi_{\pm}(x\theta)$ together where $\Psi_{+}^{*} = \Psi_{-}$, have essentially the content

$$\Psi_{\pm}(x\theta) = \{A(x), B(x), \psi(x)\} = \{0^{+}, 0^{-}, \tfrac{1}{2}^{i}\}, \qquad (13)$$

where $A_{\pm} = A \pm iB$, and the bracket on the right denotes the spin and parity. (The parity of a Majorana spinor is necessarily pure imaginary.) The Lagrangian which emerges from the supersymmetric procedure for $\Psi_{\pm}(x\theta)$ is[3]

$$\mathcal{L} = \mathcal{L}_{k.e.} + \frac{m}{2}\bar{\psi}\psi + g\bar{\psi}(A + i\gamma_5 B)\psi + \tfrac{1}{2}F^{*}F, \qquad (14)$$

where $\mathcal{L}_{k.e.}$ is a conventional kinetic energy term for $\{A, B, \psi\}$ and

$$F = mA_{+} + gA_{+}^{2}. \qquad (15)$$

Note that the potential energy $F^{*}F/2$ is positive indefinite and zero at $A_{+} = 0$. What is so special about this Lagrangian? At first sight it appears to be a conventional Yukawa-$P(\phi^4)$ interaction for $\{A, B, \psi\}$. What is special about it is that, whereas the most general parity-invariant Yukawa-$P(\phi^4)$ Lagrangian would have eight independent parameters (three masses, three boson coupling constants and three fermion coupling constants) the Lagrangian (14) has only two independent parameters, m and g. Thus there is a huge reduction in the number of independent parameters. But why go through all this machinery in order to reduce the number of parameters? Why not arbitrarily set parameters equal in the conventional eight-parameter Lagrangian? The answer is that arbitrary relations among the parameters will not be stable, i.e. will not be maintained after renormalization. Because the relations in the supersymmetric case are derived from a symmetry which is respected by the interaction, the relationships implied by the symmetry are expected to be stable. A calculation, first of the one-loop corrections, and then to all orders, shows that these expectations are indeed realized for the Lagrangian (14).

For the scalar superfield $\Phi(x,\theta)$ the spin-parity content is at first sight larger, namely,

$$\Phi(x,\theta) = \{ A, \psi, F, G, V_\mu, \lambda, D \} \tag{16}$$

However, it turns out that the Lagrangian is such that D is a function of the other fields, and if we have gauge as well as supersymmetric invariance (supergauge invariance) then a supergauge can be chosen so that the fields A, ψ, F, G vanish[5]. Thus in the supergauge invariant case the essential content of is

$$\Phi(x\theta) \simeq \{ V_\mu, \lambda \} = \{ 1^-, \tfrac{1}{2}^i \} . \tag{17}$$

Thus a super-gauge field $V_\mu(x)$ comes accompanied by a Majorana spinor field $\lambda(x)$. Correspondingly the Lagrangian for $\Phi(x\theta)$ is relatively simple, namely,[5]

$$\mathcal{L} = -\frac{1}{4} \mathrm{tr}\, F_{\mu\nu} F^{\mu\nu} - \frac{i}{2} \mathrm{tr}\, \bar{\lambda} [\not{D}, \lambda] , \tag{18}$$

which is just an ordinary Yang-Mills Lagrangian for $V_\mu(x)$ and $\lambda(x)$. Even in the special supergauge $A=\psi=F=G=0$ the Lagrangian (18) still retains, of course, the conventional gauge invariance under $V_\mu \rightarrow S^{-1} V_\mu S + S^{-1} \partial_\mu S$. Finally one can let the Yang-Mills superfield $\Phi(x,\theta) \simeq \{ V_\mu, \lambda \}$ interact with the chiral scalar superfield. One then obtains a Lagrangian of the form[5]

$$\mathcal{L} = -\frac{1}{4}\,\mathrm{Tr}\,F_{\mu\nu}F^{\mu\nu} - \frac{i}{2}\,\mathrm{Tr}\,\bar{\lambda}\,[\not{D},\lambda]$$

$$-\frac{1}{2}\bar{\psi}\not{\partial}\psi - \frac{1}{2}(D_\mu A_+)^\dagger (D_\mu A_+)$$

$$+ e\{i\bar{\lambda}_\alpha (A^\dagger + i\gamma_5 B^\dagger)\tau_\alpha \psi\} + h.c.\} + \frac{e^2}{2}\{A^\dagger \tau_\alpha B + h.c.\}^2 \qquad (19)$$

$$+ \frac{m}{2}\bar{\psi}\psi + g\bar{\psi}(A + i\gamma_5 B)\psi + \frac{1}{2}F^\dagger F\,.$$

where $\Psi_\pm(x,\theta) \simeq \{A, B, \psi\}$ is a multiplet of chiral scalar superfields, $A_+ = A + iB$, $D_\mu = \partial_\mu + ieV_\mu$, and τ_α are the group generators.

Balance-Sheet.

Having described the basic principles, we must now consider the advantages and disadvantages of supersymmetry. First let us consider the credit side.

Apart from the general elegance and beauty of this symmetry, there are five specific areas in which it is advantageous, or at least interesting. The areas can be specified as follows:

(i) Renormalization

(ii) Spontaneous Symmetry Breaking

(iii) Yang-Mills Theory (Unified Gauge Theory)

(iv) Asymptotic Freedom

(v) Mixing of Lorentz and Internal Symmetry.

We discuss these briefly in turn.

(i) The renormalization properties constitute perhaps the most striking feature of supersymmetry. Cancellations occur at almost every step so that the actual renormalization is reduced to a minimum. For example, for the chiral scalar field Lagrangian (14) there are no quadratic and linear divergences, and there is only <u>one</u> logarithmic divergence[3][6]. Furthermore, this logarithmic divergence serves as the renormalization constant for all the masses and all the wave-functions, both fermions and bosons. (Thus Kallen's prediction that at least one of the renormalization constants must be infinite is just barely fulfilled!) Another result is that for supersymmetry the vacuum expectation value of any n-point function is zero[7]. This means in particular that the Lagrangian need not be normal-ordered.

(ii) There are two kinds of spontaneous symmetry breaking that one can consider, namely the spontaneous breaking of supersymmetry itself, and the spontaneous breaking of internal symmetry by supersymmetry. One finds that the spontaneous breakdown of supersymmetry is a relatively rare occurrence[8][9], whereas the spontaneous breakdown of an internal symmetry, triggered by supersymmetry, happens frequently[4][9]. Thus with respect to spontaneous symmetry breaking, supersymmetry resembles a diamond, which itself is hard to cut, but which serves to cut glass.

(iii) There is a natural way to combine Yang-Mills gauge invariance and supersymmetry into a supergauge invariant theory, and indeed the required supergauge Lagrangian is just that given above in (18) and (19). Furthermore, if the internal symmetry of this Lagrangian is spontaneously broken (which frequently happens as discussed in (ii)) we may, by proper choice of group and representation[4][9], pick up masses for all but an abelian set of the Yang-Mills fields. In that case we obtain an infra-red convergent supergauge theory, or in other words, a supersymmetric unified gauge theory. This result is not trivial because, unlike the conventional Yang-Mills theory, supergauge theory completely determines the <u>form</u> of the potential (see equation (14)) leaving only the choice of group and representation free.

(iv) It is now well-known that, in contrast to abelian fields, a Yang-Mills field, either in self-interaction or in interaction with fermions, is asymptotically free[10]. That is, the renormalized coupling constant tends to zero as the scale parameter tends to infinity. However, it is also known that, in general, a Yang-Mills field in interaction with boson fields is not, in general, asymptotically free[10]. The reason is that renormalizability requires the introduction of a second (quartic) coupling constant for the boson field, and the latter constant is not, in general, asymptotically free. It turns out that the renormalization properties of supersymmetry are such that this problem goes away, and a supersymmetric Yang-Mills field in interaction with either one or two super matter fields (which necessarily include bosons as well as fermions) is asymptotically free[11].

(v) Finally, supersymmetry can be used to obtain a nontrivial mixing of Lorentz and internal symmetry[4][9]. The idea is very simple. Given a superfield $\Phi(x\,\theta)$ there are two ways to introduce an internal symmetry as follows:

Trivial (direct product) way: $\Phi(x\,\theta) \to \Phi^a(x\theta)$

Non-trivial way: $\Phi(x\,\theta) \to \Phi(x\,\theta^a)$

where a is the internal symmetry index. In the second case the expansion of $\Phi(x\,\theta)$ becomes

$$\Phi(x\,\theta^a) = A(x) + \bar{\theta}^a \psi_a(x) + \ldots \qquad (20)$$

so that the fields of different spin have different internal spin also. Furthermore, the relativistic and supersymmetric transformation laws remain compatible,

$$\left(U(\Lambda)\Phi\right)(x\,\theta_a) = \Phi(\Lambda^{-1}x,\, S^{-1}\theta_a), \qquad (21)$$

$$\left(U(\varepsilon_a)\Phi\right)(x_\mu \theta_a) = \Phi\left(x_\mu + \tfrac{i}{2}\bar{\varepsilon}_a \gamma_\mu \theta_a,\; \theta_a + \varepsilon_a\right). \qquad (22)$$

This result establishes that, in principle at any rate, Lorentz and internal symmetry can be successfully combined. However, for the moment, the success remains at the level of principle, because it has been shown[12] that under rather general and plausible assumptions (20) is actually the most general combination of Lorentz and internal symmetry that can be constructed, and it so happens that the spin-isospin correlations obtained from (20) are not found experimentally. Note that (20) combines only the spin and the internal symmetry. There is no mass-breaking either for the supersymmetry or the internal symmetry.

Finally, we must come to the debit side of the balance sheet. The sad fact is that in spite of its intrinsic beauty, supersymmetry has not yet found any useful application. There are three basic reasons for this. The first is the one mentioned above, that the spin-isospin correlation for supersymmetry is far from experiment, but there are two more profound difficulties. These difficulties stem from the fact that supersymmetry forces fermions and bosons to behave in a similar manner. The first difficulty then is that the fermions and bosons in a supersymmetric multiplet have the same mass. Apart from the case of the neutrino and photon, which both have zero mass, this result is in manifest disagreement with experiment. There would be no great problem if we could have a spontaneous breakdown of supersymmetry, since the breakdown would allow different fermion and boson masses to emerge. But as we have mentioned above, a spontaneous breakdown of supersymmetry is a relatively rare event. The second difficulty is that either the fermion number is not conserved (or, more exactly, is conserved only modulo two) or the boson number is conserved. Neither of these alternatives agrees, of course, with experiment.

I should hasten to add that these difficulties are not completely insurmountable, in the sense that they have already been overcome in particular models[(4)(8)(9)(13)]. Indeed one might reverse the argument, as Iliopoulos has suggested, and use mass and particle-number breaking as criteria for selecting models. However, so far no systematic way of overcoming these two difficulties has emerged, and until it does, or until some of the particular models which overcome these difficulties become realistic in other respects, the experimental identification of superfields will present a serious difficulty. Perhaps the

MONOPOLE THEORIES WITH STRINGS AND THEIR APPLICATIONS TO MESON STATES*

A. P. Balachandran, R. Ramachandran,†
J. Schechter and Kameshwar C. Wali
Department of Physics
Syracuse University, Syracuse, New York
and
Heinz Rupertsberger
Institut fur Theoretische Physik der
Universitat Wien, Wien, Austria

* Supported in part by the U.S.Atomic Energy Commission (ERDA)

† Department of Physics, Indian Institute of Technology, Kanpur, India (permanent address)

immediate future of supersymmetry is to go on ice, as Yang-Mills theory did from 1954 to 1967, until one knows how and where to use it. Perhaps it will never become useful experimentally, but serve as a simple model, analogous to the Lee model(14), on which ideas can be tested. Even in that limited context, I think that the intrinsic beauty and elegance of supersymmetry will serve to keep interest in it alive for some time to come.

References

(1) P. Ramond, Phys. Rev. D3, 2415 (1971)

(2) Y. Iwasaki, K. Kikkawa, Phys. Rev. D8, 440 (1973)
A. Neveu, J. Schwartz, Nucl. Phys. B31, 86 (1971)
J.-L. Gervais, B. Sakita, Nucl. Phys. B34, 633 (1971)

(3) J. Wess, B. Zumino, Nucl. Phys. B70, 39 (1974); Phys. Letters, 51B, 239 (1974)

(4) A. Salam, J. Strathdee, Trieste Preprint IC/74/42; Phys. Rev. D11, 1521 (1975)

(5) J. Wess, B. Zumino, Nucl. Phys. B78, 1 (1974)
A. Salam, J. Strathdee, Phys. Letters 51B, 353 (1974)
S. Ferrara, B. Zumino, B79, 413 (1974)

(6) J. Iliopoulos, B. Zumino, Nucl. Phys. 76B, 310 (1974)

(7) B. Zumino, Nucl. Phys. B89, 535 (1975)

(8) P. Fayet, J. Iliopoulos, Phys. Letters 51B, 461 (1974)
P. Fayet, Nucl. Phys. (in press)

(9) L. O'Raifeartaigh, Phys. Letters 56B, 41 (1975); Nucl. Phys. (in press)

(10) D. Politzer, Physics Reports 14, No. 4 (1974)

(11) B. Zumino, Proc. XVIIth Int. Conf. on HEP (London 1974)

(12) R. Haag, J. Lopuszánski, M. Sohnius, Nucl. Phys. 88B, 257 (1975)

(13) A. Salam, J. Strathdee, Nucl. Phys. 87B, 85 (1975)
D. Grosser, Nucl. Phys. 92B, 120 (1975)

(14) T. D. Lee, Phys. Rev. 95, 1329 (1954).

Dirac's formulation of the monopole theory modified by an additional mass term for the gauge field has been considered as a possible simplified model for quark binding.[2,3,4,5] We consider the Hamiltonian formulation of the theory resulting from this action. The original (zero mass gauge field) theory is also discussed and its Hamiltonian is shown to be essentially the same as that of the two potential formalisms.[6,4] In this case, the coordinates of the string are absorbed into what turn out to be the physically meaningful variables for the particles and the field. In the massive case, the string does play a significant role and gives rise to a static linear potential and a Yukawa potential between the monopoles. Such a potential has also been found by Nambu and others[2,3,4,5] and may lead to an acceptable model for interactions of quarks.

The theory as formulated above, however, contains infinities when the gauge field is massive. We also discuss methods for the consistent regularization of these infinities. In this way we are led to an action which is the same as that suggested by previous authors.[7,3] We show that the expression for the energy of the modified action still has infinities unless the mass of the gauge field is infinite. Thus the regularization procedure is incomplete when the gauge field has finite mass. Applications of the regularized model to charmonium and other meson states are discussed. In particular, we show that the strength of the potential which varies as the distance between the monopoles can be related to the universal Regge slope parameter α'.[8] This relation is in good agreement with the phenomenological analysis of the newly discovered resonances $\psi(3105)$, $\psi(3695)$ and $\psi(4170)$[9] if the monopoles are identified with charmed quarks and the ψ's with bound states of such quarks and their anti-particles.

Full details of the research summarized above may be found in Ref. 10.

References

1. P.A.M Dirac, Phys. Rev. 74, 817 (1948).
2. Y.Nambu, The Johns Hopkins Workshop on Current Problems in High Energy Theory. (ed. G.Domokos and S.Kövesi-Domokos) p. 1 (1974).
3. Y.Nambu, Phys. Rev. D10, 4262 (1974).
4. A.P.Balachandran, H.Rupertsberger and J.Schechter, Syracuse University preprints SU-4205-37 and SU-4205-41 (1974) and Phys. Rev. (in press).
5. A.Jevicki and P.Senjanvirc, Phys. Rev. 11, 860 (1975).
6. J.Schwinger, Phys. Rev. 144, 1087 (1966); 151, 1048 (1966); 151, 1055 (1966); D.Zwanziger, Phys. Rev. D3, 885 (1971).
7. A.O.Barut and G.L.Bornzin, Nucl. Phys. B81, 477 (1974). See also references contained therein.
8. This result was briefly reported in A.P.Balachandran, R.Ramachandran, J.Schechter, Kameshwar C. Wali and H. Rupertsberger, "Monopole Strings and Charmonium", Syracuse University preprint SU-4205-47 (1975) and Orbis Scientiae, University of Miami, Coral Gables, Florida (January, 1975) (to be published).
9. B.J.Harrington, S.Y.Park and A.Yildiz, Phys. Rev. Letters 34, 168 (1975); E.Eichten, K.Gottfried, T.Kinoshita, J.Kogut, K.D.Lane and T.M.Yan, Phys. Rev. Letters 34, 369 (1975);

K.Jhung, K.Chung and R.S.Willey, University of Pittsburgh preprint (1975); J.F.Gunion and R.S.Willey, University of Pittsburgh preprint (1975).

10. A.P.Balachandran, R.Ramachandran, J.Schechter, Kameshwar C.Wali and H.Rupertsberger, Syracuse University preprints SU-4206-53 and SU-4206-54 (April, 1975) (to be published).

QUARKS AND THE POINCARE GROUP : SU(6) x SU(3) AS A CLASSIFICATION GROUP FOR BARYONS

F. BUCCELLA*, A. SCIARRINO**,+ P. SORBA+

Recently [1], the study of the representations of the Poincaré group $\underset{\sim}{P}$ for mesons considered as free quark - antiquark systems has allowed to give a simple interpretation of the transformation between current quark and constituent quark [2] on one hand, and to specify the classification group for mesons on the other hand. Indeed, the mixing operator between the $SU(6)_W^{curr}$ algebra of light-like charges and the $SU(6)_W^{class}$ algebra of classification appears as a Wigner's rotation arising from the reduction into irreducible unitary representations (I.U.R.) of the product of two I.U.R. of P . Moreover, the L-S coupling considered in this reduction provides a justification for the use of $SU(6)_q \times SU(6)_{\bar{q}} \times O(3)$ as the classification group for collinear mesonic states.

Here*, we propose to use the same techniques for baryons considered as three free quarks systems, i.e. to consider a baryonic state as a state of an I.U.R. of P. In such a reduction, one succeeds to factorize out, in some sense, the quark spin dependence, and is left with a function which depends only on the quark momenta. From the energy momentum conservation, this function is defined on a five dimensional surface Γ^5 which is topologically equivalent to the five dimensional sphere S^5 . So after deforming the surface Γ^5 into S^5 , the group SU(3) which acts transitively on S^5 is naturally introduced.

A complete set of commuting observables carrying the property of "democracy", i.e. observables which treat the three particles symmetrically is then obtained.

In this way we find again the three Wigner's rotations, each of them acting on one current quark and giving a constituent quark.

Moreover this democratic treatment leads us to propose as a group of classification for collinear baryonic states the group SU(6) x SU(3) , the O(3) orbital part of the meson classification group being enlarged, for baryons, to a SU(3) group, which will be called the "generalized orbital momentum" (G.O.M.) group : $SU(3)_{GOM}$.

. Let us note immediately that this result is not surprising since to classify baryons, physicists have already introduced a new quantum number [4] n = 0,1,2,... to those of the classification group $SU(6) \times O(3)_L$: the states $|n, L, L_z\rangle$ are in this context states of the representations (n,0) of the SU(3) harmonic oscillator group [5] [6]. In fact a harmonic oscillator type formalism

* This note constitutes a condensed version of the work mentioned in Ref. [3].

will appear naturally in our investigations. But it is worth to stress that the group $SU(3)_{GOM}$ comes out simply in the framework of a relativistic treatment of three free quarks, while in Ref.[5] quarks interact via a non relativistic harmonic oscillator potential. In Ref.[6], one begins with a relativistic treatment of a harmonic oscillator, but afterwards the timelike excitations are neglected. So here we propose a relativistic kinematical treatment whose results are very similar to those of the dynamical model of Ref. [5] [6] for the classification of low lying baryons.

THREE PARTICLES STATES AND THE POINCARE GROUP

The reduction of the product of three representations of $\underset{\sim}{P}$ has been studied by several authors (see in particular Ref.[7]).

Here, we propose a generalized L-S coupling reduction which treats the three particles symmetrically.

A three particle state will be denoted :

(1) $$\left| [p_1][p_2][p_3]\ \sigma_1\,\sigma_2\,\sigma_3 \right\rangle = \bigotimes_{i=1}^{3} \left| [p_i]\ \sigma_i \right\rangle$$

where $|[p_i]\ \sigma_i\rangle$ is a state of the irreducible representations (m_i, s_i) of $\underset{\sim}{P}$. Inside the ket $|[p]\ \sigma\rangle$, $[p]$ represents the tetrad associated to p[7]. We write also $[p] = L(p)$ a boost which sends the four vector $\mathring{p} = (m, \vec{0})$ into the four vector $p : [p]\,\mathring{p} = p$.

The main steps of our reduction can be summarized as follows :

i) A state $|[\ell]\ j\,j_2\rangle$ of an I.U.R. (M, j) contained in the product of the three I.U.R. can be written :

(2) $$\sum_{\{\sigma_i\}} \int \prod_{i=1}^{3} \frac{d^3\vec{p}_i}{2\omega_i}\, \delta\left(\ell - \sum_{i=1}^{3} p_i\right) \times \int dR\, \mathcal{D}^{j^*}_{j_2 j_2}(R) \sum_{\sigma_i'} \mathcal{D}^{s_i}_{\sigma_i' \sigma_i}(R)$$
$$\times\ \varphi_{\{\sigma_i\}}\left([\Lambda^{-1}p_1],[\Lambda^{-1}p_2],[\Lambda^{-1}p_3]\right) \left| [p_1][p_2][p_3]\ \sigma_1'\,\sigma_2'\,\sigma_3' \right\rangle$$

where

(3) $$R_i = [p_i]^{-1}\Lambda[\Lambda^{-1}p_i] \qquad \Lambda = [\ell]\,R\,[\ell]^{-1}$$

R being a rotation belonging to the little group (stabilizer) of $\overset{\circ}{P} = (M,\vec{0})$.

ii) Choosing the tetrads $[p_i]$ such that :

(4) $$[p_i] = L(p_i \leftarrow P)\,[P]$$

$L(p_i \leftarrow P)$ being the plane Lorentz transformation which carries P into a vector parallel to p_i, we obtain :

(5) $$R_1 = R_2 = R_3 = R$$

Then, we use the three angular momenta (S_1, S_2, S_3) coupling method of Chakrabarti [8], where the spin states are classified according to the eigenvalues of the (democratic) operators : $Z = (\vec{S}_1 \times \vec{S}_2) \cdot \vec{S}_3$, $\vec{S}^2 = (\vec{S}_1 + \vec{S}_2 + \vec{S}_3)^2$, S_z and $\vec{S}_1^{\,2}, \vec{S}_2^{\,2}, \vec{S}_3^{\,2}$:

(6) $$|s, \sigma, \zeta ; s_1, s_2, s_3\rangle = \sum_{\sigma = \sigma_1 + \sigma_2 + \sigma_3} A\begin{pmatrix} s & \sigma & \zeta \\ s_1 & s_2 & s_3 \end{pmatrix} |s_1, \sigma_1 ; s_2, \sigma_2 ; s_3, \sigma_3\rangle$$

iii) The spin part of the composite particles being, in some sense, factorized out, we have now to consider the particles momenta part. In order to have a covariant treatment, it is useful to define covariant momenta [7]. To this extent, we define the P-associated tetrad by introducing the vectors $n_i(P)$ $(i = 1,2,3)$ and $n_0(P) = \frac{P}{M}$ which satisfy :

(7) $$n_\mu(P) \cdot n_\nu(P) = g_{\mu\nu} \qquad (\mu, \nu = 0,1,2,3)$$
$$\det\left(n_0(P), n_1(P), n_2(P), n_3(P)\right) = 1$$

where $g_{\mu\nu} = \operatorname{diag}(+,-,-,-)$

Then, we define the 3-vector $\vec{q}^{(i)}$ $(i = 1,2,3)$ associated to the i-th particle as :

(8) $$q_j^{(i)} = n_j(P) \cdot p_i \qquad (j = 1,2,3)$$

One can easily check that :

(9) $$n_j(P) \cdot p_i = n_j(\overset{\circ}{P}) \cdot \overset{(0)}{p}_i$$

where $\sum_{i=1}^{3} \overset{(0)}{p}_i = \overset{\circ}{P}$, $P = [P]\,\overset{\circ}{P}$

and :

(10) $$\sum_{i=1}^{3} q_j^{(i)} = n_j(P) \cdot \sum_{i=1}^{3} p_i = 0 \quad (j = 1,2,3) \quad ; \quad \sum_{i=1}^{3} \sqrt{\vec{q}^{(i)2} + m_i^2} = M$$

Using the (usual) reduced three-momenta $\vec{q}_\alpha$ and $\vec{q}_\beta$, we can see that the functions $\varphi_{\{\sigma_i\}}(\Lambda^{-1}p_1, \Lambda^{-1}p_2, \Lambda^{-1}p_3)$ can be written : $\varphi_{\{\sigma_i\}}(\Lambda^{-1}P, R^{-1}\vec{q}_\alpha, R^{-1}\vec{q}_\beta)$.

iv) At P fixed, the functions $\varphi_{\{\sigma_i\}}$ are defined on the 5-dimensional closed surface Γ^5 specified by Eq.(10). This surface is topologically equivalent to the surface S^5 of a sphere in the six-dimensional Euclidean space[9]. Moreover the SU(3) functions provide a complete orthonormal set of functions Ψ_n (n denoting the SU(3) quantum numbers) on S^5 [9]. Considering a projection mapping d from Γ^5 to S^5, we can deduce a complete orthonormal set of functions on Γ^5. This set consists of the functions :

(11) $$\varphi_{\lambda\mu L L_z \omega'}(d^{-1}(X)) = W^{-\frac{1}{2}}(X)\, \Psi_{\lambda\mu L L_z \omega'}(X)$$

where X is a point of S^5, which will be specified by five angles[10] $\psi, \varphi, \alpha, \beta, \gamma$, and W is a weight function arising from the change of variables. The indices λ and μ characterize the I.U.R. of SU(3), L specifies the orbital angular momentum of the three particle system, L_z is the component of L along a fixed axis, and ω' is a breaking degeneracy label[11].

In order to still keep the orbital angular momentum among the quantum numbers, and the "democratic" treatment of the particles, we have to impose d to commute with the rotation group O(3), subgroup of SU(3), and with the permutation group $\mathcal{S}_3$ acting on the three particles.
Then the mapping d will be determined uniquely by imposing (third condition) d to conserve the angles between the particles momenta : we reobtain in this way the radial projection suggested by Dragt[9].

v) Finally a state of an I.U.R. (M, j) of $\underset{\sim}{P}$ contained in the product of three I.U.R. can be written :

(12) $$|[P]\, j, j_z\, ; s, \zeta\, ; \lambda, \mu, L, \omega'\rangle = N \int \frac{d^3\vec{P}}{2\omega_P}\, d^3\vec{q}_\alpha\, d^3\vec{q}_\beta\, \delta(P - \sum_{i=1}^{3} p_i)$$
$$\times W^{-\frac{1}{2}} \sum_{\lambda'=|\nu|}^{\lambda} \sum_{\nu=-\lambda'}^{\lambda'} \sum_{\sigma'} a_{\nu,\lambda'}^{\lambda,\mu;L,\omega'}\, \mathcal{D}^{L}_{\nu m}(R_0^{-1})\, \mathcal{D}^{\frac{\lambda'}{2}}_{\frac{\mu}{2}\frac{\nu}{2}}(\varphi, 2\psi - \tfrac{\pi}{2}, 0) \begin{pmatrix} s & L & j \\ \sigma' & m & j_z \end{pmatrix} |[p_1][p_2][p_3]\, s, \zeta, \sigma'\rangle$$

N being a normalization factor, $a^{\lambda,\mu;L,\omega'}_{\nu,\lambda'}$ coefficients insuring the orthonormality of the SU(3) functions, and R_0 being the rotation specified by the Euler angles α, β, γ already introduced.

vi) In the above calculations, we have used particular tetrads $[p_i]$ (see Eq.(4)). In the general case, instead of the states $|[p_i]\ \sigma_i\rangle$ we have to consider the states :

$$\mathcal{D}^{s_i}_{\sigma_i \tau_i}\left([p_i]^{-1} L(p_i \leftarrow \ell)[\ell]\right) |[p_i]\ \tau_i\rangle \tag{13}$$

In this way, if we choose here the tetrad $[\ell]$ as the one appropriate to the light-like charges (see Ref.[1]), we find also for baryons the quark spin rotarions transforming a "current quark" into a "constituent quark". Let us recall that these transformations are Wigner's rotations between the quark rest frame and the baryon rest frame. In fact, one can see that we pass from a current quark state into a constituent quark state by a charge of "standard boost" $[p_i] \longrightarrow L(p_i \leftarrow \ell)[\ell]$. In terms of Wigner's states a simple change of basis has been performed ; in terms of Mackey states[12], a state has been transformed into another one which belongs to the same equivalence class[13] (i.e. the two boosts $[p_i]$ and $L(p_i \leftarrow \ell)[\ell]$ differ only by an element of the stabilizer of $\mathring{p}_i$).

SU(6) x SU(3) AND THE CLASSIFICATION OF BARYONS

So we write our baryon state :

$$|[\ell]\ j, j_z;\ s, \zeta;\ \lambda, \mu, L\ \omega';\ \beta_1, \beta_2, \beta_3\rangle =$$

$$\int \prod_{i=1}^{3} \frac{d^3\vec{p}_i}{2\omega_i} \langle [p_1]\ \tfrac{1}{2}, \Lambda_{13};[p_2]\ \tfrac{1}{2}, \Lambda_{23};[p_3]\ \tfrac{1}{2}, \Lambda_{33} | [\ell]\ j, j_z;\ s, \zeta;\ \lambda, \mu, L, \omega';\beta_1\beta_2\beta_3\rangle \tag{14}$$

$$|[p_1]\ \tfrac{1}{2}\ \Lambda_{13};\beta_1\rangle_q\ |[p_2]\ \tfrac{1}{2}\ \Lambda_{23};\beta_2\rangle_q\ |[p_3]\ \tfrac{1}{2}\ \Lambda_{33};\beta_3\rangle_q$$

where β_i $(i=1,2,3)$ stands for the internal SU(3) indices of the i-th quark : this $SU(3)_{int}$ is a subgroup of the SU(6) group whose representation $\underline{6}$ describes spin $1/2$ quark states.

Then, in analogy with Ref.[1], the group SU(6) x SU(3) appears as a good group for

the classification of baryons, instead of the generally used group $SU(6) \times O(3)_L$. The O(3) orbital momentum group is thus embedded into a larger group SU(3) we shall call "generalized orbital momentum" (G.O.M.) group. The quantum number L, provided from the reduction of $SU(3)_{GOM}$ with respect to O(3), is coupled with the quantum number S of the total quark spin in the same way as in the case of the classification group SU(6) x O(3), in order to give the total angular momentum J.
Using once more Ref.[1], one can show that $SU(6) \times SU(3)_{GOM}$ can be used not only for classifying baryons at rest, but also for collinear baryonic states.

A last problem remains : the baryonic wave functions must be completely symmetric in the internal $SU(3)_{int.}$ and Poincaré quantum numbers. The introduction of the analogous of the Chakrabarti's operator for SU(3)

$$\Upsilon = f^{ijk} \lambda_{i_1} \lambda_{j_2} \lambda_{k_3} \qquad (15)$$

$-f^{ijk}$ being the usual antisymmetric SU(3) coefficients- and the study of the symmetry properties of its eigenfunctions, allow to solve in a simple way this problem[3].

After calculations, we find, for small values of λ and μ (characterizing an irreducible representation of $SU(3)_{GOM.}$):

i) $\lambda = \mu = 0$: the corresponding states build up a 56,L = 0 of SU(6) x O(3);

ii) $\lambda = 1, |\mu| = 1$: the corresponding states build up a 70,L = 1 of SU(6) x O(3);

iii) $\lambda = 2, \mu = 0$: the corresponding states build up a 56,L = 2 of SU(6) x O(3) and a 20,L = 1 of SU(6) x O(3);

iv) $\lambda = 2, |\mu| = 2$: the corresponding states build up a 70,L = 2 ⊕ 0 of SU(6) x O(3).

This scheme presents similarities with that of the symmetric harmonic oscillator model[5][6], as one might have expected, but there are also meaningful differences. In the quoted model, the quantum number n can be compared to λ in $SU(3)_{GOM}$. Then we remark that the only difference between these two models, for λ (or n) ≤ 2, appears for $\lambda = 2$: we do not find in our model the multiplet 56,L = 0 as in Ref.[6]. Moreover, we can note that the use of this SU(6) x O(3) multiplet 56,L = 0, n = 2 for classifying known baryons is not at all clear.

Let us add that a harmonic oscillator treatment of $SU(3)_{GOM}$ is given in Ref.[3].

It is worth to remark that the $SU(3)_{GOM}$ group would also appear in a non relativistic three free quarks treatment analogous to the one proposed here, the Poincaré group being then replace by the Galilei one. Of course, this is not surprising at all since we have done a projection mapping from Γ^5 to S^5 which is just the surface of the non relativistic case. However, in such a study, the Melosh's transformation appears no more, the corresponding Wigner's rotations becoming the identity.

As a conclusion, let us emphasize once more that the results provided by the above specified relativistic kinematics treatment of quarks are quite similar to the ones given by the dynamical harmonic oscillator quark model. So the problem of the behaviour of quarks inside hadrons is once more raised !

REFERENCES

[1] F. BUCCELLA, C.A. SAVOY, P. SORBA
Lettere al Nuovo Cim. 10, 455 (1974).
Proceedings of the 3rd International Colloquium on Group Theoretical Methods (Marseille 1974).

[2] H.J. MELOSH
Phys. Rev. D.9, 1095 (1974).

[3] F. BUCCELLA, A. SCIARRINO, P. SORBA
Preprint Marseille 75/P.729 (May 1975).

[4] R.H. DALITZ
in "Proceedings of the Inter. Conference on Symmetries and Quarks Models" Gordon and Breach, New York (1970).

[5] D. FAIMAN, A.W. HENDRY
Phys. Rev. 173, 1720 (1968).

[6] R. FEYNMAN, M. KISLINGER, F. RAVNDAL
Phys. Rev. D.3, 2706 (1971).

[7] P. MOUSSA, R. STORA
in "Methods in Subnuclear Physics"
Herceg-Novi Summer School (1966), Gordon and Breach.

[8] A. CHAKRABARTI
Ann. Inst. Henri Poincaré 1, 301 (1964).

[9] A. DRAGT
J. Math. Phys. 6, 533 (1965), and
J. Math. Phys. 6, 1621 (1965).

[10] R.C. WHITTEN, F.T. SMITH
J. Math. Phys. 9, 1103 (1968).

[11] G. RACAH
Rev. Mod. Phys. 21, 494 (1968).

V. BARGMANN, M. MOSHINSKY
Nucl. Phys. 18, 967 (1960), and
Nucl. Phys. 23, 177 (1961).

B.R. JUDD, W. MILLER Jr., J. PATERA and P. WINTERNITZ
J. Math. Phys. 15, 1787 (1974),
and references therein.

[12] A.P. BALACHANDRAN, J. NILSON, L.O'RAIFEARTAIGH
Nucl. Phys. B.49, 221 (1972).

[13] P. SORBA
Thèse de Doctorat d'Etat, Marseille (1974).

75/P.742

JULY 1975

* Istituto di Fisica dell'Università - Roma (Italy)

** Fellow of C.N.R. (Italy) - On leave of absence from Istituto di Fisica Teorica dell'Università - Napoli (Italy)

+ Centre de Physique Théorique, C.N.R.S. Marseille

POSTAL ADDRESS : Centre de Physique Théorique - C.N.R.S.
31, chemin Joseph Aiguier
13274 MARSEILLE CEDEX 2 (France)

WAVE EQUATIONS FOR EXTENDED HADRONS

by

W. Drechsler

Max-Planck-Institut für Physik und Astrophysik, München, Fed.Rep.of Germany

Abstract: A formalism describing extended hadrons is presented using generalized wave functions defined on a fiber bundle constructed over space-time. The structural group of the bundle is taken to be the (4+1) de Sitter group acting as a group of motion in a locally defined space of constant curvature [the fiber] possessing a radius of curvature of the order of one Fermi. A gauge theory of strong interaction is formulated in terms of the geometry in such a de Sitter fiber bundle. This geometric description does not require the existence of any constituents for hadrons and leads to three basic nonlinear wave equations of integro-differential type for the hadronic matter wave function.

To set up a formalism capable of describing extended hadronic structures we propose the use of a fiber bundle constructed over space-time as the basic underlying geometric framework relevant for a differential geometric discription of hadronic matter at small distances [1]. The idea is that in analogy to the deviation from a flat Minkowski space-time to the curved Riemannian space-time of general relativity due to the presence of matter influencing the geometry at cosmological i.e. large distances there exists also a deviation from a flat space-time geometry at small distances, i.e. in the immediate vicinity of a hadron. The basic constituting units of matter are hadrons*) which appear in nature as extended structures of definite mass and spin possessing form factors characterized by a length parameter of the order of one Fermi. However, the inner dynamics and structure of hadronic states is at the present time still obscure despite the numerous models constructed recently with the aim to describe hadrons as bound states of certain presumed existing point-like constituents like quarks or partons carrying various quantum numbers. To speak, however, of true constituents would require that one is able to separate these constituent parts from the compound states with a reasonably small amount of energy compared to the rest energy of these compound structures**). If this is not the case the notion of a constituent loses its meaning. One has to admit that until today - even in the range of ISR energies - no experimentalist has seen an isolated quark or parton emerging from the interaction region of a high energy collision. When two hadrons collide a whole spectrum of new, short lived, excited hadrons appears. There is, however, up to the present time no direct evidence for the presence of any hadronic subunits or constituents. We think that the theory has to recognize these experimental findings at an early stage in developing a theoretical formalism for the description of hadrons.

In the following we should like to present a framework describing extended hadronic structures without the necessity to introduce any constituents. We base our description on a higher dimensional space, in fact a fiber

---

*) We do not consider leptons at this level of the description. Later we shall briefly indicate what role leptons could play in a geometrical framework constructed to describe hadronic states.

**) Compare W. Heisenberg [2].

space constructed over space-time, and represent hadronic matter by a generalized wave function defined in a world possessing the geometry of such a fiber bundle. To be specific we shall associate with each space-time point $x \epsilon V_4$ a local accompanying four-dimensional space $V_4'(x)$ of constant curvature, i.e. a de Sitter space, having a radius of curvature R of the order of one Fermi*). The local $V_4'(x)$ is, moreover, tangent at x to the base space V_4 representing curved space-time.

The fiber bundle

$$T^R(V_4) = \bigcup_{x \epsilon V_4} V_4'(x) \qquad (1)$$

will be called the de Sitter bundle constructed over space-time since it is locally the topological product of space-time and a de Sitter space $V_4'(x)$, the latter representing the local fiber F_x erected over the point x. The structural group G of the bundle is the $L_{4,1} \equiv SO(4,1)$ de Sitter group acting transitively on the fiber, which is a space isomorphic to the coset space $L_{4,1}/L_{3,1}$. Since the dimensions of base space and fiber are both four, the tangent spaces to space-time and to the fiber are isomorphic to one another allowing thereby the fiber $F_x = V_4'(x)$ to be tangent to space-time at x. A connexion on a fiber bundle possessing this property is called a Cartan Connexion [3].

We now want to study a generalized wave motion on the space $T^R(V_4)$ defined by (1). To this end a generalized wave function $\psi(x,\xi)$ is introduced which is a factorizable function of a space-time variable x and an internal variable $\xi \epsilon V_4'(x)$**) possessing, furthermore, definite representation proper-

---

*) For the discussion in this talk we shall assume R to play the role of a constant phenomenological parameter fixed from the beginning to a value around 10^{-13}cm. It is an interesting possibility to allow R to vary in an x-dependent way in response to the hadronic matter present which will, however, not be considered here.

**) The local de Sitter space $V_4'(x)$ can be embedded in a local five-dimensional pseudo Euklidean space $E_5(x)$. In $E_5(x)$ the $V_4'(x)$ is represented by the hypersurface $\xi^a\xi_a = -R^2$ (one-shell hyperboloid) with $\xi^a = (\xi^k, \xi^5)$; k = 0,1,2,3 denoting the coordinates in $E_5(x)$ and $\xi_a = \eta_{ab}\xi^b$; a = 0,1,2,3,5, with diag $\eta_{ab} = (1,-1,-1,-1,-1)$. The de Sitter group $L_{4,1}$ corresponds to a hyperbolic rotation in $E_5(x)$ leaving the hypersurface $\xi^a\xi_a = -R^2$ invariant.

ties with respect to the local Lorentz group in V_4 as well as the structural de Sitter group operating in the fiber (compare Fig. 1).

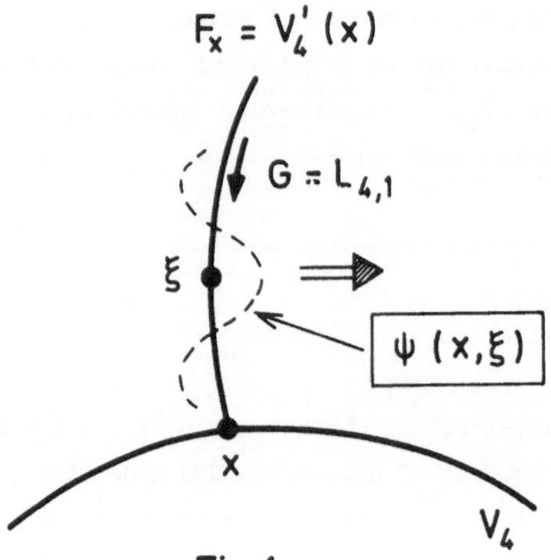

Fig. 1

Concerning the motivation for using the de Sitter structure in the fiber of the bundle $T^R(V_4)$ [which, by the way, reduces in the limit $R\to\infty$ to the familiar tangent bundle, $T(V_4)$, constructed over space-time*)] we mention the following points:

(i) The fiber F_x over x is a four-dimensional space of constant curvature being associated with a length parameter related to strong interaction (although only in a phenomenological way). Moreover, $F_x = V_4'(x)$ is <u>tangent</u> to the base space at the point x of contact. Thereby the internal variable ξ - or rather its image $\tilde{x}$ obtained by stereographically projecting the de Sitter hyperboloid onto the common tangent plane in x - acquires the status of a <u>relative coordinate</u>, i.e. $\tilde{x}$ is measured with respect to x in T_x**) (see Fig. 2, where for

---

*) The de Sitter group, i.e. the group of motion in $V_4'(x)$, contracts in the limit $R\to\infty$ to the Poincaré group, the group of motion in Minkowski space.

**) Actually the fiber could also be a space with dimension bigger than four and the contact at the point x be made through a subspace of the tangent space to the fiber.

ease of drawing the local de Sitter hyperboloid is represented as a circle).

(ii) The ten parameter de Sitter group $L_{4,1}$ which appears as a gauge group in this formalism contains besides the Lorentz subgroup a four parameter family of transformations which correspond to translations in the limit of infinite R. One could therefore say that this de Sitter bundle formalism contains some internal "translational" gauge degrees of freedom.

(iii) The Casimir operators of the de Sitter group mix mass and spin in a way reminiscent of the situation found in the Regge framework. One finds a mass formula [4] connecting mass and spin according to the relation $m^2 = m_o^2 + \frac{1}{R^2} j\,(j+1)$. There are no isospin or hypercharge dependent terms present in this formula which are known to be required to represent the masses of the observed hadrons. Thus one has later to extend the structural group of the bundle to include also generators providing charge-type quantum numbers. Here, however, we shall first focus attention on the differential geometric description of matter wave functions defined on the bundle $T^R(V_4)$ in order to obtain some insight into the inner dynamics of neutral hadronic objects when formulated in this geometrical language.

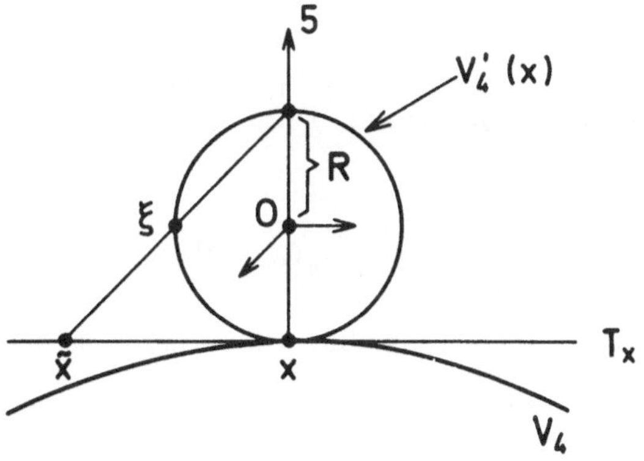

Fig. 2

To be specific we shall from now on assume that the hadronic wave function $\psi(x,\xi)$ is a bispinor, i.e. transforms as a Dirac spinor with respect to the local Lorentz transformations in the base space V_4 and as a [four component] de Sitter spinor with respect to the internal de Sitter transformations operating in the fiber $V_4'(x)$. This means that hadrons are assumed to be de Sitter spinors, i.e. having de Sitter spin $j_{d.S.} = 1/2$. Certain hadronic states could also be represented as de Sitter vectors or tensors possessing an essentially similar nonlinear internal dynamics compared to the situation for de Sitter spinors being treated below. A de Sitter scalar, on the other hand, would behave qualitatively different as far as the dynamics in the fiber are concerned. A wave function $\psi(x,\xi)$ corresponding to $j_{d.S.} = 0$ does in fact not couple to the bundle connexion which in this geometrical description is interpreted to contain the strong interaction effects.

After these remarks we write down the Lorentz and de Sitter gauge transformation properties of a bispinor quantity

$$\psi(x,\xi) = \left(\psi^{AA'}(x,[e_i(x)],\xi,[e_a(\xi)])\right);A,A' = 1,2,3,4 \tag{2}$$

with A being a conventional Dirac spinor index and A' representing the de Sitter spinor index. Furthermore, $[e_i(x)];i = 0,1,2,3$ denotes a local frame of reference at the point x in the base space, whereas $e_a(\xi);a = 0,1,2,3,5$, with $\xi^a e_a(\xi) = o$, represents a local frame of reference in $\xi\epsilon V_4'(x)$. Actually $\psi(x,\xi)$ is a spinorial wave function defined [as far as the de Sitter degrees of freedom are concerned] on the associated de Sitter spinor frame bundle which is related to the principal fiber bundle called the associated de Sitter frame bundle, $L^R(V_4)$, by replacing there the structural group $L_{4,1}$ by the covering group $\tilde{L}_{4,1} \equiv USp(2,2)$ which is identical to the intersection of the groups $U(2,2)$ and $Sp(4,C)$ *). The de Sitter frame bundle

$$L^R(V_4) = \{x,\xi,e_a(\xi)\,|\,\xi^a\xi_a = -R^2,\xi^a e_a(\xi) = 0\} \tag{3}$$

could be represented pictorially by Fig. 3 where again we draw the de Sitter

*) For more details we refer to Ref. [1].

hyperboloid for convenience as a circle.

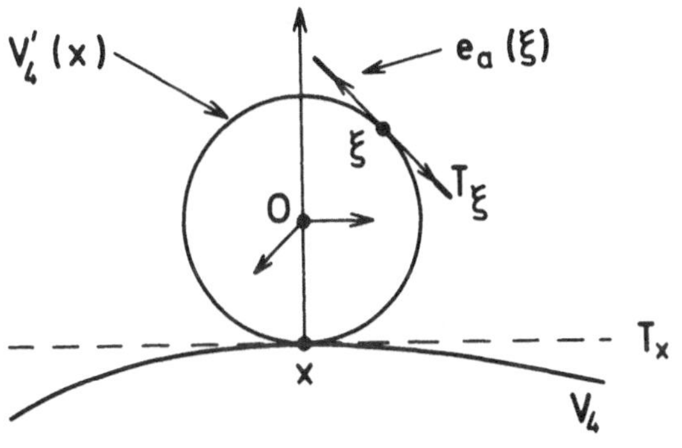

Fig. 3

The bispinor $\psi(x,\xi)$ defined on the de Sitter bundle space transforms now under Lorentz gauge transformations [i.e. changing the local Lorentz frame in $x \epsilon V_4$ in an x-dependent way] as

$$\psi'^{AA'}(x,[e_i'(x)],\xi,[e_a(\xi)]) = [S(x)]^A_B \psi^{BA'}(x,[e_i(x)],\xi,[e_a(\xi)]) \tag{4a}$$

with $S(x) = S(\Lambda(x)) = S(\bar{\omega}_{ik}(x))$ being the 4x4 Dirac representation of the Lorentz group having parameters $\bar{\omega}_{ik}(x) = -\bar{\omega}_{ki}(x)$; $i,k = 0,1,2,3$, and $e_i'(x) = [\Lambda^{-1}(x)]^k_i \, e_k(x)$ with $\Lambda(x) \epsilon L_{3,1}$. In addition $\psi(x,\xi)$ behaves under de Sitter gauge transformations in the following way:

$$\hat{\psi}^{AA'}(x,[e_i(x)],\xi',[e_a'(\xi')]) = [\hat{S}(x)]^{A'}_{B'} \psi^{AB'}(x,[e_i(x)],\xi,[e_a(\xi)]) \tag{4b}$$

with $\hat{S}(x) = \hat{S}(A(x)) = \hat{S}(\hat{\omega}_{ab}(x)) \; \epsilon USp(2,2)$ being the 4x4 spinor representation of the (4+1) de Sitter group possessing the ten x-dependent parameters $\hat{\omega}_{ab}(x) = -\hat{\omega}_{ba}(x)$; $a,b = 0,1,2,3,5$ and $e_a'(\xi') = [A^{-1}(x)]^b_a e_b(\xi)$, $\xi'^a = [A(x)]^a_b \xi^b$ with $A(x) \epsilon L_{4,1}$ where*)

$$[A^{-1}(x)]^a_b \gamma^b = \hat{S}(x)\gamma^a \hat{S}^{-1}(x) \tag{5}$$

---

*) Identical covariant and contravariant indices a or b appearing in a formula are summed over 0,1,2,3,5; identical co- and contravariant indices i,j,k are summed over 0,1,2,3.

defines the homomorphism $USp(2,2) \mapsto L_{4,1}$ and $\gamma^a = (\gamma^i, \gamma^5)$, obeying *) $\gamma^{a\dagger} = \gamma^o \gamma^a \gamma^o$ and $\{\gamma^a, \gamma^b\} = 2\eta^{ab} 1$, being the five anticommuting Dirac γ-matrices. The generators of the group USp(2,2) are given by*) $M^{ab} = \frac{i}{4} [\gamma^a, \gamma^b]$ with M^{ij} generating the Lorentz subgroup and the M^{5i} generating the special de Sitter transformations [de Sitter boosts] corresponding to translations in the limit $R \to \infty$.

The next problem is to define a Cartan connexion on the de Sitter bundle space or, more exactly, a connexion on the associated de Sitter frame bundle. In particular we are interested in a spinor connexion since we want to define a de Sitter gauge invariant differentiation process for a spinor quantity defined on $T^R(V_4)$. An infinitesimal connexion, $\Gamma^R_{(x)}$, on the spinor frame bundle (called a linear connnexion**)) is defined as a one-form on V_4 with values in the Lie algebra of the structural group of the bundle [5], i.e.

$$\Gamma^R(x) = \frac{1}{2} \omega^R_{ab}(x) \, M^{ab} \tag{6}$$

with

$$\omega^R_{ab}(x) = \omega^k(x) \Gamma^R_{kab}(x) \tag{7}$$

where $\omega^k(x) = \lambda^k_\mu(x) dx^\mu$; $k = 0,1,2,3$ [with $\lambda^R_\mu(x)$ being the Vierbein fields] is a local basis of one-forms in V_4 and $\Gamma^R_{kab}(x) = -\Gamma^R_{kba}(x)$; $k = 0,1,2,3,$; $a,b = 0,1,2,3,5$ is a set of 40 connexion coefficients called here for short the de Sitter rotation coefficients in analogy to the familiar case in general relativity. There a spinor connexion is defined as a one-form on V_4 with values in the Lie algebra of the structural group of the Lorentz spinor frame bundle [being the covering group of the Lorentz group], i.e. by [6]

$$\Gamma(x) = \frac{1}{2} \omega_{ik}(x) \, M^{ik} \tag{8}$$

$$\omega_{ik}(x) = \omega^j(x) \, \Gamma_{jik}(x) \tag{9}$$

with $\Gamma_{jik}(x) = -\Gamma_{jki}(x)$; $i,j,k = 0,1,2,3$, denoting the familiar Ricci rotation coefficients.

---

*) {,} denotes, as usual, the anticommutator, [,] the commutator.

**) We shall simply call it a connexion in the following.

It is now easy to write down a two-fold gauge invariant absolute derivative of a bispinor quantity $\psi(x,\xi)$ [disregarding for simplicity electromagnetism]. In matrix notation one has:

$$D\psi(x,\xi) = [d + i\Gamma(x) + i\Gamma^{R}(x)]\ \psi(x,\xi)$$
$$= \omega^{k}(x)[\partial_{k} + i\Gamma_{k}(x) + i\Gamma^{R}_{k}(x)]\psi(x,\xi) \tag{10}$$

with ∂_k denoting the Pfaffian derivative and $\Gamma_k(x) = \frac{1}{2}\Gamma_{kij}(x)\ M^{ij}$ affecting the unprimed spinor index of $\psi(x,\xi)$ and $\Gamma_k(x) = \frac{1}{2}\Gamma^{\ \ R}_{kab}(x)\ M^{ab}$ affecting the primed spinor index of $\psi(x,\xi)$. The operator D is de Sitter gauge invariant in the sense that

$$\hat{S}(x)D\hat{S}^{-1}(x) = \hat{D} = d + i\Gamma(x) + i\hat{\Gamma}^{R}(x) \tag{11}$$

where

$$\hat{\Gamma}^{R}(x) = \hat{S}(x)\Gamma^{R}(x)\hat{S}^{-1}(x) - i\hat{S}(x)d\hat{S}^{-1}(x) \tag{12}$$

with analogous formulae holding true for the Lorentz gauge transformations $S(x)$.

$\Gamma(x)$ in (10) describes long range gravitational effects affecting the Lorentz spinor character of the wave function $\psi(x,\xi)$ whereas the quantities $\Gamma^{\ \ R}_{kab}(x)$, contained in $\Gamma^{R}(x)$, are regarded here to represent the short range strong interaction effects in a geometrical description based on this fiber bundle formalism. Since we are primarily interested in the way the strong interaction phenomena manifest themselves in this framework we shall from now on disregard long range gravitational fields and go over to a flat space-time, i.e. take $V_4 = M$ (Minkowski space) as the base space of the de Sitter bundle*) using, furthermore, Greek indices as labels in Minkowski space [writing, for example, $\Gamma^{R}(x) = dx^{\mu}\Gamma^{R}_{\mu}(x) = dx^{\mu}\frac{1}{2}\Gamma^{\ \ R}_{\mu ab}(x)\ M^{ab}$].

There is an internal - or de Sitter - curvature quantity of supposedly strong interaction origin associated with $\Gamma^{R}_{\mu}(x)$ which is defined by

---

*) For a more careful discussion of this flat space limit see Sect. VIIa of ref. [1].

$[\partial_\mu = \partial/\partial x^\mu;\ \mu = 0,1,2,3]$

$$\mathcal{R}^R_{\mu\nu}(x) = \partial_\mu \Gamma^R_\nu(x) - \partial_\nu \Gamma^R_\mu(x) + i\,[\Gamma^R_\mu(x),\ \Gamma^R_\nu(x)] \tag{13}$$

obeying the transformation rule

$$\hat{\mathcal{R}}^R_{\mu\nu}(x) = \hat{S}(x)\,\mathcal{R}^R_{\mu\nu}(x)\,\hat{S}^{-1}(x) \tag{14}$$

As in general relativity this curvature field is constrained by Bianchi identities which are most concisely expressed through the relation

$$D\Sigma^R(x) = d\Sigma^R(x) + i\,[\Gamma^R(x),\ \Sigma^R(x)] = 0 \tag{15}$$

where

$$\begin{aligned}\Sigma^R(x) &= \frac{1}{2}\,\Omega^R_{ab}(x)M^{ab} = \frac{1}{4}\,dx^\mu \wedge dx^\nu\, R^R_{\mu\nu ab}(x)M^{ab}\\ &= \frac{1}{2}\,dx^\mu \wedge dx^\nu\, \mathcal{R}^R_{\mu\nu}(x)\end{aligned} \tag{16}$$

is a spinorial curvature two-form. In (16) $R^R_{\mu\nu ab}(x)$ represents the de Sitter analogue of the curvature tensor in general relativity [with the first two indices referring to Minkowski space and the last two referring to the internal de Sitter space], and $\Omega^R_{ab}(x)$ is the curvature two-form of de Sitter type defined on $T^R(M)$. Therefore, even in the case of a de Sitter bundle constructed over Minkowski space one expects a non-commutativity of covariant derivatives

$$D_\mu = \partial_\mu + i\Gamma^R_\mu(x) \tag{17}$$

when applied to wave functions $\psi(x,\xi)$ of de Sitter spinor type, which [in the absence of electromagnetic fields] is given by the relation:

$$[D_\mu,\ D_\nu] = i\mathcal{R}^R_{\mu\nu}(x) \tag{18}$$

The $\mathcal{R}^R_{\mu\nu}(x)$ are expected to be nonzero whenever a hadronic wave function $\psi(x,\xi)$ - representing hadronic matter - is not identically vanishing in the fiber above the point x. Moreover, the domain within which the $\mathcal{R}^R_{\mu\nu}(x)$ are expected to be nonzero is related to the distance R characterizing the de Sitter bundle. Physically speaking the $\mathcal{R}^R_{\mu\nu}(x)$ are expected to be nonzero inside hadrons. However, the curvature quantities $\mathcal{R}^{R}_{\mu\nu}(x)$ will be considered to be derived quantities when the primary object - the matter wave function $\psi(x,\xi)$ itself - is known or determined from an equation of motion which will have to be a nonlinear equation since geometry [represented through quantities like $\Gamma^{R}_{\mu}(x)$ and $\mathcal{R}^{R}_{\mu\nu}(x)$] and matter [represented by $\psi(x,\xi)$] influence each other at small distances.

Let us, however, first discuss a linear wave equation for $\psi(x,\xi)$ treating the de Sitter bundle connexion $\Gamma^{R}(x)$ as determined by the geometry, i.e. given from outside. This means that the de Sitter rotation coefficients $\Gamma^{R}_{\mu ab}(x)$ will first be regarded as external strong interaction gauge potentials. It is then easy to write down a de Sitter gauge invariant equation of Dirac type for a wave function defined on $T^{R}(M)$:

$$\gamma^{\mu}[\partial_{\mu} + i\Gamma^{R}_{\mu}(x)]\psi(x,\xi) = -im\psi(x,\xi) \tag{19}$$

We first notice that in this equation base space and fiber are only connected via the spin degrees of freedom of $\psi(x,\xi)$, i.e. via $\gamma^{\mu}\Gamma^{R}_{\mu}(x)$ which is a matrix in Dirac and de Sitter spin space. Besides the operators M^{ab} contained in the $\Gamma^{R}_{\mu}(x)$ in (19) there occur no de Sitter operators like*) $L_{ab}(\xi) = i(\xi_a\partial_b - \xi_b\partial_a)$ representing an orbital motion taking place in the fiber. However, there is indeed an equation of motion for a four component spinor $\phi(\xi)$ defined in a V_4' of constant curvature involving the operators $L_{ab}(\xi)$. Dirac wrote it down long ago in connection with cosmology [7],**)

$$\frac{1}{2R}\gamma^{a}\gamma^{b}L_{ab}(\xi)\phi(\xi) = (\mu + \frac{2i}{R})\phi(\xi) \tag{20}$$

The most surprising property of this equation is the appearance of an imaginary contribution of order $\frac{1}{R}$ to the mass μ. This contribution disappears in the limit $R\to\infty$ when (20) goes over into the conventional Dirac equation in flat Minkowski space. It is suggestive to demand that on the local fiber the wave function $\psi(x,\xi)$ should satisfy the same equation (20) with μ representing some internal mass parameter [a hadronic mass scale], i.e.

---

*) $\partial_a = \partial/\partial\xi^{a}$; $a = 0,1,2,3,5$.

**) Compare also Gürsey and Lee (ref. [8]).

$$\frac{1}{2R}\gamma^a\gamma^b L_{ab}(\xi)\psi(x,\xi) = (\mu + \frac{2i}{R})\psi(x,\xi) \qquad (21)$$

Without pursuing the discussion of the differential equations (19) and (21) further here*) - in particular with respect to a factorized form for $\psi(x,\xi)$, i.e. separating the spinor properties of $\psi(x,\xi)$ in Minkowski and local de Sitter space - we now turn to various possible nonlinear wave equations for $\psi(x,\xi)$ derived from (19).

The idea is that the $\Gamma_\mu^R(x)$ can only be effective as strong interaction gauge potentials, i.e. can only be present in a one particle theory, provided the matter wave function $\psi(x,\xi)$ is nonzero in the fiber above the point x. In other words this means that as a result of a nonvanishing $\psi(x,\xi)$ the geometry reacts with a setting up of a bundle connexion $\Gamma_\mu^R(x)$ [or a set of gauge potentials $\Gamma_{\mu ab}^R(x)$] which are constructed from the $\psi(x,\xi)$ in a bilinear fashion involving an integration over the local fiber. The question then is: can one construct from the wave function $\psi(x,\xi)$ a set of de Sitter rotation coefficients possessing the correct dimension, reality properties and de Sitter transformation character [see (12)] required for the de Sitter bundle connexion coefficients. The answer is that there are in fact <u>three</u> possible connexions $\Gamma_\mu^{R(L)}(x)$, $\Gamma_\mu^{R(L')}(x)$ and $\Gamma_\mu^{R(M)}(x)$ induced by the matter field $\psi(x,\xi)$ which are labelled by a superscript (L), (L') or (M), respectively. To each form for $\Gamma_\mu^R(x)$ belongs a certain integral or normalization constraint to be obeyed by $\psi(x,\xi)$ as a function of ξ in order that the inhomogeneous transformation property (12) for $\Gamma^R(x)$ is guaranteed**). The three possible sets of connexion coefficients are:

$$\Gamma_{\mu ab}^{R(L)}(x) = \kappa^{(L)}(x)\cdot\int\bar{\bar{\psi}}(x,\xi)(1\otimes\mathbf{L}_{\mu ab}(\xi))\psi(x,\xi)d\mu(\xi) \qquad (22a)$$

---

*) For (19) and (21) to be compatible with each other the operators ∂_μ and $L_{ab}(\xi)$ are required to commute according to the relation $[\partial_\mu, L_{ab}(\xi)] = \Gamma_{\mu a}^{Rc}(x)L_{cb}(\xi) - \Gamma_{\mu b}^{Rc}(x)L_{ca}(\xi)$.

**) We like to remark in passing that we do not intend to interprete the ξ-dependence of $\psi(x,\xi)$ in terms of probabilities since this would bring us back to the postulate of point-like constituents moving inside hadrons. This assumption, however, we wanted to abandon for the above-mentioned reasons.

$$\Gamma^{R(L')}_{\mu ab(x)} = \kappa^{(L')}(x)\cdot \int_{V_4'(x)} \bar{\bar{\psi}}(x,\xi)\,(1\otimes \mathbf{L}'_{\mu ab}(\xi))\,\psi(x,\xi)\,d\mu(\xi) \tag{22b}$$

$$\Gamma^{R(M)}_{\mu ab(x)} = \kappa^{(M)}(x)\cdot \int_{V_4'(x)} \bar{\bar{\psi}}(x,\xi)\,(1\otimes \mathbf{M}_{\mu ab})\,\psi(x,\xi)\,d\mu(\xi) \tag{22c}$$

with*)

$$\mathbf{L}_{\mu ab}(\xi) = 1\cdot\frac{1}{2}[\vec{L}_{ab}(\xi)\vec{\partial}_\mu + \overleftarrow{\partial}_\mu \overleftarrow{L}_{ab}(\xi)] \tag{23a}$$

$$\mathbf{L}'_{\mu ab}(\xi) = 1\cdot\frac{1}{2}[\vec{L}_{ab}(\xi)\vec{\partial}_\mu - \overleftarrow{\partial}_\mu \overleftarrow{L}_{ab}(\xi)] \tag{23b}$$

$$\mathbf{M}_{\mu ab} = M_{ab}\cdot\frac{1}{2}[\vec{\partial}_\mu - \overleftarrow{\partial}_\mu] \tag{23c}$$

The associated normalization constraints read:

$$\kappa^{(L)}(x)\cdot \int_{V_4'(x)} \bar{\bar{\psi}}(x,\xi)\,(1\otimes M^{a'b'})\,[\vec{L}_{ab}(\xi)-\overleftarrow{L}_{ab}(\xi)]\,\psi(x,\xi)\,d\mu(\xi) = 2i(\delta^{a'}_{a}\delta^{b'}_{b}-\delta^{b'}_{a}\delta^{a'}_{b}) \tag{24a}$$

$$\kappa^{(L')}(x)\cdot \int_{V_4'(x)} \bar{\bar{\psi}}(x,\xi)\,(1\otimes M^{a'b'})\,[\vec{L}_{ab}(\xi) + \overleftarrow{L}_{ab}(\xi)]\,\psi(x,\xi)\,d\mu(\xi) = 2i(\delta^{a'}_{a}\delta^{b'}_{b}-\delta^{b'}_{a}\delta^{a'}_{b}) \tag{24b}$$

$$\kappa^{(M)}(x)\cdot \int_{V_4'(x)} \bar{\bar{\psi}}(x,\xi)\,(1\otimes[M^{a'b'}M_{ab} + M_{ab}M^{a'b'}]\,\psi(x,\xi)\,d\mu(\xi) = 2i(\delta^{a'}_{a}\delta^{b'}_{b}-\delta^{b'}_{a}\delta^{a'}_{b}) \tag{24c}$$

---

*) $\vec{L}_{ab}(\xi)$ and $\vec{\partial}_\mu$ act to the right on $\psi(x,\xi)$, and $\overleftarrow{L}_{ab}(\xi)$ and $\overleftarrow{\partial}_\mu$ act to the left on $\bar{\bar{\psi}}(x,\xi)$ [the latter being defined below].

The functions $\kappa^{(L)}(x)$, $\kappa^{(L')}(x)$, and $\kappa^{(M)}(x)$ are obtained from (24a-c), respectively, by taking the trace with respect to a,a' and b,b'. The result is, using furthermore (21):

$$\kappa^{(L)}(x) = -\frac{i10}{\eta^R(x)} \tag{25a}$$

$$\kappa^{(L')}(x) = \frac{20}{R\mu\eta^R(x)} \tag{25b}$$

$$\kappa^{(M)}(x) = \frac{i4}{\eta^R(x)} \tag{25c}$$

where

$$\eta^R(x) = \int_{V_4'(x)} \bar{\bar{\psi}}(x,\xi)\psi(x,\xi)d\mu(\xi). \tag{26}$$

Here and in the previous integrals $d\mu(\xi) = \frac{R}{|\xi^5|}d\xi^o\wedge d\xi^1\wedge d\xi^2\wedge d\xi^3$ is the invariant measure on the coset space $L_{4,1}/L_{3,1}$ isomorphic to $V_4'(x)$ *), and $\bar{\bar{\psi}}(x,\xi) = \psi^\dagger(x,\xi)\gamma^o\otimes\gamma^o$ where one γ^o-factor is associated with the Dirac spinor index and the other γ^o-factor with the de Sitter spinor index of $\psi(x,\xi)$. The same direct product notation is used in (22a-c) and (24a-c). With the help of the expressions (22a-c) one finally obtains now three basic, by construction gauge invariant, nonlinear equations for the bispinor wave function $\psi(x,\xi)$ defined on the de Sitter bundle constructed over Minkoswi space [together with the corresponding de Sitter gauge invariant normalization constraints (24a-c)].

---

*) The space V_4' is noncompact in the ξ^o-direction. Correspondingly the integrals of the type (26) and similar ones appearing above are formally divergent. We, therefore, have to interpret the measure on V_4' either as a normalized measure given by $d\mu_N(\xi) = d\mu(\xi)/\int_{V_4'(x)} d\mu(\xi)$ or require the functions $\psi(x,\xi)$ to be of bounded support on V_4' [remember that the points at infinity in V_4' transform into themselves under hyperbolic rotations in E_5 leaving the hyperboloid $\xi^a\xi_a = -R^2$ and the cone $\xi^a\xi_a = 0$ invariant]. We, however, point out that in the final nonlinear equations (27a-c) only ratios of integrals over the local fiber will occur. We shall, therefore, adhere for simplicity to the formal expressions based on the measure $d\mu(\xi)$ without, moreover, specifying the support properties of $\psi(x,\xi)$ in the variable ξ in detail.

$$\gamma^\mu \partial_\mu \psi(x,\xi) + \frac{5}{\eta^R(x)} \Big[\int_{V_4'(x)} \bar{\bar{\psi}}(x,\xi')(1 \otimes \mathbf{L}_{\mu ab}(\xi'))\psi(x,\xi')\, d\mu(\xi') \Big] \cdot \gamma^\mu M^{ab} \psi(x,\xi) = -im\psi(x,\xi) \tag{27a}$$

$$\gamma^\mu \partial_\mu \psi(x,\xi) + \frac{10i}{R\mu\eta^R(x)} \Big[\int_{V_4'(x)} \bar{\bar{\psi}}(x,\xi)(1 \otimes \mathbf{L}'_{\mu ab}(\xi'))\psi(x,\xi')\, d\mu(\xi') \Big] \cdot \gamma^\mu M^{ab} \psi(x,\xi) = -im\psi(x,\xi) \tag{27b}$$

$$\gamma^\mu \partial_\mu \psi(x,\xi) - \frac{2}{\eta^R(x)} \Big[\int_{V_4'(x)} \bar{\bar{\psi}}(x,\xi)(1 \otimes \mathbf{M}_{\mu ab})\psi(x,\xi')\, d\mu(\xi') \Big] \cdot \gamma^\mu M^{ab} \psi(x,\xi) = -im\psi(x,\xi) \tag{27c}$$

The nonlinear terms in these integro-differential equations for $\psi(x,\xi)$ represent the de Sitter analogue of a spin-orbit coupling (LM) and (L'M) [in (27a) and (27b), respectively], and of a spin-spin coupling (MM) [in (27c)] for the internal motion taking place in the fiber. The reason for the appearance of two different spin-orbit couplings is due to the nonhermitean nature of the operator $\gamma^a\gamma^b L_{ab}(\xi)$ occurring in (21). We finally remark that the wave equations (27a-c) - which are expected to determine each a hadronic mass spectrum - are still differential equations with respect to space-time since the integrations only refer to the internal variable ξ varying in the local fiber above the space-time point x.

R E F E R E N C E S

[1] W. Drechsler, Wave Equations on a de Sitter Fiber Bundle, to be published in Fortschritte der Physik, Vol. 23, 1975;

[2] W. Heisenberg, Talk presented at the Spring Meeting of the Deutsche Physikalische Gesellschaft, München, March 1975;

[3] Ch. Ehresmann, Colloque de Topologie (espaces fibré), Bruxelles, 1950, p. 29;

[4] A.O. Barut and A. Böhm, Phys. Rev. 139, 1107 (1965);

[5] A. Lichnerowicz, Théorie globale des connexions et des groupes d'holonomie, Edizioni Cremonese, Roma 1962;

[6] D.R. Brill and J.A. Wheeler, Rev. Mod. Phys. 29, 465 (1957);

[7] P.A.M. Dirac, Ann. of Math. 36, 657 (1935);

[8] F. Gürsey and T.D. Lee, Proc. Nat. Acad. Sc. 49, 179 (1963).

COVARIANCE PRINCIPLE AND COVARIANCE GROUP IN PRESENCE OF EXTERNAL E.M. FIELDS

N. Giovannini

Institute for Theoretical Physics, University of Nijmegen,

Toernooiveld, Nijmegen, The Netherlands

Abstract:

The definition of (kinematical) covariance, by means of the properties and the equivalence relations of inertial systems, can give, as is well known, very useful information on the admissible equations of motion and on the group-theoretical concept of an elementary particle (which is defined as an irreducible quantum-mechanical system obeying such an equation). Here we consider the case where an external (classical) e.m. field is present. In this talk we show that for such a system, the relativistic covariance operator group is in general no longer homomorphic to the covering group of the Poincaré group (as in the free particle case), but to contain it only as a factor group. This larger covariance group can be derived on the basis of some simple physical assumptions, independently of any equation of motion.

Its (projective unitary-antiunitary) irreducible representations have been all determined and classified. Some conclusions are drawn concerning the properties of the corresponding covariant equations of motion and a group theoretical definition of an elementary particle in interaction with such a field is proposed (The special case of zero field reduces of course to the known results of Wigner). Some aspects of this approach lead to a possible solution of the so-called a-causality troubles for particles with spin equal to or larger than 1.

Introduction

The group theoretical definition of an elementary particle as a continuous projective unitary irreducible representation of the Poincaré group P is a well known and successful concept, as introduced in the celebrated paper of Wigner [1] . In this frame, the Poincaré group plays the role of the covariance group of special relativity, i.e. embodies the basic postulate of the theory that physics should be the same in any reference frame which is an inertial system. The elementary particles are then characterised by the values of their spin and their squared mass and it is also possible, by pure group theoretical methods, to tackle successfully the description of the time evolution of the states representing such particles, i.e. to derive covariant equations of motion. This is of course all well known, although perhaps not always clearly enough stated.

Let us now introduce a (classical) electromagnetic field and consider a charged particle moving in it. It has often been shown in the past that in most situations the so called "external" approximation, i.e. neglecting the influence of the particle on the field is a very good and useful one. For that purpose, the usual procedure is to introduce a covariant (pseudo-) vector field (the potential) in the free equation of motion by minimal coupling, the displacement operators P_μ being replaced by $\Pi_\mu = P_\mu + \frac{e}{c\hbar} A_\mu(x)$ (e the charge of the particle). Again the results of the theory are very satisfying, at least for spin 0 and $\frac{1}{2}$ particles (we shall discuss later the serious difficulties encountered in the higher spin cases [2-4]).

Physics however depend on fields and not on potentials and should thus be invariant with respect to any change in field components due to a covariant change of reference frame. Introducing a potential in the formalism induces some arbitrariness: first, to a given field does not correspond one potential but a whole class of them, hence not an equation of motion but an infinite (physically undistinguishable) class of them. Second, as a consequence of this, there is a certain choice left for the definition of the transformation law of the potential that we have introduced. Therefore one may require the potential to transform as a covariant (pseudo-) vector field under the Poincaré group but not the modified equation of motion to be Poincaré covariant. Poincaré covariance has only to hold for the (infinite) class of equations corresponding to a given field and not for each element of that class separately. This implies incidentally also that gauge transformations of the second kind become then covariant transformations for this class, too, this giving rise to difficult to handle infinite dimensional groups [5] . This is of course not new, and a natural alternative could be to develop a formalism without introducing a potential, but based only on fields, as it has been done successfully for Dirac particles (see e.g. [4] p. 21). A generalization of such a formalism to higher spins is not known. The reason is probably due to the too fundamental role of gauge invariance, even in the external approximation frame.

In this talk we choose a way inbetween, using the advantages of these two formalisms, i.e. we use potentials, but we get rid of the arbitrariness described above by fixing in some convenient way the gauge, for any given field. In this way it is possible to build up group theoretically a theory of elementary particles in interaction with an external field, by construction first of the relevant covariance operator group, independently of any equation of motion, and by the analyse then of the representations of this group. It is of course not possible here to go into details of this work. This will be done in subsequent publications [6-7] . We just would like in this talk to sketch the fundamental ideas, the way we constructed the relevant covariance group and indicate some possibilities opened by our results. We offer then our's apologies for the then unavoidable lacks of this discussion.

1. Sketch of the approach

Let us begin by considering a simple example, namely the problem of a charged massive (relativistic) particle (say of spin $\frac{1}{2}$) moving in an external constant uniform field F. In the usual procedure the (free) Dirac equation is then modified as described in the introduction : the displacement operator P_μ is replaced by $\Pi_\mu = P_\mu + \frac{e}{c\hbar} A_\mu$, with e the charge of the particle, A a potential of F. In order to understand better what does this change conceptually mean, let us first go further, in the usual way. It is quite clear that one cannot expect that the physical results (states, matrix elements, selection rules etc) will be different for two observers related by a translation transformation in space or time, since the field remains then unchanged, whereas the equation of motion cannot have this symmetry, the potential being necessarily x-dependent. It is however possible [8] to define an operator combining the translation with a gauge transformation which does commute with the equation of motion, since necessarily $(T_a \cdot A)_\mu(x) \equiv A_\mu(x-a) = A_\mu(x) + \partial\chi_a(x)$ for some gauge function $\chi_a(x)$ so that one can, on A, define pairs: $\{\chi_a, a\} \cdot A_\mu(x) \overset{def}{=} (T_a \cdot A)_\mu(x) - \partial\chi_a(x) = A_\mu(x)$. The resulting group of transformations has been shown to be a nontrivial extension of $\mathbb{R}$ (the constant gauges) by $\mathbb{R}^4$ (the translations), this because it is in this case not possible to fix the arbitrary constants of the gauge transformations such that $\chi_a(x) + T_a\chi_{a'}(x)$ becomes equal to $\chi_{a+a'}$ $\forall a, a' \in \mathbb{R}^4$ and the relevant symmetry group, of large use for the description of the solution (see [9]) is thus <u>not</u> a subgroup of the Poincaré group P.

Symmetry (equality of physical systems) is however necessarily a particular case of covariance (equivalence of physical systems) so that the relevant covariance group cannot be the Poincaré group we started with by taking the Dirac equation. We have thus to construct, in a similar way as for the symmetry just described, the relevant covariance group and ask ourselves: is the Dirac equation (and other free particle equations) with minimal coupling covariant with respect to this new group? The answer will be yes for spin 0 and ½ (Klein-Gordon and Dirac) but no for example for spin 3/2 (Rarita-Schwinger equation). We shall see that there is a relation with the troubles arising from the higher spin equations in the external approximation.
How is now defined the covariance property? Let us, for simplicity, consider a scalar equation described by a differential operator $\mathcal{O}(x,\pi F)$ on an Hilbert space $\mathcal{H}$ (to be specified later on) of functions $\psi(x,\pi F)$, with $x \in \mathcal{M}(4)$ the Minkovski space, and which depends on the field F via a potential $A \equiv \pi F$, with π an arbitrary (but fixed) choice of gauge. The states $\psi(x,\pi F)$ are thus solutions of the equation

$$\mathcal{O}(x,\pi F)\,\psi(x,\pi F) = 0 \tag{1.1}$$

Let now P_g be the substitution operator on the x-variable corresponding to $g \in P$. We have in general, since covariance under Poincaré is only true up to a possible gauge transformation as we saw,

$$\begin{aligned} P_g\,\mathcal{O}(x,(\pi F)(x))\,P_g^{-1} &= \mathcal{O}(g^{-1}x,\ (\pi F)(g^{-1}x)) \\ &= \mathcal{O}(x,\pi(gF) + \partial\chi_g(\pi F,x)) \end{aligned} \tag{1.2}$$

for some gauge transformation $\chi_g(\pi F,x)$. We construct now for each Poincaré transformation g (since Maxwell equations are invariant under P, each g must induce a covariance transformation) an operator $U_{m(g)}$ with

$U_{m(g)} = U_{\chi_g}(\pi F,x) \cdot P_g$ and

$$\begin{aligned} U_{m(g)}\,\mathcal{O}(x,\pi F)\,U_{m(g)}^{-1} &= U_{\chi_g}\,\mathcal{O}(x,\pi(gF) + \partial\chi_g(\pi F,x))\,U_{\chi_g}^{-1} \\ &= \mathcal{O}(x,\pi(gF)) \end{aligned} \tag{1.3}$$

getting so rid of the arbitrariness described in the introduction. The equation (1.3) is nothing else than the definition of the covariance property of the operator $\mathcal{O}$ under the group $\mathbb{M} = \{m(g)\}$ whose $U_{m(g)}$ do form a representation, which we shall choose irreducible to define $\mathcal{O}$.
Our purpose is just to derive this group $\mathbb{M}$ first and then, from the structure of its irreducible representations, obtain information about the covariant

(i.e. satisfying (1.3)) equations of motion. Since this group $\mathbb{M}$ is obviously completely (up to an uninteresting possible kernel) defined by its action on the potential space we construct it as follows. We choose a fixed map π as above for each field and construct pairs of gauge and Poincaré transformations m(g) in such a way that the following diagram

$$\begin{array}{ccc} F & \xrightarrow{\ \pi\ } & A \\ \downarrow g & & \downarrow m(g) \\ gF & \longrightarrow & \pi(gF) \end{array} \qquad (1.4)$$

is commutative $\forall\ g \in P$. In other words we construct a representation of $\mathbb{M}$ on a fixed cross-section of the potentials of a given external e.m. field and of its Poincaré transforms. The resulting operator group $\underline{Q}(\pi F)$ has been shown in [6] to have the following properties

(i) it appears in the following exact sequence of groups

$$1 \longrightarrow \mathbb{R}(\pi F) \longrightarrow \underline{Q}(\pi F) \longrightarrow P \longrightarrow 1,\ \underline{f},\ \psi \qquad (1.5)$$

where $\mathbb{R}(\pi F)$ is an additive group of real functions of F, independent of x which is generated by the factor set (of functions) $\underline{f}\ (g,g')\ (\pi F)$, and $\psi\ (g)\ \varphi(\pi F) \overset{\text{def}}{=} \pm\ \varphi\ (\pi(g^{-1}F)),\ \forall \varphi \in \mathbb{R}(\pi F)$ (the -sign taking into account of the antiunitary character of transformations containing time reversal). The factor set is thereby defined by

$$\underline{f}\ (g,g')\ (\pi F) \overset{\text{def}}{=} \chi_g\ (\pi(g^{-1}F),x) + \psi(g)\ \chi_{g'}\ (\pi(gg')^{-1}\ F,x) -$$

$$- \chi_{gg'}\ (\pi(gg')^{-1}\ F,x)$$

and it has been calculated in [6], using a particular convenient map π, to be equal to (for $g = (a,\Lambda)$, $g' = (a',\Lambda')$ a,a' = 4-translations, Λ, Λ' Lorentz transformations)

$$\underline{f}\ \ (g,g')\ (\pi F) = \tfrac{1}{2}\ (gg'\ F^{(o)})_{\varrho\sigma}\ \ (\Lambda a')^{\varrho}\ \ \ (a)^{\sigma} \qquad (1.6)$$

where $F^{(o)}$ is defined a s follows: we assume that the Fourier transform $\hat{F}(k)$ of F(x), with respect to the 4-dimensional Lebesgue measure exists (in the sense of distributions) and define $\hat{F}^{(o)}(k) = \hat{F}(o)$ if k = 0 and 0 else. $F^{(o)}$ is then obtained by Fourier transforming back this pseudo-tensor distribution $\hat{F}^{(o)}(k)$, and is assumed to consist only in a constant uniform field. It is easy to see that even with these restrictions a very large class of (inhomogeneous) fields is covered. It is also possible to get rid of these conditions but this will not be done here. $F^{(o)}$ will be called the constant uniform (c.u.) part of the field.

(ii) $\underline{Q}\ (\pi F) \cong \underline{Q}\ (\pi F + \partial\xi\ ) = \underline{Q}\ (\pi(gF))$

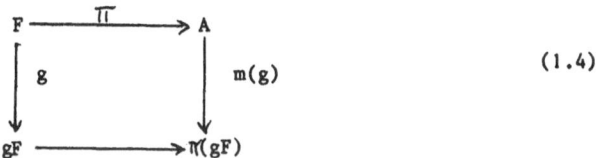

for ξ any gauge function, and all $g \in P$, so that the isomorphism class of this operator group does not depend on the particular choice of gauge we have made, nor on the choice of a particular observer, and this is of course necessary for consistency.

2. The abstract covariance group $\mathbb{M}$

It is now possible, by letting F vary in the considered class of fields to derive the abstract covariance group $\mathbb{M}$: it follows from (1.6) that the space $\mathbb{R}(\pi F)$ is spanned by the 6-dimensional (antisymmetric contravariant 4x4 tensor) space $T \wedge T \overset{\mathrm{def}}{=} \mathbb{B}$ of linear functions on c.u. e.m. fields. A basis thereof is given by the elements $\{ E_{\mu\nu} = e_\mu \wedge e_\nu \, , \ \mu,\nu = 0,1,2,3 \}$ with $\{ (e_\mu)^\nu = \delta_\mu{}^\nu \}$ a basis of the Minkovski space (and an element $B \in \mathbb{B}$ is then given by $B = B^{\mu\nu} E_{\mu\nu}$, $B^{\mu\nu} \in \mathbb{R}$). The group $\mathbb{M}$ may be then shown to appear in the following commutative diagram of exact sequences (with splitting vertical extensions):

$$\begin{array}{ccccccc} \mathbb{B} & \rightarrowtail & K & \twoheadrightarrow & U & , & A(a,a')\ ,\ \varphi \\ \| & & \downarrow & & \downarrow & & \\ \mathbb{B} & \rightarrowtail & \mathbb{M} & \twoheadrightarrow & P & , & A(g,g')\ ,\ \varphi \\ & & \downarrow & & \downarrow & & \\ & & L & = & L & & \end{array} \qquad (2.1)$$

with U the 4-dimensional translation group, L the Lorentz group, K a 10-dimensional nilpotent Lie group and with factor sets

$$A(g,g')^{\sigma\varsigma} = \tfrac{1}{2} \ ((\Lambda a')^\sigma \wedge \ a^\varsigma) \qquad (2.2)$$

and $A(a,a')$ the restriction of (2.2) on U x U. The action φ is defined as usual on contravariant 4x4 tensors. The Lie algebra of M has now the following structure: denoting the infinitesimal generators $M_{\mu\nu}$ for a rotation in the μ-ν plane of space-time, Π_μ for a translation along e_μ, and $\mathbb{F}_{\mu\nu}$ for the element $E_{\mu\nu}$ we have (with $g_{oo} = -g_{ii} = -1$):

$$\begin{aligned} [M_{\mu\nu}, M_{\varsigma\sigma}] &= g_{\mu\nu} M_{\varsigma\sigma} + g_{\nu\sigma} M_{\mu\varsigma} - g_{\nu\varsigma} M_{\mu\sigma} - g_{\mu\sigma} M_{\nu\varsigma} \\ [M_{\mu\nu}, \mathbb{F}_{\varsigma\sigma}] &= g_{\mu\nu} \mathbb{F}_{\varsigma\sigma} + g_{\nu\sigma} \mathbb{F}_{\mu\varsigma} - g_{\nu\varsigma} \mathbb{F}_{\mu\sigma} - g_{\mu\sigma} \mathbb{F}_{\nu\varsigma} \qquad (2.3) \\ [\Pi_\mu , \Pi_\nu] &= \mathbb{F}_{\mu\nu} \\ [M_{\mu\nu} , \Pi_\sigma] &= g_{\mu\sigma} \Pi_\nu - g_{\nu\sigma} \Pi_\mu \end{aligned}$$

all other commutators vanish. This Lie algebra is actually not unknown and has even a (very adapted) name, as proposed first by S.L. Glashow (see Stein [10] : the Maxwell Lie-algebra (see also [11] and [12] for further details). It was derived in an actually much more specific context, since all these papers do consider only the problem of a constant uniform field and hence obtain the generator $\mathbb{F}_{\mu\nu}$ as the eigenvalue it takes then (see [7]) in a given c.u. field, namely (up to a constant) the field itself. Further this Lie-algebra is there constructed from the Klein-Gordon or Dirac equation with minimal coupling. This shows incidentally since we have made no use of such an equation that these equations characterize representations of our group, i.e. are covariant under $\mathbb{M}$. Our derivation and interpretation are then clearly completely different and our goals somehow more ambitious and we therefore have explained in [7] why we could not share completely the point of view of these authors.

We refer also to this paper [7] for a complete description of all projective unitary/antiunitary irreducible representations (short PUAIR) of $\mathbb{M}$ which characterize the various covariant equations of motion and hence the (group theoretical) notion of "elementary particle in presence of an external e.m. field", obviously defined as a set of states in a separable Hilbert space which carries a PUAIR of $\mathbb{M}$. It turns out also from our results in [7] that, as soon as the c.u. part $F^{(o)}$ of the field vanishes, this definition reduces to the well known one of Wigner [1] . In this talk we shall only show how these representations give a possible solution of the so called troubles in the external approximation. Let us therefore restrict ourselves to the following problem:

3. A charged particle of "spin" 3/2 in a c.u. magnetic field $\vec{B}$

It is well known that the spin of a particle is a characterization of a PUAIR of the covariance group. It has thus to be "redefined" since, because of the external approximation, this covariance group has changed. We have shown in [7] that, in the case of a c.u. magnetic field, particles do not "remember the spin they had when they were free" as usually implicitely assumed but have an intrinsic integer or half integer spin j of "helicity-type" (i.e. with only 1 component and with only two polarisation states $\pm$ j, and this is actually what is observed physically). As a consequence, a free 3/2 spin representation does, when this external field is present, split in two sub-representations of the new covariance group $\mathbb{M}$: one characterized by the "helicity" spins $\pm$ 3/2 and the other by $\pm \frac{1}{2}$. It is then possible to show that (with a gyromagnetic factor of 1/s) this new "particle" is covariantly described by a direct sum of two Dirac equations, one for each subrepresentation. That this equation is not equivalent to the usual known ones can be seen by the following: it has been shown by Velo and Zwanziger [2-4] that the 3/2 (free-) spin equations with minimal coupling in an external c.u. magnetic field (such as the Rarita-Schwinger equation for instance) are a-causal in the sense that the propagators do not vanish for space-like vectors, whereas the Dirac (as the Klein-Gordon) equation is free of acausality and thus so is our equation, too. This is a quite strong hint that the origin of these pathological difficulties lies in an incorrect use of covariance and in the consequences thereof.

Let us remark also that the covariant equations of motion are then not necessarily analogous for various kinds of fields. However, because Poincaré covariance just requires such a similitude for fields related by Poincaré transformations this is as consistent as the fact that a positive or a massless free particle do not necessarily obey similar equations.

We refer to [6-7] for further details and for a less qualitative description of this problem of covariance within the external approximation.

All what we wanted to say here, was to show that this external approximation is conceptually more far reaching than perhaps expected and to show that a systematical treatment of this covariance problem does open very promising possibilities.

References

[1] E.P. Wigner, Ann. of Math. 40, 149 (39).

[2] G. Velo and D. Zwanziger, Phys. Rev. 186, 1337 (69).

[3] G. Velo and D. Zwanziger, Phys. Rev. 188, 2218 (69).

[4] A.S. Wightman,"Troubles in the external field problem for invariant wave equation" ed. Gordon and Breach (1971).

[5] U. Cattaneo and A. Janner, J. Math. Phys. 15, 1155 (74) and 15, 1166 (74), G. Rideau, Phys. Scripta, 9, 163 (74).

[6] N. Giovannini, Covariance in presence of external e.m. field, preprint (june 1975), (to be published).

[7] N. Giovannini, Elementary particles in external e.m. fields, preprint (june 1975), (to be published).

[8] A. Janner and T. Janssen, Physica 53, 1 (71).

[9] L.C. Chen and T. Janssen, Physica 77, 290 (74).

[10] E.M. Stein in "High energy and elementary particles", Trieste, 1965, IAEE Vienne, p. 563.

[11] H. Bacry, Ph. Combe and J.L. Richard, N. Cim. 70A, 267 (70).

[12] R. Schrader, Fort. der Phys. 20, 701 (72)

DYNAMICAL SU(3) MODEL FOR STRONG INTERACTIONS AND ψ PARTICLES

M.Hongoh
Centre de Recherches Mathematiques
Université de Montréal
Montréal, Québec, Canada

Abstract :

We propose as a mechanism for the internal excitation of the hadrons a simple model based on the group embedding $SU(3) \rightarrow SU(3.1)$. Ex cited states of the quark (antiquark) belong to the representation $(\ell 0)$ $((0\ell))$ of the maximal compact subgroup SU(3). The scheme leads to a definite predictions for the vector meson spectrum produced in the e^+e^- - annihilation. The SU(3.1) interpretation of ψ particles is briefly discussed.

The discovery of massive particles[1] $\psi(3105)$, $\psi'(3695)$ and a new member called ?(4.1)[2] has invited us to further theoretical speculations. Extremely narrow widths of ψ and ψ' strongly indicate the existence of a new degree of freedom which may or may not be a charm quantum number. The purpose of this report is to draw physicists' attention to the dynamical SU(3)[3] scheme and its predictions for the e^+e^- annihilation. For the sake of clarity, we focus our attention onto the internal symmetry. Spins and the spatial part of the entire wave function may be furnished later. The model is then based on the group embedding SU(3) $\rightarrow$ SU(3.1) which is one of the two dynamical groups for the 3-dimensional harmonic oscillator[4]. The role of the dynamical group for the hydrogen atoms is well known[5]; the energy levels of the hydrogen atoms can be changed by the generalized shifting operators of the dynamical group O(4.2). The fact that SU(3) is the invariance group for the 3-dimensional harmonic oscillator is also well known, but perhaps it should be emphasised here again. The highly degenerate spectrum of the 3-dimensional harmonic oscillator with the degeneracy $\frac{1}{2}(n+1)(n+2)$ can be described completely by the totally symmetric (antisymmetric) representations $(n0)$ $((0n))$ of SU(3)[6]. Increasing (decreasing) the total number of quanta by one, we obtain the next energy level of the harmonic oscillator $\hbar\omega(n \pm 1 + \frac{3}{2})$. This, on the other hand, corresponds to changing the total number of quarks or antiquarks by one. Since the SU(3) representation $(\ell 0)$ $((0\ell))$ can be realized by the quark (antiquark) basis of polynomials of degree ℓ, the next energy levels of the oscillator correspond to the representation $(\ell \pm 1,0)$ $((o,\ell \pm 1))$, i.e. higher SU(3) representations $(\ell 0)$ and (0ℓ) are obtained in such a way that the higher energy levels of the 3-dimensional harmonic oscillator are reached by the dynamical group. Besides exceptional groups[7], SU(3) can be embedded in only two rank 3 groups, SU(3.1) and SP(3.R).[8] Let us take the non-hermitian Cartan-Weyl basis, $E_{\alpha\beta}$ that satisfy the commutation relation and the hermiticity conditions,

$$[E_{\alpha\beta}, E_{\gamma\delta}] = \delta_{\alpha\delta} E_{\gamma\beta} - \delta_{\beta\gamma} E_{\alpha\delta}, \qquad \alpha\beta\gamma\delta = 1234 \tag{1a}$$

$$(E_{ij})^\dagger = E_{ji}, \quad (E_{i4})^\dagger = -\underline{E_{4i}}, \quad (E_{\alpha\alpha})^\dagger = E_{\alpha\alpha} \tag{1b}$$

Among 15 generators of SU(3.1)[9] six shifting operators and two diagonal generators span the maximum compact subgroup SU(3),

$$\sqrt{2}\,F_i^{+} = E_{jk}\,,\quad \sqrt{2}\,F_i^{-} = E_{kj} \qquad ;\ i\,j\,k = 1.2.3.$$

$$H_1 = \tfrac{1}{2}(E_{11} - E_{22})\,,\quad H_2 = \tfrac{1}{3}(E_{11} + E_{22} - 2E_{33}) \tag{2}$$

There exist six generalized shifting operators and one diagonal generator,

$$\sqrt{2}\,G_i^{+} = i\,E_{i4}\,,\quad \sqrt{2}\,G_i^{-} = i\,E_{4i}$$

$$H_3 = \tfrac{1}{4}(E_{11} + E_{22} + E_{33} - 3E_{44}) \tag{3}$$

Introducing $F_i^3 = \frac{1}{2}(E_{jj} - E_{kk})$, the commutation relations are

$$[F_i^{+}, F_i^{-}] = F_i^{3}\,,\quad [F_i^{3}, F_i^{\pm}] = \pm F_i^{\pm} \tag{4a}$$

from which we see that $\{F^{+}, F^{-}, F^{3}\}_i$ span I,U,V spin of SU(3). Similarily with $G_i^3 = \frac{1}{2}(E_{ii} - E_{44})$, we have

$$[G_i^{+}, G_i^{-}] = G_i^{3}\,,\quad [G_i^{3}, G_i^{\pm}] = \pm G_i^{\pm} \tag{4b}$$

$\{G^{+}, G^{-}, G^{3}\}_i$ span three non-independent, non-compact subgroups which are locally isomorphic to SU(1.1). Note that $(F_i^{+})^{\dagger} = +F_i^{-}$ and $(G_i^{+})^{\dagger} = -G_i^{-}$. The linear sum of the eigenvalues of three diagonal generators leads to the generalized Nishijima-Gell-Mann relation,

$$\tilde{Q} = I_3 + \tfrac{1}{2}Y + Z(\ell)\,, \tag{5}$$

where Z is the eigenvalue of H_3, which depends on the SU(3) symmetry label ℓ.[10] Denoting the general unitary representation of SU(3.1) by D(IYZ)[11], we take the lowest representation $D^{\pm}(00Z^{\pm})$ $(Z^{+} \geqslant +\frac{3}{4},\ Z^{-} \leqslant -\frac{3}{4})$ which correspond to the quark-antiquark tower, respectively (Fig. 1). The SU(3.1) IR then decomposes into an infinite direct sum of totally symmetric or anti-symmetric representation of SU(3) without multiplicity. Each SU(3) representation is equally separated along the Z axis and related to each other by the matrix elements of the generalized shifting operators, $G_0^{\pm} = \sum_i G_i^{\pm}$,

$$G_0^{\pm}|I_3 Y Z\rangle = \sum_{I_3', Y'} \langle I_3' Y' Z\pm 1|\,G_0^{\pm}\,|I_3 Y Z\rangle\,|I_3' Y' Z\pm 1\rangle. \tag{6}$$

The normalized quark, antiquark basis are[12]

$$|I_3 Y Z\rangle_Q = \left[(I+I_3)!(I-I_3)!(I-\tfrac{3}{2}Y)!\right]^{-\frac{1}{2}}\left[(-\theta-1)!\right]^{\frac{1}{2}}\cdot p^{I+I_3}\, n^{I-I_3}\, \lambda^{I-\frac{3}{2}Y}\, \gamma^{\theta}, \tag{7a}$$

$$|I_3 Y Z\rangle_{\bar{Q}} = \left[(I+I_3)!(I-I_3)!(I+\tfrac{3}{2}Y)!\right]^{-\frac{1}{2}}\left[(-\tau-1)!\right]^{\frac{1}{2}}\cdot \bar{p}^{I+I_3}\, \bar{n}^{I-I_3}\, \bar{\lambda}^{I+\frac{3}{2}Y}\, \bar{\gamma}^{\tau}, \tag{7b}$$

where $\theta = I - \frac{1}{2}Y - \frac{4}{3}Z$ and $\tau = I + \frac{1}{2}Y + \frac{4}{3}Z$. We associate with $(\ell 0)$ $((0\ell))$ the the ℓ-th excited states of quarks (antiquarks) and as a meson structure the bound state of them, i.e. mesons $\sim \bar{Q}^{\ell} Q^{\ell'} \delta_{\ell\ell'}$. Then they belong to the product representation $(\ell 0)$ x (0ℓ). For $\ell=1$ we obtain everything we know from the eight-fold way, i.e. the familiar meson nonet. For $\ell=2$ we have

$$(22) + (11)^{*} + (00)^{*}. \tag{8}$$

Once expressed in terms of the quark-antiquark basis (7), the starred representations in (8) carry an extra power of the general SU(3) scalars, $S \sim \frac{1}{\sqrt{3}}(p\bar{p} + n\bar{n} + \lambda\bar{\lambda})$. Note that SU(3) representations multiplied by powers of S are usually ignored in order to make the representation space orthogonal. If we assign[13] $\rho'(1600)$ and $\omega(1675)$ to the $(11)^*$ in (8), we obtain for the SU(3) scalar mentioned above, $M_S \sim 0.8$ GeV. At this stage it is tempting to speculate a chain structure for lower resonances (Fig.2). That is, if states for the $(22)^*$ representation from $\ell=3$ chain exist, the mass differences from ψ's would likely be $O(M_S)$, and possibly there are broad resonances like ρ'. In fact, $\frac{1}{2}(M_{\psi} + M_{\psi'}) + M_S = 4.2$ GeV. This might suggest that the observed broad resonance at $M \sim 4.1$ GeV with the width, $200 \sim 250$ MeV belong to the degenerate state of $(22)^*$ from $\ell=3$. However, it should be remembered that the the present scheme tells us nothing about radial or orbital excitation except equal mass separations for lower resonances at the exact symmetry limit. In order to include these effects, an appropriate spatial group must be introduced. Right now the experimental evidences are not clear for ρ'', ω'', ϕ'', like resonances which may correspond, respectively, to the $(11)^{**}$ and $(00)^{**}$ representation of $\ell=3$ chain; they could be broad in width and lower in intensity. Assuming the one photon exchange dominance in the present e^+e^--annihilation experimant, we should expect 1^- vector mesons with $I_3 = 0$, $I = 0, 1$ only.

Using the notation $\left|\begin{matrix} p\,q \\ Y\, I\, I_3\end{matrix}\right\rangle$, candidates for ψ's are therefore[14],

$$B_2 \sim \left|\begin{matrix} 2\,2 \\ 0\,1\,0\end{matrix}\right\rangle = \frac{1}{\sqrt{10}}\left\{-pp\bar{p}\bar{p} + nn\bar{n}\bar{n} + 2p\lambda\bar{p}\bar{\lambda} + 2n\lambda\bar{n}\bar{\lambda}\right\}, \tag{9a}$$

$$D \sim \left|\begin{matrix}22\\000\end{matrix}\right\rangle = \frac{1}{\sqrt{30}}\left\{pp\bar{p}\bar{p} - pn\bar{p}\bar{n} + nn\bar{n}\bar{n} - 3p\lambda\bar{p}\bar{\lambda} + 3n\lambda\bar{n}\bar{\lambda} + 3\lambda\lambda\bar{\lambda}\bar{\lambda}\right\}, \quad (9b)$$

$$g_3 \sim \left|\begin{matrix}22\\020\end{matrix}\right\rangle = \frac{1}{\sqrt{6}}\left\{pp\bar{p}\bar{p} + 2pn\bar{p}\bar{n} + nn\bar{n}\bar{n}\right\}. \quad (9c)$$

Among several possibilities the following assignment is perhaps the most attractive one; ψ(3105) and ψ'(3695) belong, respectively, to the $I = 1$ and $I = 0$ members of the 27-dimensional SU(3) representation (22). In each excited level of SU(3), assuming the symmetry breaking of the order of the Gell-Mann-Okubo type, we obtain $M(g_3) \sim 1.9$ GeV from (9). Observed narrow widths of the ψ particles may be attributed to an approximate SU(3.1) invariance of the strong interaction, i.e. $|\Delta\ell| \neq 0$ selection rule (this is suggested by (5)). Assuming $\pi\pi$ channel dominance, the order of magnitude estimation for $|\Delta\ell| \neq 0$ suppression gives:

$$\frac{G^2(\ell=2)}{G^2(\ell=1)} = \frac{\Gamma(\psi\to\pi\pi)}{\Gamma(\rho\to\pi\pi)}\left(\frac{M_\psi}{M_\rho}\right)^2 \frac{(\text{Phase space } \rho)}{(\text{phase space } \psi)} \simeq 1.3 \times 10^{-4} \quad (10)$$

where $G(\ell)$ is the coupling strength defined by the effective hamiltonian $\mathcal{H} = G(\ell)\,\vec{\psi}_\lambda(\vec{\varphi} \times \partial_\lambda \vec{\varphi})$. Note that in our assignment the observed suppression of of the order $\sim 10^{-2}$ for the process $\psi' \to \psi\pi\pi$, compared with the typical strong decay $\rho' \to \rho\pi\pi$, is due to the G-parity-violating decay, and it should be compared more appropriately with the process $\omega \to \rho\pi\pi$, which is, however, energetically forbidden. For leptonic decays the third diagonal generator does not correspond to an independent current like the charm current in the SU(4)scheme[15], since Z in (5) depends on the SU(3) symmetry label. Here we simply assume that the electromagnetic current, $J_\mu \sim F^3_\mu - \frac{1}{\sqrt{3}} F^8_\mu$ decouples from the possible ℓ-conserving current[16]. Then we should expect the ratio for the coupling strength of ψ's into lepton pairs, $f_{\psi'}^{-2} : f_\psi^{-2} \simeq f_\omega^{-2} : f_\rho^{-2}$. Ignoring a possible mixing of the states, we obtain

$$\frac{\Gamma(\psi' \to \bar{\ell}\ell)}{\Gamma(\psi \to \bar{\ell}\ell)} = \left(\frac{f_{\psi'}}{f_\psi}\right)^2 \frac{M_{\psi'}}{M_\psi} \sim 0.13\,, \quad (11)$$

which should be compared with the experimental value ~ 0.4. These results encourage us to make further predictions for higher excited states. We already know that the simple non-relativistic quark model gives a surprisingly good estimate for the low-lying hadron spectra and for some interactions among them. Furthermore, we expect that the relativistic effect, if any[17], will be minimum for the e^+e^--annihilation. Therefore, we justify ourselves in making the 0-th order estimate by taking only the bare quark masses into account, and here we do not assume any particular form for the interaction between two quarks. Then the bare mass of the ℓ-th excited state quark and the threshold energy of the ℓ-th level mesons[18] are approximately:

$$M_\ell = \ell \mu_0 \tag{12a}$$

$$E_\ell = 2D(\ell 0)M_\ell = \ell(\ell+1)(\ell+2)\mu_0 \tag{12b}$$

Taking $\mu_0 \sim 150$ MeV, we obtain $E_1 = 0.9$, $E_2 = 3.6$, $E_3 = 9$ GeV, for whic h $R(q^2 \to \infty) = \sum_i Q_i^2$ is $\frac{2}{3}$, 4, 14, respectively. In conclusion, we should expect at least two more ψ-like narrow resonances (even narrower) at $M \sim 9$ GeV. They could be associated with broad resonances from the $\ell = 4$ chain, and between E_2 and E_3 no more sharp resonances will be seen, but some relatively broad resonances may be observed corresponding to the higher harmonics states. As a final remark we want to point out that although the e^+e^- annihilation experiments are informative, we should be aware of their limitations; for example, g_3 which is the $I = 2$ member of the (22) representation will never be observed unless it can be produced by the 2-photon exchange processes.

REFERENCES AND FOOTNOTES

1. J.J.Aubert et al.,Phys. Rev. Letters 33, 1404 (1974);
   J.E.Augustin et al., Phys. Rev. Letters 33, 1406 (1974);
   C.Bacci et al., Phys. Rev. Letters 33, 1408 (1974);
   G.S.Abrams et al., Phys. Rev. Letters 33, 1453 (1974).
2. J.E.Augustin et al., Phys. Rev. Letters 34, 764 (1975).
3. M.Hongoh, University of Montreal preprint CRM-505, May 1975.
4. A.O.Barut, Phys. Rev. 139, B1433 (1965);
   R.C.HWA and J.NUYTS, Phys. Rev. 145,1188 (1966);
   P.Budini, Nuovo Cimento 44A, 363 (1966).
5. A.O.Barut, Dynamical Groups and Generalized Symmetries in Quantum Theory (University of Canterbury, Christchurch, New Zealand, 1972).
6. H.J. Lipkin, Lie Groups for Pedestrians,2nd Edition (North-Holland, Amsterdam 1966).
7. M.Gunaydin and F.Gursey, J. Math. Phys. 14, 1651 (1973).
8. Only a finite part of the spectrum, depending on the partition parameter, can be embedded in the compact group SU(4), and by SP(3.R) spectrum changes by the step of two. R.C.Hwa and J.Nuyts, Ref. 4.
9. R.M.Santilli, Nuovo Cimento 51, 89 (1967).
10. With $(\ell 0)$ $((0\ell))$ being the general symmetric (antisymmetric) representation of SU(3), $\frac{4}{3}(\ell - Z)$ $(\frac{4}{3}(\ell + Z))$ is an invariant in general for a given UR of SU(3.1). In fact it is a negative real number as is seen from the Gel'fand pattern, R.M.Santilli, Ref.9.
11. We are interested, of course, only in the discrete unitary representation of SU(3.1) here.
12. States similar to expression (7) are also considered by Bacry and Chang in connection with the relativistic 3-dimensional harmonic oscillator. They correspond to the coherent states in the cms. See H.Bacry, the 3rd International Colloquium on Group Theoretical Methods in Physics, Marseille , 1974. This point will be discussed in detail in a forthcoming paper, M.Hongoh, in preparation.
13. In Ref.3 these heavier mesons belonging to the starred representations are called the higher harmonics states; they have the same SU(3) quantum number as the mesons of the original multiplet.
14. g_3 is the $I_3 = 0$, $I = 2$ member of the (22) representation, therefore, it cannot be a candidate for ψ's, nevertheless listed here for completeness.
15. S.L.Glashow,J.Iliopoulos and L.Maiani Phys. Rev. D2, 1285 (1972).

16. From this point of view, we do not consider that the evidence for $\Delta S = -\Delta Q$ currents recently reported by E.G.Cazzoli et al. immediately concludes the existence of charmed baryons. E.G.Cazzoli et al., Phys. Rev. Letters 34, 1125 (1975).
17. In diffractive excitations and also in deep inelastic scatterings, we should expect the Lorentz contraction effect on the extended hadronic matters.
18. For example, ψ's are $\ell = 2$ level mesons in the present scheme.

FIGURE AND TABLE CAPTIONS

Fig. 1: The lowest states of $D^{+}(00Z^{+})$ and $D^{-}(00Z^{-})$.

Fig. 2: Equal mass-separation for lower resonances.

Table 1: Non-vanishing matrix elements for the generalized shifting operator.

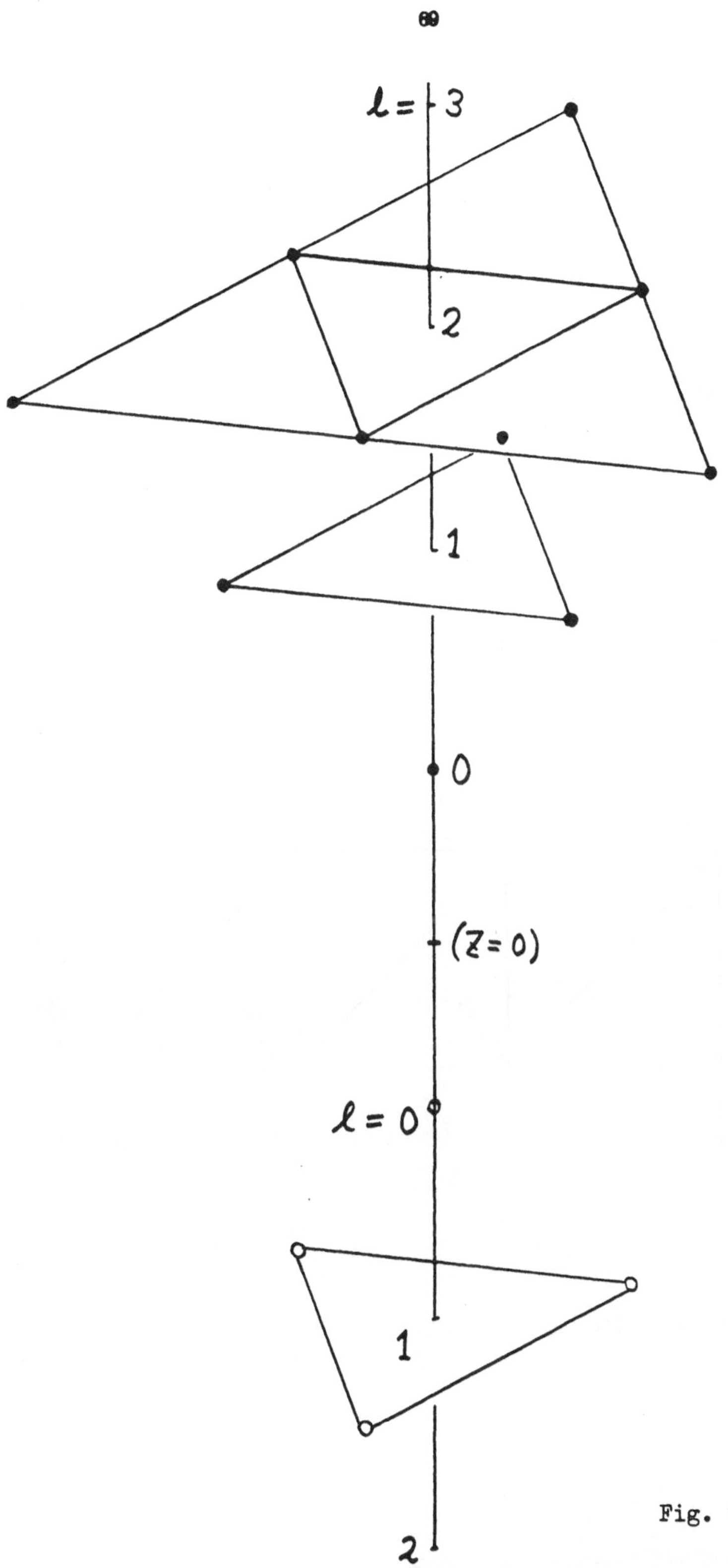

Fig. 1

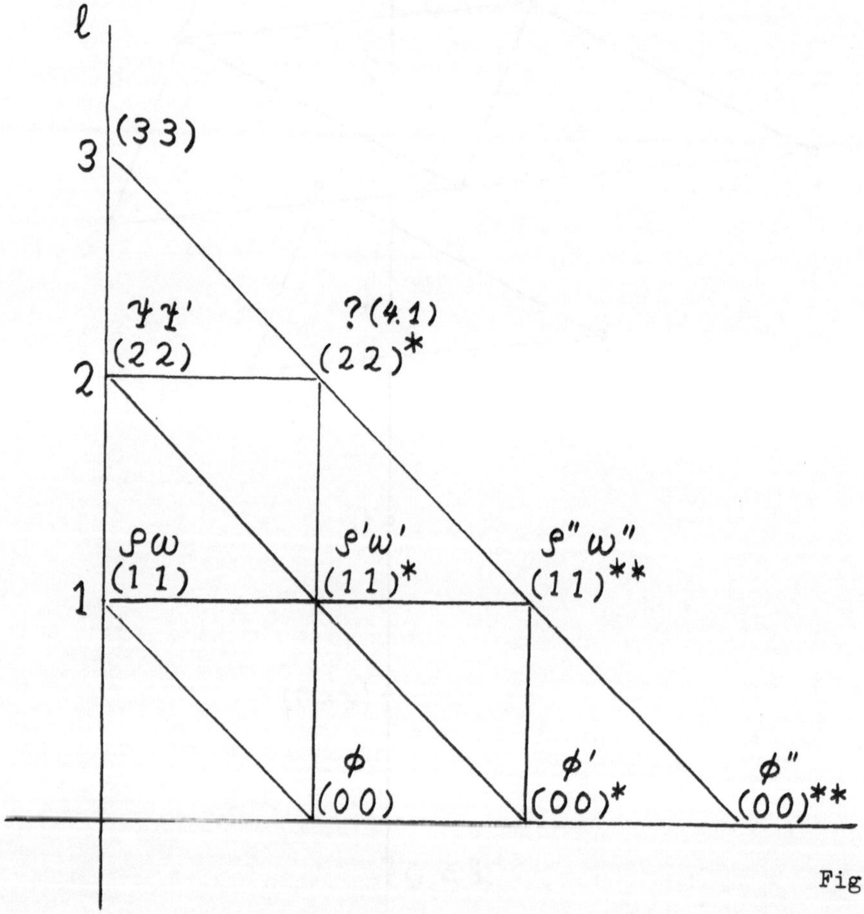

Fig. 2

TABLE 1

Non-vanishing matrix elements	Quark State	Anti-Quark State
$\left\langle I_3 + \frac{1}{2}, Y + \frac{1}{3}, Z+1 \middle\| G_o^+ \middle\| I_3, Y, Z \right\rangle$	$-((I+I_3+1)(-I + \frac{1}{2}Y + \frac{4}{3}Z)/2)^{\frac{1}{2}}$	$((I+I_3)(-I + \frac{1}{2}Y + \frac{4}{3}Z+1)/2)^{\frac{1}{2}}$
$\left\langle I_3 - \frac{1}{2}, Y - \frac{1}{3}, Z-1 \middle\| G_o^- \middle\| I_3, Y, Z \right\rangle$	$((I+I_3)(-I + \frac{1}{2}Y + \frac{4}{3}Z-1)/2)^{\frac{1}{2}}$	$((I+I_3-1)(-I + \frac{1}{2}Y + \frac{4}{3}Z)/2)^{\frac{1}{2}}$
$\left\langle I_3 - \frac{1}{2}, Y + \frac{1}{3}, Z+1 \middle\| G_o^+ \middle\| I_3, Y, Z \right\rangle$	$-((I-I_3+1)(-I + \frac{1}{2}Y + \frac{4}{3}Z)/2)^{\frac{1}{2}}$	$((I-I_3)(-I + \frac{1}{2}Y + \frac{4}{3}Z+1)/2)^{\frac{1}{2}}$
$\left\langle I_3 + \frac{1}{2}, Y - \frac{1}{3}, Z-1 \middle\| G_o^- \middle\| I_3, Y, Z \right\rangle$	$((I-I_3)(-I + \frac{1}{2}Y + \frac{4}{3}Z-1)/2)^{\frac{1}{2}}$	$((I-I_3-1)(-I + \frac{1}{2}Y + \frac{4}{3}Z)/2)^{\frac{1}{2}}$
$\left\langle I_3, Y - \frac{2}{3}, Z+1 \middle\| G_o^+ \middle\| I_3, Y, Z \right\rangle$	$-((I-\frac{3}{2}Y+1)(-I + \frac{1}{2}Y + \frac{4}{3}Z)/2)^{\frac{1}{2}}$	$((I-\frac{3}{2}Y)(-I + \frac{1}{2}Y + \frac{4}{3}Z+ 1)/2)^{\frac{1}{2}}$
$\left\langle I_3, Y + \frac{2}{3}, Z-1 \middle\| G_o^- \middle\| I_3, Y, Z \right\rangle$	$((I-\frac{3}{2}Y)(-I + \frac{1}{2}Y + \frac{4}{3}Z-1)/2)^{\frac{1}{2}}$	$((I-I_3-1)(-I + \frac{1}{2}Y + \frac{4}{3}Z)/2)^{\frac{1}{2}}$

LOCAL AND GLOBAL EQUIVALENCE OF PROJECTIVE REPRESENTATIONS

Henk Hoogland *, Theoretical physics, University of Nijmegen

1. Introduction

From the classical work of Wigner and Bargmann [1,2] it is well known that a physical system, which is invariant under a symmetry group of transformations of space-time, is described in quantum mechanics by an (anti) unitary projective representation of that group in a Hilbert space. Classifying irreducible unitary projective representations of symmetry groups is almost a routine procedure in theoretical physics nowadays and as a result the very concept "elementary particle" has eventually been associated by theoreticians with the abstract notion of equivalence class of irreducible unitary projective representations of the symmetry group of the universe (space-time) in which the particle exists.

Long before this abstract description of quantum mechanical particles, there existed already the usual description of particle states by means of wave functions obeying appropriate equations of motion.
It appeared that the solutions of the wave equation for a free particle span a Hilbert space in which the projective representation corresponding to that particle can be realised in a natural way. Thus, fortunately, the two approaches fit well together and the wave equations got a firm group theoretical basis after having been derived from classical mechanics by heuristic arguments.

Nevertheless, these two approaches are not physically equivalent, although this seems to be partly forgotten now. If we read Bargmann and Wigner [3] we see that there are two problems: projective representations in the same equivalence class may possibly

a) not look like each other (even if they are physically equivalent)

b) not be physically equivalent (even if they look like each other).

Here we are mainly concerned with the second problem which in fact says that it is not sufficient to determine all equivalence classes of irreducible unitary projective representations of a symmetry group. There is more work to be done, viz. to distinguish the physically inequivalent representations within each class.

* Postal address: Faculteit Wis- en Natuurkunde, Toernooiveld, Nijmegen, Holland

A substantial part of this work has been done elsewhere [4] and we refer to that paper for more detailed information. We give here the main concepts and results.

One of the results is that for the Poincaré and Galilei groups equivalence of projective representations indeed implies physical equivalence, so it is quite natural to forget the whole problem. However, there exist other symmetry groups of different universes for which projective equivalence does not imply physical equivalence. To be specific: we will give examples of projective representations, describing a free particle resp. a particle in a non-zero external field, which are projectively equivalent.

An analogous situation occurs in classical mechanics, where projectively equivalent representations of a symmetry group may give rise to inequivalent gauge functions, hence to inequivalent Lagrangians. The work of Lévy-Leblond [5] in this field has acted as a guide line for our investigations and especially the mathematical concept "superequivalence of group exponents" which is essential for our work, has been introduced by him. For the notions of unitary continuous projective representation (UCPR) and group exponents we refer to Bargmann [2] .

2. Locally operating UCPRs and gauge matrices

Here we introduce the special kind of UCPRs that we are interested in.

- Definition: A locally operating UCPR of a symmetry group G is a UCPR that has the following two characteristics:
  a) the Hilbert space in which it operates is spanned by complex (spinor) wave functions $\psi_\sigma(x)$ on space-time X, obeying a given equation of motion.
  b) the unitary operators U(g) work locally on these wave functions, i.e. there exists a nonsingular matrix function $A(g;x)$ on G x X such that

$$(U(g)\psi)_\sigma(g\circ x) = \sum_{\sigma'=1}^{n} A_{\sigma\sigma'}(g;x)\,\psi_{\sigma'}(x) \qquad (1)$$

Condition b) says that the value of the transformed function $U(g)\psi$ in the transformed event $g\circ x$ is related to the value of the original function ψ in the original event x, under linear mixing of the components, multiplied by (gauge) functions which in general may be different for all components, such that $U(g)\psi$ obeys the same equation of motion as ψ does.

The matrix $A(g;x)$ is called the gauge matrix of the locally operating UCPR. The "usual" wave functions in quantum mechanics have a transformation character (1) with a gauge matrix of the form

$$A_{\sigma\sigma'}(g;x) = e^{i\,\theta(g;x)} D_{\sigma\sigma'}(\gamma(g)) \qquad (2)$$

where $\theta(g;x)$ is an x-dependent real gauge function and $D_{\sigma\sigma'}(\gamma)$ a matrix representation of the homogeneous part Γ of the symmetry group, which leaves the origin x_o of X invariant.

However, we do not know any a priori reason why the gauge matrix should be restricted to this particular form. On the contrary, trying to prove a posteriori that any gauge matrix can be chosen in the form (2), we get the result that this is possible only for symmetry groups which have a special property (see (9)), including the Poincaré and Galilei groups.

Of course, the wave functions and the gauge matrices must have sufficient continuity, differentiability and integrability properties, but we do not specify them explicitely, because we do not want to assume specific features of the inner product and of the equation of motion.

It is not claimed that any UCPR can be realised by (1). There may exist projective equivalence classes of UCPRs which do not contain a locally operating representative. Such classes of UCPRs are not considered in this paper. From now on we implicitely mean by UCPR a locally operating UCPR.

The properties of the gauge matrix A can be divided in two classes:

1) specific properties, depending on the explicit form of the equation of motion and the inner product, to ensure that they are invariant under U(g). These properties do <u>not</u> necessarily imply unitarity of the matrix A(g;x) (although the operator U(g) is unitary). We will not use these specific properties in the sequel.
2) general properties, related with the projective nature of U(g), especially related with the group exponents. Our theory will be based on these general properties.

The basic property of the gauge matrix follows from the multiplication rule of the operators U(g):

$$A(g';\, g\cdot x)\, A(g;x)\, A^{-1}(g'g;x) = e^{i\xi(g',g)}, \quad \forall g',g \in G,\ \forall x \in X \qquad (3)$$

where ξ is the exponent of the UCPR with gauge matrix A (or shortly: ξ is the exponent of A). Remark that the left hand side of (3) must be independent of x. From a convention follows

$$A(e;x) = \mathbb{1} \qquad \forall x \in X \quad \text{(e unit element of G)}. \qquad (4)$$

From (3) follows by substitution $g' = \gamma'$, $g = \gamma$ and $x = x_o$ (origin of X)

$$A(\gamma';x_o)A(\gamma;x_o) = e^{i\xi(\gamma',\gamma)}A(\gamma'\gamma;x_o) \quad , \forall\gamma',\gamma\in\Gamma , \qquad (5)$$

hence the matrices $\{A(\gamma;x_o)\}_{\gamma\in\Gamma}$ form a continuous projective matrix representation (CPMR) of Γ , not necessarily unitary, but its factor system still has modulus one. Equivalence between CPMRs is defined in the usual way.

We use property (5) to define a new equivalence relation between UCPRs of the symmetry group.

- Definition: Two UCPRs of G are called globally equivalent if they are projectively equivalent and if their gauge matrices give rise to equivalent CPMRs of Γ .

The physical idea behind this definition is that two globally equivalent UCPRs operate on wave functions of the same "spinor type" with respect to the homogeneous part Γ of the group G. In the terminology of the introduction: globally equivalent UCPRs "look like each other".

It still may occur that two globally equivalent UCPRs describe systems which are not physically equivalent, because no restriction is put on their local properties. Hence the probability density, energy density etc. need not be related in the two systems.

Therefore we need a more restrictive equivalence concept, demanding more equivalence between gauge matrices.

- Definition: Two gauge matrices A and A' of G are called equivalent if they have the same dimension and if there exists a real continuous function ζ on G and a nonsingular matrix function S on X such that

$$A'(g;x) = e^{i\zeta(g)}S(g\circ x)A(g;x)\ S^{-1}(x) \quad , \forall g\in G, \forall x\in X \qquad (6)$$

- Definition: Two UCPRs are called locally equivalent if they are projectively equivalent and if their gauge matrices are equivalent.

It is clear that local equivalence implies global equivalence.

Two UCPRs which are projectively equivalent with an equivalence transformation S, that operates as multiplication with a matrix S(x), are obviously locally equivalent. Such an equivalence transformation is in fact a generalised gauge transformation of the second (i.e. local) kind. Now we have good reasons to expect that within a given global equivalence class, local equivalence of UCPRs coincides with equivalence from the physical point of view. On the other hand it still remains possible that one physical system can be described by wave functions with different numbers of components, so by UCPRs that are globally inequivalent (hence, locally inequivalent). This is problem a) of the introduction, but we do not go into that point.

The aim of this work is: to find all local equivalence classes of UCPRs contained in a given global equivalence class. (see § 4).

3. Centered gauge matrices

We try here to obtain gauge matrices in a simple form.

- Definition: A gauge matrix A is called centered if it has the property:

$$A(g;x_o) = A(\gamma(g);x_o) \quad , \forall g \in G \tag{7}$$

where $\gamma(g)$ is the homogeneous part of the transformation g. We choose $\forall x \in X$ a transformation h_x which transforms the origin x_o into x (convention $h_{x_o} = e$).

- For each centered gauge matrix A holds

$$A(g;x) = e^{i\xi(g;h_x)} A(\gamma(gh_x);x_o) \quad , \forall g \in G, \forall x \in X \tag{8}$$

which follows from (3) by substitution of (g,h_x,x_o) for (g',g,x), using (4).

- Proposition: Each equivalence class of gauge matrices contains a centered representative. (For proof see [4]).

We consider now symmetry groups which have the property that $\gamma(gh_x)$ is independent of x, hence

$$\gamma(gh_x) = \gamma(g) \quad , \forall g \in G, \forall x \in X \tag{9}$$

From (8) it is clear that for such groups any centered gauge matrix takes the form

$$A(g;x) = e^{i\xi(g;h_x)} A(\gamma(g); x_o) \tag{10}$$

This is precisely the usual form (2), where the gauge function $\xi(g;h_x)$ contains all x-dependence of the gauge matrix and where $A(\gamma(g); x_o)$ is a projective (!) matrix representation of the homogeneous part Γ of G. All symmetry groups that are a semidirect product of the translations with the homogeneous transformations (especially the Galilei and Poincaré groups) have property (9), so for these groups any gauge matrix is equivalent to the usual form (2), which now has been justified a posteriori for symmetry groups of this special type.

For groups that do not have property (9) (e.g. the Newton and de Sitter groups) there may exist gauge matrices A where the x-dependence occurs also in an overall gauge function only:

$$A(g;x) = e^{i\theta(g;x)} D(g) \tag{11}$$

but in general these gauge matrices are not centered, whereas the matrices D(g) are not a projective representation of Γ .

We are convinced that it can not be proved from the general properties of gauge matrices that for any symmetry group each gauge matrix is equivalent to (11), although this might follow from the specific properties in each individual case. Consequently we are bound to keep working with the concept of gauge matrix as a generalisation of gauge function.

4. Main result and example

In order to find all local classes of UCPRs contained in a given global class, special kinds of exponents and gauge matrices have to be defined. We refer to chapter 4 of reference [4] for the details. Fortunately we can state the results without all these technicalities.

A global equivalence class of UCPRs determines an (equivalence class of) exponent of G, say $\xi(g',g)$, and an (eq. cl. of) CPMR of Γ, say $D^{(n,\alpha)}(\gamma)$, where n is the dimension of the matrices and α an additional label.
The local equivalence classes contained in this global class are now given by the following set of gauge matrices:

$$A(g;x) = e^{i\xi(g,h_x) - i\zeta(\gamma(gh_x))} D^{(n,\alpha)}(\gamma(gh_x)) \qquad (12)$$

where ζ runs through a space of one-dimensional real continuous representations of Γ that can not be extended to such a representation of the whole group G.

When the symmetry group is a semi-direct product of the translations with the homogeneous transformations, then any representation of Γ can be extended to G and the global class contains only one local class, given by (12) with $\zeta \equiv 0$. Hence, for the "usual" symmetry groups (Poincaré, Galileï) physical equivalence coincides with projective equivalence. That this is not generally true for other groups shows the next counterexample.

The Newton group in a one-dimensional oscillating universe. (see reference [6]). This "non-relativistic" cosmological group operates on space-time as follows.

$$(x,t) \in X, \quad g = (b,a,v) \in G, \quad g \circ (x,t) = (g\circ x,\ g\circ t) \quad \text{with}$$

$$\left.\begin{aligned} g\circ x &= x + \frac{v}{\omega}\sin\omega t + a\cos\omega t \\ g\circ t &= t + b \end{aligned}\right\} \qquad (13)$$

Here ω is a "constant of nature": the frequency of the oscillating universe. In the limit $\omega \to 0$ we get the (inhomogeneous) Galilei group. The subgroup Γ of homogeneous transformations, leaving the origin $x_0 = (0,0) \in X$ invariant, contains all elements $\gamma(g) = (0,0,v)$ and the transformations that transform (0,0) to an arbitrary point (x,t) are $h_{(x,t)} = (t,x,0)$. The multiplication rule can be calculated easily from (13) and from this we get

$$\gamma(gh_{(x,t)}) = (0,\ 0,\ v\cos\omega t - a\omega\sin\omega t) \qquad (14)$$

This group has a one-dimensional space of (equivalence classes of) exponents $\{m\,\xi_1(g',g) \mid m \in \mathbb{R}\}$. Any UCPR can be written as $U(g) = e^{ibH}e^{-iaP}e^{ivN}$ and the generators obey

$$[H,P] = i\omega^2 N\,, \quad [H,N] = -iP\,, \quad [P,N] = -im\,. \qquad (15)$$

The Casimir operator is $2mH - P^2 - \omega^2 N^2$.

A free particle with mass $m > 0$ in this universe is described by the UCPR:

$$(U(g)\psi)(g\circ(x,t)) = e^{im\xi_1(g,h_{(x,t)})}\psi(x,t) \qquad (16)$$

with generators

$$\left.\begin{aligned} H &= i\partial_t \\ P &= -i\cos\omega t\,\partial_x + m\omega x\sin\omega t \\ N &= mx\cos\omega t + \frac{\sin\omega t}{\omega}\,i\partial_x \end{aligned}\right\} \qquad (17)$$

(The generators follow from the explicit form of $\xi_1(g',g)$, given in reference [4]). From the Casimir operator follows the equation of motion:

$$i\partial_t\psi(x,t) = \left\{\frac{(-i\partial_x)^2}{2m} + \tfrac{1}{2}m\omega^2 x^2\right\}\psi(x,t) \qquad (18)$$

Hence a free particle in this oscillating universe behaves like a particle on a spring in a Galilean universe. The gauge matrix of this UCPR is one-dimensional. The one-dimensional real continuous representations of Γ are $\zeta(\gamma) = \frac{f}{\omega^2}v$, $(f \in \mathbb{R})$ and they can not be extended to representations of G. (For convenience we took the arbitrary constant equal to f/ω^2) From (12) and (14) we get (one-dimensional) gauge matrices of UCPRs (one for each f) that are globally equivalent but locally inequivalent with (16):

$$A(g;x) = e^{im\xi_1(g',g) - i\frac{f}{\omega^2}(v\cos\omega t - a\omega\sin\omega t)} \qquad (19)$$

The generators in these UCPRs have extra terms, compared to (17): P gets an extra term $-\frac{f}{\omega}\sin\omega t$, N gets extra $-\frac{f}{\omega^2}\cos\omega t$ and H stays invariant. With these extra terms the generators still obey (15). The Casimir operator gives again the equation of motion, which now has the extra term $-fx\psi(x,t)$ compared to (18). This term is just a linear potential term in the Schrödinger equation, arising from a uniform external (one-dimensional) free field f.

So the new UCPRs, that we get from (19), describe particles with the same mass in uniform force fields in this oscillating universe. They are all locally inequivalent, but globally equivalent, hence projectively equivalent.

The de Sitter group also has the same features (see reference [4] for a one-dimensional model of this group).

References

[1] V. Bargmann and E.P. Wigner, Group Theoretical discussion of relativistic wave equations, Proc. Nat. Ac. Sci. U.S.A. 34 (1948) 211 - 223.

[2] V. Bargmann, On unitary ray representations of continuous groups, Ann of Math. 59 (1954) 1 - 46.

[3] See ref [1] , at least from p. 212 (5th line from bottom) to p. 213 (3rd line from above).

[4] H. Hoogland, On local and global equivalence of wave functions, super-equivalence of group exponents applied in quantum mechanics, Preprint Univ. of Nijmegen.

[5] J.M. Lévy-Leblond, Group theoretical foundations of classical mechanics: the Lagrangian gauge problem, Commun.Math.Phys. 12 (1964) 64 - 79.

[6] H. Bacry and J.M. Lévy-Leblond, Possible kinematics, J.Math.Phys. 9 (1968) 1605 - 1614.

INVARIANT EQUATIONS ON THE FIBRE BUNDLES

R. Kerner
Departement de Mecanique,
Universite de Paris VI
4, Place Jussieu
Paris - FRANCE

1.-Notations.

The Minkowskian space-time will be denoted by M_4, with a metric tensor $g_{ij} = \mathrm{diag}\,(+++-)$; $i,j = 0,1,2,3$. Let us construct a principal fibre bundle $P(M_4,G)$ with the base M_4 and the structural group G. The group G is supposed to be a semi-simple, compact Lie group of dimension N. The structure constants of the group G are denoted by C^a_{bc}, $a,b,.. = 1,2,..,N$, and its Cartan-Killing form is $g_{ab} = C^d_{ac}C^c_{db}$. Remark that this metric form is always positive definite for a semi-simple compact Lie group.

The fibre bundle whether trivial or not, locally can be endowed with a pseudo-Euclidean metric form given by

$$\tilde{g}_{\alpha\beta} = \mathrm{diag}\,(N+3\ +,\ -),\quad \alpha,\beta = 1,2,..,N+4$$

or symbolically

$$\tilde{g}_{\alpha\beta} = g_{ab} \oplus g_{ij} \qquad /1/$$

A connection on the fibre bundle $P(M_4,G)$ is a field of a 1-form ω with values in the Lie algebra of G. In local coordinates we can put $\omega = A^a_\alpha$. The canonical projection $\pi : P(M_4,G) \longrightarrow M_4$ will induce also the projection of the corresponding tangent spaces. In local coordinates we shall write. Moreover, locally we can always choose the coordinates in which

$$A^a_\alpha = (\delta^a_b, A^a_i) \quad \text{and} \quad d\pi^i_\alpha = (0, \delta^i_j) \qquad /2/$$

Then there is a canonical way of inducing the metric structure on $P(M_4,G)$ by means of the two metrics g_{ij} and g_{ab}, and consistent with the connection: we put

$$\tilde{g}_{\alpha\beta} = \begin{pmatrix} g_{ij} + g_{ab} A^a_i A^b_j & g_{ab} A^a_i \\ g_{ab} A^b_j & g_{ab} \end{pmatrix} \qquad /3/$$

and

$$\tilde{g}^{\alpha\beta} = \begin{pmatrix} g^{ij} & -g^{ij} A^b_j \\ -g^{ij} A^a_i & g^{ab} + g^{ij} A^a_i A^b_j \end{pmatrix} \qquad /4/$$

A^a_i is the Yang-Mills potential corresponding to the group G, and the field tensor is denoted by

$$F^a_{ij} = \partial_i A^a_j - \partial_j A^a_i + C^a_{bc} A^b_i A^c_j \qquad /5/$$

The metric /3/ in a natural way gives the metric connection on $P(M_4,G)$, with the coefficients $\Gamma^\alpha_{\beta\gamma}$. The covariant derivative with respect to this connection will be denoted by ∇_α.

2. Invariant equations.

Let us now generalize the Klein-Gordon and the Dirac equations for the case when the underlying manifold is $P(M_4,G)$ with the metric /3/ instead of the M_4. The generalization of the Klein-Gordon equation is easily obtained: we have to replace the d'Alembertian operator by the Lapla ce-Beltrami operator on $P(M_4,G)$:

$$\Box \longrightarrow \frac{1}{\sqrt{-\tilde{g}}} \nabla_\alpha \tilde{g}^{\alpha\beta} \sqrt{-\tilde{g}}\, \nabla_\beta \qquad /6/$$

Therefore the scalar field on $P(M_4, G)$ has to verify the following equation:

$$\left[\frac{1}{\sqrt{-\tilde{g}}} \nabla_\alpha \tilde{g}^{\alpha\beta} \sqrt{-\tilde{g}}\, \nabla_\beta - m^2\right] \psi = 0 \qquad /7/$$

Taking into account that $\sqrt{-\tilde{g}} = \sqrt{|-g_{ij}|\,|g_{ab}|} =$ Const, we have

$$\left[\nabla_\alpha \tilde{g}^{\alpha\beta} \nabla_\beta - m^2\right] \psi = 0 \qquad /8/$$

and if we fix the gauge in which

$$\partial^i A^a_i = 0 \qquad /9/$$

then $\nabla_\alpha \tilde{g}^{\alpha\beta}$ and

$$\left[\tilde{g}^{\alpha\beta} \nabla_\alpha \nabla_\beta - m^2\right] \psi = 0 \qquad /10/$$

In order to generalize the Dirac equation on $P(M_4,G)$ we write down

$$(\gamma^\alpha \nabla_\alpha + m)\psi = 0 \qquad /11/$$

where the matrices γ^α should verify

$$\gamma^\alpha \gamma^\beta + \gamma^\beta \gamma^\alpha = 2\tilde{g}^{\alpha\beta} \otimes Id \qquad /12/$$

so that

$$(\gamma^\alpha \nabla_\alpha - m)(\gamma^\beta \nabla_\beta + m) = \tilde{g}^{\alpha\beta} \nabla_\alpha \nabla_\beta - m^2 \qquad /13/$$

Let us first construct the Clifford algebra over $P(M_4,G)$ without connection, corresponding to the metric

$$\overset{\circ}{g}_{\alpha\beta} = g_{ab} \oplus g_{ij} \qquad /1/$$

Let us denote the generators of this algebra by $\overset{\circ}{\gamma}{}^{\alpha}$; we have:

$$\overset{\circ}{\gamma}{}^{\alpha}\overset{\circ}{\gamma}{}^{\beta} + \overset{\circ}{\gamma}{}^{\beta}\overset{\circ}{\gamma}{}^{\alpha} = 2\overset{\circ}{g}{}^{\alpha\beta} \otimes Id \qquad /14/$$

where Id denotes the identity matrix in $2^{\left[\frac{N+4}{2}\right]}$ dimensions. We remind that the dimension of the fundamental representation of the Clifford algebra over an Euclidean manifold of dimension s is $2^{[s/2]}$, where $[s/2]$ means the entire part of s/2 .

Let us also denote

$$\overset{\circ}{\gamma}{}^{\alpha}\overset{\circ}{\gamma}{}^{\beta} - \overset{\circ}{\gamma}{}^{\beta}\overset{\circ}{\gamma}{}^{\alpha} = 2\overset{\circ}{\sigma}{}^{\alpha\beta} \qquad /15/$$

Let us choose 4 matrices $\overset{\circ}{\gamma}{}^{i}$ out of the $\overset{\circ}{\gamma}{}^{\alpha}$, verifying

$$\overset{\circ}{\gamma}{}^{i}\overset{\circ}{\gamma}{}^{j} + \overset{\circ}{\gamma}{}^{j}\overset{\circ}{\gamma}{}^{i} = 2g^{ij} \otimes Id, \qquad \overset{\circ}{\gamma}{}^{i}\overset{\circ}{\gamma}{}^{j} - \overset{\circ}{\gamma}{}^{j}\overset{\circ}{\gamma}{}^{i} = 2\overset{\circ}{\sigma}{}^{ij} \qquad /16/$$

and N matrices $\overset{\circ}{\gamma}{}^{a}$ verifying

$$\overset{\circ}{\gamma}{}^{a}\overset{\circ}{\gamma}{}^{b} + \overset{\circ}{\gamma}{}^{b}\overset{\circ}{\gamma}{}^{a} = 2g^{ab} \otimes Id, \quad \overset{\circ}{\gamma}{}^{a}\overset{\circ}{\gamma}{}^{b} - \overset{\circ}{\gamma}{}^{b}\overset{\circ}{\gamma}{}^{a} = 2\overset{\circ}{\sigma}{}^{ab}$$
$$\overset{\circ}{\gamma}{}^{a}\overset{\circ}{\gamma}{}^{i} - \overset{\circ}{\gamma}{}^{i}\overset{\circ}{\gamma}{}^{a} = 2\overset{\circ}{\sigma}{}^{ai} \qquad /17/$$

It is easy to see that the matrices

$$\gamma_{\alpha} := \left\{ \gamma_a = \overset{\circ}{\gamma}_a, \ \gamma_i = \overset{\circ}{\gamma}_i + \overset{\circ}{\gamma}_a A^a_i \right\} \qquad /18/$$

verify

$$\gamma_{\alpha}\gamma_{\beta} + \gamma_{\beta}\gamma_{\alpha} = 2\tilde{g}_{\alpha\beta} \otimes Id \qquad /18a/$$

and the matrices

$$\gamma^{\alpha} := \left\{ \gamma^i = \overset{\circ}{\gamma}{}^{i}, \ \gamma^a = \overset{\circ}{\gamma}{}^{a} - \overset{\circ}{\gamma}{}^{i} A^a_i \right\} \qquad /19/$$

verify

$$\gamma^{\alpha}\gamma^{\beta} + \gamma^{\beta}\gamma^{\alpha} = 2\tilde{g}^{\alpha\beta} \otimes Id \qquad /20/$$

We have also

$$\gamma^i\gamma^j - \gamma^j\gamma^i = 2\mathring{\delta}^{ij} \qquad /20/$$

$$\gamma^a\gamma^i - \gamma^i\gamma^a = 2\mathring{\delta}^{ai} + 2A^a_k\,\mathring{\delta}^{ki} \qquad /21/$$

and

$$\gamma^a\gamma^b - \gamma^b\gamma^a = 2\mathring{\delta}^{ab} + 2A^a_i A^b_j\,\mathring{\delta}^{ij} + \left(A^a_j\,\mathring{\delta}^{bj} - A^b_j\,\mathring{\delta}^{aj}\right) \qquad /22/$$

In order to calculate explicitly the expression

$$(\gamma^\alpha \nabla_\alpha + m)\,\psi \qquad /23/$$

we have to define the covariant derivative of a spinor over $P(M_4, G)$. We define it as follows:

$$\nabla_\alpha \psi = \partial_\alpha \psi - \Gamma^\delta_{\alpha\beta}\,\gamma^\beta\gamma_\delta\,\psi \qquad /24/$$

where $\Gamma^\delta_{\alpha\beta}$ are the Christoffel symbols corresponding to the metric $\tilde{g}_{\alpha\beta}$.

After some calculus we see that all the terms non-linear in the γ-matrices do vanish, and we are left with the following equation:

$$(\gamma^\alpha\nabla_\alpha + m)\psi = \left[\mathring{\gamma}^i(\partial_i - A^a_i\partial_a) + \mathring{\gamma}^a\partial_a + m\right]\psi = 0 \qquad /25/$$

which even in the absence of the Yang-Mills potential reads

$$\left[\mathring{\gamma}^i\partial_i + \mathring{\gamma}^a\partial_a + m\right]\psi = 0 \qquad /26/$$

This gives rise to the mass-splitting with respect to the eigenvalues of the operator $\mathring{\gamma}^a\partial_a$.

3. Splitting of the product representations.

The object described by the generalized spinor has no real counterpart in Nature; as a matter of fact, it has not definite space-time properties, not being an irreducible representation of the Lorentz group; all its components are intermixed under the action of the matrices γ^α.

Whatx we can hope however, is to construct some reducible representation of the isometry group of $P(M_4, G)$ which can be obtained out of the product representations of the irreducible one, and then to decompose it into a sum of other representations, which may contain the semi-direct product of the Lorentz a nd the gauge groups. In other words, we shall try to find a decomposition of the tensor products of ψ into subspaces having definite properties under the semi-direct product.

It is quite easy to see when such the possibility exists. Let us remind some simple formulae concerning the Clifford algebras. Let us denote by $C(p, q)$ the Clifford algebra corresponding to the metric diag $(p+, q-)$. Then we have:

$$C(1,1) \otimes C(p,q) = C(p+1, q+1)$$
$$C(2,0) \otimes C(p,q) = C(q+2, p) \qquad /27/$$
$$C(0,2) \otimes C(p,q) = C(q, p+2)$$

Let us also remind that

$C(2,0) = C(1,1) = \mathrm{Mat}_2(\mathbb{R})$ = 2 by 2 real matrices

$C(0,2) = \mathbb{H}$ = Quaternions

$C(0,1) = \mathbb{C}$ = Complex numbers

$C(1,0) = \mathbb{R} \oplus \mathbb{R}$

Therefore the splitting of the representation into a simple sum will occur only if there appears C 1,0 in the decomposition. For exa mple, taking SU(2) for the gauge group we obtain:

dim SU(2) = 3; therefore the metric /1/ becomes diag(6+, 1-), and the Clifford algebra $C(6,1)$ is decomposed as follows:

$$C(6,1) = C(1,1) \otimes C(5,0) = C(1,1) \otimes C(2,0) \otimes C(0,3) =$$
$$C(3,1) \otimes C(0,3) = C(3,1) \otimes C(0,2) \otimes C(1,0) =$$
$$= C(3,1) \otimes H \otimes [R \oplus R] = [C(3,1) \otimes H] \oplus [C(3,1) \otimes H]$$

We have thus obtained the sum of the two irreducible representations of the Lorentz group $C(3,1)$ in tensor product with the quaternions. The base space for each of these irreducible representations is given by the 2-valued Lorentz spinors /in other words, the Lorentz spinors $\otimes R^2$ / To obtain the ordinary Lorentz spinors, we have to fix in each copy of H occuring in the decomposition o n e imaginary unit out of three, in order to make the translation operators commutative. Then the group of automorphisms of the base space is reduced to the semi-simple product of the Lorentz group and the SU(2)group.

It is easy to show that such a decomposition is possible only for the groups $SU(2k)$ and impossible for the case $SU(2k+1)$. The proof is based on simple arithmetics.

It would be interesting to investigate the equations and the automorphism groups arising from the application of the super-symmetry scheme for the generalized Dirac equation on $P(M_4,G)$.

GAUGE GROUPS IN LOCAL FIELD THEORY AND SUPERSELECTION RULES

F. Strocchi

Scuola Normale Superiore, Pisa, Italy

The basic difference between conservation laws arising from invariance under a finite Lie group and those arising from invariance under an infinite Lie Group or gauge group is discussed. Conserved currents associated to local gauge groups have the property that when their charges are conserved they define superselection rules. This theorem has strong physical implications on the electric charge superselection rule, the presumed baryon and lepton superselection rules, the existence of (hidden) color variables in non-abelian gauge theories and on possible explanations of the non-observability of quarks.

In general a symmetry of the Lagrangian gives rise to a conservation law.When the symmetry group is a continuous (Lie) group one has the additional property that the conserved quantity Q is generated by a local current $J_\mu(x)$: $Q = \int J_o(x)\, d^3x$. The deep implications of this property in quantum field theory have been extensively discussed in the literature (*local symmetries*)[1]. A natural question is then: is there any difference between a conservation law arising from a symmetry of the theory under a finite Lie group and one arising from a symmetry under an infinite Lie group or gauge group? In going from a finite Lie group to its associated local gauge group one does not gain any new conservation laws, one gets however an additional restriction on local current generating the conserved quantity Q.

Proposition 1 [2]. The invariance of the Lagrangian under an infinite Lie group or gauge group implies the existence of a divergenceless current $J_\mu(x)$ generating a conserved quantity Q. Moreover $J_\mu(x)$ is of the form

$$J_\mu(x) = \partial^\nu G_{\mu\nu}(x) \tag{1}$$

where $G_{\mu\nu}(x)$ is antisymmetric tensor.

The above result has deep implications in the quantum field theory formulation

Proposition 2 [3]. In a local quantum field theory conserved currents arising from local gauge invariance give rise to superselection rules if the gauge invariance is not broken. Therefore a conserved quantity Q can be associated to a local gauge invariance of the Lagrangian only if Q is superselected.

Proof. The idea of the proof is simple. It exploits the locality of the observables, a property which will not be discussed here since its connection with Einstein's causality is well known [4], together with eq.(1). For any local observable A one has

$$[Q,A] = \lim_{R\to\infty} \int_{|x|<R} [J_o(x),A]\, d^3x = \lim_{R\to\infty} \int_{|x|<R} [\partial^i G_{oi}(x),A] d^3x = 0$$

since by the locality of A and $G_{\mu\nu}(x)$, the surface at infinity does not contribute to the integral [3].

For a more detailed and rigorous proof, taking into account a careful definition of the charge Q and the gauge problem we refer to [3].

The physical implications of the above result are rather strong. The above Prop.2 provides in fact a rigorous proof of the superselection rule of the electric charge; in this case the content of Prop. 1 is the validity of the Maxwell's equations (again for a careful discussion of the gauge problem see ref.[3]).

Prop.2 provides also non trivial informations about the Yang-Mills color gauge theory suggested by Fritzsch, Gell-Mann and Leutwyler [5]. The introduction of color is forced by one the main problem related to quarks, that of quark statistics. The basic assumptions of FGML scheme are: i) the theory is of the Yang-Mills type and the color SU(3) is unbroken ii) observables are color singlets iii) physical states are color singlets. Prop.2 proves that ii) follows from i). For a detailed proof one may follow the pattern discussed in Ref.[3].

Finally we want to discuss property iii). This is related to the problem of reconciling the success of the quark model and the failure of detecting quarks. Recently much attention has been paid in finding a mechanism for keeping the quarks bounded into systems of zero triality (quark confinement). This philosophy implies that, e.g. the q-$\bar{q}$ potential does not decrease at infinity sufficiently fast to allow quarks to escape from the bounded system [6]. It seems difficult to understand the above mechanism in a local field theory, since the decrease of the potential is associated to the cluster property and in a local field theory satisfying Wightman axioms the slowest decrease at infinity is like $1/r$, [7]. We will see that the non abelian character of the gauge group allows to escape the above difficulty and that, just as in the two dimensional QED, the cluster property may fail if an infrared mechanism is acting. To show that this may happen without violating locality and that the q-$\bar{q}$ potential may behave at infinity like r^N, N non negative integer, one must spell out the basic properties of a local field theory of the Yang-Mills type.

Definition. An indefinite metric quantum field theory is a theory specified by : a) a set of fields $\{\phi_i\}$ defined as operator valued tempered distributions on a Hilbert space H , and having a common dense domain D ; b) a bounded hermitean sesquilinear form $<\cdot,\cdot>$ on H , called the metric: $<\cdot,\cdot> = (\cdot,\eta\cdot)$; c) a representation U of the Poincaré group in D , unitary with respect to $<\cdot,\cdot>$: $U^+\eta U = \eta$; d) a distinguished space $H' \subset H$ such that $<\cdot,\cdot>$ is non negative on H' and the unique translationally invariant vector Ψ_o, the vacuum, belongs to H' and it is cyclic with respect to the set of fields $\{\phi_i\}$; e) (spectral condition) the Fourier transform of the Wightman functions have support contained in $\bar{V}_+$; f) (locality) for $(x-y)^2 < 0$ one or the other of $[\phi_i(x), \phi_j(y)]_{\pm} = 0$ holds.

A local theory of the Yang-Mills type is expected to be a local quantum field theory is the sense of the Definition above.

Proposition 3. (*Cluster property*) In a quantum field theory of the type of Definition above, with a mass gap (0,M) the cluster decomposition for the Wightman functions holds

$$W(x_1,\dots,x_j,x_{j+1}+\lambda a,\dots x_n+\lambda a) - W(x_1,\dots x_j)W(x_{j+1},\dots x_n) \to 0$$

when the real positive number $\lambda \to \infty$ and a is a spacelike vector, the convergence being in S' . Moreover if $B_1(x_1)$, $B_2(x_2)$ denote two clusters

$$B_i(x_i) = \int dx_1'\dots dx_{r(i)}' \; f_i(x_1',\dots x_{r(i)}')\phi(x_1'+x_i)\dots\phi(x_n'+x_i)$$

$f_i \in \mathcal{D}$, i = 1,2 , one has, $\xi = x_1-x_2$,

$$|<\Psi_o',B_1(x_1)\; B_2(x_2)\Psi_o> - <\Psi_o,B_1(x_1)\Psi_o> <\Psi_o,B_2(x_2)\Psi_o>|$$
$$\leq C[\xi]^{-3/2}\; e^{-M[\xi]}\,|\xi|^{2N}\left(1+\frac{[\xi^o]}{[\xi]}\right)$$

where N is a suitable non negative integer and all the other notations are the same as in Ref.7.

Proof. The crucial point is to fully exploit the spectral condition and the locality property. One defines as in Ref.7

$$h_{12}(\xi) \equiv \langle\Psi_o, B_1(x_1)\ B_2(x_2)\Psi_o\rangle - \langle\Psi_o, B_1(x_1)\Psi_o\rangle\langle\Psi_o, B_2(x_2)\Psi_o\rangle$$

and similarly h_{21}. Then one may show [8] that $\tilde{h}_{12}(p)$ can be written in the form $\tilde{h}_{12}(p) = \Box_p^N F(p)$, where $F(p)$ is a bounded complex measure with the same support properties as $\tilde{h}_{12}(p)$. The same holds for $\tilde{h}_{21}(p) = \Box_p^{N'} F'(p)$ and by locality and edge of the wedge theorem one finds $N = N'$, $F(p) = F'(p)$. The Fourier transform $H(\xi)$ of $F(p)$ then satisfies all the properties of the standard cluster decomposition function [7]. Since $h(\xi) = \xi^{2N} H(\xi)$, the theorem can be easily proved by following the standard argument [7].

The above Proposition shows that a $q\bar{q}$ potential decreasing at ∞ more slowly than $1/r$ is incompatible with locality unless an infrared mechanism is present (infrared slavery).

<u>Proposition 4</u> In the case of no mass gap one has

$$|h_{12}(\xi)| \leq C[\xi]^{-2}\ |\xi|^{2N} \left[1 + \frac{[\xi^o]}{[\xi]^2} \right]$$

where N is a suitable non negative integer. So that the cluster property may fail.

Similarly statements can be proved for the spectral representation of the two point function (the analogue of the Lehman spectral representation for Yang-Mills theories) or for the JLD-representation .

<u>Proposition 5</u> The failure of the cluster property is consistent with locality if the theory is of the FGML type.

Proof. It is not difficult to show that in such a theory locality requires the indefinite metric [3]. Therefore the space time translations $U(a)$ need not to commute with the metric η . All is required in order to have translationally invariant Wightman functions is that $U(a)$ are η-unitary. Thus one falls in the class of theories of the Definition above. For any two local states the Fourier transform of $\langle\Psi, U(a)\Phi\rangle$ need not to be a bounded fast decreasing measure as in the case of unitary $U(a)$'s . The exploitation of locality leads to the results of Prop. 4.

It is important to stress the rôle of the non abelian character of the gauge group. This can be easily seen, e.g. in the two point function of the vector field

$$<\Psi_o, A_\mu^a(x)\, A_\nu^b(y)\, \Psi_o> = g_{\mu\nu}\, F^{ab}(x) + \partial_\mu \partial_\nu\, G^{ab}(x)$$

In the abelian case $F_{\mu\nu} = \partial_\mu A_\nu - \partial_\nu A_\mu$ is an observable and therefore the Fourier transform of $F^{ab}(x)$ must be a measure, so that the possible non unitary of the translations may affect only the unphysical purely gauge part $\tilde{G}^{ab}(p)$. In the non abelian case $F_{\mu\nu}^{ab}$ and $G_{\mu\nu}^{ab}$ are not observable and there is no condition of $\tilde{F}^{ab}$ and $\tilde{G}^{ab}$ which may both fail to be measures. This in fact seems to be the case, as perturbation theory suggests, due to the presence of Faddeev-Popov ghosts. In the ghost free gauges, there are problems with positivity, an indication that again $\tilde{F}^{ab}$ and $\tilde{G}^{ab}$ cannot be measures.

Another basic difference between the abelian and the non abelian case is that in the first case the local charge, i.e. the charge $Q_R = \int_{|x|<R} J^o(x)\, d^3x$ confined in a bounded region, is observable, whereas in the latter case the *local charges* Q_R^a *cannot be observable*. The proof follows from the structure of the algebra of charges $[Q^a, Q_R^b] = i\, f^{abc}\, Q_R^c$ and Prop.2. Again for a more detailed argument we refer to [3] .

REFERENCES

[1] See e.g. the review C.Orzalesi, Rev.Mod.Phys. 42, 381 (1970)

[2] R.Utyama, Phys.Rev. 101, 1597 (1956)

[3] F.Strocchi and A.S.Wightman, J.Math.Phys. 15, 2198 (1974)

[4] R.Haag and D.Kastler, J.Math.Phys. 5, 848 (1964)

[5] H.Fritsch, M.Gell-Mann and H.Leutwymer, Phys.Lett. 47B, 365 (1973)

[6] J.Kogut, talk at the APS Meeting, Sept.5-7, 1974 and references therein

[7] H.Araki, K.Hepp and D.Ruelle, Helv.Phys.Acta 35, 164 (1962)

[8] J.Bros, H.Epstein and V.Glaser, Commun. Math.Phys. 6, 77 (1967).

GEOMETRIC QUANTIZATION

THE ALGEBRAIC METHOD IN REPRESENTATION THEORY

(ENVELOPING ALGEBRAS)

Anthony JOSEPH*)

Institut des Hautes Etudes Scientifiques
91440 - BURES-sur-YVETTE
(France)

*) Permanent address : Department of Physics and Astronomy, Tel-Aviv University, Ramit Avic, Israel

THE ALGEBRAIC METHOD IN REPRESENTATION THEORY (ENVELOPING ALGEBRAS)

1 Introduction

In Lie algebra theory a basic open problem is to classify all irreducible representations (up to equivalence). For the category of finite dimensional representations the answer is known and is classical. By contrast a full classification for infinite dimensional representations appears to be impossible. For example, this is evidenced by the work of Arnal and Pinczon [2] on $s\ell(2)$ and by the work of McConnell and Robson [34] on A_1 which can be used to show [5] that the Heisenberg Lie algebra [18], 4.6.1, admits infinitely many inequivalent irreducible representations all very different from the standard one.

One way out of this difficulty is to consider only representations which integrate to unitary (or just bounded) representations of the corresponding Lie group(s). This has physical justification through Wigner's theorem. We call it the analytic approach. In it the Lie algebra plays only a subservient role.

About ten years ago, Dixmier proposed a purely algebraic way out which has since then generated a new area of mathematics called enveloping algebras. The philosophy is to find a less refined classification than equivalence classes rather than to exclude representations. We call it the algebraic approach. Except for finite dimensional representations, or for nilpotent Lie algebras, the exact relationship between these two approaches is not yet known, though one can find many analogies. Consequently in a given application one must make a definite choice as to which to use.

It is my personnel conviction that the algebraic approach has a very real and important contribution to make in physics. First of all it gives an overall framework which I believe physicists have been unsucessfully groping at for some years. Secondly, new techniques have been developed which go far beyond the relatively naive (though often intricate) computations of physicists. Some of these were taken over from situations where the analytic approach could be reduced to purely Lie algebraic questions and in this sense notable early contributors to enveloping algebras are Gelfand, Kostant and Harish-Chandra (the acknowledgement here is Dixmier's [18], p. 5). Other techniques are inspired by commutative (and non-commutative) algebra and in particular localization plays a fundamental role.

I do not suggest that either approach is superior in its physical applicability; rather that a choice between the two should be dependent on the situation concerned. Apart from the question of taste (i.e. if one prefers algebra to analysis) two further principles should be kept in mind.

1) If the Lie algebra is given through the action of a Lie group on a manifold (such as the Poincaré group acting on space-time) then the analytic approach is favoured. Yet for internal symmetries (e.g. su(3) in particle physics) where the symmetry has no well-established geometric origin, then the algebraic approach is at least of equal importance. In fact the search for "realizations" of Lie algebras (especially in the context of spectrum generating algebras [36] and chiral Lagrangians [14]) constitutes what I believe to be the primordial component of enveloping algebras in physics.

2) Roughly speaking, the analytic approach is best adapted to fundamental questions in physics, whereas the algebraic approach applies best to a computational scheme.

In the next section we give a precise definition of the algebraic approach. For "representation space", it is convenient to use the more precise term - module [18] , 1.2.1. This in particular excludes the possibility that we are talking about representations in Hilbert space, though in this context we remark that the module should be regarded as a common dense domain for the generators of the Lie algebra (in the language of say field theorists). Moreover this notion can often be made precise and then has a highly non-trivial content [22,39].

Throughout we take the base field $\underline{k}$ to be commutative and of characteristic zero. For example, $\underline{k}$ can be the real field $\mathbb{R}$ or the complex field $\mathbb{C}$, the latter being algebraically closed. $\mathbb{N}$ denotes the natural numbers and $\mathbb{N}^+$ the positive integers.

2 The Primitive Spectrum

Let $\underline{g}$ be a finite dimensional Lie algebra with enveloping algebra $U(\underline{g})$. Recall that if $\{X_i\}_{i=1}^n$ is a basis for $\underline{g}$, then $X_1^{k_1} X_2^{k_2} \ldots\ldots X_n^{k_n} : k_i \in \mathbb{N}$ is a basis for $U(\underline{g})$ and that for all $X,Y \in \underline{g}$, we have $[X,Y] = XY - YX$ in $U(\underline{g})$, where $[\ ,\ ]$ denotes the Lie bracket in $\underline{g}$. The importance of $U(\underline{g})$ derives from the fact that every $\underline{g}$ module extends (in the obvious fashion) to a $U(\underline{g})$ module. Given M a $U(\underline{g})$ module, we set $\mathrm{Ann}(M) = \{a \in U(\underline{g}) : aM = 0\}$. $\mathrm{Ann}(M)$ is a two-sided ideal in $U(\underline{g})$ and if ρ denotes the representation of $U(\underline{g})$ associated with M, then $\ker \rho = \mathrm{Ann}(M)$. If M_1, M_2 are isomorphic (i.e. equivalence of representations) as $U(\underline{g})$ modules, then $\mathrm{Ann}(M_1) = \mathrm{Ann}(M_2)$. By Wedderburn's theorem [23], Chap. 2, the converse holds in the category of finite dimensional simple $U(\underline{g})$ modules.

Yet the converse fails for infinite dimensional modules and this observation is the starting point of the algebraic approach. In more detail, set Prim $U(\underline{g})$ = { Ann(M) : M a simple $U(\underline{g})$ module}. Then the fundamental goal of the algebraic approach is to classify Prim $U(\underline{g})$ (the primitive spectrum) for each $\underline{g}$. This is less refined than classification by equivalence classes and unlike the latter is proving quite feasible.

One should admit from the start that Prim $U(\underline{g})$ is not sufficiently refined for all physical purposes. Thus we shall see that the description of a primitive ideal $I \in$ Prim $U(\underline{g})$ entails the use of differential operators. Then a canonical transformation like $x \to x$, $d/dx \to d/dx + f(x)$, does not alter I. On the other hand (if these operators are applied to a fixed space of functions) this transformation does in general alter the equivalence class of the associated representation [19], Sect. 3. Moreover this transformation has been used for example by Zak to contrast the behaviour of a free electron with one in a magnetic field [7,21]. In a simply connected region one may solve the equation $g'(x) = f(x)$ and eliminate $f(x)$ through a gauge transformation. However this is not possible in general and different $f(x)$ can correspond to different physical situations as exemplified by the Bohm-Aharonov effect [1]. Again one might recall the work of Miller [35] who shows that a quite large variety of different physical situations can be encompassed within various realizations of one or two Lie algebras and in fact within a quite small class of (primitive) ideals. Yet again the realization of the symplectic algebra sp(2n) through symmetrized quadratic polynomials in x_i, $\partial/\partial x_i$: $i = 1,2,..n$, leads by its action on the polynomial ring $\underline{k}[x_1,x_2,\dots x_n]$ to two inequivalent irreducible representations (the Weil representations) spanned by polynomials of odd and even degree. These have the same annihilator in $U(\underline{g})$ and both have enjoyed enormous popularity in physical models [32] , p. 69.

Nonetheless it should be emphasized that the description of a primitive ideal through differential operators allows one to recover practically all the representations of Lie algebras discussed in physics. In fact I know of no examples for which this is not so.

3 Goldie's Theorem and the Gelfand-Kirillov Conjecture.

There are two main problems in the study of Prim $U(\underline{g})$. One, its description as a set and secondly the description of a given primitive ideal. First we give two basic properties of the latter which highly motivates its physical interest.

Given $I \in$ Prim $U(\underline{g})$, we set $U = U(\underline{g})/I$. The structure of the factor algebra U should be considered as the generic property of the set of simple $U(\underline{g})$ modules with annihilator I and is hence of fundamental importance. In general U is not integral and so we set $S = \{ a \in U : ab = 0, b \in U \text{ implies } b = 0\}$.

Example 1 . Let I be the annihilator of a simple finite dimensional $U(\underline{g})$ module. Then $\dim U < \infty$ and for some $m \in \mathbb{N}^+$, it is isomorphic to the full matrix algebra M_m over $\underline{k}$. Then S is the set of all invertible matrices in M_m.

Application of Goldie's theorem [18], 3.6.2, shows that $\text{Fract } U = \{ a^{-1}b : a \in S, b \in U\}$ is an algebra under the obvious identifications and furthermore that Fract U is isomorphic for some $m \in \mathbb{N}^+$ to M_m over some non-commutative field K (which is an extension of the base field $\underline{k}$). More precisely we have

THEOREM -

(i). S is an Ore set for U.

(ii). $\text{Fract } U = M_m(K)$, up to isomorphism.

Observe that this assigns a number m to I known as its Goldie rank. M_m describes in the generic sense, the finite dimensional piece of the representation associated with I. If $\underline{g}$ is solvable, then m is always one [18], 3.7.2, and this generalizes Lie's theorem to the infinite dimensional case ! Now K can be thought of as carrying the infinite dimensional part of the representation and it is a remarkable fact that in the enveloping algebra context K is apparently always a Weyl field. In more detail, recall that the Weyl algebra A_n of index n (over $\underline{k}$) is isomorphic to the associative algebra generated by x_i , $\partial/\partial x_i$: $i = 1,2,\ldots n$, which physicists should consider as creation and annihilation operators (and correspondingly $\underline{k}[x_1,x_2,\ldots x_n]$ as Fock space). Set $K_n = \text{Fract } A_n = \{ a^{-1}b : a \neq 0, b \in A_n\}$. Let K be defined by (ii) above.

CONJECTURE - Suppose that $\underline{g}$ is a split algebraic Lie algebra. Then

(i). For some $n \in \mathbb{N}$, $K = K_n$, up to isomorphism.

(ii). $n = 1/2 \text{ Dim } U$.

The above conjecture, in a slightly different form, was proposed by Gelfand and Kirillov [20]. It has been established for $\underline{g}$ solvable [9, 33] and for the minimal primitive ideal (see Sect. 6) of a semisimple Lie algebra [16], Cor. 10.5. In [28] I have shown that the origin of the Weyl field derives from nilpotent action. In (ii), Dim denotes Gelfand-Kirillov dimension. This is roughly the same as transcendence degree; but we refer the reader to [11] for its exact definition and principle properties. We remark that $\underline{g}$ is always split if $\underline{k}$ is algebraically closed. The requirement that $\underline{g}$ be algebraic [8] is of a technical nature and we could equally well have extended slightly the class of possible candidates for K (c.f. [33]). Most Lie algebras in physics are algebraic.

We should like to emphasize three points.

1) A further integer, namely Dim U is associated with I. It is called the Gelfand-Kirillov dimension of I. Whilst it is an integer, it is not obviously even. However, this has been shown for $\underline{g}$ solvable (and algebraic) and for $\underline{g}$ semi-simple [11], 7.1.

2) The identity $K = K_n$ has the physical interpretation that the infinite part of the representation is carried by Fock or a Fock-like space (depending on the amount of localization necessary in the embedding of U in K_n). The conjecture and the theorem can be regarded as a precise way of saying that the matrix generalization of quantum realizations provides the general setting for Lie algebra representations. The representation space (in the generic sense) is spanned by finite component wave functions such as those encountered in the Dirac or Majorana equations.

3) It should be emphasized that the algebraic approach is not just a transition from the study of representations to the study of realizations. This would have little more than heuristic value since the latter problem is equally intractable (though some progress was made with the realization of $s\ell(2)$ in A_1 [26]). Rather one classifies all the possible relations between U relative to M and for this the description of U in terms of differential operators is particularly useful.

<u>Example</u> 2 Let $\underline{b}$ be a subalgebra of $\underline{g}$ and σ a finite dimensional irreducible representation of $\underline{b}$ of dimension m. Let ind (σ , $\underline{b} \uparrow \underline{g}$) [18], Chap. 5, denote the representation induced to $\underline{g}$ and set I = ker ind (σ, $\underline{b} \uparrow \underline{g}$). Under favourable circumstances (see Sects. 4-6) , I is a primitive ideal of Goldie rank m, and Gelfand-Kirillov dimension n, where n = dim $\underline{g}$ - dim $\underline{b}$. It is said to be an induced ideal.

We remark that the inducing construction in the algebraic context [16], Sect. 5, or [27], Sect. 3, exactly coincides with what physicists called non-linear or induced realizations of Lie algebras [14, 24] . Moreover the above decomposition of the representation into finite and infinite dimensional pieces was given [14] the physical interpretation of being the decomposition of the particle spectrum into baryonic (finite) and mesonic (infinite) parts. (This distinction arose because baryon number is conserved and for fixed baryon number, the baryon spectrum is finite. Also it is easy to give the finite part the correct statistics through the use of operators satisfying the canonical anticommutation relations). An important fact is that not all primitive ideals are obtained by induction from a finite (or even infinite) dimensional representation of a proper subalgebra (c.f. Sects. 5,6). In particular this is true of the annihilator of the Weil representations. I have succeeded in giving some new constructions for non-induced primitive ideals and I anticipate that these may have some useful physical applications.

4 Geometric Quantization and the Solvable Case

To describe Prim $U(\underline{g})$ as a set, we should like to relate it to some simple geometric object. Suppose (for technical simplicity) that $\underline{g}$ is an algebraic Lie algebra [8]. Recall that the adjoint group G acts by transposition in the dual space $\underline{g}^*$ of $\underline{g}$ and let $\underline{g}^*/G$ denote the corresponding orbit space. After Kirillov-Kostant each G orbit in $\underline{g}^*$ is a symplectic manifold [6], Chap. II, and hence in physicists' language provides a classical realization of $\underline{g}$. Through the conjecture of Sect. 3, primitive ideals are associated with quantum realizations. This suggests that one can construct a bijection between $\underline{g}^*/G$ and Prim $U(\underline{g})$ which we should naturally call (geometric) quantization. Actually this term is usually reserved for the rather specific procedure initiated by Kirillov [31] for the nilpotent case and generalized by Auslander and Kostant [3,4] to the solvable case. This was put in the enveloping algebra framework by Dixmier for the nilpotent case and by Borho, Gabriel and Rentschler [9], [18], Chap. 6, in the solvable case. We sketch the method below. For its wider context we refer the reader to Simms' article in this series.

Given $f \in \underline{g}^*$, let Gf denote the G orbit containing f. A polarization $\underline{b}$ for f is a subalgebra of $\underline{g}$ satisfying

(i). $\dim \underline{g} - \dim \underline{b} = \frac{1}{2} \dim Gf$ (recall Ex. 2).

(ii). $(f, [\underline{b},\underline{b}]) = 0$.

The set of all polarizations for f is denoted by P(f).

By (ii) the map $f : \underline{b} \to (\underline{b}, f)$ is a one-dimensional representation of $\underline{b}$ which we simply denote by f. Set $I(f,\underline{b}) = \ker \operatorname{ind}(f, \underline{b} \uparrow \underline{g})$. Ideally this is the required element of Prim $U(\underline{g})$. For $\underline{g}$ split solvable, $I(f,\underline{b})$ is in fact primitive [18], 6.1.1 ; but may also depend on the polarization $\underline{b}$ chosen. In the above case this difficulty can be removed by replacing ind by $\operatorname{ind}^{\sim}$ - namely, the twisted induced representation [18], 5.2. Indeed [18], 6.1.4,

THEOREM 1 - ($\underline{g}$ split solvable) . Let $f \in \underline{g}^*$, $\underline{b}_1$, $\underline{b}_2$ $\in P(f)$. Then $\ker \operatorname{ind}^{\sim}(f, \underline{b}_1 \uparrow \underline{g}) = \ker \operatorname{ind}^{\sim}(f, \underline{b}_2 \uparrow \underline{g})$.

From now on we take $\underline{g}$ solvable and $\underline{k}$ algebraically closed. Following Theorem 1, we set $I(f) = \ker \operatorname{ind}^{\sim}(f,\underline{b}\uparrow \underline{g}) : \underline{b} \in P(f)$. Then Theorem 2, [18], 6.1.7, asserts that all primitive ideals are obtained in this way.

THEOREM 2 - The map $f \to I(f)$ of $\underline{g}^*$ into Prim $U(\underline{g})$ is surjective.

Let $\mathfrak{G}$ denote the algebraic adjoint group of $\underline{g}$ [18], 6.1.5. (One has $\mathfrak{G} = G$ if $\underline{g}$ is algebraic). Then $I(f_1) = I(f_2)$ if $f_2 \in \mathfrak{G}\, f_1$. Extend $f \to I(f)$ to a map $\overline{I}$ of $\underline{g}^*/\mathfrak{G}$ into Prim $U(\underline{g})$ by passage to the quotient (i.e. in the obvious manner). Theorem 3, [18], 6.5.12, shows that different $\mathfrak{G}$ orbits give different ideals.

THEOREM 3 - $\overline{I}$ is a bijection.

$\overline{I}$ is continuous in a natural manner. Indeed take the Jacobson topology [18], 3.2, in Prim $U(\underline{g})$ and the Zariski topology in $\underline{g}^*$ (and the corresponding quotient topology in $\underline{g}^*/\mathfrak{G}$). We remark that the Jacobson topology is the non-commutative analogue of the Zariski topology and that the latter is generally used in algebraic geometry to discuss algebraic curves. It is coarser than the usual metric topology. For $\underline{g}$ solvable, $\overline{I}$ is continuous [18], 6.4.4 and [9], Sect. 6, and for $\underline{g}$ nilpotent, it is known to be a homeomorphism [15].

To establish the connection with the analytic approach, let $\mathcal{G}$ be the connected, simply connected real Lie group with Lie algebra $\underline{g}$ over $\mathbb{R}$ and $\hat{\mathcal{G}}$ the set of classes of irreducible unitary representations of $\mathcal{G}$. When $\underline{g}$ is nilpotent, $\underline{g}^*/ \operatorname{ad}(\mathcal{G}) \to \hat{\mathcal{G}}$ is a bijection [31]and [38], Thm. 2.10. Now in this case $\underline{g}$ is algebraic, so $\operatorname{ad}(\mathcal{G}) = \mathfrak{G} = G$. Furthermore the appropriate generalization of theorem 3 when $\underline{k}$ is not algebraically closed (and $\underline{g}$ is nilpotent) obtains by replacing Prim $U(\underline{g})$ by Rat $U(\underline{g})$, the set of rational ideals [18], 4.5.8. Then by [18], 6.2.4 and the above, Rat $U(\underline{g}) \to \hat{\mathcal{G}}$ is a bijection. Thus the algebraic and analytic approaches can be considered equivalent in the nilpotent case. For general solvable Lie groups the unitary representation theory has not yet been fully worked out. However for type 1 solvable Lie groups (which in particular includes the algebraic case) one has [3], Thm. 2, [4], Thm. V. 3.3,

$$\hat{\mathcal{G}} = \bigcup_{\mathfrak{O} \in \underline{g}^*/\operatorname{ad}(\mathcal{G})} \mathfrak{H}_{\mathfrak{O}},$$

where $\mathfrak{H}_{\mathfrak{O}}$ is the character group of the fundamental group of $\mathfrak{O}$ and has the structure of a torus of suitable dimension. Thus $\underline{g}^*/\operatorname{ad}(\mathcal{G}) \to \hat{\mathcal{G}}$ is a bijection only if all orbits are simply connected and this may fail even if $\underline{g}$ is algebraic, as exemplified by the oscillator group [41], Sect. 6. It remains to determine whether or not the algebraic approach exhibits a similar phenomenon (probably not).

5 Difficulties with Quantization in the Semisimple Case.

By contrast with the solvable case, quantization encounters many difficulties in the semisimple case. These we summarize below. For simplicity we take $\underline{k} = \mathbb{C}$.

1) Every simple Lie algebra, excepting $s\ell(n+1) : n \in \mathbb{N}^+$ has at least one orbit which does not admit polarization.

2) Theorem 1 fails. That is different polarizations of a given $f \in \underline{g}^*$ can give rise to different primitive ideals. This was remarked [10] by Borho and Rentschler for sp(4).

3) Theorem 2 fails. That is not all primitive ideals are induced. The first counterexample was given by Conze and Dixmier [17], Ex. 3. It turned out that this was exactly the annihilator of the Weil representations for sp(4). I have since then shown [25,29] that every simple Lie algebra, excluding $s\ell(n)$, admits at least one primitive ideal which is not induced. For sp(2n) my construction gives the annihilator of the Weil representations, whereas for the other cases it is entirely new both to mathematicians and to physicists. I believe it may have application to spectrum generating algebras and to the classification of nuclear states.

4) Unlike the solvable case the map $\underline{g}^*/G \to \text{Prim } U(\underline{g})$ is not always continuous [40]. In particular for $s\ell(3)$, the space of subregular orbits (those of dimension 4) form a cusp, whereas the corresponding family of primitive ideals (namely those of Gelfand-Kirillov dimension 4 and Goldie rank 1) form a less singular loop. This is illustrated below.

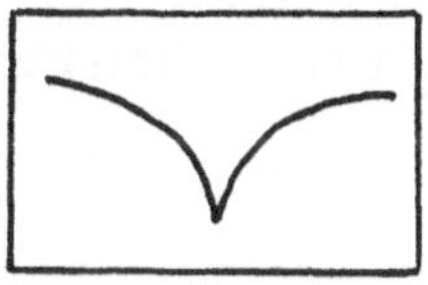

space of subregular orbits orbits.

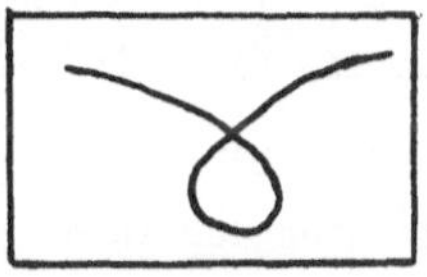

space of corresponding primitive ideals.

It is very striking that the passage from orbits to ideals exactly corresponds in this case to what algebraic geometers call a resolution of a singularity. Actually the mechanism is quite simple and derives from the $(\rho,\rho) : \rho = \frac{1}{2}$(sum of positive roots), term present in the spectrum of the Casimir invariants. This term is analogous to zero point energy and it would be very nice to find a resolution of singularity in the passage from classical to quantum mechanics. For $s\ell(3)$, Borho and I have shown [12] that the primitive ideals of higher Goldie rank form similar loops except that these have finitely many missing points where they intersect with loops of lower Goldie rank.

5) Theorem 3 fails. That is different orbits can give rise to the same ideal [10].

6) Goldie rank can be greater than one. Hence it is not sufficient to induce from a one-dimensional representation as outlined in Sect. 4. Of course for each $f \in \underline{g}^*$, one can simply induce from a finite dimensional irreducible representation σ of $\underline{b} \in P(f)$. However it can happen [12] that over a certain (finite ?) number of orbits, the resulting ideal is not primitive. Again it is not known if the Goldie rank of ker ind $(\sigma, \underline{b} \uparrow \underline{g})$ always coincides with dim σ, [18], Prob. 12.

We remark that these difficulties are not restricted to the algebraic approach and in fact become even more serious in the analytic approach.

A natural problem posed by 3) is to give a criterion for when an ideal is induced. I propose the following. First recall [27], Sect. 3, that through induction each $X \in \underline{g}$ is realized as a first-order differential operator with coefficients in $S(\underline{m})^\wedge$, where $\underline{m} = \underline{g} \ominus \underline{b}$ and $S(\underline{m})^\wedge$ denotes the formal power series completion of the symmetric algebra. (Of course this is also the essential fact behind polarization). Since each X is first order, $\underline{g}$ leaves $S(\underline{m})^\wedge$ stable. Under favourable circumstances [27], Prop. 3.5, [16], Cor. 10.5, we can replace $S(\underline{m})^\wedge$ by $R(\underline{m})$ = Fract $S(\underline{m})$ and furthermore identify $R(\underline{m})$ as a subfield of Fract $U(\underline{g})/I$. Obviously $R(\underline{m})$ is commutative. Set $L = R(\underline{m})$. Given $I \in$ Prim $U(\underline{g})$, set $U = U(\underline{g})/I$. Suppose that $L \subset U$ and let L' denote its commutant in U. Through the discussion of Sect. 3, one expects to have $L' = M_m(L)$, where m is the Goldie rank of I. In particular Dim L = Dim L' and $L = L'$ if $m = 1$. This motivates the following definition. Call a subfield L of U induced if (i) L is commutative, (ii) $[\underline{g},L] \subset L$, (iii) Dim L = Dim L'.

CONJECTURE - <u>A primitive ideal</u> I <u>is induced if and only if U admits an induced subfield</u>.

Assume that $\underline{g}$ is split solvable. Extending the results of [37] and in fact with some simplifications, I have shown that for all $I \in$ Prim $U(\underline{g})$, (i) $U(\underline{g})/I$ admits an induced field, (ii) $L = L'$ (recall that the Goldie rank equals one in this case). Combined with Theorem 2, this proves the conjecture for $\underline{g}$ solvable and $\underline{k}$ algebraically closed. An easy application of [16], Cor. 10.5, shows that the conjecture also holds for the minimal primitive ideal of a split semisimple Lie algebra (see Sect. 6). A possible generalization of quantization and its role in the Gelfand-Kirillov conjecture is discussed in [30]. (I should add that a certain primitive ideal is incorrectly identified in [30], pp. 232-233, and these pages should be ignored. For sp(4,$\mathbb{C}$) a correct description is given in [12]).

6 The Semisimple Case

In view of Sect . 5, new techniques must be devised to handle the semisimple case. Here we have space only to summarize some of the main results.

Take $\underline{g}$ simple and $\underline{k} = \mathbb{C}$. Let $\underline{h}$ be a fixed Cartan subalgebra and Δ (resp. Δ+) a choice of non-zero (resp. positive) roots. Set $\mathcal{D} = \{ \lambda \in \underline{h}^*: (\lambda,\alpha) > 0,$ for all $\alpha \in \Delta^+ \}$. $\mathcal{D}$ is called the Weyl chamber and plays a fundamental role in the representation theory of $\underline{g}$. Let $\overline{\mathcal{D}}$ denote the closure of $\mathcal{D}$ (in the metric topology). Then the walls $\overline{\mathcal{D}} \setminus \mathcal{D}$ of the Weyl chamber form a set of reflection planes which generates the Weyl group W of $\underline{g}$. Let $Z(\underline{g})$ denote the centre of $U(\underline{g})$. Given $I \in \text{Prim}\, U(\underline{g})$, then $I \cap Z(\underline{g}) = Z_\lambda$ is a maximal ideal in $Z(\underline{g})$ and after Harish-Chandra [18], 7.4.3, we can regard λ as an element of $\overline{\mathcal{D}}$. That is each $\lambda \in \overline{\mathcal{D}}$ specifies the eigenvalues of the Casimir invariants of $\underline{g}$ in some well-defined fashion, and is called the central character of I. Set $\underline{X}(\lambda) = \{ I \in \text{Prim}\, U(\underline{g}) : I \cap Z(\underline{g}) = Z_\lambda \}$. Then

1) $1 \le \text{card}\, \underline{X}(\lambda) < \infty$, for all $\lambda \in \overline{\mathcal{D}}$, [18], 8.4.4, 8.5.7 (b).

2) Order $\underline{X}(\lambda)$ by inclusion. This is not in general a total ordering; but $\underline{X}(\lambda)$ does have a unique minimal element $I_{min}(\lambda)$ and a unique maximal element $I_{max}(\lambda)$, [18], 8.4.4, 8.5.8 (a).

3) $I_{min}(\lambda) = U(\underline{g})\, Z_\lambda$, [18], 8.4.3.

4) Card $\underline{X}(\lambda) = 1$, if and only if $2\,(\lambda,\alpha)/(\alpha,\alpha) \notin \mathbb{N}^+$, for each $\alpha \in \Delta^+$, [18], 8.5.8 (a).

The sets $\{\lambda \in \mathcal{D} : 2\,(\lambda,\alpha)/(\alpha,\alpha) \in \mathbb{N}^+$, for some $\alpha \in \Delta^+\}$ are called the exceptional hyperplanes. By 3) and 4), the study of Prim $U(\underline{g})$ reduces to what happens on these planes. A similar phenomenon occurs in the analytic approach. We remark that a bounded representation gives rise to a Harish-Chandra module M [18] , p. 277, for $U(\underline{g})$. Now whereas $I_{min}(\lambda)$ is the annihilator of a Verma module $M(\lambda)$, [18] , Chap. 7, one cannot have $M = M(\lambda)$, except on the hyperplanes defined by the compact roots. This excludes the obvious interrelation. Again for a finite dimensional representation one must have $2\,(\lambda,\alpha)/(\alpha,\alpha) \in \mathbb{N}$, fot all $\alpha \in \Delta^+$.

5) Given $I \in \text{Prim}\, U(\underline{g})$, then Dim $U(\underline{g})/I$ equals the dimensional of some nilpotent orbit [11], 7.1, and is hence even (c;f; Sect. 3).

6) Let $I_1, I_2 \in \text{Prim}\, U(\underline{g})$, then $I_1 \supset I_2$ implies Dim $U(\underline{g})/I_1 \le$ Dim $U(\underline{g})/I_2$, with equality if and only if $I_1 = I_2$, [11] , 3.6.

7) By 2), 5), 6) and [16], Cor. 10.5, Dim $U(\underline{g})/I = \frac{1}{2}(\dim \underline{g} - \text{rank}\, \underline{g})$, if

and only if $I = I_{min}(\lambda)$, for some $\lambda \in \bar{\mathfrak{D}}$. Again $\text{Dim}\, U(\underline{g})/I = 0$, if and only if $\dim U(\underline{g})/I < \infty$, that is if the associated representation is finite dimensional.

8) Suppose $I = \ker \text{ind}\,(\sigma, \underline{p} \uparrow \underline{g})$, where σ is a finite dimensional representation of a parabolic subalgebra $\underline{p}$ of $\underline{g}$. Then [13], Sect. 2, $\text{Dim}\, U(\underline{g})/I = 2(\dim \underline{g} - \dim \underline{p})$.

9) By 2), 8) and [17], Prop. 2, we have card $\underline{X}(\lambda) = 2$, if λ lies on just one exceptional hyperplane [13].

Given $\lambda \in \underline{h}^*$, let $M(\lambda)$ be the Verma module with highest weight $\lambda - \rho$: $\rho = \frac{1}{2} \sum_{\alpha \in \Delta^+} \alpha$, [18], Chap. 7. (Note that the $M(w\lambda)$: $w \in W$, are not isomorphic). Now $M(\lambda)$ admits a largest non-trivial submodule L, [18], 7.1.11 (ii) and the quotient module $M(\lambda)/L$ is simple. Hence $I(\lambda) = \ker M(\lambda)/L \in \text{Prim}\, U(\underline{g})$ and if we choose $w \in W$ such that $w\lambda \in \bar{\mathfrak{D}}$, then $I(\lambda) \in \underline{X}(w\lambda)$. For each $\lambda \in \bar{\mathfrak{D}}$, let $\underline{X}^o(\lambda)$ denote the subset of all $I \in \underline{X}(\lambda)$ having the above form. Observe that card $\underline{X}^o(\lambda)$ is bounded above by the order of the Weyl group.

BASIC OPEN PROBLEM - Is $\underline{X}(\lambda) = \underline{X}^o(\lambda)$, for all $\lambda \in \bar{\mathfrak{D}}$?

10) $I_{min}(\lambda)$, $I_{max}(\lambda) \in \underline{X}^o(\lambda)$, [18], 7.6.24 and 8.4.4, and 8.5.8 (b).

Finally Borho and I have given [12] a complete description of Prim $U(\underline{g})$ for $\underline{g} = s\ell(3)$ and $sp(4)$. Space does not permit a description of the results or the methods; but I should add that many aspects of the analysis admit a straightforward generalization to arbitrary simple $\underline{g}$. Also the annihilator of the Weil representations (which has Goldie rank one) generalizes giving ideals of Goldie rank $\frac{1}{2}(\ell + 1)(2k + \ell + 2)$: $k, \ell \in \mathbb{N}$. This is a new class of primitive ideals none of which are induced. It must surely be true that such a natural generalization of the Weil representations will play an equally important role in physical models. In any case our construction is very explicit.

7 Decomposition Theory

Let M be an arbitrary $U(\underline{g})$ module and set $I = \text{Ann}(M)$. Physicists often require a decomposition theory for M especially in the discussion of coupling coefficients. Taking the algebraic approach to its logical conclusion we should attempt to do this in terms of I alone. Here there are some initial difficulties. First I can be primitive even if M is not simple. For example, the annihilator of the Verma module is always primitive [18], 8.4.4, even though the module itself is not always simple [18], Chap. 7. Also the direct sum of the Weil representations has a primitive annihilator. Again the algebraic approach does not distinguish between decomposability and reducibility. For example, $I = I_1 \cap I_2$ can mean either that $M = N_1 \oplus N_2$, where $I_j = \mathrm{Ann}(N_j) : j = 1,2.$, or that M is indecomposable and admits the submodule N with $I_1 = \mathrm{Ann}(N)$, $I_2 = \mathrm{Ann}(M/N)$. Nevertheless one does have a decomposition theory for I which generalizes a situation from algebraic geometry. Thus I can be written as a finite product of not necessarily distinct prime ideals [18], 3.1, which in the commutative case corresponds to the decomposition of the zero set of I into irreducible algebraic varieties one for each distinct prime ideal. Then, at least for $\underline{g}$ solvable [9], 10.8 and 13.4, each prime ideal can be written as (in general infinite) intersection of primitive ideals having different central character. It then remains (for physicists !) to develop a version of the Wigner-Eckart theorem.

References.

[1]. Y. Aharonov and D. Bohm, Further considerations on electromagnetic potentials in the quantum theory, Phys. Rev., 123 (1961) pp. 1511-1524.

[2]. D. Arnal and G. Pinczon, On algebraically irreducible representations of the Lie algebra of $s\ell(2)$, J. Math. Phys., 15 (1974) pp. 350-359.

[3]. L. Auslander and B. Kostant, Quantization and representations of solvable Lie groups, Bull. Am. Math. Soc., 73 (1967) pp. 692-695.

[4]. ———————, Polarization and unitary representations of solvable Lie groups, Inv. Math., 14 (1971) pp. 255-354.

[5]. S. Bamba, private communication.

[6]. P. Bernat and Coll., Représentations des groupes de Lie résolubles, Monographies Math. Soc. France, Dunod, Paris, 1972.

[7]. M. H. Boon, Representations of the invariance group for a Bloch electron in a magnetic field, J. Math. Phys., 13 (1972) pp. 1268-1284.

[8]. A. Borel, Linear algebraic groups, Benjamin, New York, 1969.

[9]. W. Borho, P. Gabriel and R. Rentschler, Primideale in Einhüllenden auflösbarer Lie-algebren, Lecture notes in mathematics, 357, Springer-Verlag, New York, 1973.

[10]. W. Borho and R. Rentschler, Primitive vollprime Ideale in der Einhüllenden von so(5,$\mathbb{C}$), Vortragsberichte, Oberwolfach 1975.

[11]. W. Borho and H. Kraft, Uber die Gelfand-Kirillov dimension, to appear.

[12]. W. Borho and A. Joseph, Primitive ideals in the enveloping algebras

of $s\ell(3,\mathbb{C})$ and $sp(4,\mathbb{C})$, to appear.

[13]. W. Borho, Berechnung der Gelfand-Kirillov-dimension bei induzierten Darstellungen, to appear.

[14]. S. Coleman, J. Wess and B. Zumino, Structure of phenomenological Lagrangians, I, II, Phys., 177 (1969) pp. 2239-2250.

[15]. N. Conze, Espace des idéaux primitifs de l'algèbre enveloppante d'une algèbre de Lie nilpotente, J. Alg., 34 (1975) pp. 444-450.

[16]. ——— , Algèbres d'opérateurs différentiels et quotients des algèbres enveloppantes, Bull. Soc. Math. France, 102 (1974) pp. 379-415.

[17]. N. Conze and J. Dixmier, Idéaux primitifs dans l'algèbre enveloppante d'une algèbre de Lie semisimple, Bull. Sci. Math., 96 (1972) pp. 339-351.

[18]. J. Dixmier, Algèbres enveloppantes, Gauthier-Villars, Paris, 1974.

[19]. ——— , Sur les algèbres de Weyl II, Bull. Sci. Math., 94 (1970) pp. 289-301.

[20]. I. M. Gelfand and A. A. Kirillov, Sur les corps liés aux algèbres enveloppantes des algèbres de Lie, I. H. E. S. publ. math., 31 (1966) pp. 5-19.

[21]. A. Grossmann, Momentum-like constants of the motion, Sect. 6. In Statistical Mechanics and field theory, Eds. R. N. Sen and C. Weil, John Wiley, New York, 1972, pp. 101-110.

[22]. Harish-Chandra, Representations of a semisimple Lie group on a Banach space I, Trans. Amer. Math. Soc., 75 (1953) pp. 185-243.

[23]. I. N. Herstein, Noncommutative rings, John Wiley, New York, 1968.

[24]. A. Joseph and A. I. Solomon, Global and infinitesimal nonlinear chiral transformations, J. Math. Phys., 11 (1970) pp. 748-761.

[25]. A. Joseph, Minimal realizations and spectrum generating algebras, Commun. math. phys., 36 (1974) pp. 325-338.

[26]. ——— , A characterization theorem for realizations of $s\ell(2)$, Proc. Camb. Phil. Soc., 75 (1974) pp. 119-131.

[27]. ——— , The Gelfand-Kirillov conjecture in classical mechanics and quantization, I. H. E. S. preprint, 1974.

[28]. ——— , A generalization of the Gelfand-Kirillov conjecture, C. N. R. S. Marseille preprint, 1975.

[29]. ——— , The minimal orbit in a simple Lie algebra and its associated maximal ideal, I. H. E. S. preprint, 1975.

[30]. A. Joseph, Realizations in classical and quantum mechanics. Proc. 3rd Int. Colloq. on Group Theoretical Methods in Physics, Eds. H. Bacry and A. Grossmann, Marseille 1974, pp. 227-235.

[31]. A. A. Kirillov, Unitary representations of nilpotent Lie groups, Uspekhi Mat. Nauk., 17 (1962) pp. 57-110 (Russian) . [Eng. transl. Russian math. surveys 1962].

[32]. H. J. Lipkin, Lie groups for pedestrians, North Holland, Amsterdam, 1965.

[33]. J. C. McConnell, Representations of solvable Lie algebras and the Gelfand-Kirillov conjecture, Proc. London Math. Soc., 29 (1974) pp. 453-484.

[34]. J. C. McConnell and J. C. Robson, Homomorphisms and extensions of modules over certain differential polynomial rings, J. Alg., 26 (1973) pp. 319-342.

[35]. W. Miller, Jr., Lie theory and special functions, Academic press, New York, 1968.

[36]. Y. Ne'eman, Algebraic theory of particle physics, Benjamin, New York, 1967, Chap. 10.

[37]. X. H. Nghiêm, Sur certains sous-corps commutatifs du corps enveloppant d'une algèbre de Lie résoluble, Bull Sci. Math., 96 (1972) pp. 111-128.

[38]. S. R. Quint, Representations of solvable Lie groups, Lecture notes, University of California, Berkeley 1972 (unpublished).

[39]. I. Segal, An extension of a theorem of L. O'Raifeartaigh, J. Funct. Analysis 1 (1967) pp. 1-21.

[40]. J. Dixmier, Idéaux primitifs complétements premiers dans l'algèbre enveloppante de $s\ell(3,\mathbb{C})$, to appear.

[41]. R. F. Streater, The representations of the oscillator group, Commun. math. phys., 4 (1967) pp. 217-236.

I. H. E. S. and Tel-Aviv - 1975.

Geometric Quantization and Graded Lie Algebras

B. Kostant

M.I.T., Cambridge, U.S.A.

Text will appear in the Proceedings of the Conference on Differential Geometrical Methods in Mathematical Physics, held in Bonn, 1 - 4 July 1975.

A SURVEY OF THE APPLICATIONS OF GEOMETRIC QUANTI ATION.

D. J. Simms, School of Mathematics, Trinity College Dublin.

One of the principal themes of this conference is geometric quantisation. This is a convenient term for a concept which, to a greater or lesser extent, has been found to be relevant to such topics as

(i) the passage from a classical dynamical system with Poisson bracket (symplectic manifold) to a quantum system (Hilbert space), see Segal (12), Kostant (5), Souriau (15), and Sniatycki (this conference),

(ii) the construction of irreducible representations of Lie groups, see Kirillov (4), Borel-Weil (13), Kostant-Auslander (6), Blattner (1),

(iii) the construction of primitive ideals in the enveloping algebra of a Lie algebra, see Dixmier (2) and Joseph (this conference),

(iv) the construction of physical systems which are elementary with respect to a prescribed symmetry group, see Souriau (15), Renouard (11), Rawnsley (10),

(v) the study of dynamical groups, see Onofri (8), Sternberg-Wolf (in progress),

(vi) coherent states for a Lie group, Onofri, Bacry (this conference),

(vii) twistor theory of general relativity, see Penrose (9), and Woodhouse (this conference),

(viii) partial differential equations, see Maslov (7), Hormander-Duistermaat (3).

The basic ingredient in geometric quantisation is the notion of a polarised symplectic manifold (M, w, F). Here M denotes a real differentiable manifold, w a non-degenerate closed differential 2-form (symplectic form) on M, and F an involutive sub-bundle of the complexified tangent bundle of M which is maximally isotropic with respect to w (polarisation of w). The quantisation process may then be described as the construction of a complex hermitian line bundle L over M with connection form α having w as curvature form, and the construction of a Hilbert space from sections of the line bundle whose covariant derivatives in the directions of F vanish. Such a line bundle can be constructed if and only if w has integral periods (quantisation condition) and is unique if, for example, M is simply connected. In this process certain functions on M have a natural representation as operators on the Hilbert space, in such a way that the Poisson bracket of functions corresponds to the commutator of operators. In this scheme a function ϕ is represented formally by the operator

$$i\hbar \nabla_{\xi_\phi} + \phi$$

where ∇_{ξ_ϕ} denotes covariant differentiation along the Hamiltonian vector field ξ_ϕ generated by ϕ.

We now describe how these ideas arise naturally in the topics listed above.

(i) Let M be the classical phase space of a dynamical system based on a configuration space X. Let w be the 2-form which equals Planck's constant times $\Sigma dp_j \wedge dq^j$ in local canonical coordinates $p_1, \ldots, p_n$, $q^1, \ldots q^n$. Let F be the polarisation spanned by the vector fields $\frac{\partial}{\partial p_j}$. The line bundle L in this case is a product $M \times \mathbb{C}$ and the connection form α equals $\Sigma p_j dq^j + \frac{1}{i}\frac{dz}{z}$ where z is the coordinate on $\mathbb{C}$. The Hilbert space is $L^2(X)$ considered as the completion of a space of functions on M. For $X = \mathbb{R}^n$ this yields the usual Schrödinger representation $p_j \to i\hbar \frac{\partial}{\partial q^j}$ and $q_j \to$ multiplication by q_j. For X an arbitrary Riemannian manifold, the Hamiltonian of a free particle is quantised as the Laplace-Beltrami operator.

(ii) Let G be a connected Lie group and let $\underline{G}^*$ be the algebraic dual of its Lie algebra $\underline{G}$. Let M be any orbit of the coadjoint action of G on $\underline{G}^*$.

Each element X of $\underline{G}$ is a linear function on $\underline{G}^*$ which restricts to a function ϕ^X on M. There is a unique symplectic structure w on M such that the Poisson bracket of ϕ^X and ϕ^Y equals $\phi^{[X,Y]}$ for all X,Y in $\underline{G}$. Let F be a polarisation of w which is invariant under G. Then G acts on the Hilbert space of the quantisation process by unitary operators. For compact groups, and for solvable groups of type I, all irreducible unitary representations can be obtained in this way. For semi-simple Lie groups the irreducible representations occurring in the Plancherelmeasure can be obtained in this way.

(iii) Let $\underline{G}$ be a Lie algebra and $\underline{G}^*$ be its algebraic dual. Each $f \in \underline{G}^*$ defines a bilinear form B_f on $\underline{G}$ by $B_f(X,y) = f([X,y])$. A subalgebra F of $\underline{G}$ is called a polarisation at f if F is maximal among vector subspaces of $\underline{G}$ on which B_f vanishes. The restriction of f to F is a 1-dimensional representation of F and we can induce to a representation of $\underline{G}$. In this way we can obtain, when $\underline{G}$ is solvable and defined over an algebraically closed field, all ideals in the enveloping algebra which are kernels of irreducible representations (primitive ideals). If we denote by M the orbit of f under the coadjoint representation of G, then B_f induces a non-degenerate skew-symmetric form on the tangent space to M at f, and this gives the symplectic structure w on M. The subalgebra F projects to a polarisation F of w. Thus we have a polarised symplectic manifold (M, w, F). When w satisfies the quantisation condition, the resulting unitary representation of G corresponds to the induced representation of $\underline{G}$.

(iv) Let G be the connected component of the Poincaré group and $\underline{G}$ its Lie algebra. The usual generators P^α, $M^{\alpha\beta}$ can be considered as linear functions on the dual $\underline{G}^*$. For fixed real s, the submanifold M_s of $\underline{G}^*$ given by

$$\tfrac{1}{2}\, \varepsilon_{\alpha\beta\gamma\delta}\, P^\beta M^{\gamma\delta} = s\, P_\alpha\ , \quad P_o > 0$$

is a 6-dimensional orbit of G in $\underline{G}^*$ and carries a natural G-invariant symplectic form w_s. The space-time translations in G generate a polarisation F_s of w_s and (M_s, w_s, F_s) is then a polarised symplectic manifold. This has been proposed by Souriau as the phase space of a classical free elementary relativistic particle of rest mass zero and spin s. Quantisation yields the usual unitary representation of G associated with such a particle in quantum mechanics.

(v) The classical phase space M of the Kepler problem is a 6-dimensional symplectic manifold on which SO(4,2) acts leaving the symplectic form w invariant. Geometric quantisation has been applied in this context using a number of different polarisations.

(vi) Let G be a compact semi-simple Lie group acting irreducibly on a finite dimensional complex vector space H. If ψ is a ray in H with isotropy group a Cartan subgroup, then the G-orbit M of ψ carries a G-invariant symplectic form w and a complex structure F. Geometric quantisation of the polarised symplectic manifold (M, w, F) gives the coherent states associated with G.

(vii) The twistor space M associated with a curved space-time by Penrose is an 8-dimensional symplectic manifold and is also a 4-dimensional complex manifold. If w is the symplectic form and if F is generated by the anti-holomorphic directions, then (M, w, F) is a polarised symplectic manifold. For flat space-time, M is complex vector space $\mathbb{C}^4$ with coordinates (Z^0, Z^1, Z^2, Z^3) and symplectic form

$$w = i(dZ^0 \wedge d\bar{Z}^2 + dZ^1 \wedge d\bar{Z}^3 + dZ^2 \wedge d\bar{Z}^0 + dZ^3 \wedge d\bar{Z}^1).$$

The 4-fold cover SU(2,2) of the conformal group acts on M leaving w and

F invariant. For fixed real s the set of U(1) orbits in the hypersurface

$$z^{o}\,\bar{z}^{2} + z^{1}\,\bar{z}^{3} + z^{2}\,\bar{z}^{o} + z^{3}\,\bar{z}^{1} \quad = \quad s$$

has the induced structure of a 6-dimensional polarised symplectic manifold (M_s, w_s, F_s). Geometric quantisation gives the holomorphic twistor functions used by Penrose to represent a mass-zero spin s particle.

(viii) In the applications to partial differential equations on a manifold X, the basic symplectic manifold is the cotangent bundle $M = T^*(X)$. For example, Rockland (this conference) considers a pseudo differential operator P such that the subset Σ of M where the principal symbol vanishes (the characteristic variety) has an induced symplectic structure. He also uses polarisations of each fibre of the conormal bundle of Σ

REFERENCES

1. R. J. Blattner: Quantisation and representation theory, A.M.S. Proceedings of Symposia in Pure Mathematics, Vol. 26, 147-165,(1973)

2. J. Dixmier: Algèbres enveloppantes, Gauthier-Villars,(1974)

3. J. J. Duistermaat: Commun. Pure Appl. Math. 27, 207, (1974).

4. A. A. Kirillov: The method of orbits in representation theory, 219-230, in Lie groups and their representations, ed. I.M.Gelfand, Adam Hilger (1975).

5. B. Kostant: Quantisation and unitary representations, Lecture notes in mathematics, Vol. 170, Springer (1970).

6. B. Kostant & L. Auslander: Polarisation and unitary representations of solvable Lie groups, Inventiones Math., 14, (1971)

7. V. Maslov: Théorie des perturbations et méthodes asymptotiques, Dunod, (1973).

8. E. Onofri: Dynamical quantisation of the Kepler manifold, Preprint, Parma 1975.

9. R. Penrose: Twistor theory, its aims and achievements, Proceedings of conference on Quantum Gravity (ed. C.J. Isham), Oxford U.P. (1975)

10. J. Rawnsley: De Sitter symplectic spaces and their quantisations, Proc. Camb. Phil. Soc. (1974) 76, 473-480.

11. P. Renouard: Variétés symplectiques et quantification, These (1969), Orsay.

12. I. Segal: Quantisation of non-linear systems. Journal of Math. Physics 1 (1960), 468-488.

13. J.-P. Serre: Représentations linéaires et espaces homogènes Kähleriens des groupes de Lie compacts. Séminaire Bourbaki, Exposé 100, (1954), Benjamin (1966).

14. D. J. Simms: Geometric quantisation of symplectic manifolds. Proceedings of International Symposium on Mathematical Physics, Warsaw 1974.

15. J.-M. Souriau: Structure des systèmes dynamiques, Dunod, (1970).

CONSTRUCTION EXPLICITE DE L'INDICE DE MASLOV. APPLICATIONS.

par Jean-Marie SOURIAU*

Abstract :

Le revêtement universel de la grassmannienne lagrangienne est plongé dans un espace numérique, ce qui permet de définir l'indice de Maslov-Arnold-Leray par une formule explicite.

Les propriétés cohomologiques de cet indice permettent de rendre transitive la transformation de Fourier ; on construit ainsi un espace de Hilbert où agissent naturellement le groupe de Heisenberg-Weyl (représentation de Schrödinger) et le groupe métaplectique (représentation de Shale-Weil).

Dans le cas d'un oscillateur harmonique, l'espace ainsi construit coïncide avec l'ensemble des solutions de l'équation de Schrödinger. Cette équation est donc explicitement intégrée ; on obtient un prolongement de la formule de Feynman qui est valable pour des durées arbitrairement grandes.

OCTOBRE 1975

*Université de Provence et Centre de Physique Théorique, CNRS, Marseille

Adresse postale : Centre de Physique Théorique - CNRS
31, chemin Joseph Aiguier
F - 13274 MARSEILLE CEDEX 2 (France)

§1 - PLANS LAGRANGIENS

Soit σ une 2-forme d'un espace vectoriel E , c'est-à-dire un tenseur covariant antisymétrique d'ordre 2 . Contractée avec deux vecteurs $X,Y \in E$, on obtient un nombre que nous noterons

$$\sigma(X,Y)$$

ou encore

$$\sigma(X)(Y)$$

ce qui a l'avantage de mettre en évidence la 1-forme

$$\sigma(X) \ : \ Y \mapsto \sigma(X)(Y)$$

et de présenter σ

$$X \mapsto \sigma(X)$$

comme une application de E dans son dual E^* . Si elle est bijective, on dit que σ est une <u>forme symplectique</u>, ou que σ donne à E une structure d'<u>espace vectoriel symplectique</u>.

Dans un espace vectoriel symplectique, la relation

(1.1) $\sigma(X)(Y) = 0$

entre deux vecteurs X et Y s'appelle <u>orthogonalité</u> ; c'est une relation symétrique ; si E' est un sous-espace vectoriel de E , l'ensemble des X orthogonaux à tous les vecteurs de E est un sous-espace vectoriel, que nous noterons orth(E') ; on a la relation

(1.2) $\dim(E') + \dim(\mathrm{orth}(E')) = \dim(E)$.

E' sera dit <u>isotrope</u> si

(1.3) $E' \subset \mathrm{orth}(E')$

c'est-à-dire si les éléments de E' sont <u>deux à deux orthogonaux</u> ; c'est le cas pour tous les espaces de dimension 1 , à cause de l'antisymétrie de σ . On appelle <u>plan lagrangien</u> tout sous-espace isotrope maximal (pour la relation d'inclusion) ; il est clair que tout sous-espace isotrope est inclus dans un plan lagrangien ; que tout plan lagrangien λ vérifie

(1.4) $$\boxed{\lambda = \mathrm{orth}(\lambda)}$$

et (grâce à (1.2)) que tous les plans lagrangiens ont la même dimension n , égale à la moitié de celle de E ; il n'existe donc que des espaces symplectiques de dimension paire.

Deux plans lagrangiens λ et μ sont dits <u>transverses</u> si

(1.5) $$\lambda \cap \mu = \{0\}$$

ce qui s'écrit aussi (à cause des dimensions de E , λ , μ)

(1.6) $$E = \lambda \oplus \mu \quad .$$

Si λ et μ sont lagrangiens transverses, nous noterons $\sigma_{\lambda\mu}$ l'application

(1.7) $$\boxed{\sigma_{\lambda\mu}(X)(Y) = \sigma(X)(Y) \qquad \forall\ X \in \lambda\ ,\ \forall\ Y \in \mu} \quad ;$$

$\sigma_{\lambda\mu}$ est une bijection linéaire de λ sur μ^* .

Soient λ , μ , ν trois plans lagrangiens <u>deux à deux transverses</u> ; alors l'application

(1.8) $$\boxed{g_{\lambda\mu\nu} = \sigma_{\lambda\mu} \circ \sigma_{\lambda\nu}^{-1} \circ \sigma_{\mu\nu}}$$

envoie μ dans son dual μ^* ; $g_{\lambda\mu\nu}$ est donc un tenseur covariant d'ordre 2 de μ , visiblement injectif ; on vérifie facilement que $g_{\lambda\mu\nu}$ est <u>symétrique</u>, donc qu'il munit μ d'une structure <u>euclidienne</u> ; nous noterons

(1.9) $$\boxed{\mathrm{sgn}(\lambda , \mu , \nu)}$$

la <u>signature</u> de $g_{\lambda\mu\nu}$, c'est-à-dire la <u>trace</u> de la matrice

$$\begin{pmatrix} 1 & & & & & \\ & 1 & & & & \\ & & \ddots & & & \\ & & & 1 & & \\ & & & & -1 & \\ & & & & & \ddots \\ & & & & & & -1 \end{pmatrix}$$

représentant $g_{\lambda\mu\nu}$ dans une base orthonormale. Il est clair que

(1.10) $$\operatorname{sgn}(\lambda, \mu, \nu) \in \{-n, -n+2, \ldots\ldots\ldots n-2, n\}$$

si $2n$ est la dimension de E ; on peut choisir λ, μ, ν pour que toutes ces valeurs soient effectivement atteintes.

Nous verrons au §4 que $\operatorname{sgn}(\lambda, \mu, \nu)$ est une <u>fonction antisymétrique</u> de ses trois arguments, et que l'on a

(1.11) $$\boxed{\operatorname{sgn}(\lambda, \mu, \nu) = \operatorname{sgn}(\rho, \mu, \nu) + \operatorname{sgn}(\lambda, \rho, \nu) + \operatorname{sgn}(\lambda, \mu, \rho)}$$

si ρ est un plan lagrangien transverse à λ, μ, ν (Leray, [III]) ; la cohomologie qui transparaît dans ces formules sera exploitée au §8 ci-dessous.

§2 - ACTION DU GROUPE SYMPLECTIQUE

Soit E un espace vectoriel symplectique de dimension $2n$, λ un plan lagrangien de E. On peut construire un plan lagrangien μ transverse à λ ; si $(S_1 S_2 \ldots S_n)$ est une base de λ, la <u>base duale</u> de S s'identifie (grâce à la dualité $\sigma_{\lambda\mu}$) à une base $(T_1 T_2 \ldots T_n)$ de μ ; il est clair que

(2.1) $$(S_1 \ldots S_n \quad T_1 \ldots T_n)$$

est une base de E, dans laquelle la matrice du tenseur σ s'écrit

(2.2) $$\left(\begin{array}{c|c} & \begin{matrix} 1 & & \\ & \ddots & \\ & & 1 \end{matrix} \\ \hline \begin{matrix} -1 & & \\ & \ddots & \\ & & -1 \end{matrix} & \end{array}\right) ;$$

on dit que (2.1) est une base canonique de E .
Si E' est un autre espace vectoriel symplectique de même dimension, l'application linéaire a qui envoie les vecteurs d'une base canonique de E sur ceux d'une base canonique de E' est évidemment un isomorphisme de la structure symplectique :

$$(2.3) \qquad a \in L(E,E') \ , \ a \text{ bijectif} \ , \ \sigma\big(a(X),a(Y)\big) = \sigma(X,Y) \quad \forall X,Y$$

l'ensemble des a vérifiant (2.3) sera noté $Sp(E,E')$, ($Sp(E)$ si $E'= E$) ; il est clair que $Sp(E)$ est un groupe, appelé groupe symplectique ; c'est un sous-groupe fermé du groupe linéaire $GL(E)$, donc un groupe de Lie ; c'est d'ailleurs un groupe semi-simple classique ; sa dimension est $n(2n+1)$.

Il résulte de la construction (2.1) des bases canoniques que $Sp(E)$ agit transitivement sur l'ensemble $\Lambda(E)$ des plans lagrangiens de E ; on constate que le stabilisateur d'un plan est un sous-groupe fermé de $Sp(E)$, dont la dimension est $n(3n+1)/2$; ce qui confère à $\Lambda(E)$ une structure de variété de dimension $n(n+1)/2$ sur laquelle $Sp(E)$ agit différentiablement ; cette variété $\Lambda(E)$ s'appelle la grassmannienne lagrangienne de E .
La construction des bases canoniques (2.1) montre aussi que la figure constituée par deux plans lagrangiens transverses est unique en géométrie symplectique. Mais des triplets lagrangiens peuvent présenter $n+1$ configurations ; en effet, si λ, μ, ν sont des plans lagrangiens deux à deux transverses dans E (resp. λ' , μ' , ν' dans E') la condition

$$(2.4) \left| \begin{array}{l} \text{Il existe } a \in Sp(E,E') \text{ tel que } a(\lambda) = \lambda' \ , \ a(\mu) = \mu' \ , \\ a(\nu) = \nu' \end{array} \right.$$

entraîne évidemment la condition

$$(2.5) \Big[\quad \operatorname{sgn}(\lambda, \mu, \nu) = \operatorname{sgn}(\lambda', \mu', \nu')$$

le calcul montre que cette condition est en fait suffisante.

§3 - REVÊTEMENT DE LA GRASSMANNIENNE LAGRANGIENNE

Considérons l'espace vectoriel complexe $\mathbb{C}^n$, muni de la structure hermitienne définie par la forme sesquilinéaire positive

(3.1) $\langle x,y\rangle = \overline{x^1}\, y^1 + \overline{x^2}\, y^2 + \ldots + \overline{x^n}\, y^n$ (x^j, y^j = coordonnées de x,y) ;

si l'on sépare la partie réelle et la partie imaginaire de $\langle x,y\rangle$:

(3.2) $\langle x,y\rangle = g(x,y) - i\,\sigma(x,y)$

on constate que σ munit $\mathbb{C}^n$ d'une structure d'espace vectoriel symplectique réel de dimension 2n ; on peut le prendre comme modèle pour un tel espace ; son groupe symplectique sera noté Sp(n) .
Le groupe unitaire U(n) est défini comme l'ensemble des

(3.3) $\{ a \in GL(n,\mathbb{C}) \,/\, \langle a(x), a(y)\rangle = \langle x,y\rangle \quad \forall x,y \in \mathbb{C}^n \}$

tout $a \in U(n)$ respecte évidemment la forme σ ; donc

(3.4) $U(n) \subset Sp(n)$;

on peut vérifier que tout élément du groupe symplectique Sp(n) s'écrit, d'une seule façon, sous la forme

(3.5) $a \circ \exp(b \circ C)$

$a \in U(n)$, b étant une matrice complexe symétrique, C la conjugaison complexe de $\mathbb{C}^n$. Ceci montre que Sp(n) est connexe et que son groupe d'homotopie est le même que celui de U(n) (nous allons constater qu'il s'agit de $\mathbb{Z}$).

(3.6) La grassmannienne lagrangienne de $\mathbb{C}^n$ sera notée $\Lambda(n)$; si $\lambda \in \Lambda(n)$, on peut choisir une base $(a_1, a_2, \ldots a_n)$ de λ qui soit orthonormale pour la structure euclidienne définie par le tenseur g (3.2) ; on a donc, $\forall$ j,k

$$g(a_j, a_k) = \delta_{jk} \quad , \quad \sigma(a_j, a_k) = 0$$

ce qui s'écrit simplement

$$\langle a_j, a_k \rangle = \delta_{jk} \quad ;$$

cette relation exprime que la matrice formée avec les colonnes a_j

$$a = (a_1 \, a_2 \ldots a_n)$$

est unitaire :

(3.7) $\left[\lambda \in \Lambda(n) \right] \Longleftrightarrow \left[\text{Il existe } a \in U(n) \, , \ \lambda = a(\mathbb{R}^n) \right]$

(3.8)

Ainsi, dans l'action de $Sp(n)$ sur la grassmannienne lagrangienne $\Lambda(n)$, $\Lambda(n)$ est orbite du sous-groupe compact connexe $U(n)$, donc elle-même une variété compacte connexe ; le stabilisateur de $\mathbb{R}^n$ dans $U(n)$ est par définition le groupe orthogonal $O(n)$; $U(n)$ est donc difféomorphe à la variété quotient $U(n)/O(n)$ (Arnold, [II]).

(3.9) De même, l'ensemble des plans lagrangiens orientés est difféomorphe à $U(n)/SO(n)$; c'est un revêtement connexe à deux feuillets de $\Lambda(n)$, la projection sur $\Lambda(n)$ consistant à "oublier" l'orientation.

Au lieu de considérer $\Lambda(n)$ comme un quotient de $U(n)$, on peut aussi la plonger dans $U(n)$([III]) ; en effet, si a et a' sont deux éléments de $U(n)$, il est clair que

$$\left[a(\mathbb{R}^n) = a'(\mathbb{R}^n) \right] \Longleftrightarrow \left[a\, C(a^{-1}) = a'\, C(a'^{-1}) \right] \quad (\ C = \text{conjugaison complexe}) ,$$

donc que l'on peut identifier $\Lambda(n)$ à l'image de $U(n)$ par l'application

(3.10) $a \longmapsto \lambda = a\, C(a^{-1})$

image qui est l'ensemble des matrices unitaires symétriques ; l'identification d'un plan lagrangien λ et d'une matrice λ est donnée par la règle

(3.11) $$[x \in \lambda] \iff [x = \lambda\, C(x)]$$

$C(x)$ étant la colonne conjuguée d'une colonne $x \in \mathbb{C}^n$; on en déduit les règles

(3.12) $$[\lambda \text{ et } \lambda' \ \underline{\text{transverses}}] \iff [\lambda - \lambda' \text{ inversible}]$$

(3.13) $$\boxed{\underline{a}(\lambda) = a\,\lambda\, C(a^{-1})} \qquad \forall a \in U(n),\ \forall \lambda \in \Lambda(n)$$

où $\underline{a}$ désigne l'<u>action</u> (3.8) d'un élément a de U(n) sur $\Lambda(n)$.
Désignons par $\widehat{U(n)}$ l'ensemble des couples

(3.14) $$(a, \varphi) \qquad a \in U(n),\ \varphi \in \mathbb{R}$$

vérifiant l'équation

(3.15) $$\det(a) = e^{i\varphi} \ ;$$

si l'on munit $\widehat{U(n)}$ de la loi de composition $\times$:

(3.16) $$(a, \varphi) \times (a', \varphi') = (aa', \varphi + \varphi')$$

$\widehat{U(n)}$ devient un groupe de Lie ; $(a,\varphi) \mapsto a$ est un morphisme de $\widehat{U(n)}$ sur U(n) , dont le noyau est le sous-groupe discret des $(I, 2k\pi)$, $k \in \mathbb{Z}$;
$\widehat{U(n)}$ est donc un <u>revêtement</u> de U(n) .
On remarque que l'application

(3.17) $$(b, \varphi) \mapsto (be^{i\varphi}, n\varphi) \qquad b \in SU(n),\ \varphi \in \mathbb{R}$$

est un isomorphisme du produit direct $SU(n) \times \mathbb{R}$ sur le groupe $\widehat{U(n)}$, donc que $\widehat{U(n)}$ est simplement connexe : $\widehat{U(n)}$ est donc <u>revêtement universel</u> de U(n).

(3.18) Grâce à la décomposition (3.5) , $\widehat{U(n)}$ pourra s'identifier à la partie du revêtement $\widehat{Sp(n)}$ située au-dessus du sous-groupe $U(n)$ de $Sp(n)$; en particulier, le générateur K du groupe d'homotopie de $Sp(n)$ s'identifie à l'élément

(3.19) $\quad K = (I\ ,\ 2\pi)$

de $\widehat{U(n)}$.

De même, si on considère la variété $\widehat{\Lambda(n)}$ des

(3.20) $\quad (\lambda, \theta) \qquad [\lambda \in \Lambda(n),\ \theta \in \mathbb{R}\ ,\ \det(\lambda) = e^{i\theta}]$

le groupe discret des L^k :

(3.21) $\quad L(\lambda, \theta) = (\lambda, \theta + 2\pi)\ ,\quad k \in \mathbb{Z}$

et la projection

(3.22) $\quad (\lambda, \theta) \longmapsto \lambda$

font de $\widehat{\Lambda(n)}$ un revêtement de $\Lambda(n)$; l'action (3.13) de $U(n)$ sur $\Lambda(n)$ se relève par l'action de $\widehat{U(n)}$ sur $\widehat{\Lambda(n)}$:

(3.23) $\quad (a, \varphi)(\lambda, \theta) = (a\ \lambda\ C(a^{-1})\ ,\ 2\varphi + \theta)$

qui est encore transitive ; on constate que le stabilisateur de l'élément $(I,0)$ de $\widehat{\Lambda(n)}$ est l'ensemble des $(a,0)$, a vérifiant $[a \in U(n)\ ,\ a = C(a)\ ,\ \det(a) = 1]$, c'est-à-dire $[a \in SO(n)]$; $\widehat{\Lambda(n)}$ est donc difféomorphe au quotient du groupe simplement connexe $\widehat{U(n)}$ par le groupe connexe $SO(n)$, donc simplement connexe ; $\widehat{\Lambda(n)}$ est donc le <u>revêtement universel</u> de
(3.24) $\Lambda(n)$; le groupe fondamental (3.21) de $\Lambda(n)$ est isomorphe à $\mathbb{Z}$ (Arnold, [II]).

§4 - INDICE DE MASLOV [I]

Si A est une matrice carrée, nous définirons le logarithme de A par la formule

$$\text{(4.1)} \qquad \boxed{\operatorname{Log}(A) = \int_{\infty}^{0} \left\{ [sI-A]^{-1} - [sI-I]^{-1} \right\} ds}$$

qui s'applique chaque fois que A ne possède pas de valeur propre négative ou nulle ; Log est une application C^{∞} qui vérifie

$$\text{(4.2)} \qquad \boxed{\exp(\operatorname{Log}(A)) = A} \qquad \text{si } \operatorname{Log}(A) \text{ existe ;}$$

d'où découle

$$\text{(4.3)} \qquad \boxed{e^{\operatorname{Tr}(\operatorname{Log}(A))} = \det(A)} \qquad ;$$

on notera que

$$\text{(4.4)} \qquad \boxed{\operatorname{Log}(A^{-1}) = -\operatorname{Log}(A)} \quad .$$

Nous définirons l'indice de Maslov $m(u,u')$ de deux points

$$u = (\lambda, \theta) \quad , \qquad u' = (\lambda', \theta')$$

de $\widehat{\Lambda(n)}$ par la formule

$$\text{(4.5)} \qquad \boxed{m(u,u') = \frac{1}{2\pi} \left[\theta - \theta' + i \operatorname{Tr}\left(\operatorname{Log}(-\lambda \lambda'^{-1})\right)\right]}$$

$m(u,u')$ existe si la matrice $-\lambda\lambda'^{-1}$ n'a pas de valeur propre négative ou nulle. Comme il s'agit d'une matrice unitaire, il suffit qu'elle n'ait pas la valeur propre -1 , c'est-à-dire que $I - \lambda\lambda'^{-1}$ soit inversible ; donc que λ et λ' soient transverses (3.12).

En utilisant (4.3) , on trouve

$$e^{2i\pi\, m(u,u')} = e^{in\pi}$$

ce qui montre que

$$(4.6)\quad \boxed{\begin{array}{ll} m(u,u') \in \mathbb{Z} & \text{si } n \text{ pair} \\ m(u,u') \in \mathbb{Z}+\frac{1}{2} & \text{si } n \text{ impair} \end{array}} \quad ;$$

toutes les valeurs permises par cette règle sont effectivement atteintes, car

$$(4.7)\quad \boxed{m(L^k(u)\ ,\ L^{k'}(u')) = k-k' + m(u,u')}$$

L étant le générateur (3.21) du groupe d'homotopie de $\Lambda(n)$.

Le groupe symplectique $Sp(n)$ agit sur $\Lambda(n)$ en conservant la transversalité des couples de plans lagrangiens ; comme $Sp(n)$ est connexe, cette action se relève en une action de son revêtement universel $\widehat{Sp(n)}$ sur le revêtement $\widehat{\Lambda(n)}$. Donnons-nous un couple de points $u,u' \in \widehat{\Lambda(n)}$ tels que $m(u,u')$ existe, donc que u et u' se projettent en des points transverses de $\Lambda(n)$; si $a \in \widehat{Sp(n)}$, $\underline{a}(u)$ et $\underline{a}(u')$ se projetteront aussi en des points transverses de $\Lambda(n)$; par suite, l'application $a \mapsto m(\underline{a}(u), \underline{a}(u'))$ envoie la variété connexe $\widehat{Sp(n)}$ dans $\mathbb{Z}$ ou $\mathbb{Z} + 1/2$; comme elle est continue, elle est constante :

$$(4.8)\quad \boxed{m(a(u)\ ,\ a(u')) = m(u, u') \qquad \forall\, a \in \widehat{Sp(E)}}$$

l'indice de Maslov est donc <u>invariant par l'action de</u> $\widehat{Sp(E)}$; sa définition (4.5) ne dépend qu'en apparence de la structure hermitienne par laquelle nous avons complété la structure symplectique de $\mathbb{C}^n$; (4.5) est en fait une <u>formule pratique de calcul</u>.

La formule

$$(4.9)\quad \boxed{m(u,u') + m(u',u) = 0}$$

est évidente sur (4.5) (utiliser (4.4)) ; quant à la formule de Leray

$$(4.10)\quad \boxed{m(u,u') + m(u',u'') + m(u'',u) = \frac{1}{2}\ \mathrm{sgn}(\lambda, \lambda', \lambda'')}$$

où $\lambda, \lambda', \lambda''$ sont les projections de u, u', u'' sur $\Lambda(n)$, elle se vérifie facilement en utilisant un choix particulier de λ, λ', λ'' correspondant à chaque signature ; elle s'étend ensuite au cas général par l'action du groupe symplectique (3.13) et de son revêtement universel (3.23). (4.9) et (4.10) impliquent immédiatement l'antisymétrie de "sgn" et la formule cohomologique (1.11) : la demi-signature apparaît comme le cobord de l'indice de Maslov.

La définition proposée ici pour l'indice de Maslov diffère d'une constante de celle de Leray (III). Indiquons comment elle se rattache à la définition originale de Maslov (I).

Une variété V , plongée dans un espace vectoriel symplectique E , est dite lagrangienne si son plan tangent est lagrangien en tout point ; on définit ainsi une application T de V dans $\Lambda(E)$ (figure 1). Maslov privilégie une direction lagrangienne particulière λ_o ; l'ensemble des $x \in V$ tels que $T(x)$ ne soit pas transverse à λ_o s'appelle contour apparent de V .

Soit F un arc de courbe tracé sur V , dont les extrémités F(0) et F(1) n'appartiennent pas au contour apparent. $T \circ F$ est une application de $[0,1]$ dans $\Lambda(E)$, qui possède un relèvement $\widehat{T \circ F}$ à $\widehat{\Lambda(E)}$. Si l'on choisit un relèvement $\hat{\lambda}_o$ de λ_o , le nombre

$$(4.11) \qquad k = m\left(\hat{\lambda}_o, \widehat{T \circ F}(1)\right) - m\left(\hat{\lambda}_o, \widehat{T \circ F}(0)\right)$$

est un entier qui ne dépend ni du choix du relèvement de $T \circ F$, ni de celui de λ_o (voir (4.7)) ; c'est l'indice de Maslov proprement dit de l'arc F . Il est nul si l'arc ne rencontre pas le contour apparent (parce qu'alors $t \mapsto m\left(\hat{\lambda}_o, \widehat{T \circ F}(t)\right)$ est une fonction continue à valeurs entières).

Si la courbe est un lacet $\left(F(1) = F(0)\right)$, la formule (4.7) montre que $\widehat{T \circ F}(1) = L^k\left(\widehat{T \circ F}(0)\right)$, L étant le générateur (3.21) du groupe d'homotopie de $\Lambda(E)$, k l'indice du lacet. Par conséquent k repère la classe d'homotopie de $T \circ F$, et ne dépend pas de λ_o ; un lacet dont l'indice de Maslov n'est pas nul rencontre donc les contours apparents attachés à toutes les directions lagrangiennes.

Si V est orientable, l'application T se relève par une application T^+ à la variété des plans lagrangiens orientables, qui est un revêtement connexe à deux feuillets de $\Lambda(E)$ (3.9) , donc identifiable au quotient de

$\widehat{\Lambda(E)}$ par L^2 . Alors l'indice de tout lacet tracé sur V est un nombre pair.

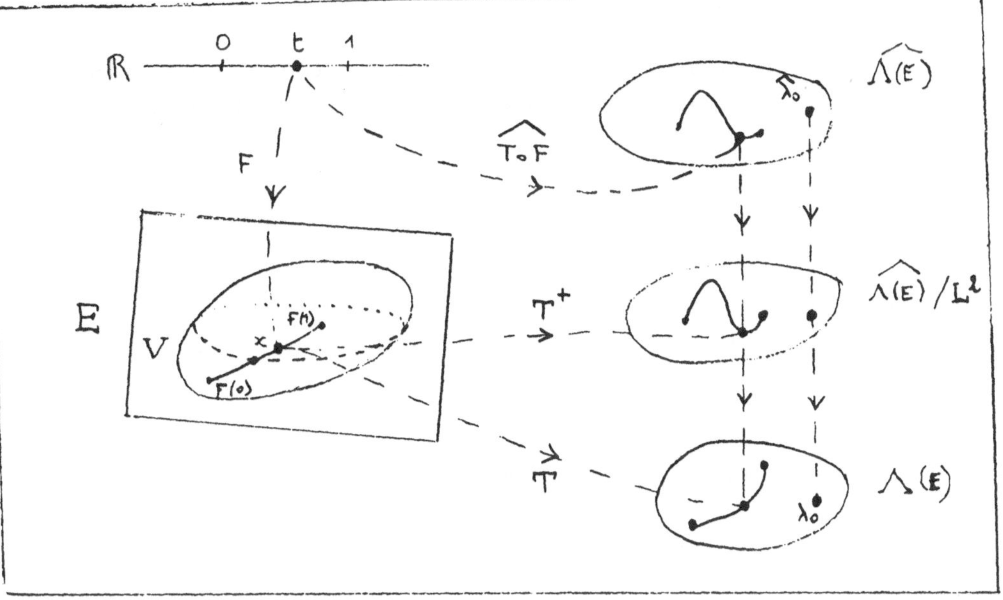

- Figure 1 -

§5 - DENSITÉS

Soit α un nombre positif, E un espace vectoriel réel de dimension n . Appelons repère toute application linéaire S de $\mathbb{R}^n$ dans E ; α-densité de E toute fonction f définie sur les repères et vérifiant

(5.1) $$f(SM) = f(S) \left|\det(M)\right|^{\alpha} \quad \text{pour toute matrice } M .$$

Les α-densités réelles forment un espace vectoriel ordonné de dimension 1 ; le produit d'une α-densité et d'une β-densité est une $(\alpha+\beta)$-densité ; la puissance β d'une α-densité positive est une $(\alpha\beta)$-densité positive.

On appellera α-densité d'une variété V tout champ continu de α-densités de l'espace tangent ; les difféomorphismes de V agissent linéairement sur les α-densités.
On sait définir l'intégrale sur V

(5.2) $$\int_V \theta$$

d'une 1-densité θ à support compact ; cette intégrale est invariante par difféomorphisme.

L'espace H_V des 1/2-densités complexes à support compact de V est muni d'une structure préhilbertienne si l'on pose

(5.3) $$\boxed{\langle\varphi,\psi\rangle_V = \int_V \overline{\varphi(x)} \times \psi(x) \qquad \forall\varphi,\psi\in H_V}$$

(5.4) Si a est un difféomorphisme de V sur une variété V' , l'image par a d'un élément de H_V est un élément de $H_{V'}$, et cette application est <u>unitaire</u> ; elle passe évidemment aux complétés $\mathcal{H}_V$, $\mathcal{H}_{V'}$. En particulier, le groupe des difféomorphismes de V sur V se représente unitairement sur H_V et $\mathcal{H}_V$.

Si V et V' sont deux variétés de dimension n et n' , un repère S de V en x et un repère S' de V' en x' définissent naturellement un repère du produit cartésien V x V' au point (x , x') ; nous le noterons $S\otimes S'$; si ψ et ψ' sont des α-densités de V et V' , il existe une α-densité de V x V' , que nous noterons $\psi\otimes\psi'$, telle que

(5.5) $$[\psi\otimes\psi'](S\otimes S') = \psi(S)\ \psi'(S') \text{ en tout point de } V \times V' ;$$

nous l'appellerons <u>produit tensoriel</u> de ψ et ψ' ; ce produit est bi-linéaire.

§6 - REPRESENTATION DE SCHRÖDINGER.

Désignons par $\mathbb{T}$ le tore (groupe multiplicatif des nombres complexes de module 1). E étant un espace vectoriel symplectique de dimension 2n , considérons la variété $Y = E \times \mathbb{T}$ parcourue par la variable

(6.1) $$\xi = (x,z) \qquad \left[x \in E\,,\ z \in \mathbb{T}\right]$$

Y peut être considérée comme un fibré principal au-dessus de E , par la projection

(6.2) $$\xi \mapsto x\,,$$

et l'action du tore

(6.3) $$Z(x,z) = (x,Zz) \qquad \left[\forall\, Z \in \mathbb{T}\,,\ \forall\,(x,z) \in Y\right]$$

Munissons Y de la 1-forme ϖ définie par

(6.4) $$\boxed{\varpi(\delta\xi) = \frac{1}{2}\,\sigma(x)(\delta x) + \frac{\delta z}{iz}}$$

Il est immédiat que la dérivée extérieure de ϖ est l'image réciproque, par la projection (6.2) de la forme σ de E , que le générateur $I(\xi)$ du tore est le vecteur vertical tel que

(6.5) $$\varpi(I(\xi)) = 1$$

Soit Quant(Y) le groupe des difféomorphismes de Y qui respectent la forme ϖ ("quantomorphismes") ; tout quantomorphisme respecte la fibration, et commute avec le tore ; il se projette donc sur E selon un difféomorphisme qui respecte σ ("symplectomorphisme") ; on définit ainsi un morphisme de groupe

(6.6) $$\mathrm{Quant}(Y) \longrightarrow \mathrm{Sympl}(E)\;;$$

ce morphisme est surjectif ; son noyau est le tore, centre de Quant(Y) ; Quant(Y) est donc une extension centrale de Sympl(E) .

Le groupe $(E,+)$ des translations de E est inclus dans $\mathrm{Sympl}(E)$; son image réciproque par le morphisme (6.6) sera appelé groupe de Heisenberg ; il agit transitivement et librement sur Y , si bien qu'il s'identifie à Y en choisissant arbitrairement son élément neutre e ; nous prendrons

$$(6.7) \qquad e = (0,1)$$

ce qui fournit sur Y la loi de groupe

$$(6.8) \qquad \boxed{(x,z) \times (x',z') = \left(x+x',\ zz'\, e^{-\frac{i}{2}\sigma(x)(x')}\right)} \ ;$$

le tore $\mathbb{T}$ est encore le centre de Y .

En choisissant une base canonique de E , on constate que l'algèbre de Lie de Y est celle des "relations de commutation" de Heisenberg ; le groupe lui-même a été introduit par Hermann Weyl.

Soit λ un plan lagrangien de E ; l'ensemble des

$$(6.9) \qquad (x,1) \quad [x \in \lambda]$$

est un sous-groupe abélien de Y , que nous noterons Y_λ ; les algèbres de Lie des Y_λ sont les sous-algèbres maximales incluses dans $\ker(\varpi)$.

Notons Y/λ la variété quotient de Y par Y_λ ; Y/λ est une variété de dimension $n+1$ sur laquelle agit Y , et en particulier $\mathbb{T}$; l'action de $\mathbb{T}$ est libre.

Puisque $\mathbb{T}$ est le centre de Y , l'espace H_λ des $\psi \in H_{Y/\lambda}$(5.3) qui vérifient la "condition de circulation"

$$(6.10) \qquad \boxed{z(\psi) = z \times \psi}$$

est invariant par l'action de Y : il constitue donc un espace de représentation unitaire du groupe de Heisenberg : c'est la représentation de Schrödinger ; nous allons chercher si l'on peut identifier les représentations de Schrödinger associées aux divers plans lagrangiens $\lambda \in \Lambda(E)$.

§7 - PAIRING

Soient $\lambda, \mu \in \Lambda(E)$, λ et μ transverses. ξ étant un point de Y , désignons par ℓ et m ses projections sur Y/λ et Y/μ (figure 2) ; l'application $\xi \mapsto \rho = (\ell, m)$ de Y dans le produit cartésien $V = [Y/\lambda] \times [Y/\mu]$ est un plongement (parce que λ et μ sont transverses).

(7.1) Soit I le générateur infinitésimal du tore agissant sur chacune des variétés Y , Y/λ , Y/μ , V ; $I(\rho) = (I(\ell), I(m))$ est l'image de $I(\xi)$ par le plongement $\xi \mapsto \rho$; par contre le vecteur $I'(\rho) = \frac{1}{2}(I(\ell), -I(m))$ est transversal à l'image de Y .

Soient $\varphi \in H_\lambda$, $\psi \in H_\mu$; $\bar{\varphi} \otimes \psi$ (5.5) est une semi-densité à support compact de V , invariante par l'action du tore (parce que φ et ψ vérifient chacune la condition de circulation (6.10)) ; si S est un repère de Y en ξ , l'application

(7.2)
$$\boxed{\omega : \quad S \mapsto [\bar{\varphi} \otimes \psi]\left(I'(\rho), \frac{\partial \rho}{\partial \xi} \circ S\right)}$$

est une semi-densité de Y , elle aussi invariante par $\mathbb{T}$.

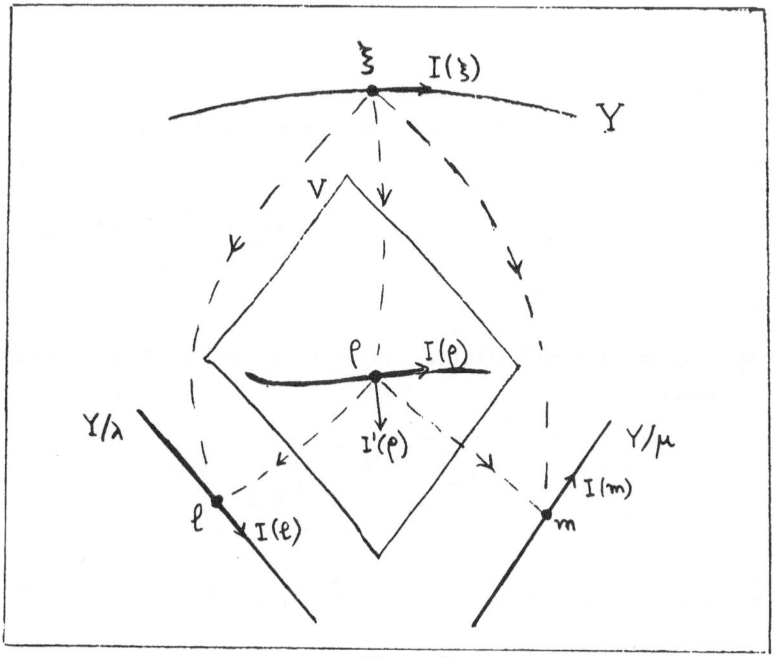

- Figure 2 -

Par ailleurs le groupe de Lie Y possède une semi-densité positive invariante ω_0 ; $\omega\,\omega_0$ est une 1-densité ; posons (cf.(5.2))

(7.3) $$\boxed{\langle \varphi, \psi \rangle_{\lambda\mu} = \int_V \omega\,\omega_0}$$;

on définit ainsi une forme sesqui-linéaire entre H_λ et H_μ , appelée "pairing" de H_λ et H_μ ; bien qu'elle ne fasse pas intervenir la structure symplectique, cette définition est équivalente à la définition originale de Kostant [IV] et Sternberg. Notons que :

Le pairing possède la symétrie hermitienne, en ce sens que

(7.4) $$\boxed{\overline{\langle \varphi, \psi \rangle_{\lambda\mu}} = \langle \psi, \varphi \rangle_{\mu\lambda}}$$;

il est invariant par l'action du groupe de Heisenberg :

(7.5) $$\boxed{\langle \underline{a}(\varphi), \underline{a}(\psi) \rangle_{\lambda\mu} = \langle \varphi, \psi \rangle_{\lambda\mu} \quad \text{si } a \in Y}$$

$a \mapsto \underline{a}$ désignant la représentation de Schrödinger (§ 6)

<u>Théorème</u> :

(7.6) Soient $\mathcal{H}_\lambda$ et $\mathcal{H}_\mu$ les hilbertiens complétés de H_λ et H_μ ; il existe une application <u>unitaire</u> $\mathcal{F}_{\lambda\mu}$ de $\mathcal{H}_\lambda$ sur $\mathcal{H}_\mu$ caractérisée par

$$\langle \varphi, \psi \rangle_{\lambda\mu} = \langle \varphi, \mathcal{F}_{\lambda\mu}(\psi) \rangle_{\mathcal{H}_\lambda} \quad \forall \varphi \in H_\lambda, \forall \psi \in H_\mu$$

Ce théorème suppose une normalisation convenable de la demi-forme invariante ω_0, à savoir

(7.7) $$\boxed{\omega_0 = [2\pi]^{-n/2} \sqrt{\ell \otimes h}}$$

ℓ étant la densité de Liouville de E , h la densité de Haar de $\mathbb{T}$;

il se vérifie en choisissant une base canonique de E associée au couple λ, μ (voir (2.1)), ce qui permet d'identifier chacun des espaces à $L^2(\mathbb{R}^n)$; on constate alors que

(7.8) $$\langle\varphi,\psi\rangle_{\lambda\mu} = \frac{1}{[2\pi]^{n/2}}\int_{\mathbb{R}^{2n}} \overline{\varphi(p)}\,\psi(q)\,e^{i\langle p,q\rangle}\,dp\,dq$$

donc que $\mathcal{F}_{\lambda\mu}$ est simplement une <u>transformation de Fourier</u> entre $\mathcal{H}_\lambda$ et $\mathcal{H}_\mu$; on peut donc considérer le pairing comme une "géométrisation" de la transformée de Fourier.

L'unitarité de $\mathcal{F}_{\lambda\mu}$ et la formule (7.4) impliquent la formule

(7.9) $$\boxed{\left[\mathcal{F}_{\lambda\mu}\right]^{-1} = \mathcal{F}_{\mu\lambda}}$$

Une question se pose alors naturellement : le pairing est-il transitif ? a-t-on $\mathcal{F}_{\lambda\mu}\circ\mathcal{F}_{\mu\nu} = \mathcal{F}_{\lambda\nu}$ si λ, μ, ν sont transverses deux à deux ? Dans le cas particulier le plus simple ($n = 1$, $\mathrm{sgn}(\lambda, \mu, \nu) = 1$), on constate, en choisissant naturellement les coordonnées, que cette question devient :

la fonctionnelle $\mathbb{F}$:

$$F(\varphi)(x) = \frac{1}{\sqrt{2\pi}}\int_{-\infty}^{+\infty} e^{\frac{i}{4}\left[x^2+y^2+4xy\right]}\,\varphi(y)\,dy$$

vérifie-t-elle $F^3 = 1$?

La réponse est <u>non</u> ; c'est seulement F^{24} qui est égal à l'identité ; un calcul élémentaire montre en effet que F^2 est le produit de la conjuguée de F (égale à F^{-1}) par l'<u>intégrale de Fresnel</u>

$$\frac{1}{\sqrt{2\pi}} \int_{-\infty}^{+\infty} e^{iy^2/2}\, dy$$

dont la valeur est $e^{i\pi/4}$; on a donc $\mathcal{F}_{\lambda\mu} \circ \mathcal{F}_{\mu\nu} = e^{i\pi/4}\, \mathcal{F}_{\lambda\nu}$; le cas général se ramène à celui-ci en choisissant une base de l'espace μ qui soit orthonormale pour la métrique $g_{\lambda\mu\nu}$ (1.8) et en la complétant canoniquement dans λ (resp. dans ν) ; ce qui conduit à la formule générale

$$\boxed{\mathcal{F}_{\lambda\mu} \circ \mathcal{F}_{\mu\nu} = e^{\frac{i\pi}{4}\,\mathrm{sgn}(\lambda,\mu,\nu)}\, \mathcal{F}_{\lambda\nu}} \tag{7.10}$$

sgn étant la fonction définie en (1.9).

Quel parti peut-on tirer de cette formule ? On peut songer à se débarrasser du terme gênant en déphasant la définition du pairing ; ce qui revient à résoudre le problème cohomologique implicitement posé par la formule (1.11) , c'est-à-dire à considérer le cocycle "sgn" comme un cobord : nous savons que ce n'est globalement possible qu'en passant au revêtement, et que la solution est fournie par l'indice de Maslov.

§8 - ESPACE DE SCHRÖDINGER

Soit donc $\widehat{\Lambda(E)}$ le revêtement universel de la grassmannienne lagrangienne $\Lambda(E)$, P la projection de $\widehat{\Lambda(E)}$ sur $\Lambda(E)$ (voir le §3).

A tout $u \in \widehat{\Lambda(E)}$, nous associerons l'espace de Hilbert $\mathcal{H}_{P(u)}$, que nous pourrons noter $\mathcal{H}_u$; l'ensemble des couples

$$(u, \psi) \qquad \left[u \in \widehat{\Lambda(E)},\ \psi \in \mathcal{H}_u \right]$$

peut être considéré comme un fibré hilbertien de base $\widehat{\Lambda(E)}$ (figure 3).

Nous dirons que u et v sont transverses si les plans lagrangiens P(u) et P(v) le sont ; nous poserons alors

$$F_{uv} = e^{-\frac{i\pi}{2} m(u,v)} \mathcal{F}_{P(u),\, P(v)} \tag{8.1}$$

m étant l'indice de Maslov (4.5), $\mathcal{F}$ la "transformation de Fourier" définie en (7.6).

Il est clair alors que

$$\text{(8.2)} \quad \begin{cases} F_{uv} \text{ est une application unitaire de } \mathcal{H}_v \text{ sur } \mathcal{H}_u \\ F_{uv}^{-1} = F_{vu} \\ F_{uv} \circ F_{vw} = F_{uw} \end{cases}$$

si u, v, w sont deux à deux transverses ; la définition (8.1) a été choisie pour assurer la dernière de ces identités, grâce à la formule de Leray (4.10).

Théorème :

(8.3) On peut prolonger l'application $(u,v) \longmapsto F_{uv}$ à tous les couples de $\widehat{\Lambda(E)}$ (transverses ou non) de façon que les formules (8.2) restent valables : ce prolongement est unique.

Etablissons d'abord deux lemmes :

(8.4) Il existe une partie Θ de $\Lambda(E)$ telle que

- Deux éléments distincts t, t' de Θ sont transverses ;
- Pour toute partie finie $(u_1, u_2, \ldots, u_p)$ de $\widehat{\Lambda(E)}$, il existe $t \in \Theta$ qui est transverse à $u_1, u_2, \ldots, u_p$.

Il suffit évidemment de choisir une telle partie dans $\Lambda(E)$ et de relever arbitrairement chacun de ses éléments. En supposant que E est l'espace $\mathbb{C}^n$, on choisira l'ensemble des matrices zI ,$[z \in \mathbb{T}]$ (avec l'identification (3.11) des plans lagrangiens à des matrices). Les propriétés (8.4) résultent immédiatement de (3.12) , qui implique

$$\left[\lambda \text{ transverse à } zI\right] \Leftrightarrow \left[\lambda \text{ n'admet pas la valeur propre } z\right].$$

C.Q.F.D.

(8.5) Pour tout couple $u,v \in \widehat{\Lambda(E)}$, il existe une application Φ_{uv} telle que

$$\Phi_{uv} = F_{ut} \circ F_{tv} \qquad [\forall t \in \Theta \ , \ t \text{ transverse à } u \text{ et } v].$$

Il suffit de montrer, si t et $t' \in \Theta$, t et t' étant transverses à u et v , que

$$F_{ut} \circ F_{tv} = F_{ut'} \circ F_{t'v}$$

ceci résulte des formules

$$F_{ut'} = F_{ut} \circ F_{tt'}, \quad F_{t'v} = F_{t't} \circ F_{tv} \quad , \quad F_{tt'} = F_{t't}^{-1}$$

valables en raison de (8.2) parce que t et t' sont transverses (8.4).

C.Q.F.D.

Il est alors élémentaire de vérifier que Φ_{uv} est le prolongement unique cherché de F_{uv} ; exemple : quels que soient u, v, w, il existe $t \in \Theta$ transverse à u, v, w (8.4) ; on a alors

$$\Phi_{uv} \circ \Phi_{vw} = F_{ut} \circ F_{tv} \circ F_{vt} \circ F_{tw} = F_{ut} \ F_{tw} = \Phi_{uw}$$

ce qui vérifie la dernière des formules (8.2).

C.Q.F.D.

Théorème :

(8.6)	$F_{uu} = I_{\mathcal{H}_u}$	$\forall u \in \widehat{\Lambda(E)}$

(8.7)	$F_{L^k(u),\,L^{k'}(u)} = i^{k'-k}\, I_{\mathcal{H}_u}$	$\forall\, k,k' \in \mathbb{Z}$

L étant le générateur (3.21) du groupe d'homotopie de $\Lambda(E)$.

Il suffit de choisir t transverse à u, d'écrire $F_{uu} = F_{ut} \circ F_{tu}$, $F_{L(u)u} = F_{L(u)t} \circ F_{tu}$, d'utiliser la définition (8.1) de F, la formule (7.9) et les propriétés (4.9), (4.7) de l'indice de Maslov.

C.Q.F.D.

(8.8) - La formule (8.7) montre que $F_{L^{4k}(u),\,L^{4k'}(u)} = I_{\mathcal{H}_u}$,

donc qu'il n'est pas nécessaire d'utiliser le revêtement universel de $\Lambda(E)$ pour parvenir au résultat : on peut si l'on veut se contenter d'utiliser la "variété de Maslov", revêtement à 4 feuillets de la grassmannienne lagrangienne (et par conséquent revêtement à 2 feuillets de la variété des plans lagrangiens orientés) ; mais ce n'est pas indispensable.

- Nous pouvons maintenant utiliser ces résultats pour <u>trivialiser</u> le fibré hilbertien de la figure 3, en identifiant toutes ses fibres à une fibre-type $\mathcal{H}_E$, que nous appellerons "espace de Schrödinger" ; en effet (7.13) et (7.16) montrent que la relation $\sim$:

$$(8.9) \qquad \left[(u,\psi) \sim (u',\psi') \right] \iff \left[\psi' = F_{u'u}(\psi) \right]$$

est une équivalence, et que la structure hilbertienne du quotient $\mathcal{H}_E$ définie par l'unitarité de l'application

$$(8.10) \qquad \psi \mapsto \text{classe } (u, \psi)$$

est indépendante de u.

- La relation (7.5) indique que les "opérateurs de Schrödinger" $\underline{a}$ commutent avec les $\mathcal{F}_{\lambda\mu}$, donc avec les F_{uv}, ce qui montre que l'on peut <u>directement définir la représentation de Schrödinger sur l'espace</u> $\mathcal{H}_E$ par la formule

$$(8.11) \qquad \underline{a}\,(\text{classe } (u,\psi)) = \text{classe } (u, \underline{a}(\psi))$$

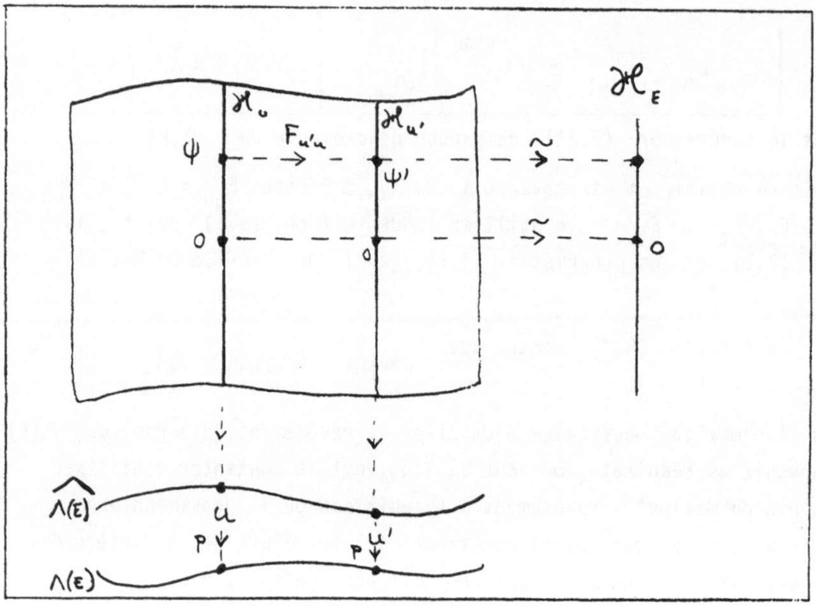

- Figure 3 -

§9 - REPRESENTATION METAPLECTIQUE

La variété Y a été munie, au §6 , d'une structure de "variété quantique" définie par la forme ϖ (6.4) et d'une structure de groupe de Lie (6.8) . Il est facile de trouver les automorphismes simultanés de ces deux structures : ce sont les images du <u>groupe symplectique</u> Sp(E) , agissant sur Y selon la règle

(9.1) $\underline{a}(x,z) = (a(x),\ z)$ $\quad [a \in Sp(E),\ x \in E,\ z \in \mathbb{T}]$

- Si a désigne un automorphisme d'un groupe de Lie G , H un sous-groupe fermé de G , il est clair que a définit un difféomorphisme de la variété G/H sur la variété G/a(H) par la formule

(9.2) $a(\text{classe}_H(x)) = \text{classe}_{a(H)}(a(x)) \qquad \forall x \in G$

En appliquant ce procédé au cas $G = Y$, $H = Y_\lambda$, $a \in Sp(E)$ (notations du §6), on voit que (9.2) définit un difféomorphisme de la variété Y/λ sur la variété $Y/a(\lambda)$, donc une application unitaire de H_{Y_λ} sur $H_{Y_{a(\lambda)}}$ (5.4) ; a étant un quantomorphisme, commute avec le tore (6.6) , si bien qu'il applique le sous-espace H_λ (défini par la condition de circulation (6.10)) sur l'espace $H_{a(\lambda)}$; cette application unitaire passe aux complétés $\mathcal{H}_\lambda$, $\mathcal{H}_{a(\lambda)}$.

Il résulte de la construction du pairing que a dédouble fonctoriellement la figure 2 - donc que

(9.3) $$\langle \underline{a}(\psi), \underline{a}(\varphi) \rangle_{a(\lambda), a(\mu)} = \langle \psi, \varphi \rangle_{\lambda\mu}$$

ce qui se traduit (Cf. (7.6)) en

(9.4) $$\mathcal{F}_{a(\lambda), a(\mu)} = \underline{a} \circ \mathcal{F}_{\lambda\mu} \circ \underline{a}^{-1}$$

Si donc nous choisissons un élément b du groupe $\widehat{Sp(E)}$, revêtement universel de $Sp(E)$ qui agit sur le revêtement $\widehat{\Lambda(E)}$ de $\Lambda(E)$, et si nous désignons par $\Pi(b)$ sa projection sur $Sp(E)$, nous aurons (Cf. (8.1))

(9.5) $$F_{b(u)\, b(v)} = \underline{\Pi(b)} \circ F_{uv} \circ \underline{\Pi(b)}^{-1}$$

ce qui montre que $\widehat{Sp(E)}$ <u>agit unitairement sur l'espace de Schrödinger</u> $\mathcal{H}_E$ selon la formule

(9.16) $$b(\text{classe } (u, \psi)) = \text{classe } \big(b(u), \pi(b)(\psi)\big)$$

c'est la <u>représentation de Shale-Weil</u> ([V]).

Cette représentation n'est pas fidèle ; considérons en effet le générateur $K = (I, 2\pi)$ du groupe d'homotopie de $Sp(n)$ (3.19) ; son action sur $\widehat{\Lambda(n)}$, donnée par la formule (3.23) , est

$$(\lambda, \theta) \mapsto (\lambda, \theta + 4\pi)$$

elle coïncide donc avec celle de L^2 (3.21) . Il résulte alors de (8.7) que

(9.7) $\underline{K}(\psi) = -\psi$

donc que K^2 appartient au noyau de la représentation : la représentation de Shale-Weil est donc une représentation unitaire de $Sp(E)/K^2$, c'est-à-dire du revêtement à deux feuillets de $Sp(E)$, appelé groupe métaplectique (d'où son nom de représentation métaplectique).

- On remarquera que le groupe métaplectique et le groupe de Heisenberg engendrent un produit semi-direct, extension par $[\mathbb{Z}/2\mathbb{Z}]\times\mathbb{T}$ du groupe des symplectomorphismes affines de E , et que celui-ci se représente sur l'espace de Schrödinger $\mathcal{H}_E$; cette "représentation de Weil-Weyl" contient les deux précédentes.

§10 - APPLICATION A L'OSCILLATEUR HARMONIQUE

On appelle oscillateur harmonique à n dimensions un point q mobile dans un espace euclidien de dimension n , les équations du mouvement dérivant du lagrangien

(10.1) $\frac{1}{2} m \left\| \frac{dq}{dt} \right\|^2 - v(q)$

où l'énergie potentielle v est une forme quadratique positive.

On peut choisir une base orthogonale où cette forme est décomposée en carrés :

(10.2) $v(q) = \frac{m}{2} \sum_{j=1}^{n} \omega_j^2 q_j^2 \qquad \omega_j > 0$

ce qui fournit n "mouvements propres" de l'oscillateur, de périodes respectives $2\pi / \omega_j$.

La linéarité des équations du mouvement et le calcul des crochets de Lagrange montrent que l'ensemble E des mouvements possède une structure d'espace vectoriel symplectique ; elle s'identifie à celle de $\mathbb{C}^n$ (§3) si :

(10.3) 1°) on a choisi des unités de longueur, masse, temps telles que la masse m soit égale à 1 et la constante de Planck à 2π ;
2°) on désigne par p_j les variables $\dfrac{dq_j}{dt}$ et on pose

(10.4)
$$x = \begin{pmatrix} \dfrac{p_1}{\sqrt{\omega_1}} + i\sqrt{\omega_1}\, q_1 \\ \cdots\cdots\cdots \\ \dfrac{p_n}{\sqrt{\omega_n}} + i\sqrt{\omega_n}\, q_n \end{pmatrix} \in \mathbb{C}^n$$

les valeurs des p_j et q^j étant prises à la date $t = 0$.

Nous nous proposons de construire et d'interpréter l'<u>espace de Schrödinger</u> associé $\mathcal{H}_E$ (§8).

Faisons quelques remarques préalables :

(10.5) Si on désigne par a_τ l'opération qui consiste à <u>retarder</u> un mouvement d'une durée arbitraire τ ,

$$\tau \longmapsto a_\tau$$

est un morphisme du groupe additif $\mathbb{R}$ dans le groupe symplectique Sp(E) ; ici, dans le groupe unitaire, car

$$a_\tau = \begin{pmatrix} e^{-i\omega_1\tau} & & \\ & \ddots & \\ & & e^{-i\omega_n\tau} \end{pmatrix} \quad ■$$

(10.6) - t étant une date arbitraire, les mouvements dans lesquels le point q <u>passe à l'origine à l'instant</u> t forment un <u>plan lagrangien</u> λ_t , et l'on a

$$\lambda_{\tau+t} = a_\tau(\lambda_t)$$

comme on a visiblement $\lambda_o = \mathbb{R}^n$, on a donc $\lambda_t = a_t(\mathbb{R}^n)$; d'où avec l'identification matricielle (3.10)

$$\lambda_t = a_t\, C(a_t^{-1}) = a_{2t}$$

■

(10.7) - Les applications $\tau \mapsto a_\tau$, $t \mapsto \lambda_t$ se relèvent respectivement aux revêtements universels $\widehat{Sp(E)}$, $\widehat{\Lambda(E)}$ par

$$\tau \mapsto b_\tau \ , \qquad t \mapsto u_t$$

de sorte que

$$b_\tau \circ b_{\tau'} = b_{\tau+\tau'} \quad ; \quad u_{\tau+t} = b_\tau(u_t) \ ;$$

on pourra choisir

$$b_\tau = (a_\tau, -[\omega_1+\ldots+\omega_n]\tau), \quad u_t = (a_{2t}, -2[\omega_1+\ldots+\omega_n]t)$$

■

(10.8) - L'<u>indice de Maslov</u> $m(u_t, u_{\tau+t})$ ne dépend que de τ ; en effet $m(u_t, u_{\tau+t}) = m(b_t(u_o), b_t(u_\tau)) = m(u_o, u_\tau)$ (formule (4.8)) ; la définition (4.5) donne alors

$$m(u_t, u_{\tau+t}) = \frac{1}{2\pi}\Big[2(\omega_1+\ldots+\omega_n)\tau + i\ \mathrm{Tr}(\mathrm{Log}(-a_{-2\tau}))\Big] \ ;$$

en utilisant le résultat auxiliaire

$$\mathrm{Log}(-e^{i\alpha}) = i\Big[\alpha - \pi - 2\pi\ \mathrm{Ent}(\alpha/2\pi)\Big] \qquad \forall \alpha \in \mathbb{R} - 2\pi\mathbb{Z}$$

où Ent désigne la partie entière, il vient

$$m(u_t\,,\,u_{\tau+t}) = \frac{n}{2} + \sum_{j=1}^{n} \mathrm{Ent}(\omega_j \tau/\pi)$$;

on voit en particulier que u_t et $u_{\tau+t}$ sont transverses si τ n'est pas demi-période d'un mouvement propre. ■

(10.9) La variété Y/λ_t peut se repérer par les variables

$$q_1, q_2, \ldots, q_n\,, \qquad \zeta = z\, e^{\frac{i}{2}\sum_j p_j q_j}$$

les valeurs des q_j et p_j étant prises à l'instant t ; un élément de l'espace $\mathcal{H}_{\lambda_t}$ s'écrit

$$\Psi(q_1, \ldots q_n)\,\zeta\,\rho$$

ρ étant une demi-densité invariante sur $\mathbb{R}^n \times \mathbb{T}$, Ψ étant à carré sommable.

Alors un élément de l'espace de Schrödinger $\mathcal{H}_E$ pourra être considéré comme une classe d'équivalence de couples $(x_t\,,\Psi_t)$ pour la relation $\sim$ (8.9) ; en effectuant le calcul du pairing (Figure 2) et en utilisant l'indice de Maslov (10.8), on constate que cette relation s'écrit

$$\Psi_{\tau+t}(q_1,\ldots q_n) = \prod_{j=1}^{n} \sqrt{\frac{\omega_j}{2\pi\,|\sin(\omega_j\tau)|}}\; e^{\frac{i\pi}{2}\left[\frac{1}{2} + \mathrm{Ent}\left(\frac{\omega_j\tau}{\pi}\right)\right]} \times \int_{\mathbb{R}^n} \Psi_t(q'_1 \ldots q'_n)\, e^{\frac{i}{2}\sum_j \frac{\omega_j}{\sin(\omega_j\tau)}\left[2q_jq'_j - [q_j^2 + q_j'^2]\cos(\omega_j\tau)\right]}\, dq'_1 \ldots dq'_n \tag{10.10}$$

chaque fois que u_t et $u_{\tau+t}$ sont transverses, c'est-à-dire quand

τ n'est pas une demi-période.

Nous savons donc que $\mathcal{H}_\varepsilon$ est l'ensemble des fonctions des variables $t, q_1, \ldots, q_n$, qui sont à carré sommable en $q_1, \ldots, q_n$ pour chaque valeur de t, et qui vérifient cette équation (10.10) ; nous savons d'ailleurs qu'elles constituent un espace de Hilbert, le carré de leur norme étant

$$(10.11) \qquad \int_{\mathbb{R}^n} \left| \psi(q_1, \ldots, q_n) \right|^2 \, dq_1 \ldots dq_n$$

qui est indépendante de t .

- En effectuant quelques dérivations sous le signe $\int$, le lecteur se convaincra sans trop de peine (voir Bérenguier [VII]), que les solutions de (10.10) vérifient l'<u>équation de Schrödinger</u>

$$(10.12) \qquad \frac{1}{2} \Delta \psi - v(q) \, \psi = i \frac{\partial \psi}{\partial t}$$

qui se trouve donc <u>explicitement intégrée</u> (10.10).

- Cette formule est d'ailleurs un prolongement de celle de Feynman ([VI]) qui s'applique seulement au cas où $|\tau|$ est inférieur à la plus petite des demi-périodes (alors l'indice de Maslov se simplifie en $\frac{n}{2} \operatorname{sgn}(\tau)$).

- On peut interpréter (10.10) de la façon suivante : un retard de τ sur une fonction d'onde ψ la transforme en $b_\tau(\psi)$, b_τ étant l'élément de $\widehat{Sp(E)}$ calculé en (10.7) , b_τ agissant sur $\mathcal{H}_\varepsilon$ par la <u>représentation métaplectique</u> (9.6) .

Utilisons cette remarque pour calculer $\psi_{\tau+t}$ dans un cas <u>non transverse</u> : traitons le cas d'un oscillateur isotrope (toutes les fréquences propres étant égales) ; prenons pour τ la demi-période $\frac{T}{2} = \frac{\pi}{\omega}$. On trouve facilement (en utilisant (10.5), (10.7), (3.21))

$$(10.13) \qquad a_{T/2} = -I \; , \; b_{T/2} = (-I, -n\pi) \; , \; u_{t+T/2} = b_{T/2}(u_t) = L^{-n}(u_t)$$

d'où, par application de (9.6) , (8.9), (8.7)

$$\psi_{t-T/2} = i^{-n}\, \psi \circ [-I]$$

ce qui peut encore s'écrire

(10.14) $$\psi_{t+T/2}(q_1, \ldots, q_n) = i^n\, \psi_t(-q_1, -q_2, \ldots, -q_n)$$

Cette propriété des solutions de l'équation de Schrödinger (10.12) peut se vérifier directement sur chaque solution stationnaire (la vérification fait intervenir l'énergie du point 0 et la parité des polynômes d'Hermite). Elle entraîne évidemment

(10.15) $$\psi_{t+T} = [-1]^n\, \psi_t \qquad ;$$

lorsque la dimension n est impaire, les solutions de l'équation de Schrödinger ont une période double de celle des mouvements classiques.

- REFERENCES -

[I] V.P. MASLOV

Théorie des Perturbations et Méthodes Asymptotiques, Dunod (1972)

[II] V. ARNOLD

Journal d'Analyse Fonctionnelle, n°1, p. 1. Traduction française : Supplément à la traduction française du livre de Maslov.

[III] J. LERAY

Communication Colloque de Rome, Janvier 1973, Acta Matematica, vol. XIV.

[IV] B. KOSTANT

Communication Colloque de Rome, Janvier 1973 Acta Matematica, vol. XIV.

[V] A. WEIL

Sur certains groupes d'opérateurs unitaires, Acta Matematica (1964), p. 143-211.

[VI] R.P. FEYNMAN

Quantum Mechanics and Path Integrals, Mc Graw Hill (1965).

[VII] G. BERENGUIER

Géométrie et Quantogéométrie de l'Oscillateur Harmonique, Thèse de 3e cycle, Université de Provence, Mars 1975.

TWISTOR THEORY AND GEOMETRIC QUANTIZATION

Nicholas Woodhouse

Department of Mathematics, King's College, London

Introduction

In this talk, I shall show how Penrose's twistor formalism [4] arises through the application, with the aid of a few geometrical tricks, of the Kostant-Souriau geometric quantization theory [3,10,11] to massless spinning particles in Minkowski space. This is not the usual way of introducing twistors, still less does it reflect the historical development of the subject, but it does have the advantage of showing up some of the similarities in the fundamental ideas of the two theories.

However, in spite of these similarities, it is important to realize that their motivations are very different. The ultimate aim of geometric quantization is the construction of a unified theory of the irreducible unitary representations of connected Lie groups by first geometrizing and then generalizing the physicist's concept of quantization. When applied to simple physical systems, such as those invariant under the Poincaré group, geometric quantization results in the synthesis within a geometric framework of various well understood techniques from conventional quantum mechanics; it does not incorporate any new physical ideas.

In his twistor theory, on the other hand, Penrose is trying to develop a new formalism for relativistic quantum field theory and, eventually, to lay the foundations for a quantum theory of gravity. Moreover, Penrose has often stressed that he is looking for a formalism which only works in four dimensional space-time. In a sense, the existence of such a formalism would explain the dimension and signature of the real world.

But, even allowing for these differences in outlook, there are a number of practical benefits which can be derived from a comparison of the two theories:

1) If one tries to construct a manifestly conformally invariant theory of massless particles by applying geometric quantization to the conformal

group C(1,3) - or, rather, to its fourfold cover, SU(2,2) - one runs into a number of difficulties (which I shall describe later). Penrose's twistor contour integration techniques provide a way of circumventing these difficulties and it is possible that, when translated into a suitable form, these techniques will lead to a friutful generalization of the Kostant-Souriau theory.

2) Kostant and Souriau's geometric formulation of standard quantum mechanics is ideally suited to answering the question: How much of twistor theory is an elegant restatement of old ideas, and how much is new physics?

3) Geometric quantization is often difficult to work with in practice: one is frequently forced to rely on the introduction of special coordinate systems. However, by using some of the tricks suggested by twistor theory, it is possible to quantize massless and massive particles in a covariant way and thus to obtain an example of a completely geometric application of the theory.

Notation

The notation used here for the $SL(2,\mathbb{C})$ spinor calculus is essentially the same as that described by Penrose [6] and Pirani [8] . Capital Roman letters are used for spinor indices (which run over 0,1) and lower case Roman letters for space-time vector and tensor indices (which run over 0,1,2,3); primed indices are used to denote conjugate spinors. The Einstein range and summation conventions are used throughout.

The correspondence between a vector Y^a and its spinor equivalent $Y^{AA'}$ is given explicitly, in <u>any</u> proper orthochronous Lorentz frame, by

$$Y^a \leftrightarrow Y^{AA'} = \begin{bmatrix} Y^{00'} & Y^{01'} \\ Y^{10'} & Y^{11'} \end{bmatrix} = \frac{1}{\sqrt{2}} \begin{bmatrix} Y^0 + Y^1 & Y^2 + i\,Y^3 \\ Y^2 - i\,Y^3 & Y^0 - Y^1 \end{bmatrix} \qquad (1)$$

Spinor indices are raised and lowered with the Levi-Civita symbols

$$\epsilon^{AB} = \begin{bmatrix} 0 & 1 \\ -1 & 0 \end{bmatrix} \qquad \epsilon_{AB} = \begin{bmatrix} 0 & 1 \\ -1 & 0 \end{bmatrix} \qquad (2)$$

and their complex conjugates $\epsilon^{A'B'}$ and $\epsilon_{A'B'}$ (these are all $SL(2,\mathbb{C})$ invariant). Thus, for example,

$$\chi^A = \epsilon^{AB}\chi_B, \ \chi_B = \epsilon_{AB}\chi^A \ \text{ and } \ \bar{\chi}_{B'} = \epsilon_{A'B'}\bar{\chi}^{A'}. \tag{3}$$

This is consistent with the usual convention for raising and lowering space-time indices since $\epsilon_{AB}\epsilon_{A'B'}$ is the spinor equivalent of the space-time metric g_{ab}. Finally, the flat spinor connection $\nabla_{AA'}$ is given in Lorentz coordinates $\{x^a\}$ by

$$\sqrt{2}\,\nabla_{AA'} = \begin{bmatrix} \partial_0 + \partial_1 & \partial_2 - i\partial_3 \\ \partial_2 + i\partial_3 & \partial_0 - \partial_1 \end{bmatrix} \qquad \partial_a = \frac{\partial}{\partial x^a} \tag{4}$$

(In the Battelle convention [6], spinor and tensor indices are regarded as abstract labels indicating the type of the geometric object to which they are attached. Thus, for example, Y^a is actually a vector, rather than the components of a vector, and (1) can be rewritten: $Y^a = Y^{AA'}$. Though this convention will not be used explicitly, it can be used to reinterpret all the equations below as relations between geometric objects, rather than the components of geometric objects.)

Massless Particles: Canonical Formalism

In classical relativistic mechanics, the kinematical variables of a massless particle with helicity $s \geqslant 0$ can be represented by a position vector X^a (relative to some origin O) and two future pointing null vectors I^a and J^a (normalized so that $I_a J^a = 1$). In terms of these, the momentum and angular momentum are given by

$$p_a = I_a \quad \text{and} \quad M^{ab} = -s\,\epsilon^{abcd} I_c J_d + X^a I^b - X^b I^a, \tag{5}$$

the form of M^{ab} being fixed by the condition that the spin vector

$$s_a = \tfrac{1}{2}\epsilon_{abcd}\, p^b M^{cd} \tag{6}$$

should be parallel to the momentum (ϵ_{abcd} is the alternating tensor).

Before quantizing this system, it is necessary to construct the classical phase space: in practical terms, this means finding a suitable expression for the symplectic 2-form of the system in terms of the variables X^a, I^a and J^a. In this search, there are two guiding principals:

1) The system is to be an elementary relativistic system: this means that the Poincare group P must act transitively on the phase space as a group of canonical transformations. By the Kostant-Kirillov-Souriau theorem [3], therefore, the phase space must be locally isomorphic (as a P-symplectic space) with an orbit in the dual of the Poincare Lie algebra.

2) The physical variables p_a and M^{ab} are to have their usual interpretation as generators of P.

The first implication of these is that each point in the phase space is determined by the values of p_a and M^{ab} alone, or, after a little calculation, that one must identify $(\tilde{X}^a, \tilde{I}^a, \tilde{J}^a)$ and (X^a, I^a, J^a) whenever

$$\tilde{X}^a = X^a + Z^a, \quad \tilde{I}^a = I^a \quad \text{and} \quad \tilde{J}^a = J^a + \epsilon^{abcd} I_b J_c Z_d - \tfrac{1}{2} Z^b Z_b I^a \qquad (7)$$

for some Z^a such that $Z^a I_a = 0$. The resulting manifold is six dimensional and has topology $\mathbb{R}^4 \times S^2$. (The need for this identification reflects the fact that even classically a massless spinning particle is not localizable: it occupies an entire null hyperplane.)

It is then not hard to show that one, and hence the only, symplectic form on M_s which gives the correct Poisson brackets for p_a and M^{ab} is

$$\sigma = s\epsilon_{abcd} I^a J^b dI^c \wedge dJ^d - dX^a \wedge dI_a \qquad (8)$$

(see Souriau [10], p.190). (As a 2-form on the nine dimensional (X^a, I^a, J^a)-space, σ is degenerate. However, the vectors in this space which annihilate σ are precisely those which generate the identification (7). Thus σ projects into a nondegenerate 2-form on M_s.)

Prequantization

The first stage in the prequantization of (M_s, σ) is to replace I^a and J^a by two spinors o^A and ι^A chosen so that

$$I^a \leftrightarrow o^A \bar{o}^{A'}, \quad J^a \leftrightarrow \iota^A \bar{\iota}^{A'} \quad \text{and} \quad o_A \iota^A = 1. \qquad (9)$$

(This is possible since I^a and J^a are null, so that $I^{AA'}$ and $J^{AA'}$ are singular.) If then

$$\pi_{A'} = \bar{o}_{A'} \quad \text{and} \quad \omega^A = s\iota^A + i X^{AA'} \bar{o}_{A'} \tag{10}$$

the identification (7) becomes $(\tilde{X}^a, \tilde{\omega}^A, \tilde{\pi}_{A'}) \simeq (X^a, \omega^A, \pi_{A'})$ whenever

$$\tilde{X}^a = X^a + Z^a, \quad \tilde{\pi}_{A'} = e^{it}\pi_{A'} \quad \text{and} \quad \tilde{\omega}^A = e^{it}\omega^A \tag{11}$$

for some $t \in \mathbb{R}$ and for some real Z^a such that $Z^{AA'}\bar{\pi}_A \pi_{A'} = 0$.

Thus a point of M_s can be fixed by specifying the pair $(\omega^A, \pi_{A'})$: the corresponding values of X^a are then given as the solutions of the linear equation

$$(\omega^A - i X^{AA'} \pi_{A'})\bar{\pi}_A = s \tag{12}$$

and the values of the momenta are given explicitly by

$$p_a \leftrightarrow \bar{\pi}_A \pi_{A'} \quad \text{and} \quad M^{ab} \leftrightarrow i(\omega^{(A} \bar{\pi}^{B)} \epsilon^{A'B'} - \epsilon^{AB} \bar{\omega}^{(A'} \pi^{B')}). \tag{13}$$

The four complex (eight real) dimensional vector space in which ω^A and $\pi_{A'}$ are independent variables is called <u>twistor space</u> (denoted T). A twistor (that is, an element of T) can be represented either as a pair $(\omega^A, \pi_{A'})$ of spinors or as a quadruple $Z^\alpha = (Z^1, Z^2, Z^3, Z^4)$ where

$$(Z^1, Z^2) = (\omega^0, \omega^1) \quad \text{and} \quad (Z^3, Z^4) = (\pi_{0'}, \pi_{1'}). \tag{14}$$

(Again, the index α can be interpreted as an "abstract" index.)

The twistors which correspond to points in M_s are those which lie in the surface $G_s \subset T$ given by

$$g(Z^\alpha) = Z^\alpha \bar{Z}_\alpha = \omega^A \bar{\pi}_A + \pi_{A'} \bar{\omega}^{A'} = 2s \tag{15}$$

where $\bar{Z}_\alpha = (\bar{\pi}_A, \bar{\omega}^{A'})$ is the <u>Hermitian conjugate</u> of Z^α. (The map

$$g : T \longrightarrow \mathbb{R} : Z^\alpha \longmapsto Z^\alpha \bar{Z}_\alpha = \omega^A \bar{\pi}_A + \bar{\omega}^{A'} \pi_{A'} \tag{16}$$

defines a pseudo-Hermitian metric on T of signature $(+,+,-,-)$.) Because the freedom available in the choice of the phases of o^A and ι^A, the

projection*

$$pr : G_s \longrightarrow M_s : Z^\alpha = (\omega^A, \pi_{A'}) \longmapsto (p_a, M^{ab}) \tag{17}$$

defined by (13) is not one-to-one and, in fact, $pr(Z^\alpha) = pr(Y^\alpha)$ whenever $Z^\alpha = e^{it} Y^\alpha$ for some $t \in \mathbb{R}$.

Now if, in T, one introduces the symplectic 2-form

$$\sigma = i\,(dZ^\alpha \wedge d\overline{Z}_\alpha) \tag{18}$$

(which gives T the structure of a pseudo-Kähler manifold) then the Hamiltonian vector field generated by $g : T \longrightarrow \mathbb{R}$ is

$$X_g = i\,(Z^\alpha \frac{\partial}{\partial Z^\alpha} - \overline{Z}_\alpha \frac{\partial}{\partial \overline{Z}_\alpha}) \tag{19}$$

which has closed integral curves of the form $t \longmapsto e^{it} Z_0^\alpha$. In other words,

$$M_s = G_s / X_g \,. \tag{20}$$

Moreover, the restiction of σ to G_s is degenerate and is annihilated by X_g (which is tangent to G_s), so that σ projects into a closed 2-form in M_s. A short calculation shows that this is precisely the symplectic form introduced above.

The point of this is that, while the symplectic structure of (M_s, σ) is not exact, that of (T, σ) is, since, in T,

$$\sigma = d\theta \quad \text{where} \quad \theta = \tfrac{1}{2} i (Z^\alpha d\overline{Z}_\alpha - \overline{Z}_\alpha dZ^\alpha). \tag{21}$$

This can be exploited in the prequantization of (M_s, σ), as follows:

If, in the bundle space of the trivial line bundle $G_s \times \mathbb{C} \longrightarrow G_s$, one puts:

$$\alpha = \theta + \frac{1}{2\pi i} \frac{dz}{z} \quad \text{and} \quad Y = X_g - 2\pi i (X_g \lrcorner\, \theta) \frac{\partial}{\partial z} = X_g - \frac{2is}{\hbar} . \frac{\partial}{\partial z}$$

---

*When $s = 0$, the twistors of the form $Z = (\omega^A, 0)$ must be omitted from G_s: the reason for this will be made clear later.

(where z is the coordinate in $\mathbb{C}$, and, in the units used here, $\hbar = (2\pi)^{-1}$) then α is a connection form on $G_s \times \mathbb{C}$ and the integral curves of Y are parallel and are of the form

$$t \longmapsto \left(e^{it} Z_0^{\alpha},\ e^{-2is\hbar^{-1}t} . z_0 \right) \ ; \quad Z_0^{\alpha} \in T, \quad z_0 \in \mathbb{C} . \tag{22}$$

These are closed whenever $2s = \oint \frac{1}{2\pi}\theta$ (the integral being taken around an orbit of X_{ε} in G_s) is an integral multiple of $\hbar$, in which case $L = (G_s \times \mathbb{C})/_Y$ is a Hausdorff manifold, and, in fact, a line bundle over M_s (with the projection $L \longrightarrow M_s$ making the diagram

$$\begin{array}{ccc} G_s \times \mathbb{C} & \longrightarrow & L \\ \downarrow & & \downarrow \\ G_s & \longrightarrow & M_s \end{array}$$

commute). Moreover, $Y \lrcorner \alpha = 0$ and $\pounds_Y \alpha = 0$, so that α projects onto a connection form (also denoted α) on L; it follows from the definition of α that the curvature of this connection is σ. Thus, when $2s\hbar^{-1}$ is integral, this construction is an explicit prequantization for (M_s, σ).

Quantization

The next stage is to find a polarization for T which is invariant under the action of the Poincaré group. This action is easily found from equation (10). Under translation through A^a,

$$(\omega^A, \pi_{A'}) \longmapsto (\omega^A + i A^{AA'} \pi_{A'}, \pi_{A'}) \tag{23}$$

and under the Lorentz rotation defined by $L^A{}_B \in SL(2,\mathbb{C})$,

$$(\omega^A, \pi_{A'}) \longmapsto (L^A{}_B\, \omega^B,\ \overline{M}^{B'}{}_{A'}\, \pi_{B'}) \quad \text{where} \quad M^C{}_A L^A{}_B = \delta^C_B \tag{24}$$

(Note that both transformations preserve the form $\epsilon_{\alpha\beta\gamma\delta}\, dZ^{\alpha} \wedge dZ^{\beta} \wedge dZ^{\gamma} \wedge dZ^{\delta}$ and the Kähler structure of T, and thus define elements of $SU(2,2)$.)

It follows that the real polarization F of T, spanned at each point by the vectors $\partial/\partial\omega^A$ and $\partial/\partial\overline{\omega}^{A'}$ is Poincaré invariant. It is also Lie propagated by X_{ε}.

This polarization induces a polarization $\tilde{F}$ of M_s: explicitly, $\tilde{F}$

is the projection into M_s of the distribution H on G_s defined by

$$H : Z \in G_s \longmapsto F_Z \cap T_Z(G_s) \subset T_Z(G_s)$$

where $T_Z(G_s)$ is the tangent space to G_s at Z (H is also Lie propagated by X_σ); $\tilde{F}$ is, in fact, the 'natural' polarization of M_s, in the sense that it is spanned by the generating vector fields of the translation subgroup of P.

The integral manifolds of $\tilde{F}$ are the surfaces in M_s given by $\bar{\pi}_A$ = const., $\pi_{A'}$ = const. so that the points of the factor space $M_s/\tilde{F}$ are parameterized by $\pi_{A'}$, and $\bar{\pi}_A$ (modulo phase); in other words, $M_s/\tilde{F}$ is simply the future half of the light cone in momentum space (denoted N_+).

Now, up to normalization, there is a unique Lorentz invariant volume element ν on N_+, given in coordinates by

$$\nu = \frac{1}{p_0}(dp_1 \wedge dp_2 \wedge dp_3) = i(d\bar{\pi}^A \wedge d\bar{\pi}_A \wedge (\pi^{B'} d\pi_{B'}) - \\ - d\pi^{A'} \wedge d\pi_{A'} \wedge (\bar{\pi}^B d\bar{\pi}_B)) \tag{25}$$

Thus the wave functions of the $\tilde{F}$ polarization can be written in the form

$$\psi = \tau \nu^{\frac{1}{2}} \tag{26}$$

where $\tau : M_s \rightarrow L$ is a section of L which is covariantly constant on the leaves of F (the precise meaning of the square root $\nu^{\frac{1}{2}}$ - which is not important here - is discussed in detail by Blattner [1]).

The prequantization of (M_s, σ) allows the sections of L to be realized in a particularly simple way: to be precise, any smooth function $f : G_s \rightarrow \mathbb{C}$ which is homogeneous of degree $-2s\hbar^{-1}$ in Z^α defines a section τ_f of L, which makes this diagram commute:

$$\begin{array}{ccc} G_s \times \mathbb{C} & \longrightarrow & L \\ \downarrow\uparrow f & & \downarrow\uparrow \tau_f \\ G_s & \longrightarrow & M_s \end{array}$$

Conversely, any section $\tau : M_s \rightarrow L$ can be obtained in this way from a homogeneous function $f_\tau : G_s \rightarrow \mathbb{C}$. Furthermore, τ will be covariantly constant on the leaves of F if, and only if, f_τ is covariantly constant in the directions in H (as a section of the trivial bundle $G_s \times \mathbb{C}$), that is, if, and only if, f_τ is of the form

$$f_\tau : Z^\alpha = (\omega^A, \pi_{A'}) \mapsto k_\tau(\bar{\pi}_A, \pi_{A'}) \exp(\tfrac{1}{2} \hbar^{-1}(\omega^A \bar{\pi}_A - \bar{\omega}^{A'} \pi_{A'}))$$

$$= k \ \ (\bar{\pi}_A, \pi_{A'}) \exp(i\hbar^{-1} p_a X^a) \tag{27}$$

where k_τ is homogeneous of degree $-2s\hbar^{-1}$ in $\pi_{A'}$:

$$\hbar(\bar{\pi}_A \frac{\partial k_\tau}{\partial \bar{\pi}_A} - \pi_{A'} \frac{\partial k_\tau}{\partial \pi_{A'}}) = 2s\, k_\tau \ . \tag{28}$$

In this realization, the inner product of the two wave functions

$$\psi = \tau\, \nu^{\frac{1}{2}} \quad \text{and} \quad \psi' = \tau'\, \nu^{\frac{1}{2}}$$

is given by

$$\langle \psi, \psi' \rangle = \int_{N_+} k_\tau \bar{k}_{\tau'} \cdot \nu \tag{29}$$

(since it is invariant under phase transformations of $\pi_{A'}$, $k_\tau \bar{k}_{\tau'}$ is a well defined function on N_+).

Finally, the relationship of this to the conventional quantum description of massless particles can be seen by introducing the spinor field

$$\chi_{A'B'C'\ldots} = \int_{N_+} [\pi_{A'} \pi_{B'} \pi_{C'} \ldots f_\tau] \nu = \int_{N_+} [\pi_{A'} \pi_{B'} \ldots k_\tau e^{i\hbar^{-1} p_a X^a}] \nu \tag{30}$$

(with $2s\hbar^{-1}$ indices). The integrand is again independent of the phase of $\pi_{A'}$ and $\chi_{A'B'C'\ldots}$ satisfies the zero-rest-mass field equation

$$\nabla^{AA'} \chi_{A'B'C'\ldots} = 0 \tag{31}$$

Twistor Theory

So far, nothing exceptional has been achieved. It was already known that geometric quantization leads in a straightforward way to the conventional quantum description of a free massless particle [9,10]. The twistor formalism has done little except to make the calculations simpler.

The parting of the ways between conventional physics and twistor theory proper come when one tries to repeat this analysis in a way which respects the invariance of massless under the full conformal group of Minkowski space (denoted $C(1,3)$). Except for one minor subtlety, which I shall deal with presently, the construction of the classical phase space and its prequantization are unchanged. The trouble is that the polarization intro-

above is not invariant under the action of $C(1,3)$ on T. This action is found by introducing a new geometrical realization of a twistor:

The definition of the ω^A part of a twistor Z^α, corresponding to some fixed point of M_s, depends on the choice of an origin O in Minkowski space $\mathfrak{M}$. If O is translated through Y^a then the $\pi_{A'}$ part of Z^α is unaltered, while ω^A transforms according to

$$\omega^A \longmapsto \omega^A - i Y^{AA'} \pi_{A'} \tag{32}$$

The spinor ω^A can, therefore, be regarded as a field on $\mathfrak{M}$ rather than as a fixed (but origin dependent) quantity. From (32), this field satisfies the <u>twistor</u> <u>equation</u>,

$$\nabla_{A'}{}^{(A}\omega^{B)} = 0 \tag{34}$$

and, in fact, the general solution of this is

$$(\omega^A)_x = \omega_0^A - i X^{AA'}\pi_{A'} \quad \text{where } x \in \mathfrak{M} \text{ and } \pi_{A'} \text{ is fixed.} \tag{35}$$

Here, ω_0^A is the value of ω^A at $O \in \mathfrak{M}$ and X^a is the position vector of x relative to O. Thus the map $\omega^A \mapsto (\omega_0^A, \pi_{A'})$ defines an isomorphism between T and the solution space of the twistor equation.

The point of this is that (34) is conformally invariant in the sense that if $\rho : \mathfrak{M} \to \mathfrak{M}$ is a conformal isometry, and if ω^A satisfies (34), then so does $\rho(\omega)^A$, but with $\nabla_{AA'}$ replaced by the connection $\tilde{\nabla}_{AA'}$ of the conformally transformed metric $\tilde{g}_{ab} = \rho(g)_{ab} = \Omega^2 g_{ab}$, which is related to $\nabla_{AA'}$ by

$$\tilde{\nabla}_{AA'}\chi^B = \nabla_{AA'}\chi^B + \epsilon_A{}^B \Upsilon_{A'C}\chi^C \quad \text{where } \Upsilon_a = \nabla_a \ln\Omega. \tag{36}$$

Thus if ρ fixes the origin O then its action on T is

$$\rho: Z^\alpha = (\omega_0^A, \pi_{A'}) \mapsto (L^A{}_B\omega_0^B, \bar{M}^{B'}{}_{A'}(\pi_{B'} + i U_{BB'}\omega_0^B)) \tag{37}$$

where $U_a = (\Upsilon_a)_0$ and $L^A{}_B$ is the $SL(2,\mathbb{C})$ transformation of the spin space at O induced by ρ and $M^C{}_A L^A{}_B = \delta^C_B$. Again, this transformation defines an element of $SU(2,2)$.

The subtlety mentioned above arises because, strictly speaking, the only conformal isometries which fix O are the Lorentz rotations and the dilatations. The full 15-dimensional group C(1,3) acts not on $\mathcal{M}$ but on its compactification $\overline{\mathcal{M}}$ which is obtained from $\mathcal{M}$ by attaching the 'null cone at infinity' [5]. The only role that this plays here is that, when $s = 0$, the phase space of a conformally invariant particle includes points at infinity corresponding to twistors of the form $(\omega^A, 0)$, and has topology $S^3 \times S^2 \times \mathbb{R}$ rather than $\mathbb{R}^4 \times S^2$.

One now sees that there is an obvious conformally invariant polarization for T: this is the Kähler polarization K, spanned at each point by the antiholomorphic vectors $\{\partial/\partial\overline{Z}_\alpha\}$.

As before, K induces a polarization $\tilde{K}$ on each M_s (when $s \neq 0$, this is Kähler, but when $s = 0$, $\tilde{K} \cap \overline{\tilde{K}}$ is one dimensional). The $\tilde{K}$ wave functions are of the form $\psi = \tau\mu^{\frac{1}{2}}$ where τ is a section of L, covariantly constant in the directions in $\tilde{K}$ and μ is a 3-form, orthogonal to $\tilde{K}$ and Lie propagated by the directions in $\tilde{K}$.

Again, τ can be represented by a function $f_\tau : G_s \longrightarrow \mathbb{C}$, only now f_τ must be of the form

$$f_\tau : Z^\alpha \longmapsto k_\tau(Z^\alpha)\, e^{-\hbar^{-1} Z^\alpha \overline{Z}_\alpha} \tag{38}$$

where $k_\tau : T \longrightarrow \mathbb{C}$ is holomorphic and homogeneous of degree $-2s\hbar^{-1}$ in Z^α.

This time, there is no natural choice for μ. However, if $u : T \longrightarrow \mathbb{C}$ is holomorphic and homogeneous of degree -2 then

$$\mu = u^2 \epsilon_{\alpha\beta\gamma\delta}\, Z^\alpha dZ^\beta \wedge dZ^\gamma \wedge dZ^\delta \tag{39}$$

projects from G_s to a 3-form in M_s which has the desired properties. Thus the $\tilde{K}$ wave functions can be represented symbolically in the form

$$\psi = k_\tau(u^2 \epsilon_{\alpha\beta\gamma\delta}\, Z^\alpha dZ^\beta \wedge dZ^\gamma \wedge dZ^\delta)^{\frac{1}{2}} \tag{40}$$

where $h = k_\tau u : T \longrightarrow \mathbb{C}$ is holomorphic and homogeneous of degree $-2s\hbar^{-1} - 2$. In short, there is a one-to-one correspondence between holomorphic homogeneous twistor functions of degree $-2s\hbar^{-1}-2$ and $\tilde{K}$ wave functions on M_s.

At this point, the Kostant-Souriau procedure breaks down: such functions are necessarily singular, and, even allowing singularities, there are no square integrable wave functions.

Penrose's solution to this problem begins with the observation that if $h : T \to \mathbb{C}$ is holomorphic and homogeneous of degree $-2s\hbar^{-1}-2$ then the space-time field

$$(\varphi_{A'B'C'\ldots})_x = \oint \pi_{A'}\pi_{B'}\pi_{C'} \ldots h(-iX^{EE'}\pi_{E'},\pi_{E'})\pi^{F'}d\pi_{F'} \quad (41)$$

(where X^a is the position vector of x) satisfies the zero-rest-mass field equation (31). The integral here is around any closed one dimensional contour in the $\pi_{A'}$ space which avoids the singularities of h. Because of the homogeneity of h, the integrand is a closed 1-form, so that the integral depends only on x and the cohomology class of the contour (in the complement of the singularity set of h). Provided the contour varies continuously as x varies, the result will be a smooth, and, indeed, analytic, field in $\mathcal{M}$.

When s is negative, the fields have unprimed indices and are given by homogeneous holomorphic functions on the dual space T^*, so that, in place of (41) one has

$$(\varphi_{ABC\ldots})_x = \oint \pi_A\pi_B\pi_C \ldots g(\pi_E, iX^{EE'}\pi_E)\pi^F d\pi_F; \quad g : T^* \to \mathbb{C} \quad (42)$$

The idea is that, eventually, all the momentum space integrals of conventional quantum theory - such as inner products, charge integrals and Feynman diagrams - should be replaced by integrals over compact contours in twistor space (or in the product of several twistor spaces). These are necessarily finite, so there would be no need for renormalization.

For example, when $s = \frac{1}{2}\hbar$, the inner product of two fields $\varphi_{A'}$ and $\psi_{B'}$ can be given by an integral over an 8-dimensional contour in $T \times T^*$

$$\langle\varphi,\psi\rangle = (2\pi i)^{-3}\oint (Z^\alpha W_\alpha)^{-1}\, h(Z^\alpha)\, g(W_\alpha)\epsilon_{\alpha\beta\gamma\delta}\epsilon^{\pi\rho\sigma\tau}dZ^\alpha\wedge\cdots\wedge dZ^\delta\wedge dW_\pi\wedge\cdots\wedge dW_\tau \quad (43)$$

Here h generates $\varphi_{A'}$ and g generates $\overline{\psi}_B$ and the contour is fixed by the choice of contours for $\varphi_{A'}$ and $\overline{\psi}_B$. Similar expressions exist for other values of the helicity, but when $s \geq \hbar$, the contour is compact-with-boundary rather than closed.

Massive Particles

I shall end by making a few remarks about massive particles. Again, the kinematical variables can be represented by a position vector X^a and two future pointing null vectors $I^a \leftrightarrow o^A\bar{o}^{A'}$ and $J^a \leftrightarrow \iota^A\bar{\iota}^{A'}$, only now (5), (6) and (8) must be replaced by

$$p_a = 2^{-\frac{1}{2}} m(I_a + J_a) , \quad M^{ab} = -s\, \epsilon^{abcd} I_c J_d + X^a p^b - X^b p^a \tag{44}$$

$$\tilde{X}^a = X^a + r p^a , \quad \tilde{I}^a = I^a \text{ and } \tilde{J}^a = J^a , \text{ where } r \in \mathbb{R} \tag{45}$$

$$\sigma = s\, \epsilon_{abcd} I^a J^b dI^c \wedge dJ^d + m\, dI_a \wedge dX^a . \tag{46}$$

One way to quantize this system is to first form the 12-dimensional exact symplectic space (W, σ) in which X^a, p_a and $\chi^A = s^{\frac{1}{2}} (2m)^{-\frac{1}{2}} o^A$ are independent variables and in which

$$\sigma = d(p_a dX^a + i\, p_{AA'} (\chi^A d \bar{\chi}^{A'} - \bar{\chi}^{A'} d \chi^A)). \tag{47}$$

The phase space $(M_{m,s}, \sigma)$ of a particle with rest mass m and spin $s > 0$ is recovered from (W, σ) by taking the 10-surface $G_{m,s} \subset W$ on which $p_a p^a = m^2$ and $p_{AA'} \chi^A \bar{\chi}^{A'} = s$ and factoring out the two commuting Hamiltonian vector fields generated by $p_a p^a$ and $p_{AA'} \chi^A \bar{\chi}^{A'}$. As before, when $2s$ is an integral multiple of $\hbar$, the prequantization line bundle can be expressed as a factor space of $G_{m,s} \times \mathbb{C}$.

In this notation, the polarization $\tilde{F}$ used by Renouard [9] is induced from the polarization of (W, σ) spanned by the vector fields $\partial/\partial \bar{\chi}^{A'}$ and $\partial/\partial X^a$. The factor space $M_{m,s}/\tilde{F}$ is $N_{m,+} \times S^2$ where $N_{m,+}$ is the future m-mass shell in momentum space and the $\tilde{F}$-wave functions are of the form

$$\psi : (X^a, p_a, \chi^A) \mapsto \left[\varphi_{ABC\ldots} \chi^A \chi^B \chi^C \ldots e^{i\hbar^{-1} p_a X^a} \right] (\nu \wedge \chi^A d\chi_A)^{\frac{1}{2}} \tag{48}$$

where ν is the invariant volume element on $N_{m,+}$ and $\varphi_{ABC\ldots}$ depends only on p_a. The conventional quantum mechanical description is recovered by introducing the space-time field

$$\tilde{\varphi}_{ABC\ldots} = \int_{N_{m,+}} \left[\varphi_{ABC\ldots} e^{i\hbar^{-1} p_a X^a} \right] \nu \tag{49}$$

Alternatively, one can recover $(M_{m,s}, \sigma)$ from the product $T \times T = \{(Z^\alpha, Y^\alpha)\}$ of two twistor spaces with the exact symplectic form

$$\sigma = i (dZ^\alpha \wedge d\bar{Z}_\alpha + dY^\alpha \wedge d\bar{Y}_\alpha) \tag{50}$$

To be precise, $(M_{m,s}, \sigma)$ is the symplectic manifold obtained from the

11-surface in $T \times T$ given by

$$Z^{\alpha}\bar{Z}_{\alpha} = s, \quad Y^{\alpha}\bar{Y}_{\alpha} = -s, \quad Z^{\alpha}\bar{Y}_{\alpha} = 0 \quad \text{and} \quad M = \tfrac{1}{2}m^2 \tag{52}$$

where

$$M = (I^{\alpha\beta}\bar{Z}_{\alpha}\bar{Y}_{\beta})(I_{\alpha\beta}Z^{\alpha}Y^{\beta}) \quad \text{and} \quad I^{\alpha\beta} = \begin{bmatrix} \epsilon^{AB} & 0 \\ 0 & 0 \end{bmatrix} \tag{53}$$

by factoring out the commuting Hamiltonian vector fields generated by $Z^{\alpha}\bar{Z}_{\alpha}$, $Y^{\alpha}\bar{Y}_{\alpha}$ and M (in Dirac's terminology, $Z^{\alpha}\bar{Y}_{\alpha} = 0$ is a first class constraint, that is, σ restricts to a nondegenerate 2-form on the 14-surface $Z^{\alpha}\bar{Y}_{\alpha} = 0$). Explicitly,

$$Z^{\alpha} = e^{ir}(\tfrac{1}{2}s\lambda^{-1}\iota^{A} + ix^{AA'}(\lambda\bar{o}_{A'}), \lambda\bar{o}_{A'}) \tag{54}$$

$$Y^{\alpha} = e^{it}(\tfrac{1}{2}s\lambda^{-1}o^{A} + ix^{AA'}(\lambda\bar{\iota}_{A'}), \lambda\bar{\iota}_{A'}) \tag{55}$$

where $\lambda = m^{\frac{1}{2}}2^{-\frac{1}{4}}$ and $r, t \in \mathbb{R}$.

This representation of $(M_{m,s}, \sigma)$ is not unique: the choice made for the right hand sides in (52) can be changed, subject to the constraint that the momentum and angular momentum defined by Z^{α} and Y^{α} should add to give the momentum and angular momentum of the massive particle. The transformations of Z^{α} and Y^{α} which preserve this constraint form a classical 'internal symmetry' group isomorphic with $SU(2) \times \tilde{E}(2)$ [4, 7] ($\tilde{E}(2)$ is the double covering of the Euclidean group $E(2)$).

This time, the natural choice for the polarization in $T \times T$ is the Kähler polarization spanned by the vector fields $\partial/\partial\bar{Z}_{\alpha}$ and $\partial/\partial\bar{Y}_{\alpha}$. This choice results in a representation of quantized massive particles by holomorphic functions on $T \times T$ of fixed homogeneities in their two arguments. Again, the space-time fields are obtained by contour integration.

Acknowledgments

Most of the ideas presented here were formed during the dialogue which has been taking place in Oxford between the twistor theorists and those working on geometric quantization, and it would be futile to attempt to trace specific ideas to particular individuals. However, I should mention the names of Keith Hannabuss and Alan Carey (who will be publishing a paper on this subject shortly), George Sparling, Paul Tod and, of course, Roger Penrose himself.

The correspondence between the symplectic structure of twistor space and that of the phase space of a zero spin zero mass particle was first observed by Crampin and Pirani [2].

I acknowledge with thanks the support of the SRC.

References

1) R.Blattner: in: Proceedings of Symposia in Pure Mathematics, Vol XXV (Amer. Math. Soc., Providence, 1974).
2) M.Crampin and F.A.E.Pirani: in: Relativity and Gravitation: eds: Ch.G. Kuper and A.Peres (Gordan and Breach, London, 1971).
3) B.Kostant: in: Lectures in Modern Analysis: ed: C.T.Taam: Lecture Notes in Mathematics, 170 (Springer, Heidelberg, 1970).
4) R.Penrose: in: Quantum Gravity: eds: C.Isham, R.Penrose and D.Sciama (Clarendon Press, Oxford, 1975).
5) R.Penrose: in: Group Theory in Non-Linear Problems: NATO Advanced Study Institute, series C: ed A.O.Barut (Reidel, 1971).
6) R.Penrose: in: Battelle Rencontres, 1967: eds: C.M.DeWitt and J.A.Wheeler (Benjamin, New York, 1968).
7) Z.Perjes: Twistor Variables in Relativistic Mechanics (preprint, Budapest, 1974).
8) F.A.E.Pirani: in: Lectures on General Relativity, Brandeis Summer Institute, Vol I, 1964: eds: S.Deser and K.W.Ford (Prentice Hall, Englewood Cliffs N.J., 1965).
9) P.Renouard: Thesis (Paris, 1969)
10) J-M.Souriau: Structures des Systemes Dynamiques (Dunod, Paris, 1970).
11) D.J.Simms and N.M.J.Woodhouse: Lectures on Geometric Quantization (to be published).

Footnote: After this talk, Professor Kostant suggested that, in the conformally invariant case, the Kostant-Souriau theory could be saved by constructing the quantum Hilbert space from certain cohomology groups associated with the prequantization line bundle and the antiholomorphic polarization. Unfortunately, this does not work since the wave functions would then be represented by products of $\frac{1}{2}$-forms with holomorphic forms on T satisfying $\mathcal{L}_{X_g}\beta = -2is\hbar^{-1}\beta$, where $\beta = \beta_{\alpha\beta\ldots}dZ^\alpha \wedge dZ^\beta\ldots$, and the singularities would still be present.

QUANTISATION AS DEFORMATION THEORY,

F.J. BLOORE AND M. ASSIMAKOPOULOS,
Department of Applied Mathematics
and Theoretical Physics
the University Liverpool, England

1. Notation

We shall use the same notation and concepts as in our talk [1] of last year to this colloquium namely:

M = Riemannian configuration manifold, metric tensor g, local coordinates $q^1 \dots q^n$.

$T^{(s)}M$ = space of real fully symmetric contravariant tensor fields S on M, with valence v(S) = s.

$C_s(S) = S^{i_1,\dots i_s}(q)p_{i_1}\dots p_{i_s}$ = homogeneous function on phase space T^*M associated with S.

$[S,T] \in T^{(s+t-1)}M$ = Schouten concomitant, related to Poisson bracket by $\{C_s(S), C_t(T)\} = - C_{s+t-1}([S,T])$.

$A = \bigoplus_{s=0}^{\infty} T^{(s)}M$ = Graded Lie algebra of sequences of symmetric tensors, with Schouten concomitant as Lie product.

H = Hilbert space of wave functions Ψ on M.

2. Quantisation and deformation theory

Any quantisation scheme associates each $C_s(S)$ with a Hermitian linear operator $Q_s(S)$ on H. For example we could take,

$$\left.\begin{array}{ll} \varphi \in T^{(0)}M, & (Q_0(\varphi)\Psi)(m) = \varphi(m)\Psi(m), \quad m \in M \\ X \in T^{(1)}M, & (Q_1(X)\Psi)(m) = \tfrac{1}{2}(-iX^i\nabla_i + \text{conjugate})\Psi(m) \\ U \in T^{(2)}M, & (Q_2(U)\Psi)(m) = \tfrac{1}{2}(-U^{i_1 i_2}\nabla_{i_1}\nabla_{i_2} + \text{conjugate} - U^{i_1 i_2}{}_{;i_1 i_2})\Psi(m) \\ S \in T^{(s)}M, & (Q_s(S)\Psi)(m) = \tfrac{1}{2}((-i)^s S^{i_1 \dots i_s}\nabla_{i_1}\dots\nabla_{i_s} + \text{conjugate})\Psi(m), \ s>2 \end{array}\right\} \quad (1)$$

where ∇_i is the covariant derivative for the Riemannian connection on M. All quantisation schemes agree with this one to leading order. With the scheme above, we find that the commutator

$$[Q_s(S),Q_t(T)] = -i(Q_{s+t-1}([S,T]) + Q_{s+t-3}(F_1(S,T)) + Q_{s+t-5}(F_2(S,T)) + \dots) \quad (2)$$

The new Lie product

$$-C([S,T]') = \{C_s(S), C_t(T)\}' = -C_{s+t-1}([S,T]) - C_{s+t-3}(F_1(S,T)) - \cdots$$

is a deformation [2] of the original Poisson bracket or Schouten concomitant. All quantisation schemes furnish such deformations.

The map $F_1 : T^{(s)}M \times T^{(t)}M \to T^{(s+t-3)}M$ is a cocycle of order 2 in the Lie algebra cohomology of A . If

$$\eta : T^{(s)}M \to T^{(s-2)}M$$

is a cochain of order 1 in A , and we alter some given quantisation scheme Q in (2) to another one, Q' , related to Q by

$$Q'_s(S) = Q_s(S) + Q_{s-2}(\eta(S))$$

then the new commutation relations are

$$[Q'_s(S), Q'_t(T)] = -i\left(Q'_{s+t-1}([S,T]) + Q'_{s+t-3}((F_1 - d\eta)(S,T)) + \cdots\right).$$

It is well know that F_1 is not exact, for any configuration manifold M ; there is no quantisation scheme in which all commutators are the quantisations of the corresponding Poisson brackets. So the problem in this formulation is which scheme to choose. Different schemes have different equations of motion. We shall scrutinize these to see if there is a "best" scheme.

3. Time development

In classical mechanics, for a system with Hamiltonian $\frac{1}{2}g^{ij}p_i p_j \equiv C_2(\frac{1}{2}g^{-1})$ we have

$$\frac{d}{dt} C_s(S) = \{C_s(S), C_2(\tfrac{1}{2}g^{-1})\} = C_{s+1}([\tfrac{1}{2}g^{-1}, S]).$$

In quantum mechanics, the eq. (2) is an equal time commutation relation for operators $Q_s(S)$ whose time development is given by

$$\frac{d}{dt} Q_s(S) = i\left[Q_2\left(\tfrac{1}{2}g^{-1}\right), Q_s(S)\right].$$

Tensors K which satisfy $[\tfrac{1}{2}g^{-1}, K] = 0$ are called Killing tensors, [3]; for such a K, $C(K)$ is a constant of the classical motion. It is natural to hope that $Q(K)$ will be a corresponding constant of the quantal motion, and that such a result is true in only one quantisation scheme. We present the results of a study of this question.

4. Results

No general result in this direction is known. We report here some results for special cases.

J. Underhill and S. Taraviras (communication to this Colloquium) have found that for manifolds M which are Riemannian spaces of constant curvature, there is a quantisation scheme Q in which $[Q(g^{-1}), Q(K)] = 0$ whenever $[g^{-1}, K] = 0$. Thus for such manifolds F_1 is exact when restricted to the arguments (g^{-1}, K). In our own work we examine whether a scheme Q' exists in which one has a stronger condition, namely that for all S,

$$[Q'(g^{-1}), Q'(S)] = -i\,Q'([g^{-1}, S]),$$

or equivalently whether F_1 is exact when restricted to the arguments (g^{-1}, S). Because it is difficult to do better, we restrict our consideration to tensors U of valence 2 and to potential functions $\eta(U)$ of the form

$$\eta(U) = \eta_{ab}U^{ab} + \eta_{abc}U^{ab;c} + \eta_{abcd}U^{ab;cd}. \qquad (3)$$

For the sake of uniqueness we take

$$\eta_{ab} = \eta_{ba} \qquad (4)$$

$$\eta_{abc} = \eta_{bac} \qquad (5)$$

$$\eta_{abcd} = \eta_{bacd} = \eta_{abdc} \qquad (6)$$

The equations

$$F_1(g^{-1}, X) = d\eta(g^{-1}, X) \tag{7a}$$

$$F_1(g^{-1}, U) = d\eta(g^{-1}, U) \tag{7b}$$

yield several conditions on the tensors $\eta_{ab}, \eta_{abc}, \eta_{abcd}$ namely, (11)-(18),(21)-(24). We have not succeeded in classifying the general solutions to these equations, for dim $M > 3$. If dim $M = 2$ there is a unique solution for $\eta(U)$. If dim $M = 3$, there is no solution unless M has constant curvature. For any space of constant curvature, there is a unique solution for $\eta(U)$. This does not really extend the result of Underhill and Taraviras since they deal with all valences of U simultaneously whereas we restrict consideration to $v(U) = 2$. The uniqueness of a solution η to (7) is unusual in cohomology theory. One can usually add a coboundary, $\eta \to \eta + d\theta$ at will. However here the 1-cochain η must map $T^{(s)}M$ into $T^{(s-2)}M$ and no exact 1-cochain has this property.

5. Calculations

With the scheme Q given by eq. (1), it is straight forward to compute, for $v(X) = 1$, $v(U) = 2$,

$$F_1(g^{-1}, X) = -\tfrac{1}{4}\,\mathrm{div}[g^{-1}, \mathrm{div}\, X] = -\tfrac{1}{2}\Delta\, \mathrm{div}\, X \tag{8}$$

$$F_1(g^{-1}, U) = V([g^{-1}, U]) + B(g^{-1}, U). \tag{9}$$

Here V is a vector field depending only on $[g^{-1}, U]$ and so may be eliminated by a redefinition of Q_3. We will ignore it. The vector field B is given by [4],

$$B^j(g^{-1}, U) = \tfrac{2}{3}\left(R^a{}_b\, U^{bj} - U^a{}_b\, R^{bj}\right)_{;a}.$$

This is not a function of $[g^{-1}, \mathcal{U}]$ and must be accounted for by a suitable redefinition of Q_2, by means of η.

In components the equation

$$F_1(g^{-1}, X) = d\eta(g^{-1}, X) = -[X, \eta(g^{-1})] - \eta([g^{-1}, X])$$

(note that $\eta(X) = 0$) becomes, with the form (3) for η,

$$\tfrac{1}{2} X^a{}_{;ab}{}^b = X^a \eta^b{}_{b;a} + 2\eta_{ab} X^{a;b} + 2\eta_{abc} X^{a;bc} + 2\eta_{abcd} X^{a;bcd} \tag{10}$$

for all vector fields X. We may equate the coefficients of $X^{a;(bcd)}$ (the bracketed indices are symmetrised), to get

$$\eta_{a(bcd)} = \tfrac{1}{4} g_{a(b} g_{cd)} . \tag{11}$$

We define f_{abcd} by

$$\eta_{abcd} = \tfrac{1}{4} g_{ab} g_{cd} + f_{abcd} \tag{12}$$

so that

$$f_{abcd} = f_{bacd} = f_{abdc} \quad , \quad f_{a(bcd)} = 0 \tag{13}$$

from which one may deduce

$$f_{abcd} = f_{cdab} \quad , \quad f_{(abc)d} = 0 . \tag{14}$$

Thus f_{abcd} has as many independent components as the curvature tensor R_{abcd}. When we equate the coefficients of $X^{a;(bc)}$ in (10) we obtain

$$\eta_{a(bc)} = 0 . \tag{15}$$

Eqs. (5) and (15) imply

$$\eta_{abc} = 0 . \tag{16}$$

Substitution of (11), (12) and (16) in (10) gives

$$X^a \eta^b{}_{b;a} + 2\eta_{ab} X^{a;b} + \tfrac{2}{3} f_{abcd} \left[2(R^{cb}{}_e{}^a X^e)^{;d} + R^{db}{}_e{}^a X^{e;c} + R^{db}{}_e{}^c X^{a;e} \right] = 0 .$$

Equating the coefficients of $X^{a;b}$ and X^{a} yields, after some work

$$f_{abc\alpha} R_{\beta}{}^{abc} = f_{abc\beta} R_{\alpha}{}^{abc} = \frac{3}{2} \gamma_{\alpha\beta} \tag{17}$$

$$f_{abcd;\alpha} R^{dabc} = 0 . \tag{18}$$

One solution to equations (13,14,17,18) is any constant multiple of

$$\overset{\circ}{f}_{abcd} \equiv -\frac{1}{8}\left(2 g_{ab} g_{cd} - g_{ac} g_{bd} - g_{ad} g_{bc}\right) . \tag{19}$$

The analysis of the equation

$$F_1(g^{-1}, U) = d\gamma(g^{-1}, U) \tag{20}$$

using eq. (9) is similar but much more complicated. It involves derivatives of U up to third order, and gives, after some work the following conditions

$$f_{abcd;e} = 0 , \tag{21}$$

$$\left(\gamma_{ab} + \frac{1}{4} R_{ab}\right)_{;c} = 0 , \tag{22}$$

$$-\frac{1}{8} R_{jmnp} = R_{j}{}^{b}\left(h_{bnpm} - h_{bpnm}\right) + R_{j}{}^{ab}{}_{p} h_{abmn} - R_{j}{}^{ab}{}_{n} h_{abmp} \tag{23}$$

where

$$f_{abcd} = \overset{\circ}{f}_{abcd} + h_{abcd} . \tag{24}$$

The eq. (18) now follows from (21).

The tensor field f_{abcd} looks to be grossly overdetermined. We conclude with a short analysis of these equations for the case when M is indecomposable. One may then use Eisenhart's theorem [5] to deduce from (22) that

$$\gamma_{ab} = -\frac{1}{4}\left(R_{ab} + k g_{ab}\right)$$

where k is some constant. The equations (17) and (24) then give

$$h_{abc\alpha} R_{\beta}{}^{abc} = -\frac{3}{8} k g_{\alpha\beta} . \tag{25}$$

Further, since

$$h^{c}{}_{cab;e} = 0$$

by (21) and (24), we may write

$$h^{c}{}_{cab} = \mu g_{ab} \tag{26}$$

where μ is some constant. The (mn) -trace of eq. (23) then gives

$$R_{jp}(1-4\mu) + 3k g_{jp} = 0 .$$

Thus if M is not an Einstein space, then $k = 0$ and $\mu = \frac{1}{4}$. We have not succeeded in analysing the equations further in their generality.

If we insist that

$$h_{abcd} = \lambda \overset{\circ}{f}_{abcd}$$

should satisfy eq. (23), then we find from eq. (23) that, for $\dim M > 2$,

$$R_{j(mn)p} = \overset{\circ}{R}_{j(mn)p}$$

where $\overset{\circ}{R}$ is the curvature tensor for a space of constant curvature. This implies [6] that M is a space of constant curvature. Thus (20) holds for any space of constant curvature, for $U \in T^{(2)}M$.

If $\dim M = 2$, then solving the equations explicitly gives, uniquely,

$$f_{abcd} = \frac{3}{4} \overset{\circ}{f}_{abcd} .$$

Now let M be an indecomposable 3-manifold. All three-dimensional Einstein spaces have constant curvature. Suppose M is not an

Einstein space. Then eq. (26), in the form

$$g^{ab} h_{abcd} = \frac{1}{4} g_{cd}$$

provides 6 equations for the 6 independent components of h. The determinant of the coefficients turns out to be $(\det g)^{-1}$ so the solution is unique, and by inspection is

$$h_{abcd} = -\frac{1}{2} \overset{0}{f}_{abcd} .$$

So M has to be of constant curvature after all!

References

[1] F.J. Bloore, Proc. 3rd int. coll. on group theo. methods in phys. p.104.

[2] M. Lévy-Nahas, J. Math. Phys. 8 (1967) 1211.

[3] P. Sommers, J. Math. Phys. 14 (1973) 787.

[4] F.J. Bloore, Report to Colloque CNRS: Geometrie Symplectique et Physique Mathematique, Aix en Provence 1974, p82

[5] L.P. Eisenhart, Trans. Am. Math. Soc. 25 (1923) 297.

[6] Kobayashi+Nomizu "Foundations of differential geometry" (Wiley, 1963) Vol. 1, Chap. V.

RELATIVISTIC CANONICAL SYSTEMS: A GEOMETRIC APPROACH TO THEIR SPACE-TIME STRUCTURE AND SYMMETRIES.

K. Drühl, Max-Planck-Institut
D-813 Starnberg, Postfach 1530

1. The general theory of classical canonical systems.

A broad class of classical physical systems may be described by using the theory of finite dimensional canonical manifolds. In this approach the variables of a given system are represented by differentiable functions on some underlying manifold M. These functions form a commutative R - algebra $\mathfrak{A}$ under pointwise algebraic operations, and a Lie algebra under the Poisson bracket:

1. i) $\{A, B\} = \Lambda(dA, dB)$

ii) $= \mathfrak{Z}_{dA}(B)$ in $\mathfrak{A}$ for A, B in $\mathfrak{A}$.

The Poisson bracket is antisymmetric, satisfies the Jacobi identy and gives rise to an antisymmetric tensorfield Λ, which we assume to be nondegenerate. We note that both the differentiable structure and the canonical structure on M are completely defined by the algebra $\mathfrak{A}$ and its bracket 1) [1,2]. In analyzing these systems we may therefore avoid any direct reference to the underlying manifold M, and use the algebra of variables only. This approach is of certain physical interest, since the algebra $\mathfrak{A}$ and its bracket structure may be expected to have most direct analogues in corresponding quantum theories. (see for example the contributions on geometric quantization in this volume). Furthermore problems of covariance under change of coordinates cannot arise, and - surprisingly enough - many calculations become much simpler than they would be by using local charts.

Vectorfields (i.e. derivations on $\mathfrak{A}$) like ξ_{dA} as defined by 1.ii) are called canonical. They leave the tensorfield Λ invariant and satisfy the commutation relations:

$$2. \qquad [\xi_{dA}, \xi_{dB}] = \xi_{d\{A,B\}} .$$

The differential forms dA,dB may in general be replaced by any two closed forms.

A general framework like this does not yet contain any reference to space-time concepts. We may however introduce such concepts by postulating certain structural properties of the algebra of variables $\mathfrak{A}$; again without having to use the underlying manifold explicitly. In this contribution we shall only state the main assumptions and results; details may be found in reference [2] .

We shall illustrate the basic ideas of this approach by a simple example. Take for M the real plane with global coordinates Q, P. Let the bracket be given by the following relations, and specify the dynamical law by a certain vectorfield $Y = \xi_{dH}$ with Hamiltonian H:

$$3. \quad \text{i)} \quad \{P, Q\} = -\{Q, P\} = 1$$

$$\text{ii)} \quad \{Q, Q\} = \{P, P\} = 0$$

$$\text{iii)} \quad H = \frac{1}{2}(P^2 + Q^2) \quad ; \quad Y = P\frac{\partial}{\partial Q} - Q\frac{\partial}{\partial P}$$

We may consider the set of submanifolds Q = const as a configuration manifold. The corresponding variables depend on Q only and from a subalgebra $\mathfrak{A}_o$:

$$4. \quad \text{i)} \quad A = a(Q) \quad \text{in } \mathfrak{A}_o , \quad a \text{ differentiable;}$$

$$\text{ii)} \quad Y(A) = P \cdot \frac{\partial a}{\partial q}(Q) .$$

Variables linear in P as in 4.ii) may then be interpreted as momenta, since they map $\mathfrak{A}_o$ to itself under the bracket:

$$\text{iii)} \qquad B = b(Q).P, \quad A = a(Q) \ : \ \{B, A\} = b(Q)\,\frac{\partial a}{\partial q}\,(Q) \ .$$

The basic structural features of this example then are the existence of a certain subalgebra $\mathfrak{A}_o$, satisfying appropriate bracket relations as in 4.ii), iii) . Let us formalize these features in the following set of axioms:

<u>Kinematical Axioms:</u> Let $(M, \{..,..\}, Y)$ be a canonical system, where $\{..,..\}$ is the Poisson bracket on the algebra $\mathfrak{A}$ of variables over M , and Y is a canonical vectorfield. There exists a subalgebra $\mathfrak{A}_o$ of $\mathfrak{A}$ such that:

K.i. $\mathfrak{A}_o$ is a maximal commutative Lie subalgebra under $\{..,..\}$;

K.ii. $\mathfrak{A}_o \cap Y(\mathfrak{A}_o) = (0)$, $Y(\mathfrak{A}_o) = (Y(A)$, A in $\mathfrak{A}_o)$;

K.iii. $\mathfrak{A}_o \cup Y(\mathfrak{A}_o)$ defines the differentiable structure on M ;

K. iv. $\{Y(\mathfrak{A}_o), \mathfrak{A}_o\} \subset \mathfrak{A}_o$, i.e. $\{Y(A), B\}$ is in $\mathfrak{A}_o$ for A, B in $\mathfrak{A}_o$;

K. v. for any A in $\mathfrak{A}_o$ the vectorfield ξ_{dA} generates a group.

The example we have given above is easily seen to satisfy these axioms. We observe that as a consequence of K.i) and 2) the functions in $\mathfrak{A}_o$ define a commutative Lie algebra $\mathcal{L}_o$ of canonical vectorfields on M . The starting point of our analysis then is to study the set of maximal integral submanifolds of $\mathcal{L}_o$, and the differentialbe structure on any such submanifold. Furthermore one may prove that all axioms remain valid if we replace the space of functions $Y(\mathfrak{A}_o)$ by the $\mathfrak{A}_o$ - module $\mathfrak{A}_1$ it generates (containing all functions which are locally finite linear combinations of elements from $Y(\mathfrak{A}_o)$ with coefficients in $\mathfrak{A}_o$). Using mainly the Frobenius integrability theorems [4, Chap. 1.8] and the properties of functions in $\mathfrak{A}_o$ and $\mathfrak{A}_1$ as stated in K.ii. - K.v. we obtain the following results [2] :

Theorem 1.

i) Through any point p in M there passes a unique maximal integral submanifold p^* of $\mathcal{L}_o$ of dimension n, where $2n = \dim M$. The corresponding subspace of the tangentspace at p is maximal isotropic with respect to the bilinear form defined by Λ. On any such p^* functions from $\mathfrak{A}_1$ define a linear n-dimensional space of functions which separates the points and determines the differentiable structure. Each such function ranges over all of R, hence p^* is diffeomorphic to the vectorspace R^n.

ii) Let $\mathfrak{V}$ be a basis of open neighbourhoods for M. For any V in $\mathfrak{V}$ denote by V^* the set of integral manifolds passing through V: $V^* = (p^*; p \text{ in } V)$. Then $\mathfrak{V}^* = (V^*; V \text{ in } \mathfrak{V})$ satisfies all axioms for a basis of neighbourhoods on the set M^* of maximal integral manifolds. The resulting topological space is a connected Hausdorff space with a countable basis. Any function A in $\mathfrak{A}_o$ defines a unique continuous function A^* on M^*.

iii) The algebra $\mathfrak{A}_o^* = (A^*; A \text{ in } \mathfrak{A}_o)$ defines a differentiable structure on M^* and is equal to the algebra of all differentiable functions on the resulting manifold N. To any vectorfield X on N there corresponds a unique function P_X in $\mathfrak{A}_1$ and vice versa, such that:

$$\{P_X, A\}^* = X(A^*), \quad A \text{ in } \mathfrak{A}_o .$$

Under the identification of $\mathfrak{A}_o$ and $\mathfrak{A}_o^*$ the map $P : X \longrightarrow P_X$ is an $\mathfrak{A}_o$ - linear isomorphism. There exists a unique diffeomorphism φ mapping the cotangent bundle $L^*(N)$ over the quotient manifold N onto the manifold M such that:

$A \circ \varphi = A^o$ is constant on the cotangent spaces,

$P_X \circ \varphi = P_X^o$ is the standard function associated to X in $L^*(N) = P_X^o(q, \omega_q) = \omega_q(X_q)$.

If the axiom K.v. is given up, similar though weaker results may be obtained. In particular the integral submanifolds p^* may be diffeomorphic to any covering manifold of some open submanifold of R^n. The topological space obtained in Theorem 1.ii. will not in general be Hausdorff, however the pair (M^*, $\mathfrak{A}_o^*$) will still enjoy all other properties of a manifold [3]. Leaving aside these generalizations we may henceforth restrict our attention to systems of the form ($L^*(N)$, $\{..,..\}$, Y).

A complete classification of canonical systems satisfying the kinematical axioms is then given by the following theorem [2] .

<u>Theorem 2.</u> Let ($L^*(N)$, $\{..,..\}$, Y) satisfy the kinematical axioms with subalgebra $\mathfrak{A}_o$ constant on the cotangent spaces.

i) There exists a closed differential 2-form ω_F on N such that the Poisson bracket is given by:

α) $\{A, B\} = 0$

β) $\{P_X, A\}^* = X(A^*)$

γ) $\{P_X, P_{X_1}\} = P_{[X, X_1]} + F(X, X_1)$

δ) $F(X, X_1)^* = \omega_F(X, X_1)$.

where A^*, B^* resp. X, X_1 are variables resp. vectorfields on N .

ii) There exists a nowhere degenerate symmetric tensorfield g and a closed differential 1-form η on N such that:

α) $g(dA^*, dB^*) = \{Y(A), B\}^*$

β) $Y = \zeta_{dT} + Y^o$, where :

γ) $2T(q, \omega_q) = g(\omega_q, \omega_q)$ is the function on $L^*(N)$ corresponding to g, and:

δ) $Y^o(A) = 0$, $Y^o(P_X)^* = \eta(X)$.

Conversely let a tensorfield g and differential forms ω_F, η on N be given with these properties. Then ($L^*(N)$, $\{\ldots,\ldots\}$, Y) with bracket defined by i. α - δ) and vectorfield Y defined by ii. β - δ) satisfies the kinematical axioms.

Let us illustrate Theorem 2 by choosing a suitable chart (U; Q^i) on N and the associated chart (V; Q^i, P_k) on $L^*(N)$. In this chart we obtain:

5. i) α) $\{Q^i, Q^k\} = 0$

β) $\{P_k, Q^i\} = 1$ for $i = k$ and vanishes otherwise

γ) $\{P_k, P_i\} = f_{ik}(Q)$

δ) $\frac{\partial}{\partial q^l} f_{jk}(Q) + \text{cycl} = 0$.

ii) $$Y = \zeta_{dH}, \quad H = \frac{1}{2} \sum_{ik} g^{ik}(Q)\, P_i\, P_k + w(Q),$$

where: $$\frac{\partial w}{\partial q^i} = \eta_i, \text{ and } \eta = \sum_i \eta_i\, dQ^i .$$

A physical interpretation is obtained if we assume the points in $L^*(N)$ to describe individual states of the system, moving along the integralcurves of Y in the course of dynamical evolution. We leave it as an easy exercise to write down the differential equations for these curves in the given chart. This"phase space interpretation" contains in particular all systems of non-relativistic particles, moving on quasi-geodesic lines in the configuration space N of arbitrary Riemannian structure. Their interactions and external fields are then described by the "scalar"- and "tensor"- potentials η and ω_F.

However a different interpretation is obtained if we consider relativistic systems. These may be singled out by an additional set of axioms, which postulate essentially the existence of time-like variables and a certain uniqueness condition for them.

The tensorfield g then is of signature $(1, n-1)$, and we must interpret the quotient N as the space-time of general relativity. The classical fields, namely gravitation and electromagnetism, are described by g and ω_F, and we must consider $\mathfrak{A}$ as a classical algebra of geometrical observables. For details and the notion of states appropriate for such an interpretation we refer to reference [2].

2. Relativistic canonical systems and their symmetries.

Let us consider the case of relativistic canonical systems as sketched above. In this case the form η vanishes, and Y is globally Hamiltonian. The structure of these systems may then be completely classified in terms of the space-time manifold N, the gravitational field tensor g and the electromagnetic differential form ω_F. What is the notion of symmetries appropriate for such systems?

Let φ be a canonical transformation on $L^*(N)$ leaving Y invariant:

6. $$Y^{\phi} = \phi \circ Y \circ \phi^{-1} = Y \; ; \quad \phi(F) = F \circ \varphi \quad \text{for } F \text{ in } \mathfrak{U} .$$

It is not difficult to check that for any such transformation the system will again satisfy all kinematical axioms with subalgebra $\phi(\mathfrak{U}_0)$ replacing the original $\mathfrak{U}_0$. However $\phi(\mathfrak{U}_0)$ will in general be different from $\mathfrak{U}_0$. We call ϕ a symmetry of the system, if it permutes $\mathfrak{U}_0$:

7. $$\phi : \mathfrak{U}_0 \longrightarrow \mathfrak{U}_0 \; ; \qquad Y^{\phi} = Y .$$

Since φ is canonical (i.e. leaves the bracket invariant) it follows that φ permutes the integral submanifolds of $\mathcal{L}_0$, and hence induces a diffeomorphism φ^0 on N.

Theorem 3. Let ϕ be a symmetry of the relativistic system $(L^*(N), \{\ldots,\ldots\}, Y)$.

Then:

i) $\phi^0 [g] = g$

ii) $\phi^0 [\omega_F] = \omega_F$,

where g and ω_F are defined as in Theorem 2, and ϕ^o corresponds to the induced diffeomorphism φ^o on N. The symbols in i) and ii) denote the corresponding transformed quantities.
Conversely let φ^o be a diffeomorphism on N satisfying i) and ii) Then there exists a unique symmetry ϕ of the corresponding system defined by:

iii) $\phi(A)^* = \phi^o(A^*)$, $\phi(P_X) = P_{X\phi_o}$.

We realize then that for space-times admitting no isometries there exist no symmetries of the corresponding system.

On the other hand for systems admitting large groups of symmetries it is often possible to select from $\mathfrak{A}$ a finite-dimensional Lie algebra of functions which contains the Hamiltonian and generates all of $\mathfrak{A}$ (in the sense of differential geometry). Let us take as an example the system defined by N equal to Minkowski space with metric g and vanishing ω_F. The Poincaré group $\mathfrak{P}$ acts as a group of isometries on N, and hence as a group of symmetries on the system. Choosing global coordinates Q^i on N, we may write down the vectorfields on N corresponding to the infinitesimal generators of $\mathfrak{P}$. These define functions in $\mathfrak{A}_1$:

8. i) $$P_i \ , \quad \hat{T} = \sum_{i,k} t^i{}_k \, P_i \, Q^k \ ,$$

where $T = (t^i{}_k)$ is a skew-symmetric matrix with respect to g. From Theorems 2. and 3. we then easily obtain the following bracket relations:

9. i) $$\{Q^i, Q^k\} = 0$$

ii) $$\{P_i, Q^k\} = \delta^k_i \qquad \text{(Kronecker symbol)}$$

iii) $$\{\hat{T}, Q^i\} = \sum_k t^i{}_k \, Q^k$$

$$\text{iv)} \quad \{P_i, P_k\} = 0$$

$$\text{v)} \quad \{P_k, \hat{T}\} = \sum_i t_k^{\ i} P_i$$

$$\text{vi)} \quad \{\hat{T}, \hat{T}_1\} = \widehat{[T, T_1]}$$

$$\text{vii)} \quad \{H, Q^i\} = \sum_k g^{ik} P_k$$

$$\text{viii)} \quad \{H, P^i\} = \{H, \hat{T}\} = 0$$

In this way we obtain a 16-dimensional Lie algebra of functions (including the constant function) on $L^*(N)$, which is sufficient to define all bracket relations required in the kinematical axioms. We wish to emphasize however that for general space-times no such choice of a unique finite-dimensional subspace of functions is possible, and one has to use the general formulation given above.

References:

1. Jost, R.: Poisson brackets (An unpedagogical lecture), Rev. Mod. Phys. 36, 572-579 (1964)

2. Drühl, K.: On the space-time interpretation of classical canonical systems, preprint Max-Planck-Institut Starnberg, May 1975.

3. Drühl, K.: Relativistic canonical systems, preprint Max-Planck-Institut Starnberg, February 1975

4. Hermann, R.: Differential geometry and the calculus of variations, New York, London: Academic Press 1968

Propagators in Quantum Mechanics on Multiply Connected Spaces

Norman E. Hurt

Introduction.

In several recent papers [4], [5], [6], [11], [15] quantum mechanics and quantum mechanical propagators on multiply connected spaces have been discussed. The quantum mechanics of a free particle moving on the Riemannian manifold M = G, a simple Lie group, was examined by J. S. Dowker in [5]. In this paper he showed that the quasi-classical approximation is "exact" and that the propagator may be calculated either by "the sum over classical paths" or by the stationary state method. In a later paper [6] Dowker suggested a natural framework (following [11]) for quantum mechanical propagators on multiply connected, homogeneous spaces.

In this paper we indicate the interconnections of Dowker's formulation and the yoga of the Selberg trace formula. The Selberg trace formula plays a very fundamental role in a broad cross section of mathematics. Below we demonstrate its vitality in modern physics.

§1. Quantum Mechanical Propagators.

Various compact and noncompact homogeneous manifolds occur in the world of mathematical physics: e.g., the spherical top, the particle with spin, the classical energy level of the Kepler problem, etc. This leads naturally to the consideration of homogeneous manifolds of the form $M = G/H$, where initially G is a separable, locally compact unimodular group with compact isotropy subgroup $H = \{g \in G \mid gm_o = m_o\}$. Let $C_c(G)$ denote the associative algebra over $\mathbb{C}$ of continuous functions with compact support, under the multiplication $(f_1 * f_2)(g) = \int_G f_1(gx^{-1}) f_2(x)dx$. The associated space of

spherical functions, $C_c(H\backslash G/H)$, is a subalgebra of $C_c(G)$. If the convolution algebra of spherical functions is commutative, then (G,H) is a Gelfand pair.

The quantum mechanical propagator for a homogeneous Riemannian manifold $M = G/H$, with Laplace-Beltrami operator Δ, is denoted by $K_{G/H}(t;x,y)$. It is characterized by the equations

$$(i \frac{\partial}{\partial t} + \frac{1}{2} \Delta_x) K_{G/H}(t;x,y) = 0 \text{ for } t > 0$$

and

$$\psi(x) = \lim_{t\to 0^+} \int_{G/H} K(t;x,y)\psi(y)dy$$

for ψ in a suitable class of functions on M. What is essential here is that the propagator is a two-point invariant function:

$$K_{G/H}(t;gxH,gyH) = K_{G/H}(t;xH,yH),$$

which follows from the invariance of $\Delta_{G/H}$ by the action of G on $M = G/H$: $G \times M \to M$, $(g,xH) \to gxH$. Any two-point invariant function, e.g., $K_{G/H}$, defines a rotationally symmetric function on M, i.e., $f(hm) = f(m)$ for all h in H. Viz,. $f(m) = K(t;m,m_o)$ (or $m = zH \to K_{G/H}(t;zH) = K_{G/H}(t;y^{-1}xH,H)$). And conversely. Furthermore, any rotationally symmetric function f(m) may also be considered as a function on G, viz., $\tilde{f}(g) = f(m)$ if $gm_o = m$. It is easily checked that $\tilde{f}$ is H-bi-invariant on G; and conversely, every spherical function on G defines a r.s. function f on M.

Under the Gelfand pair hypothesis, there is a special class of spherical functions on G which are "orthogonal and span" the space of all spherical functions. These spherical functions are the elementary or zonal spherical (z.s.) functions. The z.s. functions can be characterized in several ways: e.g., ϕ in $C(H\backslash G/H)$ is a z.s. function iff $\phi(e) = 1$ and $f*\phi = \lambda_f \phi$ for all f in $C_c(H\backslash G/H)$.

The relation to representation theory is well-known also. There is a one-one correspondence between the set of positive definite (p.d.) z.s. functions on G and the set $\hat{G}(1)$ of equivalence classes of class one representations of G The notion of spanning is in the sense of a Fourier expansion of a suitable spherical function f:

$$f(g) = \int_{\hat{G}(1)} \hat{f}(\lambda)\phi_\lambda d\hat{\mu}(\lambda)$$

where $\hat{f}(\lambda) = \int_G f(g)\phi_\lambda^* dg$.

In the case G is compact and so $\hat{G}$ is discrete, then for every f in $L^2(H\backslash G/H)$ there is a Fourier series expansion

$$f = \sum_{\lambda\in\hat{G}(1)} f(\lambda)\phi_\lambda$$

where $d(\lambda)\hat{f}(\lambda) = \int_G f(g)\phi_\lambda^*(g)\, dg = \int_{G/H} f\phi_\lambda^* = \int_{H\backslash G/H} f\phi_\lambda^*$; and if U_ϕ is the (class one) representation associated to the p.d.z.s. function ϕ, then $\phi(gH) = \int_H \chi_{U_\phi}(g\cdot h)\, dh$ where χ is the character of the representation U_ϕ of G on $\mathcal{H}(U_\phi)$.

Since $K_{G/H}(t;m,m_o)$ gives a spherical function on G, we might consider an expansion in terms of the z.s. functions on G, viz.,

$$K_{G/H}(t;m,n) = \int_{\hat{G}(1)} \phi_\lambda(g)d\hat{\mu}_t(\lambda),\ gn = m,$$

$$\text{or} = \sum_{\lambda\in\hat{G}(1)} \phi_\lambda(g)F_\lambda(t)$$

if G is compact. The function $F_\lambda(t)$ is the <u>spectral</u> <u>density</u>.

When M = G is a compact connected Lie group, it is easy to calculate the spectral density. Knowing $\Delta_G\chi_\lambda = h_\lambda\chi_\lambda$, then $F_\lambda(t) = d(\lambda)\exp(-ih_\lambda t)$ (where $F_\lambda(0) = \lim_{t\to 0^+} \int_G K_G(t;g)\chi_\lambda(g)^*\, dg = \chi_\lambda(e) = d(\lambda)$).

Thus $K_G(t;g) = \sum_{\lambda\in\hat{G}(1)} d(\lambda)\exp(-ih_\lambda t)\chi_\lambda(g)$, (i.e., eq. (19) in [5]).

As sketched by Dowker [6] if $K_G(t;g)$ is the propagator on G, then $K_{G/H}(t;zH) = \int_H K_G(t;zh)\,dh$. Viz., if $\pi: G \to M = G/H$ is the natural projection then $\Delta_M(f)\circ\pi = \Delta_G(f\circ\pi)$. Thus $(\Delta_M K_M)(gH) = [\Delta_G \int_H K_G(t;\cdot,h^{-1})\,dh](g) = [\int_H \Delta_G K_G(t;\cdot,h^{-1})\,dh](g) = -i\,\partial/\partial t\,[\int_H K_G(\cdot)\,dh] = -i\,[\partial/\partial t\, K_M](gH)$; and $\lim_{t\to 0^+} \int f\circ\pi(g)\, K_G\,(t;g)\,dg = f\circ\pi(e) = f(H)$. Thus $K_M(t;gH) = \sum_{\lambda\in\hat{G}(1)} d(\lambda)\exp(-ih_\lambda t)\phi_\lambda(gH)$, or $K_M(t;xH,xH) = \sum_{\lambda\in\hat{G}(1)} d(\lambda)\dim Z\,(H(\lambda))\exp(-ih_\lambda t)$ (1.1)

where $Z(H(\lambda)) = \{v\in H_\lambda \mid U_\lambda(h)v = v$ for all h in H$\}$, where $H(\lambda)$ is the representation space of λ, $d(\lambda) = \dim H(\lambda)$.

We now turn our attention to the relation between the propagator K_M on $M = G/H$ and the propagator $K_{\mathcal{M}}$ on $\mathcal{M} = \Gamma\backslash G/H$, where Γ is a discrete subgroup of G. The best way to look at this question is as follows. Suppose the propagator is known only over the subspace γm_o where γ varies over Γ. Then what is the relation of the spectral density of the propagator we observe to that of the unobserved propagator on the whole space M? The connection is given by the Selberg trace formula which we review next.

§2. The Selberg Trace Fromula.

Let Γ be a closed subgroup of finite index of a separable locally compact group G. Let L be a finite dimensional unitary representation of Γ. The induced representation U^L acting on $H(U^L)$, [12]. Let χ_L, resp χ_{U^L}, denote the associated characters $g\to \mathrm{Tr}L(g)$, resp. $g\to \mathrm{Tr}U^L(g)$. For f in $L^1(G)$, define U^L_f by $(U^L_f\,\alpha_1,\alpha_2) = \int_G f(g)(U^L(g)\,\alpha_1,\alpha_2)dg$ for α_i in $H(U^L)$. Then $\mathrm{Tr}(U^L_f) =$

$\int_G \chi_{U^L}(g)f(g)\, dg$ can be computed:

$$\mathrm{Tr}(U^L_f) = \frac{1}{\mathcal{O}(\Gamma\backslash G)} \sum_{\tilde{y}\in\Gamma\backslash G} \int_\Gamma f(y^{-1}\gamma y)\chi_L(\gamma)\, d\gamma$$

$$\text{or } \int_{\Gamma\backslash G} [\int_\Gamma f(y^{-1}\gamma y)\chi_L(\gamma)\, d\gamma] \times d\nu(\tilde{y}) \quad (*)$$

If U^L decomposes discretely as $U^L = \sum_{\lambda} \oplus_G n(\lambda)\, U^\lambda$ with multiplicity $n(\lambda)$, then

$$\mathrm{Tr}(U^L_f) = \sum_{\lambda\in\hat{G}} n(\lambda)\ \mathrm{Tr}\, U^\lambda_f = \int_{\hat{G}} \mathrm{Tr}\, U_f\, d\hat{\mu}(M) \quad (**)$$

The Selberg trace formula (STF) in its first version is that (*) = (**). These hypotheses are met when Γ is a discrete subgroup of G such that $\Gamma\backslash G$ is compact. The induced representation U^L is a discrete direct sum of IUR of G, each occurring with finite multiplicity $n_\Gamma(\lambda,L)$; so $[U^L] = \sum_{\lambda\in\hat{G}} n_\Gamma(\lambda,L)\lambda$ where $\hat{G}$ is the set of equivalence classes of IUR's. If f is an <u>admissable</u> function--i.e., (a) the series $\sum_{\gamma\in\Gamma} f(y^{-1}\gamma x)L(\gamma)$ converges absolutely, uniformly on compacts of G × G, to a continuous function F(x,y,L) and (b) U_f is of trace class, then the STF holds

$$\sum_{\lambda\in\hat{G}} n_\Gamma(\lambda,L)\ \mathrm{Tr}\, U^\lambda_f = \int_{\Gamma\backslash G} \mathrm{Tr}\, F(x,x,L)\, d\dot{x} = \sum_{\gamma\in C_\Gamma} \chi_L(\gamma)\ \mathrm{Vol}(\Gamma_\gamma\backslash G_\gamma)\ J_\gamma(f)$$

where C_Γ is a complete set of representatives in Γ of G-conjugacy classes of elements of Γ, G_γ is the centralizer of γ in G, $\Gamma_\gamma = \Gamma\cap G_\gamma$, $J_\gamma(f) = \int_{G_\gamma\backslash G} f(x^{-1}\gamma x)\, dx_\gamma$

The connection of the STF with class one representations is as follows. A C_1-<u>spherical</u> function f has the property that $Ⓗ_\lambda(f) = \mathrm{Tr}\, U^\lambda_f = 0$ unless λ is of type one in which case $Ⓗ_\lambda(f) = \hat{f}(\lambda)$. Thus if f is also admissable and if $\sum_{\lambda\in\hat{G}(1)} n_\Gamma(\lambda,L)\hat{f}(\lambda)$ converges absolutely then the Selberg-Tamagawa trace formula (STTF) states that

$$\sum_{\lambda\in\hat{G}(1)} n_\Gamma(\lambda,L)\hat{f}(\lambda) = \sum_{\gamma\in C_\Gamma} \chi_L(\gamma)\ \mathrm{Vol}(\Gamma_\gamma\backslash G_\gamma)\ J_\gamma(f).$$

§3. Back to Propagators.

In the situation above where Γ is a discrete subgroup of G with $\Gamma\backslash G$ compact, Γ acts on G/H by left translation and the quotient space $\Gamma\backslash G/H$ is compact. As above with $L = I$, $L_2(\Gamma\backslash G/H) = \sum \oplus \mathcal{H}(\lambda)$ where the Laplacian $\Delta \in \mathcal{D}(G/H)$ acts on $\mathcal{H}(\lambda)$ by the scalar $h_\lambda(\Delta)$. Formally the sum $\sum_{\lambda\in\hat{G}(1)} d(\lambda)\exp(-ith_\lambda(\Delta))$ is the trace of the operator $\exp(-it\Delta)$ on $L_2(\Gamma\backslash G/H)$. Of course $\exp(-it\Delta)$ for $-\infty < t < \infty$ is a group with distributional kernel $\sum_\lambda \sum_{j=1}^{d(\lambda)} \exp(-ith_\lambda(\Delta))\phi_{\lambda j}(x)\overline{\phi_{\lambda j}(y)}$, where $\{\phi_{\lambda j}\}$, $j=1,\ldots,d(\lambda)$ is an ONB for $\mathcal{H}(\lambda)$.

If the Euclidean case is mimicked one expects that on $L_2(H\backslash G/H)$ the operator $\exp(-it\Delta)$ has an integral kernel $k(t;y^{-1}x) = g(it;y^{-1}x)$ where $g_t(m)$, m G/H, is the fundamental solution of the heat equation $\Delta u = \partial u/\partial t$ on G/H. The analogous integral operator on $L_2(\Gamma\backslash G/H)$ should then be obtained by "periodizing" k by "wrapping it around" Γ G/H along the orbits of Γ on G/H, i.e., (for L nontrivial) $K_{\Gamma\backslash G/H}(t;y,x) = \sum_{\gamma\in\Gamma} L(\gamma)k_t(y^{-1}\gamma x)$ with trace being $\sum_{\gamma\in\hat{G}(1)} n_\Gamma(\lambda,L)\exp(-ith_\lambda(\Delta)) = \int_{\Gamma\backslash G/H} K(t;x,x)dx$. (Requiring the wave functions to transform according to $\psi(\gamma m) = L(\gamma)\psi(m)$, one is led to $\int_M k(t;x,y)\psi(y) = \int_{\Gamma\backslash M} K(t;x,y)\psi(y)$ where $K(t;x,y) = \sum_{\gamma\in\Gamma} L(\gamma)k(t;x,\gamma y)$ as observed by Selberg [16]. Later this was the motivation for eq. (5') of [6] and [11].)

This leads naturally to the conjecture that K_t has a trace $\int_{\Gamma\backslash G/H} K_t(x,x)dx =$

$$\sum_{\lambda\in\hat{G}(1)} n_\Gamma(\lambda,L)\exp(-ith_\lambda(\Delta)) = \int_{\Gamma\backslash G/H} \sum_{\gamma\ \Gamma} \chi_L(\gamma)k_t(x^{-1}\gamma x)dx = \sum_{\gamma\in\Gamma} \chi_L(\gamma) \int_F k_{G/H}(t;x^{-1}\gamma x)dx \tag{3.1}$$

where F is a measurable fundamental domain for Γ in G/H. In this generalized STF we emphasize that everything is understood only formally.

The conjecture crystallizes the status of the remarks in [4], [5], [6], [11], and elsewhere. The evidence which leads us to expect a GSTF to be true is as follows.

Of course (3.1) reduces to (1.1) in the case Γ = id and G compact. Furthermore, it was observed by Schulman [15] for SU(2) and Dowker [5] for a compact Lie group G that the propagator is formally

$$K_G(t;x) = \sum_{\substack{\text{classical} \\ \text{paths}}} k(t;\gamma x)$$

where k(t;exp H) satisfies the "radial equation"

$$i\frac{\partial}{\partial t} j - \left(\sum_{k=1}^{n} \frac{\partial^2}{\partial h_k^2} + 4\pi^2 |\rho|^2\right) j = 0$$

for H regular, where: T is the maximal torus in G; the Lie algebra of T is $\mathcal{T}$ with basis $H_1,\ldots,H_n$; $\rho = \frac{1}{2}\sum_{r=1}^{\frac{N-n}{2}} \theta_r$ where θ_r are the positive roots of G (N = dim G); $|\cdot|$ is the Cartan-Killing norm; $j(H) = \Pi_r\,[\exp(\pi i\theta_r(H) - \exp(-\pi i\theta_r(H))]$ for H in $\mathcal{T}$; $\mathcal{T}'$ is the set of regular (i.e., $j(H) \neq 0$) points of $\mathcal{T}$; $H = \sum_k h_k H_k$. This fits precisely in our framework as hinted by Dowker. Viz., the propagator K(t;x,y) on G satisfies $K(t;x,y) = K(t;y^{-1}x,e) = K(t;y^{-1}x)$; and the map $z \to K(t,z)$ is invariant by inner automorphisms of G. Thus it is determined by the restriction to the maximal torus $T\subset G$ and K(t;exp) satisfies the radial equation for $h = \exp H \in T'$. The positive roots θ_r of G are linear forms on $\mathcal{T}$ which take integral values on a discrete subgroup Γ of $\mathcal{T}$. The function

$$T' \ni h \to k_T(t;h) = \exp\frac{(4\pi^2 i\,|\rho|^2\,t)}{(it)^{n/2}} \sum_{\gamma\in\Gamma} \frac{[L(h)\,\exp(-i|h|^2/4t)](h)}{j(h)}$$

is a W-invariant solution of the radial equation (where W is the Weyl group with order $|W|$). Then $\int_G K(t;g)\,dg = \frac{1}{|W|}\int_T |j(u)|^2 \times [\int_G K(t;gug^{-1})\,dg]\,du =$

$\frac{1}{\mathrm{Vol}(T/\Gamma)|W|}\int_{T/\Gamma}\sum_{\gamma\in\Gamma} |j(h+\gamma)|^2\, k_T(h+\gamma)\,dh$. This is given by (16-(18) in [5].

(Note that $|\rho|^2 = R/6$ where R is the scalar curvature of G.)

For the heat equation on these spaces, the analogous formulae have been proven in [1], [7], and elsewhere.

In the simple case that $G = R$, $H = \{e\}$, and $\Gamma = Z$, the GSTF reduces to the following analogue of the Poisson-Jacobi (theta) formula

$$\sum_{n\in Z} e^{-itn^2} = \sqrt{\pi/t}\sum_{m\in Z} \exp[\frac{i(2\pi m)^2}{4t}]\exp(\frac{-i\pi}{4}) \qquad (3.2)$$

where n^2 is the spectrum of $-\Delta$ on $S_1^1 = R/Z$ and $2\pi m$ is the length of the iterated closed geodesic (length 2π on S_1^1).

Colin de Verdiere has generalized the classical Poisson-Jacobi formula

$$\sum_{n\in Z}\exp(-n^2/z) = \sqrt{\pi z}\sum_{m\in Z}\exp(-\pi^2 m^2 z) \qquad \mathrm{Re}\, z > 0$$

for certain Riemannian manifolds with negative sectional curvature to

$$\sum_{k\geq 0}\exp(-\lambda_k/z) = \sum_{\ell\, L^+\cup\{0\}} f_\ell(z)\exp(-z\ell^2/4) \quad (F)$$

where F denotes the techniques of nonlinear Fourier transform, $0 = \lambda_0 < \lambda_1 \leq \lambda_2 \leq \cdots$ are the eigenvalues of the Laplacian $-\Delta$ on the compact connected Riemannian manifold M, L = set of lengths (and their opposites) of periodic geodesics on M. Note also Chazarain's formula [2].

The author is unable at present to prove the GSTF in the most general case. (Perhaps via techniques of Nelson and Ray it has been suggested.) However in the case in [11], $G = R$, $H = \{e\}$ and $\Gamma = Z$, the GSTF can be proven. The dynamics of the situation here is the fixed axis rigid rotator with Lagrangian $L = \frac{I}{2}\dot{\phi}^2$, $0 \leq \phi < 2\pi$, on $M = SO(2)$; a CONB for the wave functions is given by $\{\psi_m(\phi) = \frac{1}{\sqrt{2\pi}} e^{im\phi}\}$, $m \in Z$, with $E_m = m^2/2I$. Then $K(t,\phi) = \frac{1}{2\pi}\sum \exp(i\phi n)\exp(-in^2/2\gamma)$ $= \frac{1}{2\pi}\Theta_3(\frac{\phi}{2}, -\frac{1}{2\gamma\pi})$ where $\gamma = I\hbar t$ and $\Theta_3(z,t) = \sum_{n\in Z} e^{i\pi tn^2} e^{2inz}$. For $\mathrm{Im}\, t > 0$, the Poisson summation formula gives

$$K(t,\phi) = (\frac{\gamma}{2\pi i})^{1/2} \sum_{n\in Z} e^{i\gamma(\phi-2n\pi)^2/2} = \sum_{\gamma\in\Gamma} k(t;\phi+\gamma)$$

where k is the free particle propagator. The GSTF is just the case $\phi = 0$.

<u>Theorem</u>. The GSTF is true for the case $G = R$, $H = \{e\}$ and $\Gamma = Z$, i.e., (3.2) holds in the sense of Wiener's fourier transform.

The proof is straight forward.

§4. <u>Geodesics and Propagators</u>.

As noted above the philosophy in Schulman, Dowker and elsewhere is to express the propagator as a "sum over all classical paths." If M is a compact Riemannian manifold, then each closed path g (distinct from the identity) of $\Pi_1(M)$ corresponds to a closed geodesic γ_g of class g whose length is minimal among the closed curves of the same class as g. If M is of negative sectional curvature then there is only one closed geodesic of each homotopy type and every closed geodesic is so obtained. So there is a biunique correspondence between closed geodesics and nontrivial elements of $\Pi_1(M)$, or between the free homotopy classes of closed paths and the set C_Γ of conjugacy classes of elements of Γ.

In the situation above we have Γ a discrete torsion-free subgroup of Lie group G with Γ G compact. Then $M = \Gamma$ G/H is a compact Riemannian with simply connected covering space G/H and $\Gamma \simeq \Pi_1(M)$. From these remarks, if M has negative sectional curvature the GSTF(3.1) is modified by writing

$$K_{\Gamma\backslash G/H}(t;y,x) = \sum_{\substack{\text{"all closed} \\ \text{geodesics } \gamma\text{"}}} k(t;y^{-1}\gamma x)$$

A large class of manifolds of negative sectional curvature in this form are $\Gamma\backslash G/H$ where G is a noncompact connected simple Lie group of R-rank one and finite center, H is a maximal compact subgroup of G, Γ is a discrete subgroup of G acting freely on G/H; and G/H is a rank one symmetric space of noncompact type. This is an extremely interesting case for then the length spectrum (lengths of the periodic geodesics γ_g and their multiplicities) is determined by the (harmonic) spectrum of Δ on M. (Huber [10], Atiyah and Duistermaat (to appear), Gangolli (to appear)) and a "generalized" length spectrum plus $\mathrm{Vol}(\Gamma\backslash G)$ determines the (harmonic) spectrum.

§5. STF and Geometric Quantization.

As we know, an important object in geometric quantization is the quantized Hilbert space associated with a Kählerian polarization F, $H^{\circ}(M,\mathcal{O}(E))$ where $E \to M$ is a holomorphic line bundle over (M,Ω) with the curvature of the connection on E being Ω, etc. This situation arises when $M = G/H$ is a bounded symmetric domain with cocompact Γ acting freely; then $M = \Gamma\backslash G/H$ is an algebraic manifold. In this case the first version of the STF applies. E.G., if $E_\lambda \to M$ is the bundle corresponding to the holomorphic discrete series of Harish Chandra then the multiplicity of the "energy levels" for the "energy manifold" M is

$$n_\Gamma(\lambda) = \dim H^{\circ}(M,\mathcal{O}(E_\lambda)) = \sum_{\gamma} \mathrm{Vol}(\Gamma_\gamma\backslash G_\gamma)J_\lambda(\gamma) = \text{dimension}$$

of the space of automorphic forms for such a representation. (Cf. Hotta-Parthasarathy, et al.)

In general in the situation at the end of §5 if Γ has no elliptic elements $n_\Gamma(\lambda) = \mathrm{Vol}(\Gamma\backslash G)d(\lambda)$ when λ is integrable. (Cf. Langlands, Schmid, et al.)

References

[1] A.-I. Benabdallah, Bull. Soc. Math. Fr. 101 (1973), 265-283.

[2] J. Chazarain, Invent. math. 24 (1974), 65-82.

[3] Y. Colin de Verdiere, Compos. Math. 27 (1973), 83-106, 159-184.

[4] C. M. DeWitt, Ann. Inst. H. Poincare 11 (1969), 152-206.

[5] J. S. Dowker, Ann. Phys. 62 (1971), 361-382.

[6] J. S. Dowker, J. Phys. A (1972).

[7] L. D. Eskin, Amer. Math. Soc. Trans. 75 (1968), 239-254.

[8] R. Gangolli, Acta Math. 121 (1968), 151-192.

[9] I. M. Gelfand et al., Automorphic Functions and Representation Theory, (W. B. Saunders, Philadelphia, 1969).

[10] H. Huber, Math. Ann. 138 (1959), 1-26.

[11] M. Laidlaw and C. M. DeWitt, Phys. Rev. D3 (1971), 1375-1378.

[12] K. Maurin, General Eigenfunction Expansions and Unitary Representations of Topological Groups, (PWN, Warszawa, 1968).

[13] H. P. McKean, Comm. Pure Appl. Math. 25 (1972), 225-246.

[14] A. Preismann, Comm. Math. Helv. 15 (1943), 175-216.

[15] L. S. Schulman, Phys. Rev. 176 (1968), 1558-1569.

[16] A. Selberg, J. Indian Math. Soc. 20 (1956), 47-87.

[17] T. Tamagawa, J. Fac. Sci. Univ. Tokyo 8 (1960), 363-386.

ON THE QUANTISATION OF THE KEPLER MANIFOLD *

Enrico Onofri

Istituto di Fisica dell'Università di Parma, I-43100 Parma
and
Istituto Nazionale di Fisica Nucleare, Sezione di Milano.

The geometric quantisation of the Kepler manifold in any number of degrees of freedom is constructed. The Kepler manifold is the phase space of the regularized Kepler motion and is shown to be a SO(2,n)-homogeneous symplectic manifold corresponding to an extremely singular orbit in the co-adjoint representation (of dimension 2(n-1)). The quantisation is obtained by approximating this orbit by more regular ones, which are equivalent to homogeneous bounded domains of type IV. The most relevant result is that the usual quantum-mechanical hydrogen-atom model is recovered in the particular representation introduced by Fock in 1935 (SO(n)-homogeneous integral equation in momentum space).

References

FOCK, V., Z.Physik 98, 145 (1935)

BANDER, M. and ITZYKSON, C., Rev.Mod.Phys. 38(2) 330 (1966)

ONOFRI, E. and PAURI,M. J.Math.Phys. 13(4) 533 (1972)

SIMMS, D.J. "Geometric Quantisation of Energy Levels in the Kepler Problèm", Symposia Math. Vol.XIV, Acad.Press 1974.

SOURIAU, J.M., "Sur la Variété de Kepler", Symposia Math. Vol.XIV Acad.Press 1974.

*ONOFRI, E., "Dynamical Quantization of the Kepler Manifold", Università di Parma, preprint 047 (1975).

ON WAVE FUNCTIONS IN GEOMETRIC QUANTIZATION

by

Jędrzej Śniatycki

Department of Mathematics and Statistics, University of Calgary.

In the abstract formulation of quantum mechanics states are elements of an abstract hilbert space. A choice of a complete system of commuting observables yields a representation of states by wave functions. Knowledge of the classical counterparts of the observables forming a complete commuting system enables one to interpret the wave functions from the point of view of the classical phase space. However, in the process of quantization of a classical system the situation is reversed. One has only the classical phase space to begin with, and has to choose a maximal set of commuting functions on the phase space to define wave functions and their scalar product. The standard quantization procedure is possible if the classical system has a distinguished configuration space with a euclidean structure. Then, one uses cartesian coordinates of the configuration space as a complete set of commuting observables, and the wave functions form the space of square integrable complex functions on the configuration space. In the case when the configuration space has no euclidean structure the quantization procedure is more difficult and the geometric nature of wave functions is more complicated. If some observables in the complete commuting set have discrete spectra the wave functions are generalized functions (distributions) and one has to be very careful in defining their scalar product, since multiplication of generalized functions is usually not possible.

The phase space of a dynamical system with n degrees of freedom can be represented by a $2n$-dimensional manifold X. The lagrange bracket divided by the Planck's constant defines a symplectic form ω on X. Dynamical variables are represented by functions on X.

A symplectic manifold (X,ω) is quantizable if ω defines an integral de Rham cohomology class. In this case there exists a complex line bundle L over X with a connection ∇ such that ω is the curvature form of ∇, and with an invariant hermitian form. Given such a line bundle, one can associate to each function on X a linear operator on the space of sections of L in such a way that the poisson bracket of two functions is associated to the commutator of the corresponding operators, divided by $i\hbar$, where $\hbar$ is the Planck's constant divided by 2π. In order to obtain a physically meaningful quantization one has to choose a complete system of commuting observables. Since one has no hilbert space of states, one does not know, a priori, which functions on X will qualify as observables. A globalized classical counterpart of the notion of a complete set of everywhere independent observables is that of a real polarization tion of a symplectic manifold. A real polarization of (X,ω) is a foliation F of X by lagrangian submanifolds, that is by n-dimensional submanifolds Λ of X, called the leaves of the foliation, such that ω restricted to Λ vanishes identically. The sections of L covariant constant along F are possible candidates for wave functions. However, there is no natural way to define a scalar product for such sections, since there is no canonical density in the space Y of all leaves of the foliation F. This is one of the reasons for the necessity to introduce a bundle N of half-forms normal to F and define wave functions as sections of $L\otimes N$ covariant constant along F. Such smooth sections exist only if the leaves of F are simply connected. In the case when the leaves of F are not simply connected one can represent wave functions as generalized sections of $L\otimes N$ covariant constant along F. Generalized sections of $L\otimes N$ can be treated as continuous linear functionals on the space of smooth sections with compact supports of the bundle $L\otimes N^{+}$, where N^{+} is the hermitian dual of N. The generalized sections of $L\otimes N$ covariant constant along F have supports in the Bohr-Sommerfeld set S defined as the union of all leaves Λ of F such that the holonomy group of the flat connection in $(L\otimes N)|\Lambda$ is trivial.

In this talk, I would like to discuss the case of a real polarization satisfying the following conditions.

Completeness: for each leaf Λ of F, the canonical flat affine connection in Λ is complete.

Local triviality: the space Y of all leaves of F admits a manifold structure such that the canonical projection $\pi: X \to Y$ is a fibration admitting local trivializations which induce affine isomorphisms of the fibres.

Under these assumptions leaves of F are isomorphic to $R^{n-k} \times T^k$, where T^k denotes a k-torus. The hamiltonian vector fields in F with closed orbits define an involutive k-dimensional distribution K contained in F. There is a unique density κ on K which is invariant under the hamiltonian vector fields in F and gives the total volume 1 to each integral manifold of K. The Bohr-Sommerfeld set S and its projection $\pi(S)$ are submanifolds of X and Y, respectively, of codimension k.

Let H_0 denote the space of smooth sections of $(L \otimes N)|S$ covariant constant along $F|S$ with supports projecting to compact sets in $\pi(S)$. It is a subspace of the space of all generalized sections of $L \quad N$ covariant constant along $F|S$ with a pre-hilbert structure defined as follows. Since L has an invariant hermitian form and N is the bundle of half-forms normal to F, there is a sesquilinear pairing of sections of $(L \otimes N)|S$ covariant constant along $F|S$ which yields a generalized density in Y with support in $\pi(S)$. The density κ on K can be used to make this pairing into a genuine density on $\pi(S)$, and the scalar product of two elements from H_0 is given by integretion of the corresponding density over $\pi(S)$. The completion H of H_0 can serve as the hilbert space of wave functions in the representation given by a polarization F.

Quantization of a function f constant along F gives an operator of multiplication by f, and the spectrum of this operator is completely determined by the range of the restriction of f to the Bohr-Sommerfeld set S. In order to quantize arbitrary functions one would have to generalize Blattner-Kostant-Sternberg kernels to the case of polarizations with not simply connected leaves.

It has been suggested by B. Kostant that in the case of polarizations with not simply connected leaves one might be able to use for representation spaces higher cohomology groups of X with coefficients in the sheaf S of germs of smooth sections of $L \otimes N$ covariant constant along F. This has been verified by R.J. Blattner, J. Rawnsley and D.J. Simms in the case of quantization of a one-dimensional harmonic oscillator in the energy representation.

Under the assumptions of completeness and local triviality, for all $m \neq k$, the m'th cohomology group $H^m(X,S)$ of X with coefficients in S vanishes. If in addition the distribution K is orientable, then $H^k(X,S)$ and the space of smooth sections of $(L \quad N)|S$ covariant constant along $F|S$ are isomorphic as modules over the ring C of complex-valued functions on X constant along F. The subgroup of $H^k(X,S)$ isomorphic to H_0 has an induced pre-hilbert structure, and its completion gives another possibility for the space of wave functions in the representation given by a polarization F. The fact that $H^k(X,S)$ and the space of smooth sections of $(L \quad N)|S$ covariant constant along $F|S$ are isomorphic C-modules implies that the quantization of functions constant along F gives equivalent operators in both representation spaces.

References:

R.J. Blattner, *Quantization and representation theory*, in Harmonic analysis on homogeneous spaces, A.M.S. Proc. Sym. Pure Math., vol. 36 (1973), pp. 147-165.

R.J. Blattner, *Pairings of half-form spaces*, to appear in proceedings of Coll. Int. du C.N.R.S. "Géométrie symplectique et physique mathématique", Aix-en-Provence, 1974.

K. Gawędzki, *Geometric quantization kernels*, to appear.

B. Kostant, *Quantization and unitary representations*, Lecture notes in Math., vol. 170 (1970), pp. 87-208, Springer, Berlin.

B. Kostant, *Symplectic spinors*, Conv. di Geom. Simp. e Fis. Mat., INDAM Rome, 1973, Symposia Math. series, Academic Press. Vol. XIV

B. Kostant, *On the definition of quantization*, to appear in proceedings of Coll. Int. du C.N.R.S. "Géométrie symplectique et physique mathématique", Aix-en-Provence, 1974.

J. Rawnsley, *De Sitter symplectic spaces and their quantizations*, to appear in Proc. Camb. Phil. Soc.

D.J. Simms, *Geometric quantisation of the harmonic oscillator with diagonalised Hamiltonian*, Proc. of 2nd. Int. Coll. on Group Theoretical Methods in Physics, Nijmegen, 1973.

D.J. Simms, *Geometric quantisation of symplectic manifolds*, Proc. of Int. Sym. on Math. Phys., Warsaw, 1974.

D.J. Simms, *Metalinear structures and a geometric quantisation of the harmonic oscillator*, to appear in proceedings of Coll. Int. du C.N.R.S. "Géométrie symplectique et physique mathématique", Aix-en-Provence, 1974.

J.-M. Souriau, *Structures des systèmes dynamiques*, Dunod, Paris, 1970.

J. Śniatycki, *Bohr-Sommerfeld quantum systems*, Proc. of 3rd. Int. Coll. on Group Theoretical Methods in Physics, Marseille, 1974.

J. Śniatycki, *Bohr-Sommerfeld conditions in geometric quantization*, Reports on Math. Phys., vol. 7 (1974) p. 127-135.

J. Śniatycki, *Wave functions relative to a real polarization*, to appear in Int. J. of Theor. Phys.

J. Śniatycki, *On cohomology groups Appearing in geometric quantization*, to appear.

A. Weinstein, *Symplectic manifolds and their lagrangian submanifolds*, Advances in Math., vol. 6 (1971), pp. 329-346.

Dynamical Prequantization, Spectrum-generating algebras and the Classical Kepler and Harmonic Oscillator Problems

Kishor C. Tripathy

Department of Physics, University of Delhi, Delhi-110007, India.

ABSTRACT: The prequantization scheme for the three dimensional classical Kepler and harmonic oscillator problems has been discussed in the light of the work of Souriau and Kostant and via the spectrum-generating algebras associated with the dynamical systems.

1. Introduction

Recently, a surge of activities on the problem of quantization of classical systems has been initiated by Souriau's programme[1] and Kostant's work on quantization and unitary representations[2].

Weyl's Ω-rule: Earlier, Weyl[3] prescribed a remarkable method of constructing phase-space representation of quantum meçhanics(i.e., a linear one-to-one map of operators in a Hilbert space into c-number functions).

If g(q,p) is a classical observable, then define the Fourier transform as

$$g(q,p) = \iint_{-\infty}^{+\infty} \gamma(\xi,\eta)\, e^{i(\xi\cdot q+\eta\cdot p)}\, d\xi\, d\eta . \quad (1.1)$$

Since the correspondence is linear, the phase-space representation can be completely specified by the operators associated with $\exp(i(\xi q+\eta p))$.

He prescribed then the Ω-rule such that

$$e^{i(\xi\cdot q+\eta\cdot p)} \longrightarrow \Omega(\xi,\eta)\, e^{i(\xi\hat{q}+\eta\hat{p})} . \quad (1.2)$$

Thus, the operator $\hat{g}(\hat{q},\hat{p})$ corresponding to g(q,p) is given by

$$\hat{g}(\hat{q},\hat{p}) = \iint_{-\infty}^{+\infty} \gamma(\xi,\eta)\, \Omega(\xi,\eta)\, e^{i(\xi\cdot\hat{q}+\eta\hat{p})}\, d\xi\, d\eta \quad (1.3)$$

We have the inverse mapping,

$$g(q,p) = \frac{1}{2\pi} \iint_{-\infty}^{+\infty} \mathrm{Tr}\left(\hat{g}\, e^{-i(\xi\hat{q}+\eta\hat{p})}\right) . \Omega(\xi,\eta)^{-1} \times \quad (1.4)$$
$$\times\, e^{i(\xi q+\eta\cdot p)}\, d\xi\, d\eta .$$

$\Omega(\xi,\eta)$ is the boundary value of an entire analytic function of ξ, η and has no zeros for real ξ, η . Further,

$$\Omega^{*}(\xi,\eta) = \Omega(-\xi,-\eta), \qquad (1.5)$$

$$\text{and} \quad \Omega(0,0) = 1 . \qquad (1.6)$$

(1.5) implies the reality condition for Ω and ensures that the real functions are mapped onto self-adjoint operators and vice-versa.Weyl's Ω-rule, however fails in general since the distribution function could be negative.

In Schrödinger approach,one resorts to formal quantization of classical generalized coordinates X_k and the canonical momenta P_k which are defined locally.If we consider $M=S^1$,the unit circle,multiplication by the angle $X= \theta$, is not an operator in the Hilbert space of periodic functions $f(\theta)= F(\theta+ 2\pi)$.This elucidates the difficulties with the formal quantization of generalized coordinates and momenta.

Dirac's work involves a map of classical dynamical variables f_i to self-adjoint,irreducible operators $K(f_i)$ with suitable domains in a Hilbert space satisfying

$$[K(f_i) , K(f_j)] = i\hbar K([f_i, f_j])$$

$$\text{and } K(1) = I .$$

In all conventional approaches,the solution of the Dirac problem is carried out by quantizing the Heisenberg algebra:(q,p,I) while no apriori guarentee is made for preserving the self-adjointness of the rest of the operator functions f(q,p).

Van Hove's prequantization scheme[4]: Let $\hbar > 0$.A prequantization scheme on a manifold $M(= R_{2n}$, the Euclidean phase-space) is a mapping f(q,p) (the C^{∞} functions of infinitesimal canonical transformations which generate one-parameter subgroups of the Lie pseudo-group of contact transformations) onto the set of self-adjoint operators in a complex, infinite dimensional, separable Hilbert space such that

$$K([g,f]) = \frac{1}{i\hbar}[K(g), K(f)],$$
$$K(1) = \mathbb{1}.$$

Van Hove's method of Euclidean prequantization fails, however, in a simple dynamical system like the Kepler problem where the Hamiltonian vector field is not 'complete' since orbits with l=0 reach the point q=0 within a finite lapse of time.

Souriau's scheme:[1] Let (M, Ω) be the symplectic manifold of a classical dynamical system, M= the state space and Ω = the symplectic closed 2-form on M. The diffeomorphism $\phi : M \rightarrow M$ is a canonical transformation if $\phi^*(\Omega) = \Omega$. Let F(M) be C^∞ real-valued functions on M. For each $f_a \in F(M)$, define the vector field X_{f_a} such that

$$d f_a = X_{f_a} \lrcorner \Omega$$

$$[f_a, f_b] = X_{f_a}(f_b), \ \forall f_a, f_b \in F(M).$$

Under the above Poisson bracket relation, the vector space F(M) becomes a Lie algebra.

Considerthe Hamiltonian dynamical system(M, Ω, H) where $H \in F(M)$ is the Hamiltonian function if it has no critical points(dH =0). The integral curves of the vector field X_H are solutions of the Hamilton's equations and generate the one-parameter group of canonical transformations. Thus, in Souriau's scheme, the vector fields X_H are complete also. The essential feature of Souriau's prequantization scheme is that it enables to construct the contact manifold Ω_{2n+1}, one dimension higher than the phase space Ω_{2n}.

2. Dynamical Prequantization [5,6,7]

We discuss in this section the prequantization scheme for the classical

Kepler and harmonic oscillator problems in the light of the Kostant-Souriau scheme and using the dynamical symmetry associated with the classical mechanical systems.

a) Let (M, Ω) be the symplectic manifold for a classical dynamical system; M= the state space and Ω=the canonical closed 2-form on M.It admits a maximal dynamical symmetry K($\mathbb{K}$, correspondingly is the Lie algebra of infinitesimal canonical transformations on M) acting transitively on each energy surface $M_E \approx K/K_0$, K_0 being the stability subgroup of some point on M_E.This implies that all the orbits of the dynamical system are diffeomorphic to one another and that the Hamiltonian is a certain function of the canonical invariants of K.

b)The vector field X_H on M generates a global action in $\Omega$of R(R= O(2) or U(1) for compact orbits, = O(1,1) for non -compact orbits).This defines the Hamilton group G_H (= R) and the Hamiltonian appears as a function of the single ~~of~~ element of the Lie algebra.

c)There exists a dynamical group(spectrum-generating group) G such that it possesses a global canonical action in Ω and contains K & G_H as subgroups.The compact and non-compact orbits correspond to different open intervals of the energy and correspondingly there exists analytic continuation within the submanifolds(energy surfaces).

Further,the elements of the Lie algebra $\mathbb{G}$ of G satisfy the classical equation of motion,

$$\frac{\partial \mathbb{G}}{\partial t} + [H, \mathbb{G}] = 0 \cdot$$

Note that the elements of $\mathbb{K}$ and $\mathbb{G}$ are independent of time as it should be.

d) We note that the construction of the canonical realisation for G provides directly Souriau's prequantization in the following sense.

The irreducible representation of G (quantal representation) is such that every eigenspace of G_H carries an irreducible representation of the 'symmetry group' K, i.e.,the Casimir operators of K commutes with G_H (irreducibility condition).

3.Construction of the spectrum-generating algebras

Let (M, Ω) be the symplectic manifold of a classical dynamical system. Ω = the symplectic closed two-form on M is such that $d\Omega=0$. Let $F(M)= \{f_1, f_2, \ldots\ldots, f_n\}$ be the C^∞ functions on M. Ω is called the phasespace for the underlying dynamical system and in the canonical co-ordinates is given by

$$\Omega = \sum_i dp_i \wedge dq_i - dH \wedge dt. \tag{3.1}$$

For $\forall\, f_a \in F(M)$, we define the covariant and the contravariant vector fields df_a and X_{f_a} respectively as

$$df_a = X_{f_a} \rfloor \Omega = (\partial f_a / \partial p_i)\, dp_i + (\partial f_a / \partial q_i)\, dq_i \tag{3.2}$$

$$\text{and } X_{f_a} = \frac{\partial}{\partial t} + \sum_i \left(\frac{\partial f_a}{\partial p_i} \frac{\partial}{\partial q_i} - \frac{\partial f_a}{\partial q_i} \frac{\partial}{\partial p_i} \right), \tag{3.3}$$

where $X_{f_a} \rfloor \Omega$ defines the contraction of Ω by X_{f_a}. Let $\{X_{f_a}\} = V(M)$. For $f_a, f_b \in F(M)$, we have

$$X_{f_a}(f_b) = \frac{\partial f_b}{\partial t} + \sum_i \left(\frac{\partial f_a}{\partial p_i} \frac{\partial f_b}{\partial q_i} - \frac{\partial f_a}{\partial q_i} \frac{\partial f_b}{\partial p_i} \right).$$
$$= \tag{3.4}$$

Under the Poisson bracket relation (3.4),the real vector space F(M) becomes a Lie algebra.The map $f_a \longrightarrow X_{f_a}$ is a Lie algebra homomorphism of F(M) into V(M) on M;

i.e.,
$$X_{\alpha f + \beta g} = \alpha X_f + \beta X_g,$$
$$X_{[f,g]} = X_f X_g - X_g X_f, \quad f, g \in F(M). \tag{3.5}$$

Consider the triplet (M, Ω, H), the Hamiltonian dynamical system. Then

$$H \to X_H = \frac{\partial}{\partial t} + \sum_i \left(\frac{\partial H}{\partial p_i} \frac{\partial}{\partial q_i} - \frac{\partial H}{\partial q_i} \frac{\partial}{\partial p_i} \right) \tag{3.6}$$

$$\text{and } dH = \sum_i \left(\frac{\partial H}{\partial p_i} dp_i + \frac{\partial H}{\partial q_i} dq_i \right) + \frac{\partial H}{\partial t} dt \tag{3.7}$$

If $H = H(p_i, q_i)$, then $\frac{\partial H}{\partial t} = 0$. Now,

$$X_H(f_a) = \frac{\partial f_a}{\partial t} + [H, f_a] \qquad (3.8)$$

Using the classical equation of motion, we have

$$[H, f_a] + \frac{\partial f_a}{\partial t} = df_a/dt = 0. \qquad (3.9)$$

Thus, $\mathbb{G} = \{ f_a,\ a = 1,2,3,\ldots\ldots,n : df_a/dt = \frac{\partial f_a}{\partial t} + [H, f_a] = 0 \}$ defines the spectrum generating algebra for the given Hamiltonian H. If $[H, f_a] = 0 = X_H(f_a)$, then we obtain the symmetry algebra $\mathbb{K}$ $\mathbb{G} = \{f_a,\ 1 \le a \le m < n : \frac{\partial f_a}{\partial t} = 0, [H, f_a] = 0\}$.

4.(a) Kepler motion [6,8]

We have $H = p^2/2 + V(q)$, $V(q) = V(q^2) = -1/q$, (4.1)

where we have used the reduced mass $\mu = 1$ and the coupling parameter $\lambda = 1$. The constants of the motion which have vanishing Poisson bracket with H are given by[8]

$$\underline{L} = \underline{q} \times \underline{p},$$

$$\underline{f} = a_1(q^2, H, l^2)\,\underline{q} + a_2(q^2, H, l^2)\,\underline{p}$$

$$= (\alpha_0'/l^2)\,\underline{L} \times \underline{A} - \alpha_2'\underline{A}, \qquad (4.2)$$

where $\underline{A} = \underline{p} \times \underline{L} - \underline{q}/q$ (the conventional Lenz-vector),

$$\underline{L} \times \underline{A} = l^2(1 - q/l^2)\,\underline{p} + (\underline{p}\cdot\underline{q}/q\,l^2)\,\underline{q}, \qquad (4.3)$$

$$a_1 = \alpha_0'(\underline{p}\cdot\underline{q}/l^2 q) - \alpha_2'(2H + 1/q), \qquad (4.4)$$

$$a_2 = \alpha_0'(1 - q/l^2) + \alpha_2'\,\underline{p}\cdot\underline{q}. \qquad (4.5)$$

α_0', α_2' are arbitrary constants depending upon H and l^2. The vectors $\underline{A}$ and $\underline{L} \times \underline{A}$ lie on the plane of the orbit and so also $\underline{f}$. For negative energy motions ($E < 0$), we have

$$\underline{f}\cdot\underline{f} = a_1 q^2 + 2a_1 a_2\,\underline{p}\cdot\underline{q} + a_2^2 p^2$$

$$= -\sigma l^2 + \sigma_0(H), \qquad (4.6)$$

where $\sigma_0(H) > l^2 \geq -1/2H$ and $\sigma = +1$. For $E \gtrless 0$, σ takes values -1 and 0 respectively. Thus, the symmetry algebra $\mathbb{K}$ is spanned by

($\underline{L}$, $\underline{f}$) whose Poisson bracket relations satisfy the Lie algebra isomorphic to O(4) for E 0.Further,the Casimir invariants are given by

$$C_1 = \underline{L}^2 + \underline{f}\,2 = \sigma_0(H),$$
$$C_2 = 0 = \underline{L} \cdot \underline{f} .$$

We note that the symmetry algebra $\mathbb{K}$ possesses the local canonical action on the energy surface $M_E \approx O(4)/\ O(2)$, i.e., M_E is a homogeneous space with stability subgroup O(2). We consider $\mathbb{G}$ to be such that the commutant of O(4) is O(2) and $\mathbb{G}$ contains O(4) $\otimes$ O(2) as subgroups.

Let $\mathbb{G}$ = O(4,2) whose elements satisfy the following equal-time Poisson bracket relations:

$$[L_{ab}, L_{cd}]_t = g_{ac} L_{bd} + g_{bd} L_{ac} - g_{ad} L_{bc} - g_{bc} L_{ad},$$

$$a,b,c,d = 1,2,3,4,5,6 \text{ and } g_{ii} = g_{44} = -g_{55} = -g_{66} = 1 ;\ i,j = 1,2,3. \quad (4.7)$$

Let us identify $L_i = \epsilon_{ijk} L_{jk}$, $f_i = L_{i4}$, $M_i = L_{i5}$,

$\Gamma_i = L_{i6}$, $T = L_{45}$, $\Gamma_4 = L_{46}$ and $\Gamma_0 = L_{56}$.

$$\text{Let } \underline{B} = A_1 \underline{q} + A_2 \underline{p} . \quad (4.8)$$

From (4.7) and (4.8), we have

$$[\underline{f}, \underline{B}]_t = S = a_1 A_2 - a_2 A_1 + \text{terms involving } \underline{p} \times \underline{p} ,\ \underline{q} \times \underline{q} \text{ and } \underline{q} \times \underline{p} , \quad (4.9)$$

$$\text{and} \quad [S, \underline{f}]_t = \underline{B} = A_1 \underline{q} + A2\, \underline{p} . \quad (4.10)$$

From (4.9) and (4.10), we obtain 6

$$A_1 = \left(\frac{u}{q}\frac{\partial a_1}{\partial H} + \frac{a_2}{q}\right)\frac{\partial S}{\partial q} + \frac{1}{l}\,(a_1 u + a_2 p^2)\frac{\partial S}{\partial l}$$

$$A_2 = \frac{u}{q}\frac{\partial S}{\partial q}\frac{\partial a_2}{\partial H} - 1/l\,(a_1 q^2 + a_2 u)\frac{\partial S}{\partial l} ,$$

$$u = \underline{p} \cdot \underline{q} . \quad (4.11)$$

Thus, $S = a_1 A_2 - a_2 A_1$

$$= \left[\frac{u}{q}\left(a_1 \frac{\partial a_2}{\partial H} - a_2 \frac{\partial a_1}{\partial H}\right) - a_2^2/q\right]\frac{\partial S}{\partial q} + (l + 1/2Hl)\frac{\partial S}{\partial l} . \quad (4.12)$$

We consider the following cases:

a) $\alpha_0' = 0, \quad \alpha_2' = \frac{1}{\sqrt{-2H}} \quad , \underline{f} = -\alpha_2' \underline{A}$

b) $\alpha_2' = 0, \quad \alpha_0'/l = 1/\sqrt{-2H} \quad , \underline{f} = 1/l\sqrt{-2H}(\underline{L} \times \underline{A})$.

For case (a) (using (4.12) and equation of motion), we have

$$S = \frac{X(H)}{\sqrt{-2H}}\left(-\frac{1+2Hq}{\sqrt{-2H}}\cos\beta - u\sin\beta\right) + \frac{Y(H)}{\sqrt{-2H}}\left(-\frac{1+2Hq}{\sqrt{-2H}}\sin\beta + u\cos\beta\right),$$

$$\beta = (-2H)^{\frac{1}{2}}(u - 2Ht). \tag{4.13}$$

(4.13) shows that S is a linear combination of two rotational scalars Γ_4 and T (say) with arbitrary coefficients X(H) and Y(H). Let us put

$$X(H) = Y(H) = (-2H)^{\frac{1}{2}}$$

and
$$\Gamma_4 = -\frac{1+2Hq}{\sqrt{-2H}}\cos\beta - u\sin\beta$$

$$T = -\frac{1+2Hq}{\sqrt{-2H}}\sin\beta + u\cos\beta . \tag{4.14}$$

Substituting for S from (4.14) in (4.11) and using the values of a_1 and a_2, we obtain after some simplification

$$\underline{M} = q\,\underline{p}\cos\beta - (-2H)^{-\frac{1}{2}}(u\,\underline{p} - \underline{q}/q)\sin\beta$$

and
$$\underline{\Gamma} = q\,\underline{p}\sin\beta + (-2H)^{-\frac{1}{2}}(u\,\underline{p} - \underline{q}/q)\cos\beta \tag{4.15}$$

Now, the equal-time commutation relation between M and gives us

$$\Gamma_0 = (-2H)^{-\frac{1}{2}}. \tag{4.16}$$

Note that is independent of time.

For case(b), we obtain the following expressions for $\underline{M}$, $\underline{\Gamma}$, Γ_4, T and Γ_0.

$$\underline{M} = \frac{1}{l}\left[(2+2Hq)\,\underline{q} - uq\,\underline{p}\right]\cos\beta + 1/l\sqrt{-2H}\left[-\frac{u}{q}(1+2Hq)\,\underline{q} + (u^2-q)\,\underline{p}\right]\sin\beta$$

$$\underline{\Gamma} = 1/l\left[(2+2Hq)\,\underline{q} - uq\,\underline{p}\right]\sin\beta - 1/l\sqrt{-2H}\left[-\frac{u}{q}(1+2Hq)\,\underline{q} + (u^2-q)\,\underline{p}\right]\cos\beta . \tag{4.17}$$

The structure for Γ_4, T, Γ_0 remains same as in the case (a).

The Casimir invariants for O(4,2) and $O(4)_{1,2,3,4}$ are given by

$$Q_2 = \tfrac{1}{2} L_{ab} L^{ab} = 0,$$

$$Q_3 = -1/48 \in^{abcdef} L_{ab}L_{cd}L_{ef} = 0,$$

$$Q_4 = 1/4\ (L_{ab} L^{bc} L_{cd} L^{da} - Q_2^2 - 8\ Q_2) = 0, \qquad (4.18)$$

$$C_1 = \underline{L}^2 + \underline{f}^2 = -1/2H = (\Gamma_o)^2,$$

$$C_2 = \underline{L}.\underline{f} = 0. \qquad (4.19)$$

We note that the eigenspace $\{ \Psi_{n l m} \}$ of $O(4)_{1,2,3,4}$ furnishes the eigenstates for Γ_o (irreducibility condition). In case of non-compact orbits i,e, ,for $\mathbb{G}_H = O(1,1)$, $\mathbb{G}$ is still $O(4,2)$; however, the symmetry algebra is $O(3,1)_{1,2,3,5}$ and $\mathbb{G}_H$ is generated by L_{46}.

4.b Harmonic oscillator[5]

We have $H = p^2/2 + q^2/2$, $m = \omega = 1$. (4.20)

In this case, $\mathbb{K} = SU(3)$ and is spanned by $\underline{L}$, T_{ij};,where

$$\underline{L} = \underline{q} \times \underline{p} \ ,$$

$$T_{ij} = T_1\, q_i q_j + T_2 (q_i p_j + q_j p_i) + T_3\, p_i p_j - 1/3\ \delta_{ij}\, T_{kk}, \qquad (4.21)$$

where T_1, T_2, T_3 are given by

$$T_1 = \varphi_1 + 2\ \underline{p}.\underline{q}\ \varphi_2 + p^2 \varphi_3 ,$$

$$T_2 = (p^2 - q^2) \varphi_2 - \underline{p}.\underline{q}\ \varphi_3 ,$$

$$T_3 = \varphi_1 - 2 \underline{p}.\underline{q}\ \varphi_2 + q^2 \varphi_3 ,$$

$$T_1 T_3 - T_2^2 = \sigma = +1. \qquad (4.22)$$

φ_1, φ_2, φ_3 are arbitrary functions of H and l^2 and satisfy

$$\varphi_1^2 + 2H \varphi_1 \varphi_3 + l^2 \varphi_3 - 4 (H^2 - l^2) \varphi_2^2 = +1,$$

$$2H\ \varphi_1 + 2l^2 \varphi_3 = \Phi(H). \qquad (4.23)$$

The Casimir operators C_2 , C_3 are given by

$$C_2 = L_i L_i + \tfrac{1}{2} T_{ij} T_{ij} = - (-2H / \sqrt{3})^2,$$

$$C_3 = (\delta L_i L_j - T_{jk} T_{ki}) T_{ij}$$
$$= -\,{}^{1}\!/_{3\sqrt{3}}\,\Phi^3(H) = (-2H/\sqrt{3})^3 \qquad \text{(special case; } \varphi_2 = \varphi_3 = 0, \varphi_1 = 1\text{)}. \tag{4.24}$$

The energy surface M_E on which the global canonical action of K is defined is given by

$$M_E \approx SU(3)/SU(2) \approx S^5, E \neq 0,$$
$$\approx SU(3)/SU(3) \approx \{o\}, E = 0, \tag{4.25}$$

where $K_o = SU(2)$ ($SU(3)$) is the stability subgroup of some point on the homogeneous space(energy surface) for $E > 0$($E = 0$). Thus, the whole phase space Ω_6 is filled by the energy surface according to

$$\Omega_6 = \{0\} \cup \{R \times S^5\} = K_0. \tag{4.26}$$

The Hamiltonian flow defines in R_6 a global action of U(1) which together with SU(3) gives the global realisation of SU(3,1).

To construct $G = SU(3,1) = \{f_a : \underline{L}, T_{ij}, \underline{P}, \underline{K}, S \mid df_a/dt = 0 = \frac{\partial f_a}{\partial t} + [H, f_a]\}$ satisfying the Poisson bracket relations:

$$[L_i, L_j] = \epsilon_{ijk} L_k$$
$$[L_i, T_{km}] = \epsilon_{ikl} T_{lm} + \epsilon_{iml} T_{kl},$$
$$[T_{ij}, T_{kl}] = L_m(\epsilon_{ikm}\delta_{jl} + \epsilon_{ilm}\delta_{jk} + \epsilon_{jkm}\delta_{il} + \epsilon_{jlm}\delta_{ik}),$$
$$[L_i, P_j] = \epsilon_{ijk} P_k,$$
$$[L_i, K_j] = \epsilon_{ijk} K_k$$
$$[T_{ij}, P_k] = K_i \delta_{jk} + K_j \delta_{ik},$$
$$[T_{ij}, K_k] = -(P_i \delta_{jk} + P_j \delta_{ik}), \tag{4.27}$$

we proceed as follows:

$$\text{Let } \underline{P} = f\,\underline{p} + g\,\underline{q}$$
$$\underline{K} = f'\,\underline{p} + g'\,\underline{q}. \tag{4.28}$$

Using (4.27) and the classica equation of motion, we finally obtain (for the special case : $\varphi_1 = 1, \varphi_2 = \varphi_3 = 0, \Phi(H) = 2H$)

$$T_{ij} = (q_i q_j + p_i p_j) - (2H/3)\, \delta_{ij},$$

$$\underline{P} = \underline{p} \cos t - \underline{q} \sin t,$$

$$\underline{K} = \underline{p} \sin t + \underline{q} \cos t,$$

$$S = 2H.$$

References.

1. J.M.Souriau, Comm.Math.Physics 1 ,374(1966).

2. B.Kostant, Lecture notes in Math. 170, 87-207(1970)(Springer-Verlag. C.T.Taam (editor)).

3. H.Weyl,Z.Phys 46,1(1927);

See also,J.R.Klauder and E.C.G.Sudarshan in Fundamentals of quantum optics(Benjamin,N.Y.,1968);

J.E.Moyal,Proc.Camb.Phil.Soc 45,99(1949);
S.P.Misra and T.S.Shankara,J.Math.Physics 9,229(1968).

4. L.Van Hove,Acad.Roy.Belg.Bull.Claase.Sci.Mem,37,610(1951).

5. R.Gupta and K.C.Tripathy,Delhi Preprint:DU/Dec/74(to be published).

6. K.C.Tripathy,R.Gupta and J.Anand,J.Math.Physics 16,1139(1975) .

7. E.Onofri and M.Pauri,J.Math.Physics 13, 533(1972);

R.Hermann, ibid,13 ,838(1972);

D.J.Simms (this proceeding).

8. K.C.Tripathy and J.Anand,Nuovo Cimento 17B,71(1973).

Weyl Quantisation on a Sphere

J. Underhill and S. Taraviras

Department of Applied Mathematics and Theoretical Physics,
University of Liverpool

Introduction Weyl's quantisation rule[1] for a non-relativistic particle assigns to the classical position and momentum q^i, p_j irreducible operators $\hat{q}^i, \hat{p}_j$ satisfying the canonical commutation relations; to the observable

$$f(q,p) = \int \varphi(x,y) \exp i(q^j x_j + p_k y^k) d^n x\, d^n y$$

it assigns the operator

$$\hat{f} = \int \varphi(x,y) \exp \tfrac{1}{2} i x_j y^j \exp i x_k \hat{q}^k \exp i y^\ell \hat{p}_\ell \, d^n x\, d^n y.$$

This rule provides a quantisation for a free particle in n-dimensional Euclidean space $\mathbb{R}^n$ in the following sense: if $H = \frac{1}{2}(p_1^2 + \dots + p_n^2)$ is the free particle Hamiltonian, then $\hat{H} = \frac{1}{2}(\hat{p}_1^2 + \dots + \hat{p}_n^2)$, and for each classical observable f,

$$[\hat{f}, \hat{H}] = i\widehat{\{f, H\}}, \tag{1}$$

($\{\}$ denotes the Poisson bracket). As a consequence there is an exact correspondence between the classical and quantum mechanical time dependence of observables, so that in particular constants of the motion correspond in the two theories.

In the Schrödinger representation $\hat{q}^i = q^i$, $\hat{p}_j = -i\frac{\partial}{\partial q^j}$ act on functions defined on configuration space $\mathbb{R}^n$. The operator $\hat{f}$ can be expressed as an integral operator whose kernel K_f is given by

$$K_f(q', q'') = (2\pi)^{-n} \int d^n p\, f(q,p) e^{-2ip_j v^j},$$

where $q^i = \frac{1}{2}(q'^i + q''^i)$, $v^i = \frac{1}{2}(q''^i - q'^i)$. With the usual Riemannian structure on $\mathbb{R}^n$, q is the midpoint of the geodesic joining q' and q''; in fact if $\gamma(t)$ is the geodesic for which $\gamma(0) = q$, $\dot{\gamma}(0) = v$, then $\gamma(1) = q''$, $\gamma(-1) = q'$. Hence we obtain a coordinate-free version of the Weyl rule: given $f \in \mathcal{F}(T^*M)$, (where $M = \mathbb{R}^n$), then

$$(\hat{f}\varphi)(q') = \int_M K_f(q', q'')\,\varphi(q'')\,d\mu(q'') \tag{2}$$

where
$$K_f(q', q'') = (2\pi)^{-n} \int d\mu(\varpi)\, f(q, \varpi)\, e^{-2i\varpi(v)}. \tag{3}$$

In these equations $d\mu(q'')$ is the Riemannian measure on M, $d\mu(\varpi)$ the corresponding measure on the cotangent space at q.

We shall be concerned with possible generalisations of Weyl quantisation to more general systems with Hamiltonians quadratic in the momenta. Such a system may be thought of as consisting of a free particle moving on a Riemannian space M, which we shall usually suppose to be the sphere S^2. Our results generalise without difficulty to the case in which M is any space of constant curvature, and there is reason to suppose that they may lead to quantisations for arbitrary symmetric spaces.

We may define $\hat{f}$ by (2) and (3) for any space for which almost all pairs (q', q'') can be joined by a unique geodesic, so that the map

$$\begin{aligned} \chi :\quad & M \times M \to TM \\ & (q', q'') \to (q, v) \end{aligned} \tag{4}$$

is defined almost everywhere on $M \times M$. However, such a quantisation will not, even in simple cases, satisfy (1). This can be checked by putting (1) in a more convenient form. Firstly, suppose that $\hat{H} = -\frac{1}{2}\Delta$ where Δ is the Laplacian on M - this can be proved for the spaces of constant curvature we consider later, and will be imposed on any quantisation we consider, together with the condition $(\widehat{f^v}\,\varphi)(q') = f(q')\,\varphi(q')$ where f is a function on M, f^v its vertical lift to T^*M.(2) Then (1) says $(\Delta'' - \Delta')\,K_f(q', q'') = 2i\,K_{\{H,f\}}(q', q'')$, where Δ' (Δ'') acts on the q' (q'') variables only. On any locally symmetric space $\Delta'' - \Delta'$ can be decomposed as

$$\Delta'' - \Delta' = \sum_i d^i D_i \tag{5}$$

where d^i, D_i are first order differential operators at (q, v) , vertical and horizontal respectively.(2) (In terms of a coordinate system about q, $D_i = \left(\frac{\partial}{\partial q^i}\right)^H$, d^i is related to the differential of the exponential map $T_qM \to M$.) Using this result it is possible to check the validity of (1) for quantisations like (2), (3). Even for spaces of constant curvature we find that (1) is not satisfied, although we can satisfy the weaker condition: if f is a constant of the motion, so is $\hat{f}$.

<u>Quantisation on the sphere</u> Let N be the complement in $S^2 \times S^2$ of the set of pairs of diagonally opposite points. Two points of N can be joined by a unique geodesic of length $< \pi R$, (R the radius of the sphere). χ maps N onto the submanifold of TS^2 consisting of points (q, v) for which $t^2 = g_q(v, v) < \frac{1}{4}\pi^2 R^2$. Let $w = v \tan\frac{t}{R} \Big/ \frac{t}{R}$ then $\eta : (q', q'') \to (q, w)$ maps N onto all of TS^2. Define

$$K^s_f(q', q'') = (2\pi)^{-2} \sec \frac{t}{R} \int d\mu(\varpi)\, f(q_r, \varpi)\, e^{-2i\varpi(w)}, \tag{6}$$

then $(\Delta'' - \Delta')K^s_f = 2i \sec^2 \frac{t}{R} K^s_{\{H,f\}}$. It follows that $[\hat{H}, \hat{f}] = 0$ iff $\{H, f\} = 0$: constants of the motion quantise into constants of the motion. We have also to check that functions on S^2, and the Hamiltonian H quantise in the correct way; this can be done. Of course in these and most other cases of interest, the integral in (6) does not converge and must be considered as a distribution on TS^2 (or, via η, on $S^2 \times S^2$), the Fourier transform of a regular distribution on T^*S^2. Hence K^s_f defines $\hat{f}$ as a symmetric operator (for real f) on the dense subspace of $\mathcal{L}^2(S^2)$ consisting of C^∞ functions.

Even this weak result, which seems to be the best one can achieve with a quantisation rule of the type of (2), (3), does not generalise to spaces of non-constant curvature. We describe below an alternative approach.

Second method Weyl's rule makes K_f an integral over the cotangent space at a single point. A more general rule is given by

$$K_f(q', q'') = \int \mathcal{K}(q, v; q_1, p_1)\, f(q_1, p_1)\, d^n q_1\, d^n p_1 .$$

On a symmetric space (1) implies

$$(d^i D_i)_{q,v} \mathcal{K} = -2i \{H, \mathcal{K}\}_{q_1, p_1},$$

(the operators on the left act only on the q, v variables, the Poisson bracket is evaluated with $\mathcal{K}$ regarded as a function of q_1, p_1).

We are concerned only with the solutions of this equation

which quantise (i) the Hamiltonian and (ii) functions on configuration space in the correct way. A brief description of a partial solution for the case of the two-dimensional sphere follows.

First, the differential equation possesses solutions depending only on certain quantities which fix the set $q, v; q_1, p_1$ up to rotations of the sphere. E.g., if Ω is a rotation which sends q into some arbitrarily chosen origin O, and v into a vector (of length t) in some fixed direction at O, and if Ω maps q_1, p_1 into Q, P, then we may assume that $\mathcal{K}(q, v; q_1, p_1) = \tan\frac{t}{R}\,\Theta(t, Q, P)$.

The equation satisfied by Θ is simplified by observing that some of the first order operators occurring in it provide a representation of $so(3)$, (the representation arising from the usual action of $SO(3)$ on T^*S^2). If Θ is expanded in terms of suitable eigenfunctions of the Casimir operator J, then the coefficients in the expansion depend only on t and on t_1 = length of p_1; those labelled by the eigenvalue $j(j+1)$ of J satisfying $2j+1$ coupled first order ordinary differential equations. For low values of j, $(1,2)$, the equations can be solved explicitly, though this seems difficult in general. The $j = 0$ coefficient is restricted only by (i) and (ii). The second of these restrictions also applies to coefficients with $j > 0$, and where the differential equations can be solved explicitly it is possible to find (many) solutions which also satisfy (ii).

For the details, let Q, $\bar{Q}$ be complex coordinates on the sphere, related to spherical polar coordinates θ, φ by

$Q = \tan\frac{1}{2}\theta\, e^{i\varphi}$, $\bar{Q} = \tan\frac{1}{2}\theta\, e^{-i\varphi}$, and let Q , $\bar{Q}$, P , $\bar{P}$ be corresponding coordinates on T^*S^2 . Then Θ satisfies

$$t_1 N \Theta = \frac{i(1+4t^2)}{2t}\Big\{\Big(t\frac{\partial}{\partial t}+1-M\Big)A + \Big(t\frac{\partial}{\partial t}+1+M\Big)\bar{A}\Big\}\Theta$$

where

$$A = i\Big(Q^2\frac{\partial}{\partial Q}+\frac{\partial}{\partial \bar{Q}} - 2\bar{P}Q\frac{\partial}{\partial \bar{P}}\Big),$$

$$\bar{A} = i\Big(\bar{Q}^2\frac{\partial}{\partial \bar{Q}}+\frac{\partial}{\partial Q} - 2P\bar{Q}\frac{\partial}{\partial P}\Big),$$

$$M = P\frac{\partial}{\partial P} - \bar{P}\frac{\partial}{\partial \bar{P}} + Q\frac{\partial}{\partial Q} - \bar{Q}\frac{\partial}{\partial \bar{Q}},$$

$$t_1 N = -i(1+Q\bar{Q})^2\Big\{P\frac{\partial}{\partial Q}+\bar{P}\frac{\partial}{\partial \bar{Q}} - \frac{2P\bar{P}}{1+Q\bar{Q}}\Big(Q\frac{\partial}{\partial P}+\bar{Q}\frac{\partial}{\partial \bar{P}}\Big)\Big\},$$

$$t_1^2 = P\bar{P}(1+Q\bar{Q})^2.$$

These operators satisfy $[A,\bar{A}] = 2M$, $[M,A] = A$, $[M,\bar{A}] = -\bar{A}$, N commutes with M, A and $\bar{A}$. The functions

$$\Phi_{jmn}(Q,\bar{Q},P,\bar{P}) = 2^n\left(\frac{1+iQ\sqrt{\bar{P}/P}}{1-iQ\sqrt{\bar{P}/P}}\right)^n\left(\frac{P+Q^2\bar{P}}{t_1}\right)^m P_{j-m}^{m+n,\,m-n}\left(\frac{-i(Q\bar{P}-\bar{Q}P)}{t_1}\right)$$

are simultaneous eigenstates of M, N and $J = A\bar{A} + M^2 - M$ with eigenvalues m , $2n$ and $j(j+1)$ respectively, ($j = 0,1,2,\dots$, m and n take on the values $-j, -j+1, \dots, j-1, j$).

If

$$\Theta = \sum_{j,m,n} \alpha_{jmn}(t,t_1)\,\Phi_{jmn}(Q,\bar{Q},P,\bar{P}),$$

then

$$2ni\,t_1\alpha_{jmn} = \frac{(1+4t^2)}{2t}\Big\{(j+m)\Big(t\frac{\partial}{\partial t}+1-m\Big)\alpha_{j\,m-1\,n} + (j-m)\Big(t\frac{\partial}{\partial t}+1+m\Big)\alpha_{j\,m+1\,n}\Big\}.$$

For low values of j, the α_{jmn} are Legendre functions whose order depends on nt_1. The conditions on the α_{jmn} arising from (i), (ii) have the forms

$$\text{(ii)} \quad \sum_{n=-j}^{j} \frac{\Gamma\left(\frac{n+j+1}{2}\right)}{\Gamma\left(\frac{n-j+1}{2}\right)} \int_0^\infty \alpha_{jmn}(r, r_1)\, t_1\, dr_1 \quad \propto \quad \delta_{m0}\, \delta(t^2),$$

$$\text{(i)} \quad \int_0^\infty t_1^3\, \alpha_{000}(r, r_1)\, dr_1 \quad \propto \quad \delta'(t^2).$$

References

(1) H. Weyl. Theory of Groups and Quantum Mechanics, 2nd Edit., Dover, 1931, p.275.

(2) K. Yano and S. Ishihara. Tangent and Cotangent Bundles, Dekker, 1973, pp. 6, 87.

CONFORMAL GROUP, QUANTIZATION, AND THE KEPLER PROBLEM

Joseph A. Wolf

§1. INTRODUCTION. This is a report on some joint work with Shlomo Sternberg. We consider a variation on geometric quantization for the orthogonal groups SO(2,n), realizing certain of their representations on the nonzero cotangent bundle of the (n-1)-sphere. Here the elliptic orbits of the Kepler problem (with collision orbits regularized) appear as SO(2)-orbits. Another viewpoint, related by a geometric Cayley transform, gives the hyperbolic orbits as SO(1,1)-orbits in the nonzero cotangent bundle of real hyperbolic (n-1)-space. This gives a correspondence between the classical bound states and the classical scattering states for the hydrogen atom.

Our group-theoretic considerations are valid with only minor changes for the unitary groups U(2,n), the special unitary groups SU(2,n), and the unitary symplectic groups Sp(2,n). While there is a connection with the harmonic oscillator, the physical interpretations are not always so clear. In any case, here I just indicate the situation for SO(2,n). Complete details will appear elsewhere.

§2. A NILPOTENT CO-ADJOINT ORBIT. Let $R^{2,n}$ denote the real vector space with standard basis $\{e_{-1}, e_0, e_1, \ldots, e_n\}$ and inner product $\langle u,v\rangle = u_{-1}v_{-1} + u_0v_0 - (u_1v_1 + \ldots + u_nv_n)$. $O(2,n)$ is the orthogonal group of $R^{2,n}$, $G = SO(2,n)$ denotes its identity component, and the alternating tensor square $\Lambda^2(R^{2,n})$ is identified with the Lie algebra $\mathfrak{g} = \mathfrak{o}(2,n)$ under

$$u \wedge v : x \longmapsto \langle x,u\rangle v - \langle x,v\rangle u .$$

Here the adjoint representation is given by $Ad(g)(u \wedge v) = gu \wedge gv$.

If $\xi \in \mathfrak{g}$ let E_ξ denote its range. If E_ξ is 2-dimensional and totally isotropic, then $\xi^2 = 0$, and E_ξ projects onto $R^{2,0} = \mathrm{span}\{e_{-1}, e_0\}$, so ξ has unique expression

$$(2.1) \qquad \xi = s(e_{-1} + p) \wedge (e_0 + q) \text{ where } \begin{cases} p,\ q \in R^{0,n} = \mathrm{span}\{e_1, \ldots, e_n\} \\ \|p\|^2 = \|q\|^2 = -1,\ \langle p,q\rangle = 0 \end{cases}$$

All such ξ form a single $O(2,n)$-orbit. Here $s = \langle \xi, e_{-1} \wedge e_0 \rangle$, and that single orbit falls into two G-orbits as $s > 0$ or $s < 0$. We will use the orbit

$$(2.2) \qquad \mathcal{V} = \{ \xi \in \mathfrak{g} \text{ as in (2.1)} : s > 0 \}.$$

The semisimple Lie algebra $\mathfrak{g}$ is identified with its dual space $\mathfrak{g}^*$ under the Killing form, and we view $\mathcal{V}$ as a (co-adjoint) orbit of G on $\mathfrak{g}^*$. That gives $\mathcal{V}$ the structure of G-homogeneous symplectic manifold.

In the notation (2.1), think of q as a point on the unit sphere $S^{n-1} = \{ x \in R^{0,n} : \|x\|^2 = -1 \}$ and sp as an arbitrary nonzero cotangent vector to S^{n-1} at q. This identifies $\mathcal{V}$ with the bundle $T^+(S^{n-1})$ of nonzero cotangent vectors to S^{n-1}. In this identification, the subgroup

$$G_1 = SO(1,n) = \{ g \in G : ge_{-1} = e_{-1} \}$$

is visibly transitive on $T^+(S^{n-1})$, and thus on $\mathcal{V}$. Furthermore $\xi = s(e_{-1} + p) \wedge (e_0 + q) \longmapsto sp \wedge (e_0 + q)$ is a bijection of $\mathcal{V}$ onto the principal nilpotent coadjoint orbit of G_1, which is

$$(2.3) \qquad \mathcal{V}_1 = \{ \xi_1 \in \mathfrak{g}_1 : \dim E_{\xi_1} = 2 \text{ and } \dim(E_{\xi_1} \cap E_{\xi_1}^{\perp}) = 1 \}.$$

$\mathcal{V}$ now carries three symplectic structures: as co-adjoint orbit of G, from the natural symplectic structure on the cotangent bundle of S^{n-1}, and from the natural symplectic structure of $\mathcal{V}_1$. Here our result is

THEOREM. The three symplectic structures on $\mathcal{V}$ coincide. In particular, the natural symplectic structure on $T^+(S^{n-1})$ is invariant under the action of $G = SO(2,n)$.

§3. ORBITS FOR THE KEPLER PROBLEM. We have $R^{2,n} = R^{2,0} \oplus R^{0,n}$ as above, and the G-stabilizer of this splitting is the maximal compact subgroup $K = SO(2) \times SO(n)$. Here $SO(n)$ acts on $\mathcal{V}$ through its usual action on the tangent bundle $T(S^{n-1})$,

$$A : s(e_{-1} + p) \wedge (e_0 + q) \longmapsto s(e_{-1} + Ap) \wedge (e_0 + Aq) ,$$

and $SO(2)$ acts by rotations, the rotation r_φ through an angle φ sending $s(e_{-1} + p) \wedge (e_0 + q)$ to

$$s(\cos\varphi\, e_{-1} + \sin\varphi\, e_0 + p) \wedge (-\sin\varphi\, e_{-1} + \cos\varphi\, e_0 + q)$$
$$= s(e_{-1} + \cos\varphi\, p - \sin\varphi\, q) \wedge (e_0 + \sin\varphi\, p + \cos\varphi\, q) .$$

On (co)-tangent vectors of length s, this rotation r_φ is geodesic flow $f_{\varphi/s}$ at time φ/s. The infinitesmal generator of the geodesic flow $\{f_t\}$ is the vector field V_H corresponding (by exterior derivative and the symplectic form) to $H = -s^2/2$, so $\{r_\varphi\}$ has infinitesmal generator that is the Hamiltonian field for $(-2H)^{1/2} = s$. Since the SO(2)-orbits are the orbits of the geodesic flow, they are the elliptic orbits of the Kepler problem with collision orbits regularized.

Similarly $R^{2,n} = R^{1,1} \oplus R^{1,n-1}$ where $R^{1,1} = \text{span}\{e_{-1},e_n\}$ and $R^{1,n-1} = \text{span}\{e_0,e_1,\dots,e_{n-1}\}$. The G-stabilizer of this splitting is a two-component group with identity component $K' = SO(1,1)\times SO(1,n-1)$, and $\mathcal{U}$ is the union of three K'-invariant sets

$$\mathcal{U}^+ = \{t(e_{-1}+p)\wedge(e_n+q) : t>0,\ p,q\in R^{1,n-1}, \|p\|^2=-1, \|q\|^2=1, p\perp q, \langle e_0,q\rangle>0\},$$

$$\mathcal{U}^0 = \{\xi\in\mathcal{U} : E_\xi \cap R^{1,1} \neq 0\}, \text{ and}$$

$$\mathcal{U}^- = \{t(e_{-1}+p)\wedge(e_n+q) : t<0,\ p,q\in R^{1,n-1}, \|p\|^2=-1, \|q\|^2=1, p\perp q, \langle e_0,q\rangle<0\}.$$

Let H_+^{n-1} (resp. H_-^{n-1}) denote the real hyperbolic (n-1)-space that is is the sheet $\langle e_0,q\rangle > 0$ (resp. $\langle e_0,q\rangle < 0$) of the mass hyperboloid $\|q\|^2 = 1$ in $R^{1,n-1}$. Then $\mathcal{U}^+$ (resp. $\mathcal{U}^-$) is identified with its bundle $T^+(H_+^{n-1})$ (resp. $T^+(H_-^{n-1})$) of nonzero cotangent vectors, $\xi = t(e_{-1} + p)\wedge(e_n + q)$ corresponding to the vector tp of length $|t|$ at q. Here SO(1,n-1) acts through its usual action by isometries and SO(1,1) acts, as before, by hyperbolic rotations proportional to the geodesic flow. So the SO(1,1)-orbits on $\mathcal{U}^\pm$ are the hyperbolic orbits of the Kepler problem.

If one interprets the SO(2)-orbits on $\mathcal{U}$ as the classical bound states for the hydrogen atom, and the SO(1,1)-orbits on $\mathcal{U}^\pm$ as the scattering states, then the Cayley transform relating SO(2) to SO(1,1) gives a sort of correspondence between those states. The geometric picture for this Cayley transform comes from noting that the sign condition on $\langle e_0,q\rangle$ identifies H_+^{n-1} with the upper hemisphere of S^{n-1} and H_-^{n-1} with the lower hemisphere:

$$t(e_{-1}+p)\wedge(e_n+q) = t\langle e_0,q\rangle\{(e_{-1}+p)\wedge(e_0 + \langle e_0,q\rangle^{-1}(q-\langle e_0,q\rangle e_0+e_n)\}.$$

Another interesting picture comes from taking p for base point and tq for (co)-tangent vector.

§4. GEOMETRIC QUANTIZATION. We turn to the question of quantizing the action of $G = SO(2,n)$ on its co-adjoint orbit $\mathcal{V} = T^+(S^{n-1})$.

The standard Kostant-Souriau quantization procedure does not work here because there is no G-invariant polarization. In effect, a result of Ozeki and Wakimoto says that any such polarization would be a parabolic subalgebra $\mathfrak{q}$ of $\mathfrak{g}_C$, a result of mine would then say $\mathfrak{q} = \mathfrak{p}_C$ for some parabolic subalgebra $\mathfrak{p}$ of $\mathfrak{g}$, and of course $\mathfrak{p}$ would necessarily have codimension n-1 in $\mathfrak{g}$. But the maximal parabolic subalgebras of $\mathfrak{g}$ are the stabilizers of null lines, which have codimension n, and the stabilizers of null planes, which have codimension 2n-1, and so $\mathfrak{p}$ does not exist.

There are several possibilities for circumventing this lack of polarizations:

(i) weaken the definition of polarization,

(ii) view $\mathcal{V}$ as a limit of polarized co-adjoint orbits,

(iii) use the Kostant-Sternberg-Blattner half-form method.

In the first approach, one takes the usual definition of invariant polarization as complex subalgebra $\mathfrak{q}$ of $\mathfrak{g}_C$, but no longer requires that $\mathfrak{q} + \bar{\mathfrak{q}}$ be an algebra; that is done implicitly in N. Woodhouse's report at this conference. In the second approach, one has a smooth family $\mathcal{V}_t$ of co-adjoint orbits with $\mathcal{V} = \mathcal{V}_0$, with representations π_t associated to $\mathcal{V}_t$ for $t \neq 0$, in such a way that one can make sense of $\pi_0 = \lim \pi_t$ and associate it to $\mathcal{V}$; in E. Onofri's report here, that is done for elliptic semisimple approximating orbits and holomorphic discrete series approximating representations, and I have a comment on this in §6 below. Sternberg and I use the third approach.

§5. HALF FORMS AND VARYING POLARIZATIONS. Let P denote the standard polarization on $T^+(S^{n-1})$; its maximal integral manifolds are the cotangent spaces with origin deleted. Then $G_1 = SO(1,n)$ is the stabilizer of P in $G = SO(2,n)$, and the G-translates of P are parameterized by the mass shell $H = \{x \in R^{2,n} : \|x\|^2 = 1\}$:

$$G/G_1 = SO(2,n)/SO(1,n) \cong SO(2,n)(e_{-1}) = H .$$

Given $x = ge_{-1} \in H$, let P_x denote the image $g(P)$. The half form method gives a family of Hilbert spaces $\mathcal{H}_x$, and nondegenerate pairings between them, stable under the action of G. Here G_1 has natural irreducible unitary representation ψ on $\mathcal{H} = \mathcal{H}_{e_{-1}}$ by standard geometric quantization using $P = P_{e_{-1}}$; in fact ψ is the principal series representation that corresponds to the trivial character on the minimal parabolic subgroup, and $\mathcal{H}$ is $L^2(S^{n-1})$. More generally, if $g \in G$ then g carries $\mathcal{H}$ to $\mathcal{H}_x$, $x = ge_{-1}$, and we pair this back to $\mathcal{H}$ using the half forms. Thus G acts on $L^2(S^{n-1})$, and this action π restricts to the representation ψ of G_1. Sternberg and I still have to clarify some technical matters with the half form pairing here.

§6. LIMIT METHOD. I'll close by exhibiting the representation π of G, corresponding to the co-adjoint orbit $\mathcal{U} = T^+(S^{n-1})$, as a limit of spherical principal series representations. This has the advantage of simplicity over Onofri's procedure with the holomorphic discrete series, but the disadvantage of obscuring the place of G_1 and $L^2(S^{n-1})$ as compared with the half form method.

Fix $\xi = (e_{-1}+e_{n-1})\wedge(e_0+e_n)\in\mathfrak{g}$. Its matrix is $\begin{pmatrix} J & 0 & -J \\ 0 & 0 & 0 \\ J & 0 & -J \end{pmatrix}$ where $J = \begin{pmatrix} 0 & -1 \\ 1 & 0 \end{pmatrix}$. Then $\eta = -\frac{1}{4}(e_{-1}-e_{n-1})\wedge(e_0-e_n)$ is another nilpotent element of $\mathfrak{g}$. It has matrix $\frac{1}{4}\begin{pmatrix} -J & 0 & -J \\ 0 & 0 & 0 \\ J & 0 & J \end{pmatrix}$, and so $h = [\xi,\eta]$ has matrix $\begin{pmatrix} 0 & 0 & I \\ 0 & 0 & 0 \\ I & 0 & 0 \end{pmatrix}$ where $I = \begin{pmatrix} 1 & 0 \\ 0 & 1 \end{pmatrix}$. Now

$$[h,\xi] = 2\xi\ , \quad [h,\eta] = -2\eta \quad \text{and} \quad [\xi,\eta] = h\ .$$

So $\{h,\xi,\eta\}$ is a standard generating triple for a split three dimensional simple subalgebra (TDS) in $\mathfrak{g}$, that is

$$h \longrightarrow \begin{pmatrix} 1 & 0 \\ 0 & -1 \end{pmatrix}, \quad \xi \longrightarrow \begin{pmatrix} 0 & 1 \\ 0 & 0 \end{pmatrix}, \quad \eta \longrightarrow \begin{pmatrix} 0 & 0 \\ 1 & 0 \end{pmatrix}$$

defines a Lie algebra isomorphism of span$\{h,\xi,\eta\}$ onto $\mathfrak{sl}(2;R)$. From this we see that

$\xi_t = \xi + th$ is semisimple with real eigenvalues for $t \neq 0$.

Let B be a minimal parabolic subgroup of G whose Lie algebra contains ξ and h, and denote

$\mathcal{U}_t = Ad(G)\cdot\xi_t$ viewed as a co-adjoint orbit,

π_t : the corresponding principal series representation $(t \neq 0)$,

φ_t : the positive definite spherical function for π_t $(t \neq 0)$.

Then the π_t, $t \neq 0$, are irreducible unitary representations of G on $L^2(G/B)$ given by formulas that depend smoothly on t , and one has

$\pi = \lim_{t \to 0} \pi_t$: unitary representation of G on $L^2(G/B)$.

Here π_t corresponds to the orbit $\mathcal{V}_t$ for $t \neq 0$, and so π corresponds to $\mathcal{V} = \mathcal{V}_0$.

One obtains the same limit with the spherical functions. For φ_t defines π_t in the standard manner when $t \neq 0$, and $\varphi = \lim_{t \to 0} \varphi_t$ is a positive definite spherical function and thus defines a limit representation π .

Departments of Mathematics,
University of California at Berkeley
The Hebrew University of Jerusalem
Tel Aviv University

BROKEN SYMMETRIES

EXCEPTIONAL GROUPS AND ELEMENTARY PARTICLES[+*]

Feza Gürsey
Yale University, Physics Department
New Haven, CT. 06520

ABSTRACT If the exceptional observables introduced by Jordan, von Neumann and Wigner are identified with charge states corresponding to internal degrees of freedom of elementary particles, then one is led to the classification of quarks and leptons by means of exceptional groups. It is shown that the groups of the E series are likely candidates and a gauge field theory based on E_7 is given as an example. The hierarchy of symmetry breaking is linked to the hierarchy of stability groups for geometries that are associated with the exceptional groups.

I. Introduction: the New World Picture

Thanks to far reaching recent developments in gauge field theories[1,2] (unification of weak and electromagnetic interactions, renormalizability of spontaneously broken gauge field theories, asymptotic freedom in non-abelian gauge theories) and momentous experimental discoveries (scaling[3] in lepton-hadron, hadron-hadron scattering and e^+e^- annihilation into hadrons; existence of weak interactions mediated by neutral currents[3]; new hadron families associated with a new quantum number[4,5]) we have now the possibility of describing electromagnetic, weak and strong interactions in a unified way: local renormalizable field theory is applicable to the whole world of elementary particles. The fields that occur in the local Lagrangian are leptons, quarks (which come in three colors and at least four flavors), gauge vector bosons which mediate interactions among these fundamental fermions and scalar mesons (Higgs fields) with self interactions and interactions with the fermions and vector bosons.

The Lagrangian of the theory is determined by Poincaré invariance, the principle of local (gauge) invariance with respect to a compact internal symmetry group G, and finally the principle of renormalizability (smooth high energy behavior). For a truly unified theory G is a simple

+Talk presented at the 4th International Colloquium on Group Theoretical Methods in Physics, Nijmegen, June 1975.

*Research (Yale Report COO-3075-124) supported in part by the U.S. Energy Research and Development Administration under Contract No. AT(11-1)-3075.

or semi-simple Lie group. It must have $SU(2)xU(1)xSU^c(3)$ as a subgroup. Here $SU^c(3)$ is the exact group of strong interactions (the color group) leading to asymptotic freedom at high energies and hence to scaling (up to logarithmic terms) in scattering involving hadrons. The SU(2)xU(1) subgroup is associated with the electromagnetic and weak parts of the interaction (including neutral currents).

The vector bosons belong to the adjoint representation of G. The fermions and the Higgs scalars may belong to other irreducible representations. Some of the Higgs fields develop non zero vacuum expectation values through the minimization of the Higgs potential which describes their self interactions. Then the vacuum is no longer symmetrical with respect to a subgroup of G and we have the phenomenon of spontaneous symmetry breaking which plays a fundamental role in all the major fields of physics as beautifully explained by Professor Michel in his contribution to this conference. The vacuum expectation values of the Higgs fields give masses to the fermions and to the vector bosons. Those vector bosons associated with the unbroken subgroup of G (color gluons and the photon) remain massless.

Three of the quarks are carriers of isotopic spin and hypercharge. A fourth one carries the new hadronic quantum number called charm. There may be other quarks associated with other flavors and new leptons corresponding to new degrees of freedom that characterize the group G. Thus the fermion multiplet must include at least 4 colored quarks (12 charge states) and 4 leptons (e, μ, ν^e and ν^μ). There[6] are strong indications that there exist another heavy charged lepton E^- and its neutral companion N^E. Then, in a unified theory the basic fermions will belong to an irreducible representation of G which has dimension $d>16$ or $d>18$. The dimension of the adjoint representation (number of parameters of G) must be at least 12 (8 for color, 4 for the weak and electromagnetic group).

For a physical interpretation we require another principle: color confinement. According to this principle the only observable scattering states are those that are invariant under the exact color group $SU^c(3)$. This includes leptons, weak bosons and the photon associated with SU(2)xU(1) and those bound states of quarks that are color singlets. These will include mesons ($q\bar{q}$), baryons (qqq), antibaryons ($\bar{q}\bar{q}\bar{q}$) and resonances or bound states of mesons and baryons. Hadrons being bound states will be extended objects held together by gluons. They may have string or bag structure. Field theory is directly applicable not to these hadrons but to their point like constituents (quarks), to leptons and gluons.

The picture of the world we have outlined has already received impressive experimental support. The evidence for the new approach has been eloquently spelled out at this conference by Professor Iliopoulos. There are some serious problems: namely, the difficulty of incorporating gravitation in our world picture (provisionally excluded by the principle of renormalizability), the absence of a rigorous mathematical justification for color confinement, the necessity for the introduction of very high mass bosons (of the order of the Planck mass) for a truly unified theory based on a simple group G, the stability of the proton, the need for too many Higgs scalars to explain the hierarchy of masses and coupling constants, the difficulty of introducing the Higgs fields as bound states (dynamical symmetry breaking) and the absence of a principle for selecting the gauge group G.

It is to this last question that I would like to address myself in this talk. I shall summarize some results of work[7] done in collaboration with Dr. M. Günaydin,[8] Dr. P. Sikivie and Dr. P. Ramond.[9]

Some examples of universal gauge field theories have been developed and studied by Georgi and Glashow[10] (SU(5)), Pati and Salam[11] (SU(4)xSU(4)), Fritzsch and Minkowski[12] (SO(10), SO(14), SU(16)xSU(16), etc.) and others. The most economical and elegant of these schemes is the SO(10) model. They all suffer however, either from proton instability or chiral anomalies in renormalization.

II. Are charges the JNW exceptional observables?

In searching for a principle that leads to the compact group G, we are trying to single out a finite Hilbert space on which the generators of G act. Are there such finite spaces that arise naturally in the classification of Hilbert spaces? An affirmative answer was given by Jordan, von Neumann and Wigner more than four decades ago.[13] Confronted by the new puzzling phenomena of the nuclear world following the discovery of the neutron and the study of β-decay they looked for all possible extensions of Quantum Mechanics which had been so successful in understanding atomic physics. They showed that the usual formulation of Quantum Mechanics in a vector space (the kets and the bras of the dual space) with observables represented by hermitian operators which act on states represented by kets is equivalent to an algebraic formulation in which both observables and states are represented by hermitian matrices, the states being now associated with projection operators

$$P_\alpha = |\alpha><\alpha| \qquad (P_\alpha^2 = P_\alpha) \tag{2.1}$$

that are in one-to-one correspondence with kets, $|\alpha>$. A state which is not a pure state is also represented by a hermitian operator, namely a density matrix

$$D = \sum_i p_i^2 |\alpha_i><\alpha_i| \ , \quad \mathrm{Tr}\ D = \sum p_i^2 = 1 \quad . \tag{2.2}$$

Instead of the expectation value $<\alpha|\Omega|\alpha>$ of a hermitian operator (observable) Ω, and its matrix elements $<\alpha|\Omega|\beta>$ we consider respectively the positive numbers

$$\mathrm{Tr}\ P_\alpha\Omega = <\alpha|\Omega|\alpha> \text{ and } \mathrm{Tr}(P_\alpha\Omega P_\beta\Omega) = |<\alpha|\Omega|\beta>|^2. \tag{2.3}$$

For states that are not pure we substitute density matrices D for the projection operators P_α. Note that in all these physically meaningful quantities the operation on hermitian matrices that is relevant is their symmetrized product since contact with experiment is made after taking the trace of matrix products. Thus, the only product which occurs in this algebraic formulation of quantum mechanics is the symmetrized (or Jordan) product of hermitian matrices which we denote by

$$A.B = \tfrac{1}{2}(AB + BA), \ (A = A^\dagger;\ B = B^\dagger) \quad . \tag{2.4}$$

This product which defines the algebra of observables is commutative but non associative, the associator being

$$(A,X,B) = -\ (B,X,A) = (A.X).B - A.(X.B) \quad . \tag{2.5}$$

In fact, if A, X, B are matrices over complex numbers, we have

$$(A,X,B) = \tfrac{1}{4}\ [X,[A,B]], \tag{2.6}$$

so that A and B are compatible (simultaneously measurable) observables if $(A,X,B) = 0$ for all X.

The automorphism groups of the Jordan algebra of nxn hermitian matrices are the orthogonal groups SO(n) for real matrices, the unitary groups SU(n) for complex matrices and the symplectic groups Sp(n,q) for nxn quaternionic matrices (hermitian with respect to quaternion conjugation). Thus we recover the classical groups and their associated Hilbert spaces.

An infinitesimal unitary transformation on X reads

$$X' = X + i[X,C] \quad . \tag{2.7}$$

If we express the antihermitian matrix in terms of two hermitian matrices A and B by $\frac{1}{4}[A,B] = iC$, then Eq. (2.7), takes the form

$$X' = X + (A,X,B) \tag{2.8}$$

expressing the infinitesimal action of the automorphism group purely in terms of the Jordan algebra. Then A and B are also observables associated with the transformation of the observable X. The finite unitary transformation on X can be written in two ways

$$X' = e^{-iC}Xe^{iC} = 1 + \frac{i}{1!}[X,C] + \frac{i}{2!}[[X,C],C] + \ldots \qquad (2.9a)$$

in terms of the Lie algebra of commutators or, alternatively

$$X' = E_{A,B}X = X + \frac{1}{1!}(A,X,B) + \frac{1}{2!}(A,(A,X,B),B) + \ldots \qquad (2.9b)$$

in terms of associators that only involve the Jordan algebra.

Now, Jordan, von Neumann and Wigner found one case in which the algebra of observables is obeyed by operators that are not matrices over an associative division algebra. They are 3x3 octonionic matrices hermitian with respect to octonionic conjugation. Segal and Sherman[14] later showed that idempotent matrices of this kind can be made to correspond to pure states and a general matrix of the form

$$X = \begin{pmatrix} \alpha & c & \bar{b} \\ \bar{c} & \beta & a \\ b & \bar{a} & \gamma \end{pmatrix}, \qquad (2.10)$$

with Tr X = 1 can represent a quantum mechanical density matrix. Here α, β, γ are real and a,b,c are octonions with the bar denoting octonionic conjugation. A determinant form can be uniquely associated with X by the formula

$$I \operatorname{Det} X = X^{\cdot 3} - X^2 \operatorname{Tr} X + \tfrac{1}{2} X \{(\operatorname{Tr} X)^2 - \operatorname{Tr} X^2\}, \qquad (2.11)$$

where I is the 3x3 unit matrix and $X^{\cdot 3} = X.X^2 \ldots$ Chevalley and Schafer[15] showed that these observables can be transformed by the generalization of the unitary transformations in the form (2.9b) with A and B matrices elements of the set (2.10) with zero trace. Thus the group has 2x26=52 parameters and is the exceptional group F_4 acting on the exceptional observable X. The generators of F_4 correspond to observable charges. The automorphism group of the octonions was already known to be the exceptional group G_2. Its infinitesimal action on the octonion x is given by

$$x' = x - \tfrac{1}{4}[[a,b],x] + \tfrac{3}{4}[a,b,x] \qquad (2.11)$$

where [a,b,x] is the completely antisymmetrical associator with respect to the octonion product which is non commutative and non associative. Here a and b are such that their scalar parts $\frac{1}{2}(a+\bar{a})$ and $\frac{1}{2}(b+\bar{b})$ vanish so that G_2 has 2x7=14 parameters. It has a maximal subgroup $SU^c(3)$ which leaves one of the 7 octonionic imaginary units invariant.

Another subgroup of F_4 that acts on X and commutes with $SU^c(3)$ is an SU(3) associated with the 3x3 matrix structure of X in which the octonion unit left invariant by $SU^c(3)$ plays the role of the imaginary unit of complex numbers. Thus, F_4 admits $SU(3)xSU^c(3)$ as a maximal subgroup. We also note that the group F_4 leaves invariant Tr X, Tr X^2 and Det X.

We have arrived at the remarkable conclusion that there exists a finite Hilbert space associated with an exceptional realization of the algebra of observables with a transformation law that involves two different SU(3) groups, one associated with octonions which has the same structure as the color group $SU^c(3)$ and the other with the SU(3) of unitary symmetry. Furthermore these observables arise only in the JNW formulation,as the ket associated with an idempotent X does not transform linearly under F_4.

Since the discovery of these exceptional observables there has been a case of the missing observables in physics as this exceptional Hilbert space could not be related to any known states in Physics. Now with the new discoveries of a rich charge structure for elementary particles, the above considerations strongly suggest the identification of the charge space with the space of exceptional observables.

The 26 and 52 dimensional representations of F_4 have the following decomposition in terms of $SU(3)xSU^c(3)$:

$$26: (8,1) + (3,3^c) + (\bar{3},\bar{3}^c) \qquad (2.12)$$

$$52: (8,1) + (1,8^c) + (6,\bar{3}^c) + (\bar{6},3^c) \; . \qquad (2.13)$$

Hence if we take G = F_4 the (26) representation unifies 8 Majorana (4 Dirac) leptons with integer charges together with 3 fractionally charged quarks and 3 antiquarks. The adjoint representation contains color singlet vector bosons together with octet colored gluons and twelve lepto quark bosons.

The other exceptional groups E_6, E_7 and E_8 act also on finite Hilbert spaces with elements that are partly real or complex numbers and partly exceptional observables. Hence if we take as our guiding principle for finding G the identification of internal degrees of freedom with exceptional observables we are uniquely led to exceptional groups. The need for charm singles out the E series as natural candidates for the universal gauge group G.

III. Examples of schemes based on E_6 and E_7. Spontaneous Symmetry Breaking and Geometry.

The smallest Hilbert space E6 acts on is provided by a pair

(X_1,X_2) of exceptional observables which can be combined in a single complex matrix J which is hermitian with respect to octonionic conjugation only. The infinitesimal E_6 transformation on this 27 dimensional representation is

$$T_{A,B,C}\, J = J + (A,J,B) + i\, C.J \quad , \qquad (3.1)$$

where A,B,C are traceless real matrices belonging to the set (2.10). Then the group has 3x26=78 parameters. J^* corresponds to an inequivalent $\overline{27}$ representation for which the transformation law is as in (3.1) with $C \to -C$.

We now introduce the symmetric Freudenthal product

$$JxK = J.K - \frac{1}{2} J \,\mathrm{Tr}\, K - \frac{1}{2} K \,\mathrm{Tr}\, J - \frac{1}{2} \mathrm{Tr}(J.K - \frac{1}{2} J \,\mathrm{Tr}\, K - \frac{1}{2} K \,\mathrm{Tr}\, J). \qquad (3.2)$$

If J_1 and J_2 transform like 27, J_1xJ_2 transforms as $\overline{27}$ and we have

$$I\, \mathrm{Det}\, J = (J \times J).J \quad . \qquad (3.3)$$

Under E_6 we have the invariants

$$I_1 = \frac{1}{2} \mathrm{Tr}\, JJ^*, \quad I_2 + iI_3 = \mathrm{Det}\, J, \quad I_4 = \frac{1}{2} \mathrm{Tr}\, \{(JxJ).(J^*xJ^*)\} \qquad (3.4)$$

If J represents Higgs scalar fields, then a linear combination of these invariants forms a Higgs potential the minimization of which gives a spontaneous breaking of E_6 into SO(10)xSO(2). This is just the gauge group considered by Fritzsch and Minkowski. This last group leaves invariant an idempotent J which obeys JxJ = 0, Tr J = 1. This has a geometric interpretation: SO(10)xSO(2) is the stability group of the geometrical object J which moves in a generalized projective space under the action of E_6. Thus, spontaneous symmetry breaking is intimately connected with the hierarchy of subgroups that are stability groups of various geometrical objects.[16]

E_7 acts in the 56 dim. space $(\alpha_1, J_1, J_2, \alpha_2)$ where J_1 and J_2 transform respectively like 27 and $\overline{27}$ of the E_6 subgroup and the complex numbers α_1 and α_2 are E_6 invariant. The parameters of E_7 can be grouped into A,B,C, namely the generators of E_6, a real parameter λ and a complex Jordan matrix K giving 133 real parameters. Under the maximal subgroup $SU(6)xSU^c(3)$ we have

$$56 = (20,1) + (6,3) + (\bar{6},\bar{3}) \qquad (3.4)$$

$$133 = (35,1) + (1,8) + (15,3) + (\overline{15},\bar{3}) \qquad (3.5)$$

giving 10 Dirac leptons (4 charged, 6 neutral), 6 colored quarks and six antiquarks for the 56. The adjoint representation has 35 color singlet vector bosons, 8 colored gluons and 90 leptoquark colored bosons. Under spontaneous breaking $SU(6) \rightarrow SU(4)xSU(2)$. Thus we have the $SU(4)xSU(2)xSU^c(3)$ decomposition.

$$56 = (4,1,1^c) + (\bar{4},1,1^c) + (6,2,1^c) + (1,2,3^c) + (4,1,\bar{3}^c) + (1,2,\bar{3}^c) \qquad (3.6)$$

We identify $(4,1,1^c)$ with left handed leptons, $(\bar{4},1,1^c)$ with right handed leptons, (6,2,1) with heavy leptons, $(4,1,3^c)$ with the colored quarks $\mathcal{P}$, $\mathcal{N}$, λ and $\mathcal{P}'$, $(1,2,3^c)$ with two new heavy quarks $\mathcal{N}'$ and λ' and the elements involving $\bar{3}^c$ with antiquarks. The charged current is taken to be a Cabibbo rotated $j^1{}_2 + j^4{}_3$. Then there are no $\Delta S = 1$ neutral currents. The ratio $R = (e^+e^- \rightarrow \text{hadrons})/*e^+e^- \rightarrow \mu^+\mu^-)$ turns out to be 4, but with one charged heavy lepton with decay products counted as hadrons becomes 5 and with the contribution of two can go up to 6. The experimental value is between 5 and 6.

The E_6 and E_7 schemes are worked out in more details in two forthcoming papers[9].

I am grateful to my colleagues at Yale and to Professor M. Gell-Mann for stimulating discussions and to Professor L. Michel for his interest and kind hospitality at IHES.

References

1. For a review, see for example S. Weinberg, Rev. Mod. Phys. 46, 255 (1974).
2. The 4 quark gauge model in reviewed in M.K. Gaillard, B.W. Lee, J.L. Rosner, Rev. Mod. Phys. 47, 277 (1975).
3. For reviews, see Proceedings of the London XVII International Conference in High Energy Physics (1974).
4. For $c\bar{c}$ states, see additional references in Ref. 2 and G.D. Feldman et al., Phys. Rev. Lett. 35, 819 (1975).
5. For dimuon events see A. Benvenuti et al., Phys. Rev. Lett. 34, 419 (1975) and 35, 1203 (1975).
6. M. Perl, Invited talk at the 1975 meeting of the Division of Particles and Fields of the American Physical Society, Seattle.
7. F. Gürsey, in Johns Hopkins University Workshop on Current Problems in High Energy Particle Theory, p.15, (Johns Hopkins University, 1974), and F. Gürsey, Algebraic Methods and Quark Structure, in International Symposium on Mathematical Problems in Theoretical Physics, Ed. H. Araki, p. 189 (Springer, 1975). For related work see A. Gamba, J. Math. Phys. 8, 775 (1967).

8. M. Günaydin and F. Gürsey, Phys. Rev. D9, 3387 (1974).
9. F. Gürsey, P. Ramond and P. Sikivie, Phys. Rev. D, to be published. Also A Universal Gauge Theory Model Based on E_6, to be published.
10. H. Georgi and S.L. Glashow, Phys. Rev. Lett. 32, 433 (1974).
11. J. Pati and A. Salam, Phys. Rev. D8, 1240 (1973).
12. H. Fritzsch and P. Minkowski, Caltech Preprint (1975), to be published in Annals of Physics.
13. P. Jordan, J. von Neumann and E.P. Wigner, Ann. Math. 35, 29 (1934).
14. J.E. Segal, Ann. Math. 48, 930 (1947), S. Sherman, Ann. Math. 64, 593 (1956).
15. C. Chevalley and R.D. Schafer, Proc. Natl. Acad. Sci. U.S., 36, 137 (1950).
16. For a review see H. Freudenthal, Advances in Mathematics, vol. I, p. 145 (1965).

A propos des brisures spontanées de symétrie

Louis MICHEL
Institut des Hautes Etudes Scientifiques
Bures-sur-Yvette (France)

C'est Goldstone qui avait été pressenti pour donner ici une revue sur le sujet des brisures spontanées de symétrie. Je suis certain que, comme moi, vous regrettez tous qu'il n'ait pas pû accepter et que j'ai eu à le remplacer.

Ayant fait une revue sur ce sujet[≠] au 1er Symposium Européen de physique mathématique à Varsovie [1] en mars 1974, et devant parler de ce sujet dans une semaine au colloque de la Société Française de Physique à Dijon [2] je ne voudrais pas me répéter ici. Cet exposé ne sera donc pas systématique; il sera incomplet et partial puisque ni Goldstone ni Higgs ne seront cités. Il vous présentera quelques idées très générales, en insistant plus sur leur aspect mathématique, et en citant simplement leurs applications physiques.

1. Un exemple historique.

Commençons par un exemple, que je crois être le premier exemple historique de symétrie brisée. Il fut découvert par Jacobi [3] en 1834 en considérant une masse m de fluide incompressible, isolé, en cohésion par son attraction gravitationnelle, et soumis à une mouvement de rotation uniforme autour d'un axe fixe (Ox_3), de vitesse angulaire constante Ω_0. Lorsque $\Omega_0 = 0$, la figure d'équilibre est une

---

≠ Le texte de la conférence de Varsovie avait d'ailleurs été distribué à tous les participants du 3ème Colloquium, l'an dernier à Marseille.

TABLEAU 1.

Masse fluide m, vitesse angulaire constante Ω_o autour de Ox_3. La surface libre du liquide incompressible est donnée par l'équation

$$\Sigma_{i=1}^{3}(x_i/a_i)^2 = 1 \qquad (1)$$

On définit les intégrales

$$A^{(k)} = a_1a_2a_3 \int_o^\infty \frac{u^k du}{\sqrt{D}}, \quad A_i^{(k)} = a_1a_2a_3 \int_o^\infty \frac{u^k du}{\sqrt{D}(a_i^2+u)},$$

$$A_{ij}^{(k)} = a_1a_2a_3 \int_o^\infty \frac{u^k du}{\sqrt{D}(a_i^2+u)(a_j^2+u)}$$

où $D = (a_1^2+u)(a_2^2+u)(a_3^2+u)$. Noter que $A_1^{(o)} + A_2^{(o)} + A_3^{(o)} = 2$.

L'équilibre hydrostatique dans le repère tournant s'écrit

$$\frac{p}{\rho} = V(\vec{x}) + \tfrac{1}{2}\,\Omega_o^2(x_1^2 + x_2^2) + C^{te} \qquad (2)$$

où p est la pression, ρ la densité : $m = 4\pi\, a_1a_2a_3\, \rho/3$ et l'énergie potentielle

$$V(\vec{x}) = \pi G\rho(A^{(o)} - \Sigma_{i=1}^{3} A_i^{(o)} x_i^2) \qquad (3)$$

G est la constante de gravitation et $\Omega = \Omega_o/\sqrt{\pi G\rho}$ est sans dimension. En écrivant que $p = 0$ sur la surface, la comparaison de (1) et (2) donne les conditions d'équilibre

$$(\Omega^2 - 2A_1^{(o)})a_1^2 = (\Omega^2 - 2A_2^{(o)})a_2^2 = -2A_3^{(o)}a_3^2 \qquad (4)$$

La première égalité peut encore s'écrire

$$(a_1^2 - a_2^2)(\Omega^2 - 2A_{12}^{(1)}) = 0 \qquad (5)$$

Solutions à symétrie axiale : $a_1^2 - a_2^2 = 0$

(il en existe deux pour chaque $\Omega < 0{,}44933\ldots$ trouvées par Mac Laurin)

Solutions à trois axes inégaux : $\Omega^2 - 2A_{12}^{(1)} = 0$

(il en existe une pour chaque $\Omega < 0{,}37230\ldots$ trouvée par Jacobi)

Le point de bifurcation satisfait $a_1^2 - a_2^2 = 0$ <u>et</u> $\Omega^2 - 2A_{12}^{(1)} = 0$ et (de (4)) $a_1^2\, a_2^2\, A_{12}^{(o)} = a_3^2\, A_3^{(o)}$ soit $\Omega = 0{,}37230\ldots$, $\frac{a_3}{a_1} = \frac{a_2}{a_1} = 0{,}582724\ldots$.

sphère; pour un certain domaine de valeur de Ω_0, il existe un équilibre statique dans le repère tournant, la surface d'équilibre du liquide étant un ellipsoïde de révolution aplati. Le Tableau 1 donne les équations, en fonction de Ω_0, satisfaites par les longueurs des axes principaux a_1, a_2, a_3 de l'ellipsoïde. Cela donne a une assez bonne approximation de l'aplatissement de la Terre. Mais ces mêmes équations ont aussi une autre famille de solutions découvertes par Jacobi. Elles sont asymétriques : $a_1 \neq a_2$, l'ellipsoïde ayant trois axes inégaux. La solution commune à ces deux familles est un "point de bifurcation". Ce qui est remarquable c'est sa généralité. D'autres familles de solutions (non statique) avec courants de convexion partent du même point. Si on utilise comme paramètre cinématique sans dimension t

$$t = \frac{\text{Energie de rotation}}{|\text{Energie potentielle}|} \tag{1}$$

le théorème du viriel impose pour tout système en interaction gravitationnelle

$$0 \leq t \leq \tfrac{1}{2} \tag{2}$$

Pour la sphère $t = 0$ ($\Omega_0 = 0$) ; le point de bifurcation apparaît pour $t = 0{,}1375$.. Si le fluide n'est pas incompressible, la densité ρ (pour les solutions statiques dans le repère tournant) est constante sur des ellipsoïdes concentriques dont la symétrie axiale disparaît pour cette même valeur de $t = 0{,}1375\ldots$, quelle que soit l'équation d'état du fluide, comme viennent de le montrer Bertin et Radicati [4] en utilisant une remarque mathématique de Roberts [5].

Quand l'ellipsoïde a trois axes inégaux, on obtient pour Ω_0 ou t fixé, à un instant donné, toute la famille de solution à partir de l'une d'elle en faisant tourner l'ellipsoïde autour de l'axe Ox_3, qui est l'axe de symétrie du problème. Cette situation est générale.

2. L'action du groupe de symétrie sur l'ensemble des solutions.

Quand un problème physique a un groupe de symétrie G, ce groupe agit sur l'ensemble M des solutions du problème. Ces points de M, invariants par G forment un sous-ensemble M^G (en général fermé) et représentent les solutions qui ont aussi la symétrie G . Chaque autre solution $m \in M$ est invariante pour un sous-groupe G_m de G (appelé petit groupe ou stabilisateur ou groupe d'isotropie de m). Nous notons $G(m)$ la famille des solutions transformées de m par G . On dit que $G(m)$ est une orbite du groupe. Si deux points m' et m sont sur la même orbite, il existe alors $g \in G$ (pas forcément unique!) tel que $m' = gm$ et on vérifie que les stabilisateurs sont conjugués i.e.

$$G_{m'} = gG_m g^{-1} \tag{3}$$

La réciproque n'est pas vraie; si deux points x et y ont des stabilisateurs conjugués (3), x et y ne sont pas nécessairement sur la même orbite, mais par définition, ils sont sur le même stratum; un stratum est ainsi l'union de toutes les orbites d'un même type. On notera $S(x)$ le stratum de x . Il y a un ordre partiel naturel sur les classes de conjugaison des stabilisateurs qui apparaissent dans l'action de G sur M : (on note $[G_x]$ celle de G_x) . $[G_x] < [G_y]$ si G_x est conjugué d'un sous-groupe de G_y .

L'ensemble M des solutions est très différent en général pour le même problème traité en mécanique classique ou quantique. Considérez par exemple le problème de Kepler dans l'espace à 3 dimensions (potentiel central en $\frac{k}{r}$) ; dans le cas classique aucune solution n'a la symétrie $O(3)$, car la nécessité d'avoir des conditions initiales détruit cette symétrie; dans le cas quantique tous les états de moment orbital $\ell = 0$ ont la symétrie sphérique. Pour les $\ell > 0$, à chaque orbite d'état pur (représentables par un vecteur d'état) on peut faire correspondre un "mélange" (représentable par un matrice densité) en intégrant sur l'orbite, ce qui est encore un état invariant.

Lorsque la symétrie est brisée, savoir laquelle des solutions d'une orbite est choisie dans le phénomène qu'on étudie, est d'un intérêt secondaire, surtout pour les symétries spatiales; dans ce cas, le choix est dû à une irrégularité (par exemple germe cristallin) ou même parfois à une fluctation statistique. Ce qui est important c'est de pouvoir prévoir en quel sous-groupe la symétrie peut être brisée (sans résoudre complètement le problème!). C'est ce qui a été fait en cristallographie à la fin du siècle dernier. Les interactions entre les ions ou les atomes constituant un cristal sont invariantes par déplacement euclidien (translation et rotation), mais le cristal (même supposé parfait et indéfini) n'est invariant que pour un sous-groupe du groupe Euclidien E(3). Il y a 230 classes de groupes cristallographiques.*) Parfois et de plus en plus, pour un problème de mécanique statistique classique ou quantique, ainsi que pour un problème de mécanique quantique on peut prévoir l'ensemble des stabilisateurs qui apparaitront pour l'action de G dans l'espace des états et donc comment peut se casser la symétrie. La revue la plus récente de cette question, dans le cadre général des C^*-algèbres a été faite par D. Kastler [6] ; je vous y renvoie . J'avais étudié avec lui [7] le cas ≠ où G est le groupe Euclidien E(3) . En plus des classes de groupes cristallographiques nous avons trouvé une infinité dénombrable de "classes de symétrie" que nous avons décrites (à une conjugaison près dans $InL_+(3,R)$)* . Bien que cette classification rejettent beaucoup de sous-groupes fermés de E(3) comme stabilisateurs d'états, il faut cependant noter que les stabilisateurs possibles forment un ouvert dans l'espace compact des sous-groupes fermés de E(3) ≠ .

Tout en restant général, nous allons particulariser un peu !

3. <u>Principe variationnel et invariance par un groupe compact</u> G .

C'est une situation qu'on rencontre assez souvent en physique. Il s'agit

---

*) Comme on ne distingue pas, de ce point de vue des symétries, les cristaux de tailles différentes, les classes de conjugaisons sont dans le groupe $InL_+(3,R)$: le groupe linéaire inhomogène à 3 dimension, pour les opérations homogènes de déterminant positif. Dans InL(3,R) il n'y aurait que 219 classes.

La réf. [1] contient un bref résumé

≠ En effet, les sous-groupes fermés d'un groupe localement compact forment un espace topologique compact cf. Bourbaki VI.8,§5, généralisant un travail de Macbeath et Swierkowski.

de trouver les extrema d'une fonction f , réelle, différentiable≠ definie sur une variété M et invariante par G . Appelons $\mathfrak{F}$ l'ensemble de telles fonctions. Si m est un extremum de f , tous les points de G(m) le sont aussi. J'ai applelé "critiques" les orbites qui sont des orbites d'extrema pour <u>toutes</u> les fonctions $\in \mathfrak{F}$. Ces orbites sont faciles à caractériser [8]. Ce sont celles qui sont isolées dans leur stratum (c'est-à-dire, il y a un voisinage qui contient l'orbite critique et qui n'en contient aucune autre du même type). Ce résultat découle du fait que le gradient d'une fonction $f \in \mathfrak{F}$ en tout point m est orthogonal à l'orbite et tangent au stratum. Le physicien qui a choisi un modèle -donc une fonction à varier - et qui a trouvé ainsi la cassure de symétrie qu'il cherchait ne doit pas croire que son modèle est vérifié si l'orbite de solutions qui lui convient est une orbite critique. Il aurait pu partir de n'importe quelle autre fonction (avec le même groupe de symétrie G) et il a simplement vérifié un théoreme général. De telles orbites critiques jouent souvent un grand rôle. C'est ce que Radicati et moi avons vérifié pour les brisures de symétrie interne des hadrons ([9], [10], [11]); pour des exemples pour le groupe SO(4), pris à la relativité générale, ou a l'hydrodynamique, voir [12].

Naturellement, les physiciens sont intéressés par la nature des extrema. Pour cela il faut former en chaque point $m \in M$,le Hessian *) d^2f de la fonction f .

En chaque point $m \in M$, le petit groupe G_m agit linéairement sur le plan tangent $T_m(M)$ par la représentation linéaire $g \to D_m(g)$. Le Hessian est invariant par G_m , i.e. $\forall_g \in G_m$, $d^2f = D_m(g)\, d^2f\, D_m(g)^T$. Cela implique en général une certaine dégénérescence de son spectre. De plus le noyau de d^2f , Ker d^2f ,

≠ Pour éviter des difficultés techniques nous les prenons C^∞ ; la plupart des résultats sont un peu plus généraux.

* Avec un choix de coordonnées x^i , les coordonnées du gradient sont $\frac{\partial f}{\partial x^i}$ et le Hessian est représenté par la matrice réelle symétrique $\frac{\partial^2 f}{\partial x^i \partial x^j}$.

est le plan tangent en m à l'orbite $G(m)$, $T_m(G(m)) \subset T_m(M)$. L'extremum est un minimum si $d^2f \geq 0$. Admettons que la fonction f dépendent de paramètres extérieurs (en thermodynamique il s'agit de p,T ; dans l'exemple de la section 1, il s'agit de t ou Ω_o , cf. [4]) . L'équilibre est donné par $df = 0$, $d^2f \geq 0$ et l'apparition d'un point de bifurcation par "Ker d^2f strictement plus grand que $T_m(G_m)$."

4. Theorie de Landau des changements de phase du 2e ordre.

La situation générale que nous venons de décrire est un cas particulier de la theorie des catastrophes de Thom [13], théorie qui a classé les types de singularités en l'absence de groupe G . Son extension à G compact est en cours. D'autre part, la théorie de Morse [14] impose des relations entre la nature des extrema de f et les nombres de Betti de M en l'absence de G . Son extension "équivariante" pour G compact est possible [15] et je suis actuellement intéressé par ses applications à la physique$^{\neq}$. Un des buts de la théorie de Landau [17], [18] est de prédire quelle brisure spontanée de symétrie peut se produire dans un cristal par un changement de phase du second ordre. C'est une excellente illustration de la situation générale que nous venons de décrire. Le groupe cristallographique G n'est pas compact, mais ses représentations irréductibles réelles sont toutes de dimension finie et orthogonales (car G a un sous-groupe invariant abélien d'index fini); f est le potentiel thesmodynamique, M serait l'Hilbert $\mathcal{L}^2(G)$ mais en fait on considère un sous-espace $\mathcal{E}$ de représentation irréductible et f est minimum à l'origine pour la phase la plus symétrique. Notons par (x,x) le produit scalaire orthogonal sur $\mathcal{E}$. Si en réduisant (sur les réels) la représentation de G sur $\mathcal{E} \otimes \mathcal{E}$ on retrouve la représentation sur $\mathcal{E}$, nous obtenons un homomorphism G-équivarient

$$\mathcal{E} \otimes \mathcal{E} \xrightarrow{\vee} \mathcal{E}$$

$\neq$ Un des premiers et bel exemple de l'application de la théorie de Morse à la physique est la remarque de Van Hove [16] sur les singularités de la densité des fréquences élastiques dans un cristal.

qui définit une algèbre symétrique # (en général non associative) sur $\mathcal{E}$: nous notons simplement $\vee(x \otimes y) = x_\vee y$. Notons Δ_x l'opérateur lineaire sur $\mathcal{E}$ défini par $\Delta_x y = x \vee y$ et remarquons que $\operatorname{tr} \Delta_x = 0$ sinon $x \to \operatorname{tr} \Delta_x$ serait une forme linéaire équivariante sur $\mathcal{E}$, ce qui est impossible si la représentation irréductible de G sur $\mathcal{E}$ n'est pas triviale. Nous pouvons alors démontrer le

Lemma. Tout polynôme du 3e degré G-invariant sur $\mathcal{E}$ ne peut être minimum qu'à l'origine.

L'invariance par G interdit au polynome d'avoir un terme linéaire en x . Le terme constant est sans importance. Soit

$$f = \frac{\alpha}{2}(x,x) + \frac{\beta}{3}(x \vee x,x) \quad , \qquad \beta \neq 0 \tag{4}$$

ce polynome. Nous calculons alors aisément

$$df = \alpha x + \beta x \vee x \tag{5}$$

$$d^2 f = \alpha I + 2\beta \, \Delta_x \tag{6}$$

L'origine $x = 0$ est un extremum. Une autre orbite d'extremum est donné par

$$x \vee x = \frac{-\alpha}{\beta} x \tag{7}$$

c'est-à-dire x est un idempotent de l'algèbre symétrique. Remarquons que lorsque f est extremum $(df = 0)$, x est vecteur propre de $d^2 f$

$$d^2 f x = \alpha x + 2\beta x \vee x = -\alpha x \tag{8}$$

D'autre part,

Par exemple, pour l'exemple de la section 1, les quadrupoles, c'est-à-dire les matrices 3×3 réelles symétriques de trace nulle forment la représentation irréductible (réelle) de dimension 5 de O(3) : $Q_1 \vee Q_2 = \frac{\sqrt{3}}{2}(Q_1Q_2 + Q_2Q_1) - I\frac{2}{\sqrt{3}}(Q_1,Q_2)$ où le produit scalaire a été défini par $(Q_1,Q_2) = \frac{1}{2} \operatorname{tr} Q_1Q_2$. Remarquons que pour tout Q , $Q \vee Q \vee Q = Q(Q,Q)$.

$$\mathrm{tr}\, d^2 f = \alpha n \quad , \quad n = \dim \mathcal{E} \qquad (9)$$

La comparaison de (8) et (9) montre que pour $x \neq 0$, $d^2 f$ a des valeurs propres de signe opposé et donc ne peut satisfaire $d^2 f \geq 0$ ce qui prouve le lemme (pour $\alpha > 0$, $x = 0$, $df = 0$, $d^2 f = \alpha I > 0$) .

Dans un développement limité du potentiel thermodynamique f au voisinage de zero, en faisant varier p,T continûment on ne pourra donc pas passer d'un minimum à zéro (phase invariante par G) a un minimum pour $x \neq 0$ si les invariants du 3e ordre ne sont pas nuls. C'est une des conditions nécessaire de la théorie de Landau.

On qualifie d'"actives" les représentations réelles irréductibles d'un groupe cristallographique G satisfaisant la théorie de Landau. Récemment Mozrzymas a trouvé des relations d'équivalence intéressantes entre représentations actives≠ [12]. La théorie de Landau s'applique en dehors de la cristallographie et même pour l'exemple de la section 1 comme l'ont montré Bertin et Radicati [4] dans la cassure de la symétrie O(2) .

5. Les idempotents des algèbres symétriques G-invariantes.

Il ne faudrait pas croire que les invariants du 3e ordre, et plus généralement, les algèbres symétriques G-invariantes qui permettent de les former ne jouent pas un grand rôle dans les brisures spontanées de symétrie. Au contraire, comme Radicati et moi l'avons constaté, les idempotents et nilpotents des algèbres symétriques interviennent dans les brisures de symétrie interne des hadrons. Le tableau 2 donne les principales relations que nous avons observées. Ces résultats ont été généralisés par plusieurs auteurs [21 à 25], dont le prochain conférencier, Prof. Gürsey.

Il est peut-être téméraire de vouloir comprendre les brisures des symé-

---

≠ Nous travaillons ensemble pour compléter explicitement les prédictions de brisures symétriques de la théorie de Landau.

TABLEAU 2 - Référence Michel et Radicati [11], voir aussi [9], [10].

Groupe de symétrie de l'algèbre des courants

$SU(3) \times SU(3)$ et P et C (P = parité, C = conjugaison de charge)

Représentation adjointe de dimension 16, notée généralement $(8,1) \oplus (1,8)$. C'est l'espace des matrices 3×3 hermitiques de trace nulle : $x^* = x$, $\mathrm{tr}\, x = 0$ pour chaque $SU(3)$ et le produit $\vee$ correspondant se définit comme pour les quadrupoles ($2^{\text{ème}}$ note de la section 4). On a encore $x \vee x \vee x = x(x,x)$.

Les charges sont les intégrales, sur tout l'espace, des courants. Les directions remarquables sont notées.

- y hypercharge
- q courant électromagnétique et charge électrique
- $c_{\pm}$ courants faibles } Cabibbo [26]
- z hypercharge faible } Cabibbo [26]

Ces directions satisfont les relations

$$y \vee y + y = 0 \ , \ q \vee q + q = 0 \ , \ c_{\pm} \vee c_{\pm} = 0 \ , \ z \vee z + z = 0$$

$$z = 2c_{+} \vee c_{-}$$

$$(y,z) = \frac{1}{\sqrt{2}}\left(1 - \frac{3}{2}\sin^2\theta\right)$$ où θ est l'angle de Cabibbo ([26]) .

La symétrie G_o des particules est inférieure à celle des courants. Elle se produit essentiellement dans la représentation notée généralement $(3,\bar{3}) \oplus (\bar{3},3)$. C'est l'espace de 18 dimensions réelles des matrices complexes 3×3 . Action du groupe: $\forall u_1 \times u_2 \in SU(3) \times SU(3)$, $x \to u_1 x u_2^*$. L'algèbre symétrique, noté $_T$ est définie par

$$(x_T x)x^* = \det x^*$$

qui satisfait l'identité :

$$(x_T x)_T (x_T x) = x \det x$$

Deux cas de brisure de symétrie remarquables de $SU(3) \times SU(3)$ sur le sous-groupe G_o:

$SU(3)$ diagonal, dans la direction x telle que $x_T x = \sqrt{\frac{2}{3}}\, x$

$SU(2) \times SU(2) \times U(1)$, dans la direction x telle que $x_T x = 0$.

Le dernier cas correspond à la symétrie chirale : la masse m_π des π est négligée. Dans le premier cas c'est la différence de masse entre les mésons π et K qui est négligée. La nature est plus complexe; elle tient des deux cas à la fois tout en étant plus proche du cas chiral.

tries internes des particules fondamentales tant que nous n'avons pas une théorie, mais il est possible que cette théorie ne soit découverte que lorsque nous aurons mieux compris le mécanisme des brisures de ces symétries internes. C'est ce qui semble se passer actuellement en interprétant ces symétries internes comme symétrie de jauge (cf. la conférence d'Illiopoulos) ou peut-être même comme super-symétrie (cf. la conférence d'O'Raifeartaigh). Peut-être même sommes nous très avancés dans cette voie comme va vous le montrer beaucoup plus éloquemment le prochain conférencier, le professeur Gürsey.

Pour terminer je vous signale un tout autre domaine où les idempotents de ces algèbres symétriques jouent le rôle essentiel dans la brisure de symétrie : C'est la théorie des bifurcations, comme vient de me l'apprendre le prétirage "Group Representation Theory and Branch Points of non linear functional Equations" (University of Minnesota) de D.H. Sattinger actuellement en visite à l'I.H.E.S.. Dans les exemples qu'il choisit il est amené à chercher les idempotents $x \vee x = \lambda x$.

REFERENCES

[1] L. Michel, Some mathematical models of symmetry breaking. Application to particle physics, (to appear in "Proceedings of 1974 Warsaw Symposium in Mathematical Physics").

[2] L. Michel, Les brisures spontanées de symétrie en physique (à paraître dans le Journal de Physique).

[3] C.G.J. Jacobi, Poggendorf Annalen de Physik und Chimie 33 (1834) 229.

[4] G. Bertin, L.A. Radicati, The bifurcation from the Mac Laurin to the Jacobi sequence as a second order phase transition. Preprint Scuola Normale Superiore (Pisa).

[5] P.H. Roberts, Astrophys. J. 136 (1962) 1108.

[6] D. Kastler, Equilibrium states of matter and operator algebras, Convegno di C^*-algebra, Roma (1975) (to be published).

[7] D. Kastler, G. Loupias, M. Mebkhout, L. Michel, Comm. math. Phys. 27, (1972) 195.

[8] L. Michel, C.R. Acad. Sc. Paris 272 (1971) 433; pour plus de détails :"Proceedings 3rd Gift seminar in Theoretical Physics", p. 49-131, Madrid 1972.

[9] L. Michel, L. Radicati, Proceedings of the fifth Coral Gables Conference, "Symmetry Principles at High Energy", p. 19, W.A. Benjamin Inc., New York, (1968).

[10] L. Michel, L. Radicati, Mendeleev Symposium, Acti Accad. Sci. Torino II Sci. Fis. Mat. Natur., p. 377-389 (1971).

[11] L. Michel, L. Radicati, Ann. of Phys. 66, 758-783 (1971).

[12] F. Pegoraro, Comm. math. phys. 42 (1975) 41.

[13] R. Thom, Modèles Mathématiques de la Morphogénèse, Collection 10/18, Union générale d'Editions, Paris 1974. Cours Enrico Fermi 1973, à publier.

[14] J. Milnor, Morse Theory, Annals of Mathematical Studies, N° 51, Princeton University Press, est probablement un des meilleurs cours sur cette théorie.

[15] A.G. Wassermann, Topology 8 (1969) 127.

[16] L. Van Hove, Phys. Rev. 89 (1953) 1189.

[17] L. Landau , Phys. 2. Sovejt. 11 (1973) 545.

[18] L. Landau, E.M. Lifschitz, Statistical Physics, § 136 (traduit du russe aux Editions Mir, Moscou).

[19] J. Mozrzymas, Preprint Instytut Fizyki Teoretycznej Uniwersytetu Wrocławskiego, n° 306.

[20] L. Michel, L. Radicati, Evolution of particle physics, p. 191 (dedicated to E. Amaldi) academic Press New York (1970).

[21] L. Abellanas, J. Math. Phys., 13, 1064 (1972).

[22] Pegoraro and J. Subba Rao, Nucl. Phys. B44, 221 (1972).

[23] C. Darzens, Ann. Phys. 76, 236 (1973).

[24] R.E. Mott, N. Phys. B84 (1975) 260.

[25] S. Eliezer, Phys. Let. 53B (1974) 86.

[26] N. Cabibbo, Phys. Rev. Lett. 10 (1963) 531.

COHERENT STATES

GEOMETRY OF GENERALIZED COHERENT STATES [+]

H. BACRY [*]

A. GROSSMANN [**]

and

J. ZAK [***]

ABSTRACT : Various attempts have been made to generalize the concept of coherent states (c.s.). One of them, due to Perelomov, seems to be very promising but not restrictive enough. The Perelomov c.s. are briefly reviewed. One shows how his definition gives rise to Radcliffe's c.s. The relationship between the usual and Radcliffe's c.s. can be investigated either from group contraction point of view (Arecchi et al.) or from a physical point of view (with the aid of the Poincaré sphere of elliptic polarizations of electromagnetic plane waves). The question of finding complete subsets of c.s. is revisited and an attempt is made to restrict the Perelomov definition.

[+] Talk given at the 4th International Colloquium on Group Theoretical Methods in Physics, Nijmegen, 1975.

SEPTEMBER 1975

[*] UER Expérimentale et Pluridisciplinaire de Marseille Luminy, and Centre de Physique Théorique, CNRS, Marseille.

[**] Centre de Physique Théorique, CNRS, Marseille

[***] Physics Dept., Technion, Haïfa

Postal Address : Centre de Physique Théorique - C.N.R.S.
31, chemin Joseph Aiguier
F - 13274 MARSEILLE CEDEX 2 (France)

I - Introduction

Coherent states (c.s.) introduced by Schrödinger [1] have been shown [2] to play an important role in Quantum Optics [3 - 5]. They have so many nice properties [6 - 7] that many attempts have been made to generalize them. The most attractive attempt is probably that of Perelomov [8] who, emphasizing the role played by the nilpotent Weyl group (also known as the Heisenberg group), defined a way of constructing systems of generalized coherent states (g.c.s.) associated with (almost) any irreducible unitary representation of any Lie group. The property of the ordinary c.s. which has been emphasized by this author in his generalization is the transitive action of the Weyl group on the set of c.s.. The definition of Perelomov will be discussed below.

Another attempt was made a few months earlier by Barut and Girardello [9] where the accent was on the fact that usual c.s. are eigenstates of an unbounded operator, namely the annihilation operator. Their g.c.s. are eigenstates of a nilpotent generator of a given semi-simple (non compact) Lie group. As already stressed by Perelomov [8] their method cannot be extended to all Lie groups, especially to compact ones.

Other systems of g.c.s. have been defined by various authors [10-15] for specific physical problems. The remarkable fact is that all these sets involve Lie groups and appear as special cases of the Perelomov definition. Apart the Weyl and oscillator [16] groups which underly the usual c.s. and the ones of the Landau electron [13], the Lie groups which have been involved are SU(2), SO(4,2) and SU(n,1) : (a) An SU(2) system of g.c.s. has been introduced by Radcliffe [10] in 1971 under the name of spin coherent states ; this system has already been investigated in many works [4, 10, 11, 17-21]. The angular momentum c.s. invented by Atkins and Dobson [11] in relation with the Schwinger [22] - Bargmann [23] approach of the roation group are closely related with the Radcliffe ones ; (b) Gürsey and Orfanidis [12] have used the conformal group to define four vector coherent states associated with four vector position and energy momentum operators ; (c) SU(n,1) sets of g.c.s. have been investigated [14,15] in the special case $n = 3$ for a covariant description of the relativistic harmonic oscillator.

In the present paper, we intend to describe the relationship between the geometric properties of different types of g.c.s. After a brief review of the Perelomov definition of g.c.s. we will show how it allows the introduction of the Radcliffe spin c.s. The connection between harmonic oscillator c.s. (h.o.c.s.) and Radcliffe's ones is investigated. The Perelomov definition is criticized and restricted in order to get a richer structure.

II - Perelomov's definition of a system of g.c.s. [8]

Definition : Let G be a Lie group and $\mathcal{H}$ the Hilbert space of an irreducible unitary continuous representation of G. Let $\hat{\mathcal{H}}$ be the projective space associated with $\mathcal{H}$ ($\hat{\mathcal{H}}$ is the set of rays of $\mathcal{H}$, i.e. the set of one dimensional subspaces of $\mathcal{H}$ referred to as the set of states). Let $\hat{\psi}$ be an arbitrary state, the set of all $g\hat{\psi}$ where $g \in G$ is called by Perelomov a system of generalized c.s.

Such systems have the following properties

i) Let H be the stabilizer of $\hat{\psi}$, that is the subgroup of G such that $H\hat{\psi} = \hat{\psi}$. Any element g of G can be written as a product $g = xh$ where $h \in H$ and $x \in G/H$. One readily sees that g.c.s. can be parametrized by the elements of the coset space G/H .

ii) The group G acts transitively on g.c.s. This means that given two g.c.s. $x_1\hat{\psi}$ and $x_2\hat{\psi}$, there exists a group transformation mapping x_1 on x_2 (one also says that the g.c.s. form a homogeneous space of G).

iii) Suppose there exists on G/H an invariant measure dx . If $|x\rangle$ denotes a normalized vector lying on the ray $x\hat{\psi}$, the operator

$$B = \int |x\rangle\langle x|\, dx \qquad (1)$$

provided it exists, is a multiple of the unit operator : $B = \lambda I$.
It follows that any element $|\psi\rangle$ of $\mathcal{H}$ can be written in the form

$$|\psi\rangle = \frac{1}{\lambda} \int dx \, |x\rangle \langle x|\psi\rangle$$

It follows, from (1) that the system of g.c.s. is complete

iii) From (1) , any wave function on G/H can be written as follows

$$\psi(y) = \langle y|\psi\rangle = \frac{1}{\lambda} \int dx \; \psi(y) \, K(y,x)$$

where $K(y,x) = \langle y|x\rangle$ is a reproducing kernel.

One easily recognizes some important properties of the ordinary c.s. when G is the nilpotent Weyl group. In fact, if we use the Perelomov definition for the Weyl group, we get an infinite number of systems of g.c.s. in which the usual system appears as a very special case. In fact, according to a famous theorem of Von Neumann [24] , the Weyl group only has one kind of continuous irreducible faithful representations. In this representation, any state $\hat{\psi}$ can be shifted in a non trivial way by transformations generated by x_i and $-i \frac{d}{dx_i}$. In other words, any state lies on a two-dimensional system of g.c.s. in the Perelomov sense. The usual c.s. are the ones which lie on the orbit of the ground state of a harmonic oscillator. It follows that the Perelomov definition of g.c.s. does not contain one of the characteristic properties of Schrödinger c.s., namely the closeness of c.s. to the classical states, a property which comes from the minimalization of the Heisenberg uncertainty relations $\Delta x_i \cdot \Delta p_i = \frac{\hbar}{2}$
Unfortunately, such a property is not easily generalizable to arbitrary Lie groups.

III - Radcliffe's c.s. from Perelomov definition

According to Perelomov's ideas, given a couple $(G, \mathcal{H})$, we have to decompose $\hat{\mathcal{H}}$ into a union of orbits [25] (homogeneous spaces) of G, each of them corresponding to a set of g.c.s. When G is the ordinary rotation group, such a decomposition has been made in [26] +. The results have a simple geometrical description we are going to recall here : first, let us define the concept of constellation.

Constellation of order n : Let $A_1, A_2, \ldots, A_k$ be k points of a manifold, with weights $\alpha_1, \alpha_2, \cdots \alpha_k$, respectively. The α 's are strictly positive integers and satisfy the relation $\alpha_1 + \alpha_2 + \cdots + \alpha_k = n$.

First Example : Any complex polynomial in one variable of degree n is associated with a constellation of order n in the complex plane (its roots) and vice-versa (if the polynomials are defined up to a non-zero factor).

Second Example : Any complex polynomial in one variable of degree $\leq n$ is associated with a constellation of order n on a two dimensional sphere (Proof : if the degree of the polynomial is m, we say that n-m roots are infinite ; then, the extended complex line is mapped on the Riemann sphere through a stereographic projection). The set of constellations of order n on the Riemann sphere will be referred to as the nth sky $S^{(n)}$.

Theorem [26]. The projective space $\hat{\mathcal{H}}$ associated with an (n+1)-dimensional Hilbert space $\mathcal{H}$ can be identified with the nth sky $S^{(n)}$.

As a consequence, finding a finite projective representation of G is equivalent to finding how G acts on the corresponding sky ++. The answer is quite simple for SU(2) : "JUST ROTATE THE SKY". Therefore, spin j

---

+ The corresponding decomposition of $\mathcal{H}$ (instead of $\hat{\mathcal{H}}$) into a union of orbits has been made by Mickelsson and Niederle [27].

++ About the action of SL(2,C) on the sky $S^{(2j)}$ associated with the representation D_{jo}, see reference [28]. I am grateful to Dr. R. Shaw for having pointed out this reference to me.

states are constellations of order $2j$ and two constellations are on the same orbit if and only if they can be brought into superposition by rotation. Two such constellations will be said to have the same shape. The classification of orbits which are present in a representation of SU(2) has been given in [26]. Let us note that many descriptions are possible + but the following one is quite simple : the state $|jm\rangle$ is represented by a constellation of order $2j$ with one point at the North pole with multiplicity $2m$ and one point at the South pole with multiplicity $2j - 2m$. The operators $J_\pm$ act in a very elementary way on such states. It is clear in this scheme that the states $|jm\rangle$ and $|j\ -m\rangle$ lie on the same orbit.

According to Perelomov's definition, any system of g.c.s. is given by an orbit and thus characterized by a shape of constellation. The system which has been introduced by Radcliffe [10] is the one of "collapsed" constellations that is the one which contains the state $|jj\rangle$. Therefore any Radcliffe c.s. can be labelled by spherical coordinates $\Omega = (\theta, \varphi)$ or by a complex number z. It follows that a spin c.s. can take the value $z = \infty$ in contradistinction with the h.o.c.s. The orbit of Radcliffe c.s. is sometimes referred to as the Bloch sphere [4].

The main properties of spin c.s. have been established in [10, 17, 20]. Let us mention some of them

$$|z\rangle = (1 + |z|^2)^{-j} e^{zJ_+} |j\ -j\rangle \qquad (2)$$

$$\langle z'|z\rangle = \left[\frac{(1 + \bar{z}'z)}{(1+|z'|^2)(1+|z|^2)} \right]^j \qquad (3)$$

In the (θ, φ) notation, one gets $z = \tan\frac{\theta}{2} e^{-i\varphi}$; it follows that

+ Due to the transitive action of U(n) on $S^{(n-1)}$ any state can be represented by a given constellation.

$$\langle \theta' \varphi' | \theta \varphi \rangle = \left(\cos \frac{\Theta}{2}\right)^{2j} \tag{4}$$

where Θ is the angle between the two corresponding radii on the Bloch sphere S (two orthogonal states are opposite on S) . The completeness relation reads

$$\int d\Omega \, |\Omega\rangle\langle\Omega| = \frac{4\pi}{2j+1} \tag{5}$$

where $d\Omega$ is the usual rotationnally invariant measure.

Remarks : i) The complex parametrization of the Radcliffe c.s. is intimately related to the Riemann sphere used by Vilenkin [29] in his construction of the SU(2) representations.

ii) SU(2) is generally used in physical problems involving two level systems. In the case of the polarization space of the electron, $\hat{\mathcal{H}}$ is a sphere which can be readily embedded in the ordinary space because the rotation group acts in an obvious way on it with an obvious interpretation. In the case of the polarization space of the photon, $\hat{\mathcal{H}}$ is the Poincaré sphere but rotations of this sphere are not related with the physical rotations of the photon states. Isospin and quasi-spin states also correspond to abstract spheres.

iii) According to the work of reference [26] , the sky representation can be used for any finite-dimensional Hilbert space. The Bloch sphere [4] corresponds to symmetrized states of N identical coherent two-level atoms. A generalization of the Bloch sphere for the description of non coherent identical systems appears to be possible with the aid of the constellation concept.

IV - Connection between spin c.s. and h.o.c.s.

Radcliffe [10] has described a relationship between his spin c.s. and the c.s. of the harmonic oscillator in one dimension. It has been shown in [17] that this relationship is better understood with the aid of a group contraction [30 , 31] . Moreover, it follows from the work by Atkins and Dobson [11] , that another relationship can be found between Radcliffe c.s. and the c.s. of the two-dimensional h.o. This is closely related to the Schwinger [22]- Bargmann [23] way of studying the SU(2) group. We intend to show here how this approach can be given a physical interpretation with the Poincaré sphere of elliptic polarizations of an electromagnetic plane wave.

1) The Poincaré sphere and the angular momentum c.s.

Poincaré [32] has shown that every elliptic polarization of an electromagnetic plane wave (propagating in a given direction) is represented by a point on a sphere [+] . A modern group theoretical approach of the Poincaré sphere geometry would be as follows. Let

$$H = \frac{p_x^2 + p_y^2 + x^2 + y^2}{2} \tag{6}$$

be the (classical) Hamiltonian of the two-dimensional h.o. It can also be written

$$H = \bar{z}_+ z_+ + \bar{z}_- z_- \tag{7}$$

where

$$z_\pm = \frac{1}{2}\left[(x - iy) \pm i(p_x - ip_y)\right] \tag{8}$$

Since it is a two degrees of freedom problem, a complete set of constants of the motion must contain four classical observables. If we discard the phase and energy, the complex number $z = z_+ / z_-$ uniquely define a solution, a polarization. It is clear that z belongs to the extended complex line. With the aid of a stereographic projection, we are led to the Poincaré sphere.

Now, it is clear from Eq(7) that H is invariant under SU(2) . Therefore SU(2) must act on the Poincaré sphere. The SU(2) generators

[+] North (resp.South) hemisphere corresponds to right (resp. left) polarizations ; the latitude angle 2I is such that $\cos I = (A^2-B^2)/(A^2 + B^2)$ where A and B are the half axes lengths of the ellipse ; the longitude angle is 2φ where φ is the angle of the main axis with a given direction in the polarization plane.

are

$$J_3 = \frac{1}{2}(xp_y - yp_x)$$

$$J_1 = \frac{1}{4}(p_x^2 - p_y^2 + x^2 - y^2) \qquad J_2 = \frac{1}{2}(xy + p_x p_y) \tag{9}$$

which are constants of the motion since $[\vec{J},H] = 0$. We readily note that $J^2 = \frac{1}{4}H^2$ and that $\vec{J}$ defines exactly one point on the Poincaré sphere of radius $H/2$ with the very meaning indicated in the last footnote. In other words, the knowledge of $\vec{J}/H$ determines uniquely the shape and the orientation of the ellipse.

The quantum mechanical approach is quite analogous : we define the annihilation operators as in (8)

$$a_{\pm} = \frac{1}{2}\left[(x - iy) \pm i(p_x - ip_y)\right] \tag{10}$$

and the corresponding (Hermitian conjugate) creation operators $a^+_{\pm}$. We get

$$H = a_+^{\dagger} a_+ + a_-^{\dagger} a_- \tag{11}$$

$$[a_{\varepsilon}, a_{\varepsilon}^{\dagger}] = \delta_{\varepsilon\varepsilon'} \tag{12}$$

Then, the $\vec{J}$ operators expressed in terms of a and a^+ are exactly the ones Schwinger [22] introduced in his study of the SU(2) group. If $z_{\pm}$ are the eigenvalues of $a_{\pm}$, we see how we go from h.o.c.s. z_+, z_- to the spin c.s. $|z\rangle$ just by defining equivalence classes

$$|z_+ z_-\rangle = |\lambda z_+, \lambda z_-\rangle \tag{13}$$

each equivalence class defining a spin c.s. In the Poincaré interpretation, two harmonic oscillator motions are equivalent if they correspond to the same polarization (that is if their corresponding ellipses have same eccentricity and orientation)$^+$.

$^+$ Another interesting property is the following one : the operator a^+_+ (resp. a^+_-) can be interpreted as the creator of a point at North (resp. South) pole of the Poincaré sphere. Therefore $J_+ = a^+_+ a_-$ raises a point from South to North pole and $J_- = a^+_- a_-$ does the opposite (see [26]).

Résumé

Two-dimensional h.o.c.s.	$\longleftrightarrow$	spin c.s.
Bargmann-Schwinger study of SU(2)	$\longleftrightarrow$	Vilenkin study of SU(2)
Electromagnetic plane wave	$\longleftrightarrow$	elliptic polarization

2) Group contraction of the rotation group into the oscillator group

A set of Lie algebras $G(\alpha, \beta)$ of dimension 4 which has been extensively investigated by Miller [33] are intimately related with special functions. These Lie algebras have the following commutation brackets

$$\begin{cases} [J_3, J_1] = i J_2 \\ [J_3, J_2] = -i J_1 \\ [J_1, J_2] = i\alpha J_3 + i\beta E \\ [E, \vec{J}] = 0 \end{cases} \tag{14}$$

These real Lie algebras are the ones of $SO(3) \times \mathbb{R}$ for $\alpha > 0$, $SO(2,1) \times \mathbb{R}$ for $\alpha < 0$, $E(2) \times \mathbb{R}$ for $\alpha = \beta = 0$ and Osc(1) for $\alpha = 0$, $\beta \neq 0$. Here E(2) denotes the Euclidean group in the two dimensional space, Osc(1) is the oscillator group [16] and $\mathbb{R}$ is the one dimensional translation group. Miller [33] has shown that these Lie algebras are related through contraction processes [30, 31] in the following way :

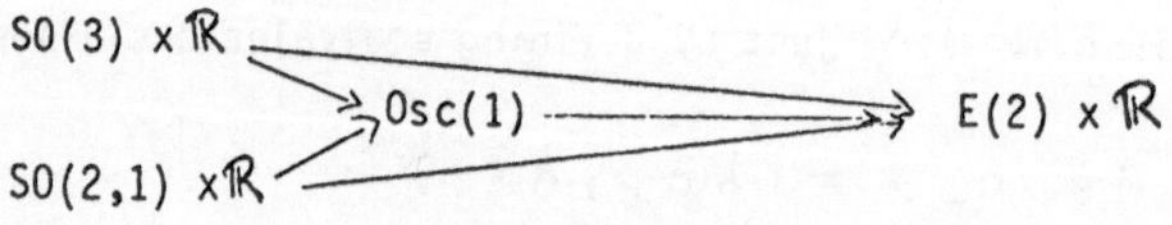

each arrow denoting a contraction.

Arecchi et al. [17] have stated that the contraction from $SO(3) \times \mathbb{R}$ to Osc(1) transform the spin c.s. into the h.o.c.s. This statement is true but the proof presented by these authors is incomplete. Our purpose is to give an exact derivation of the contraction by using unitary representations of the real groups under consideration, i.e. $SO(3) \times \mathbb{R}$ and Osc(1) .

We start from the Lie algebra of $SO(3) \times \mathbb{R}$ with generators $\vec{J}$ and E satisfying (14) with $\alpha = 1$ and $\beta = 0$. We perform the following change of basis

$$\begin{cases} H = J_3 + \left(c + \frac{1}{2} - \frac{1}{2c}\right) E \\ A_{1,2} = \frac{1}{\sqrt{2c}} \; J_{1,2} \\ F = E \end{cases} \tag{15}$$

The commutation rules read :

$$\begin{cases} [H \; , A_{\pm}] = \pm A_{\pm} \\ [A_+, A_-] = \frac{1}{c} H - \left(1 + \frac{1}{2c} - \frac{1}{2c^2}\right) F \\ [F \; , H] = 0 \end{cases} \tag{16}$$

For $c = \frac{1}{2}$, the change of basis is the identity one. When we make c going to infinity, we get the Osc(1) Lie algebra as a contracted Lie algebra. Obviously, it would be possible to perform this contraction by use of a simpler parametrization than (15) . However the one we chose is convenient for the study of c.s.

So far, we have investigated the contraction on the Lie algebra. Let us now see what we get for the representation. We start from the Vilenkin description [29] by polynomial of degree $\leq 2j$

$$\begin{cases} J_3 = z \frac{\partial}{\partial z} - j \\ J_+ = - z^2 \frac{\partial}{\partial z} + 2jz \\ J_- = \frac{\partial}{\partial z} \\ E = 1 \end{cases} \tag{17}$$

Let us renormalize by setting

$$z = \mathfrak{Z} / \sqrt{2j} \tag{18}$$

This corresponds to a redefinition of Radcliffe's states (Eq.2)

$$|y\rangle = \left(1+\frac{|y|^2}{2j}\right)^{-j} e^{y/\sqrt{2j}\, J_+} |j,-j\rangle \tag{19}$$

With this change of normalization, Eqs.(15) and (17) give us

$$\begin{cases} H = y\frac{\partial}{\partial y} - j + c + \frac{1}{2} - \frac{1}{2c} \\ A_+ = -\frac{1}{\sqrt{jc}}\, y^2 \frac{\partial}{\partial y} + \sqrt{\frac{j}{c}}\, y \\ A_- = \sqrt{\frac{j}{c}}\,\frac{\partial}{\partial y} \\ F = 1 \end{cases} \tag{20}$$

If we now make $c = j$ and <u>then</u> make j going to infinity, we readily obtain the usual h.o.c.s. of the group Osc(1) , namely

$$H = y\frac{\partial}{\partial y} + \frac{1}{2} \qquad A_+ = y \qquad A_- = \frac{\partial}{\partial y} \qquad F = 1 \tag{21}$$

Let us now give a more rigorous description of what we have just arrived at. Let $\mathcal{B}$ be a Bargmann space [7] and let P_j be the projector on the subspace of polynomials of degree less than or equal to $2j$. Each set of the following operators

$$\begin{cases} H^{(j)} = \left(y\frac{\partial}{\partial y} + \frac{1}{2} - \frac{1}{2j}\right) P_j \\ A_+^{(j)} = \left(-\frac{1}{2j}\, y^2 \frac{\partial}{\partial y} + y\right) P_j \\ A_-^{(j)} = \frac{\partial}{\partial y}\, P_j \\ F^{(j)} = P_j \end{cases} \tag{22}$$

defines an irreducible representation of the Lie algebra generated by $H^{(j)}$, $A^{(j)}$ and $F^{(j)}$. When j goes to infinity, we get the Bargmann representation of the group Osc(1).

Remarks : i) The change (18) from z to ζ corresponds to giving to the $2j^{th}$ sky a radius $\sqrt{2j}$.

ii) The question arises how to define a Radcliffe c.s. as a function of ζ instead of the ket defined in (19) . The answer follows from the identity

$$e^{\bar{\zeta}'[-\frac{1}{2j}\zeta^2\frac{\partial}{\partial\zeta}+\zeta]}\,1 = \left(1+\frac{\bar{\zeta}'\zeta}{2j}\right)^{2j} \tag{23}$$

In fact, since 1 is the function associated with the state $|\,j\ -j\,\rangle$, one readily sees that the Radcliffe c.s. (in the ζ variable) corresponds to the function

$$= R^{(j)}_{\bar{\zeta}'}(\zeta) = \left(1+\frac{|\zeta'|^2}{2j}\right)^{-2j}\left(1+\frac{\bar{\zeta}'\zeta}{2j}\right)^{2j} \tag{24}$$

which, when j goes to infinity provide us with the usual c.s. functions

$$R^{(\infty)}_{\bar{\zeta}'}(\zeta) = e^{-\frac{|\zeta'|^2}{2}}\,e^{\bar{\zeta}'\zeta} \tag{25}$$

iii) A set of fundamental invariants of the algebra (16') is given by F and

$$Q_j = \frac{H^2}{2j} - \left(1+\frac{j-1}{2j^2}\right)HF + \frac{1}{2}(A_+A_- + A_-A_+) \tag{26}$$

We readily see that Q_∞ is the invariant of the group Osc(1) +

Summary

In F_1 , the Bargmann spaces of entire functions of ζ , one can define a sequence of representations D_j of the SU(2) x $\mathbb{R}$ Lie algebra on an increasing sequence of subspaces P_jF_1 (representations given by Eqs.(15) and (22) with c = j). When j tends to infinity, this Lie algebra contracts into that of Osc(1) . The operators (22) which are bounded for

---

+ The Casimir operator of SU(2) , the eigenvalues of which are j(j+1) is given by $Q_j' = 2j\,Q_j + (j+\frac{1}{2}-\frac{1}{2j})^2\,F^2$.

j finite tend to the unbounded operators (21) in the following sense : if f belongs to H domain, $\lim_{j\to\infty} \| H^{(j)} f - Hf \| = 0$. Moreover the Radcliffe c.s. $R^{(j)}_{\bar{\xi}'}(\xi)$ of Eq.(24) tends to the usual c.s. (25) , i.e.

$$\lim_{j\to\infty} \| R^{(j)}_{\bar{\xi}'} - R^{(\infty)}_{\bar{\xi}'} \| = 0$$

V - Completeness of subsets of g.c.s.

Any system of g.c.s. being complete, it is natural to look for some complete subset. Such a question has already been answered for the usual c.s. by Von Neumann [25] , by Bargmann et al [34] and Perelomov [35]. In this special case an interesting complete set of c.s. which has been investigated, is generated by a discrete subgroup of the Weyl group, namely the group of discrete translations of a lattice in phase space

$$e^{imbx}\, e^{inap} \tag{27}$$

where x and p are position and momentum operators, m and n are integers and a and b are related by the condition $ab = 2\pi$.

It is therefore natural to look, following Perelomov [8] , for complete subsets of g.c.s. which are orbits of some subgroup of the group under consideration. In the case of g.c.s. associated with the Weyl group, a necessary and sufficient condition has been given in [36] for a state to generate a complete set under the lattice group (27).

For the case of spin c.s. , it can be readily seen that any subset of 2j+1 spin c.s. of spin j is complete, i.e. form a (nonorthogonal) basis of the representation space. The proof is as follows : a spin c.s. $|z_1\rangle$ has components of the form [26] :

$$\begin{vmatrix} 1 \\ \alpha_1 z_1 \\ \alpha_2 z_2^2 \\ \vdots \\ \alpha_{2j} z^{2j} \end{vmatrix}$$

Consider 2j+1 such states : $|z_1\rangle$, $|z_2\rangle$, ... $|z_{2j+1}\rangle$. For these c.s. to be independent, it is necessary that the determinant

$$\begin{vmatrix} 1 & 1 & 1 & \cdots & 1 \\ \alpha_1 z_1 & \alpha_1 z_2 & \alpha_1 z_3 & & \alpha_1 z_{2j+1} \\ \alpha_2 z_1^2 & \alpha_2 z_2^2 & \alpha_2 z_3^2 & & \alpha_2 z_{2j+1} \\ \alpha_{2j} z_1^{2j} & \alpha_{2j} z_2^{2j} & \alpha_{2j} z_3^{2j} & & \alpha_{2j} z_{2j+1}^{2j} \end{vmatrix}$$

does not vanish. It is readily seen that this determinant is different from zero if and only if all the z_i's are distinct. This proves the statement.

VI - Conclusion

The Perelomov definition of g.c.s. is only based on the transitivity property and no physical justification has been given for that. Moreover, according to this definition, any state is coherent. This is not very satisfactory and it is desirable not only to justify the need of an orbit but also to restrict the definition by using some physical argument. The most physical argument which is used for usual c.s. is probably the closeness of h.o.c.s. to classical states, a property which is expressed by the minimalization of the Heisenberg inequality. Unfortunately, we do not know how to generalize the Heisenberg uncertainty principle to all Lie groups but it is clear that if we were able to express the closeness of g.c.s. to classical states, the g.c.s. would be parametrized by coordinates in the phase space of the system. A phase space is a particular case of a symplectic manifold and it has already been shown [37-41] how non trivial symplectic manifolds are naturally involved in the description of classical relativistic or non relativistic elementary systems. In this case, the symplectic manifolds are canonically related to the Poincaré and Galilei group by the Kostant-Souriau theorem [42] . Let us underline that the transitivity property only expressed the elementary character of the classical system under consideration.

It follows from our discussion that, whatever is the way of introducing the concept of closeness to classical states, the orbit of g.c.s. must be a symplectic one. Let us examine how strong is the restriction for an orbit to be a symplectic one in the case of the SU(2) group. According to the Kostant-Souriau theorem [42] , the only symplectic homogeneous spaces of a Lie group G are the orbits of G on the dual vector space of the Lie algebra. It is quite simple to see that the only symplectic homogeneous space of SU(2) is the sphere S_2 (as a coset space it is SO(3)/SO(2)). According to [26] the only states which have SO(2) as stability subgroup are the states of type $|jm\rangle$ with $m \neq 0$. On the $2j^{th}$ sky there are $j + \frac{1}{2}$ or j such orbits according to j is half integral or integral. It is quite remarkable that the restriction of symplecticness only select a <u>finite</u> number of orbits among an infinity.+ Obviously, the Radcliffe choice is the most natural one.

The restricted definition we proposed is unfortunately not so successful in the case of the Heisenberg-Weyl or oscillator group, because we are still left with an infinite number of symplectic orbits. However, in the case of the Galilei group, it can be shown, for a spinless particle, that one of possible systems of g.c.s. would be of the form

$$e^{i\frac{\vec{p}^2}{2m}\tau}\, e^{i\vec{p}\vec{a}}\, f((\vec{p}-\vec{k})^2)$$

+ Except in the case $j = \frac{1}{2}$ for which the projective space is a single orbit.

where τ , $\vec{a}$, $\vec{k}$ are parameters. We immediately note that the ordinary c.s. belong to this kind. Similar g.c.s. could be defined for relativistic particles with the aid of the Poincaré group[+]. One of the most promising sets of g.c.s. seems to be the twistor space [43] .

[+] All symplectic manifolds invariant under the Poincaré group have been classified in [41].

- REFERENCES -

[1] E. SCHRÖDINGER
Naturwissenchaften 14 , 664 (1926)

[2] R.J. GLAUBER
Phys. Rev. 131 , 2766 (1963)

[3] J.R.KLAUDER and E.C.G.SUDARSHAN
Fundamental of Quantum Optics (Benjamin 1968)

[4] M. NUSSENZWEIG
Introduction to Quantum Optics (Gordon and Breach, 1973)

[5] F. ROCCA
in the present proceedings

[6] P. CARRUTHERS and M.M. NIETO
Rev. Mod. Phys. 40 , 411 (1968)

[7] V. BARGMANN
Comm. Pure Appl. Math. 14, 187 (1961)

[8] A.M. PERELOMOV
Commun.Math.Phys. 26, 222 (1972)

[9] A.O.BARUT and L. GIRARDELLO
Commun.Math.Phys. 21, 41 (1971)

[10] J.M. RADCLIFFE
Journ. Phys. A4, 313 (1971)

[11] P.W. ATKINS and J.C. DOBSON
Proc. Roy. Soc. A 321, 321 (1971)

[12] F. GÜRSEY and S. ORFANIDIS
Phys. Rev. D7, 2414 (1973)

[13] W.G. TAM
in Proceedings 1st. Colloquium on Group Theoretical Methods in Physics, Marseille 1972 (H. Bacry, Editor), and references therein.

[14] H. BACRY and N.P. CHANG
Phys. Lett. B 44, 286 (1973)

[15] H. BACRY
in Proceedings 3rd International Colloquium on Group Theoretical Methods in Physics, Marseille 1974 (H. Bacry and A. Grossmann, Eds.)

[16] R. STREATER
Commun.Math.Phys. 4 , 217 (1967)

[17] F.T. ARECCHI, E. COURTENS, R. GILMORE and H. THOMAS
Phys. Rev. A6 , 2211 (1972)

[18] J. KUTZNER
Phys. Lett. A 41, 475 (1972)

[19] E.H. LIEB
Commun.Math.Phys. 31, 327 (1973)

[20] J. BELLISSARD and R. HOLTZ
J. Math.Phys. 15 , 1275 (1974)

[21] W. WITSCHEL
J. Phys. A7, 1847 (1974)

[22] J. SCHWINGER
in L.C. BIEDENHARN and VAN DAM Eds., Quantum Theory of Angular Momentum (Academic Press, 1965)

[23] V. BARGMANN
Rev. Mod. Phys. 34 , 300 (1962)

[24] J. VON NEUMANN
Mathematical Foundations of Quantum Mechanics, (Princeton 1955)

[25] L. MICHEL
in the present proceedings

[26] H. BACRY
J. Math. Phys. 15, 1686 (1974)

[27] J. MICKELSSON and J. NIEDERLE
Commun.Math.Phys. 16, 191 (1970)

[28] F.A.E. PIRANI
in Lectures at Brandeis Summer School 1964, vol. I (Prenctice Hall, 1965)

[29] N.J. VILENKIN
Special Functions and the Theory of Group Representations (American Math. Society, 1968), chapter III, section 2

[30] E. INÖNU and E.P. WIGNER
Proc. Nat. Acad. Sci. 39, 510 (1953)

[31] E. SALETAN
J. Math. Phys. 2, 1 (1961)

[32] H. POINCARE
Théorie mathématique de la lumière (vol. 2, Paris, 1892)

[33] W. MILLER
Lie Theory and Special Functions (Acad. Press 1968)

[34] V. BARGMANN, P. BUTERA, L. GIRARDELLO and J.R.KLAUDER
Reports on Math. Phys. 2 , 221 (1971)

[35] A.M. PERELOMOV
Theoreticheskaya i Matematicheskaya Fizika, 6, 213 (1971)

[36] H. BACRY, A. GROSSMANN and J. ZAK
Phys. Rev. B (to appear)

[37] H. BACRY
Classical Hamiltonian for a Spinning Particle (unpublished)

[38] H. BACRY
Commun.Math.Phys. 5, 97 (1967)

[39] H. BACRY
in Proceedings 1st International Colloquium on Group Theoretical Methods in Physics, Marseille 1972 (H. Bacry, Editor)

[40] R. ARENS
Commun.Math.Phys. 21, 139 (1971)

[41] R. ARENS
J.Math.Phys. 12, 2415 (1971)

[42] D. SIMMS
Talk given at the present colloquium and references therein.

[43] N. WOODHOUSE
Talk given at the present colloquium.

Coherent States for Boson Systems in Quantum Field Theory and Statistical Mechanics.

F.ROCCA, Université de Nice, France.

The concept of coherent states, first introduced as quantum states of a single oscillator, is now generalized in different ways. In group theory, generalized coherent states have been introduced for unitary irreducible representations of any Lie group. In quantum theory and statistical mechanics generalized coherent states have been constructed in representations of canonical commutation relations which are inequivalent to the usual Fock one.

Coherent state representations of Lie groups will be considered in other talks. I should want to present a review of coherent states for quantum systems of bosons, in a systematic way : an unique and sufficiently general formalism is outlined, the coherent states are defined in this formalism, then the different situations, from the single oscillator to the non-Fock representations are considered, with, in each case, examples of applications in concrete physics.

1. Definitions and Formalism.

Let us consider a Bose system and let H be the one-particle Hilbert space. The dimensionality of H is the number of degrees of freedom of the system. We denote by (,) the complex scalar product on H and by σ its imaginary part; σ is a symplectic form on H.

Weyl system.

A Weyl system ($\mathcal{H}$, W) on H consists of a complex Hilbert space $\mathcal{H}$ with scalar product $\langle \mid \rangle$ and a map f $\rightarrow$ W(f) of H into the unitary group of $\mathcal{H}$ such that :

i) $W(f)\, W(g) = W(f + g)\, e^{-i\sigma(f,g)}$ (1)

$W(0) = \mathbb{1}$ (Weyl relations)

ii) $\lambda \in \mathbb{R} \rightarrow W(\lambda f)$ is weakly continuous for each $f \in H$.
A cyclic Weyl system ($\mathcal{H}$, W, Ω) is a Weyl system admitting a cyclic vector $\Omega \in \mathcal{H}$.

State.

A state is a functional $E : H \rightarrow \mathbb{C}$ satisfying :

i) $E(0) = 1$ (2)

ii) $$\sum_{j,k=1}^{n} c_j^* c_k \, E(f_k - f_j) \, e^{i\sigma(f_j, f_k)} \geqslant 0 \qquad (3)$$

$\forall c_j \in \mathbb{C}$, $\forall f_j \in H$, $j = 1, 2, \ldots, n$; $\forall n$

iii) $\lambda \in \mathbb{R} \rightarrow E(f + \lambda g)$ is continuous, $\forall f, g \in H$.

Given a Weyl system ($\mathcal{H}$, W) and a unit vector $\Phi \in \mathcal{H}$, the functional E_Φ defined by $E_\Phi(f) = \langle \Phi | W(f) | \Phi \rangle$ is a state. Conversely by the Gelfand-Naimark-Segal construction, there is, for any state E, a cyclic Weyl system ($\mathcal{H}_E$, W_E, Ω_E) such that

$$E(f) = \langle \Omega_E | \, W_E(f) \, | \Omega_E \rangle. \qquad (4)$$

Field operator.

For a given state E, the field operator A(f), $f \in H$, (unbounded operator acting on $\mathcal{H}_E$) is the infinitesimal generator of the group $\lambda \rightarrow W_E(\lambda f)$:

$$W_E(f) = e^{iA(f)} \qquad (5)$$

Creation and annihilation operators $A^{\pm}(f)$, $f \in H$, are defined by :

$$A^{\pm}(f) = \frac{1}{2} (A(f) \mp i \, A(if)) \qquad (6)$$

and satisfy the usual commutation relations

$$\left[A^{\pm}(f), A^{\pm}(g) \right] = 0$$

$$\left[A^{-}(f), A^{+}(g) \right] = (f,g) \qquad (7)$$

Fock state.

A well-known and very useful state is the Fock state E_0, defined by

$$E_0(f) = e^{-1/2\,(f,f)} \tag{8}$$

We shall note $(\mathcal{H}_0, W_0, \Omega_0)$ the corresponding cyclic Weyl system and $A_F(f) = A_F^+(f) + A_F^-(f)$ the field operator. Recall some properties of the Fock state :

i) $A_F^-(f)\,|\Omega_0\rangle = 0 \qquad \forall f \in H \qquad (\,|\Omega_0\rangle$: vacuum vector)

ii) W_0 is irreducible $\Leftrightarrow$ E_0 is a pure state

iii) $f \longrightarrow W_0(f)$ is continuous with respect to the strong $\mathcal{H}$-operator topology and the H-norm topology.

Coherent states [1].

Let H' the algebraic dual of H, considered as a symplectic real space with the form σ . For each $G \in H'$, we define the coherent state E_G by

$$E_G(f) = e^{-1/2(f,f) + i\,G(f)} = E_0(f)\, e^{iG(f)} \tag{9}$$

The transformation of states $E_0 \longrightarrow E_G$ is a gauge transformation. Particular cases with direct physical interest will be now reviewed, in connection with the dimensionality of H and with the continuity properties of the form G.

2. Dim H = 1 : coherent states of the one-dimensional oscillator.

In this case each element of H is of the form αe, $\alpha \in \mathbb{C}$, $(e,e) = 1$ and each $G \in H'$ is determined by some $\beta \in \mathbb{C}$ such that

$$G_\beta(\alpha e) = \beta\alpha^* + \beta^*\alpha \tag{10}$$

The definition (9) is now written :

$$E_\beta(\alpha e) \equiv E_\beta(\alpha) = e^{-\frac{1}{2}|\alpha|^2 + i(\beta\alpha^* + \beta^*\alpha)} \tag{11}$$

It is immediate, by the Von Neumann unicity theorem, that the corresponding cyclic Weyl system is unitarily equivalent to the Fock one :

$$E_\beta \longleftrightarrow (\mathcal{H}_0, U_\beta W_0 U_\beta^{-1}, \Omega_0) = (\mathcal{H}_0, W_0, \Omega_\beta)$$

with, following (4) :

$$E_\beta(\alpha) = \langle \Omega_\beta | W_0(\alpha) | \Omega_\beta \rangle$$

thus

$$e^{-1/2|\alpha|^2 + i(\beta\alpha^* + \beta^*\alpha)} = \langle \Omega_\beta | e^{i(\alpha^* a + \alpha a^+)} | \Omega_\beta \rangle \tag{12}$$

The solution of (12) in the representation space $\mathcal{H}_0$ is well-known :

$$|\Omega_\beta\rangle = e^{\beta a^+ - \beta^* a}|\Omega_0\rangle = e^{-1/2|\beta|^2} \sum_n \frac{\beta^n}{\sqrt{n!}} |n\rangle \tag{13}$$

The coherent vector states $|\Omega_\beta\rangle$ (usually quoted "coherent states" in standard quantum mechanics [(2)]) have some very attractive properties which are now briefly summarized :

a) They are eigen-vectors of the annihilation operator :

$$a|\Omega_\beta\rangle = \beta|\Omega_\beta\rangle \tag{14}$$

b) The coherent states are quantum states most closely approximating classical states : for them the uncertainly relation $\Delta p . \Delta q \geqslant \hbar/2$ has its minimum value (the equality sign holds). This corresponds to a Gaussian localization in both the coordinate and momentum spaces.

c) The coherent vector states $|\Omega_\beta\rangle$ give rise to a "resolution of the identity" :

$$\frac{1}{\pi}\int |\Omega_\beta\rangle\langle\Omega_\beta|\, d^2\beta \;=\; \sum_n |n\rangle\langle n| \;=\; \mathbb{1}_{\mathcal{H}_0} \tag{15}$$

$$(d^2\beta = d(\mathrm{Re}\,\beta)\, d(\mathrm{Im}\,\beta))$$

They are not orthogonal to each other:

$$\langle\Omega_\beta|\Omega_\alpha\rangle = \exp\left\{-\frac{1}{2}|\beta|^2 - \frac{1}{2}|\alpha|^2 + \beta^*\alpha\right\} \tag{16}$$

so that they form an over-complete linear dependent set. Actually, it is possible to find countable subsets which are complete, in particular the subset of all coherent states $|\Omega_\beta\rangle$ for which $\beta = \sqrt{\pi}\,(n + im)$, n and m being arbitrary integers [3]. Owing to the usual position-momentum decomposition $a = \frac{1}{\sqrt{2}}(q + ip)$ these states appear as lattice states in phase space.

d) Using (15), an arbitrary vector $|\Psi\rangle \in \mathcal{H}_0$ can be expanded :

$$|\Psi\rangle = \frac{1}{\pi}\int d^2\beta\, |\Omega_\beta\rangle\langle\Omega_\beta|\Psi\rangle \tag{17}$$

The coefficients $\langle\Omega_\beta|\Psi\rangle = \Psi(\beta)$ satisfy the equation

$$\Psi(\beta) = \frac{1}{\pi}\int d^2\alpha\, \langle\Omega_\beta|\Omega_\alpha\rangle\, \Psi(\alpha) \tag{18}$$

so that $\langle\Omega_\beta|\Omega_\alpha\rangle$, given by (16), acts as a reproducing kernel.

Furthermore the function $\Psi(\beta)$ is shown from (13) to be of the form

$$\Psi(\beta) = e^{-1/2|\beta|^2} f(\beta^*)$$ where f is an entire function.

This establishes the close connection between the coherent-state representation and the representation in the Segal-Bargmann space [4].

e) Every density matrix ρ in $\mathcal{H}_0$ can be approximated in the trace-norm topology by density matrices which are diagonal in coherent states [5] :

$$\| \rho - \rho_n \|_1 \underset{n\to\infty}{\longrightarrow} 0$$

with

$$\rho_n = \int P_n(\alpha)\, |\Omega_\alpha\rangle\langle\Omega_\alpha|\, d^2\alpha \tag{19}$$

$(P_n)_n$ is a sequence in $\mathcal{S}(\mathbb{R}^2)$.

Driven oscillator.

The ground state of the hamiltonian $h = \frac{1}{2}(p^2 + q^2 - 1)$, describing an harmonic oscillator (with unit mass and angular frequency, $\hbar = 1$) is the vacuum $|\Omega_0\rangle$: $h|\Omega_0\rangle = 0$. If we introduce c-number driving terms α_1 and α_2, we get :

$$h' = \frac{1}{2}(p^2 + q^2 - 1) - \alpha_1 p - \alpha_2 q.$$

Then, the ground state of the driven oscillator is the coherent state $|\Omega_\alpha\rangle$, $\alpha = \frac{1}{\sqrt{2}}(\alpha_1 + i\alpha_2)$, with :

$$h'|\Omega_\alpha\rangle = -|\alpha|^2\,|\Omega_\alpha\rangle \tag{20}$$

Model of ideal monomode laser.

The simpler model of a monomode laser is obtained supposing that the current distribution in the cavity acts as a classical source which excites the single mode. Then the laser can be considered as a driven oscillator. It is furthermore necessary to average over the phase of the oscillator, since, at optical frequencies, we have no certainty about the initial phase. Thus, the laser is described in this model by the density matrix

$$\rho = \frac{1}{2\pi}\int_{-\pi}^{+\pi} |\Omega_{|\alpha|e^{i\theta}}\rangle\langle\Omega_{|\alpha|e^{i\theta}}|\, d\theta \tag{21}$$

In counting experiments, the counting rates involve the mean value $\langle N\rangle$ of the particle number $N = a^+a$, and the probability P_n to have n photons in the field state. With the density matrix (21), we obtain :

$$\langle N\rangle = \mathrm{Tr}\{\rho\, a^+a\} = |\alpha|^2 \tag{22}$$

$$P_n = \mathrm{Tr}\{\rho\, |n\rangle\langle n|\} = \frac{\langle N\rangle^n}{n!}\, e^{-\langle N\rangle} \tag{23}$$

The Poisson law (23) is experimentally verified for well-stabilized monomode lasers. More realistic models, including phase diffusion, noise and quantization of atoms, which involve again coherent states, can be found in the literature [6].

Note to end this part that the coherent states in the case Dim H = n are a direct generalization of the previous ones, without new interesting features, since the Von Neumann unicity theorem holds again. In physical applications, the multi-mode lasers, with mode coupling phenomena, are described using these coherent states.

3. Dim H = ∞ and continuous G : coherent states in Quantum Field Theory.

Let us suppose that, in the definition (9) of coherent states, the real linear form G is a continuous element of H'. Then, we can find in H an element g such that

$$G(f) = 2\,\sigma(f,g) \tag{24}$$

Conversely, each $g \in H$ defines a coherent state E_g by

$$E_g(f) = e^{-1/2\,(f,f) + 2i\,\sigma(f,g)} \tag{25}$$

The corresponding Weyl system is again equivalent to the Fock one :

$$E_g \longleftrightarrow (\mathcal{H}_0, U_g W_0 U_g^{-1}, \Omega_0) = (\mathcal{H}_0, W_0, \Omega_g)$$

with

$$|\Omega_g\rangle = \exp\{-i A_F(g)\}|\Omega_0\rangle \tag{26}$$

since

$$\langle\Omega_0| e^{iA_F(g)} e^{iA_F(f)} e^{-iA_F(g)} |\Omega_0\rangle = e^{-1/2(f,f) + 2i\sigma(f,g)} \tag{27}$$

Some properties of the one-mode case can be easily rewritten :

i) the coherent vector states $|\Omega_g\rangle$ are eigenvectors of the annihilation part of the field operator :

$$A_F^-(f)\,|\Omega_g\rangle = (f,g)\,|\Omega_g\rangle \tag{28}$$

ii) The set $\{ |\Omega_g\rangle,\ g \in H \}$ is total in the Fock space $\mathcal{H}_0$;

iii) The mean value of the particle number is :

$$\langle \Omega_g | N_F | \Omega_g \rangle = \| g \|^2 \tag{29}$$

where $N_F = \sum_i A_F^+ (f_i) A_F^- (f_i)$, for any basis $(f_i)_i$.

Optical coherence.

The coherence properties of an electromagnetic field(in the radiation gauge)are expressed from the correlation functions [(7)] :

$$G^{(n)}_{j_1,\dots,j_{2n}} (x_1,\dots,x_{2n}) = \mathrm{Tr}\Big\{ \rho \prod_{i=1}^{n} A^+_{j_i}(x_i) \prod_{\ell=n+1}^{2n} A^-_{j_\ell}(x_\ell) \Big\}$$

$$x = (\vec{x}, t)$$

The field is said to be completely coherent if these functions factorize in the form :

$$G^{(n)}_{j_1,\dots,j_{2n}} (x_1,\dots,x_{2n}) = \prod_{i=1}^{n} g^*_{j_i}(x_i) \prod_{\ell=n+1}^{2n} g_{j_\ell}(x_\ell)\ , \ \forall n \tag{30}$$

where $\vec{g} \in L$, one-photon Hilbert space. Indeed the functions $G^{(n)}$ are directly connected with experimental measurements involving n photodetections and the condition (30) expresses the statistical independence of these ones. For example, for n = 1, the condition (30) implies the maximal visibility of the pattern in an interference experiment, and for n = 2 the absence of "bunching effect" in the Hanbury-Brown and Twiss experiment[(7)].

The factorization condition (30) is satisfied if the field is in a coherent state :

$$\rho = |\Omega_{\vec{g}}\rangle\langle \Omega_{\vec{g}}| \quad , \ \vec{g} \in L$$

since, from (28) :

$$\begin{aligned} A_j^- (x) |\Omega_{\vec{g}}\rangle &= A_F^- (D_x^+ \vec{e}_j) |\Omega_{\vec{g}}\rangle \\ &= g_j(x) |\Omega_{\vec{g}}\rangle \end{aligned} \tag{31}$$

$(D_x^+ (y) = D^+ (x - y))$

4. Dim H = ∞ and not continuous G : non-Fock coherent states.

If the form G in (9) is not continuous, the Weyl system corresponding to E_G cannot be equivalent to the Fock one (the propertie iii) of the Fock state is evidently wrong for such a E_G). The previous properties of Fock coherent states, which were essentially stated in the representation Hilbert space $\mathcal{H}_0$, are in general lost : it is necessary to completely reconstruct the cyclic Weyl system from a given form G and to study separately each case. This work is not a formal one since the concrete physical situations are actually very often described by non-Fock states, in particular for photons. We give two examples.

Radiation field from external currents (8).

We consider in the radiation gauge an electromagnetic field interacting with an external current. For accelerated charged particles, the Fourier transform on the mass-shell of the transverse part $\vec{j}(\vec{k})$ behaves like $\frac{1}{|\vec{k}|}$ when $|\vec{k}|$ goes to zero and so the L-norm of $\vec{j}$ is infinite :

$$(\vec{j},\vec{j}) = \int \frac{d^3\vec{k}}{2|\vec{k}|} \vec{j}^*(\vec{k}) . \vec{j}(\vec{k}) = \infty \tag{32}$$

However, there exists in L a dense subspace $\mathcal{A}$ such that $(\vec{f},\vec{j})$, $\vec{f} \in \mathcal{A}$, is finite. We can restrict ourselves to define states as functionals on $\mathcal{A}$ without physical alterations. Working in the Heisenberg picture, the out and in-operators are related by :

$$\vec{a}_{out}(\vec{k}) = \vec{a}_{in}(\vec{k}) + i\,\vec{j}(\vec{k}) \tag{33}$$

that is :

$$\vec{a}_{out}(\vec{k})\,|\Omega_0\rangle = i\,\vec{j}(\vec{k})\,|\Omega_0\rangle \tag{34}$$

if we start with the vacuum : $\vec{a}_{in}(\vec{k})\,|\Omega_0\rangle = 0$.
The out-state is then easily determined :

$$\begin{aligned} E_{\vec{j}}(\vec{f}) &= \langle\Omega_0|\, e^{iA_{out}(\vec{f})}\,|\Omega_0\rangle \\ &= e^{-1/2(\vec{f},\vec{f}) + 2i\sigma(\vec{f},\vec{j})} \end{aligned} \tag{35}$$

$E_{\vec{j}}$ is a coherent state in the sense of (9) but it is not a Fock coherent state since $\vec{f} \to (\vec{f},\vec{j})$, $\vec{f} \in \mathcal{A}$, $\vec{j} \in L$, is not continuous with respect to the norm in L.

The corresponding cyclic Weyl system is $(\mathcal{H}_0, W_{\vec{j}}, \Omega_0)$ with :

$$W_{\vec{j}}(\vec{f}) = W_0(\vec{f})\, e^{2i\,\sigma(\vec{f},\vec{j})} \tag{36}$$

One can directly prove on (36) that the gauge transformations of the first kind cannot be unitarily implemented in the space $\mathcal{H}_0$, i.e. there does not exist a weakly continuous group of unitaries such that :

$$W_{\vec{j}}(e^{it}\,\vec{f}) = U_t\, W_{\vec{j}}(\vec{f})\, U_t^{-1} \tag{37}$$

Consequently no particle number operator (infinitesimal generator of U_t) can be found in the coherent state $E_{\vec{j}}$. All questions concerning measurements connected with the particle number (mean-value, counting probability, etc...) are then meaningless in this situation. This is the ultimate reason why the (incorrect) use of the Fock representation in describing these measurements leads to infrared divergences.

The mathematical motivation of the introduction of a cut-off (corresponding to the experimental accuracy) is the necessity to recover rigorously Fock coherent states of section 4 to obtain non divergent results.

Phase problem.

When we consider the one-mode equation :

$$a|\Omega_\beta\rangle = \beta\,|\Omega_\beta\rangle = |\beta| e^{i\theta}\,|\Omega_\beta\rangle \tag{38}$$

it is tempting to introduce the polar decomposition $a = |a|\, e^{i\,\theta_{op}}$ and to look at θ_{op} as the "phase operator", so that the coherent state $|\Omega_\beta\rangle$ would be a state with the well-defined phase θ. Despite some successes in physical semi-phenomenological applications, this last point of view is theoretically wrong : θ_{op} is not self-adjoint, and, more generally, it is impossible to find in this one-mode case a self-adjoint operator θ_{op} satisfying the canonical commutation relation [9] :

$$[N, \theta_{op}] = i \tag{39}$$

Phase states, with existence of phase operator, have been recently constructed [10]. They are states of infinitely extended systems with finite particle density. These non-Fock states are obtained using the well-known procedure in Statistical Mechanics of thermodynamical limit on box systems.

We perform now such an operation on Fock coherent states to investigate their connection with phase states.

Let us start with $H = \mathcal{D}(\mathbb{R}^3)$ and suppose the boson system is confined in a volume V. The natural Weyl system is then the Fock one and the coherent vector states are, using (26) :

$$|\Omega_{g_V}\rangle = \exp\{-i A_F(g_V)\} |\Omega_0\rangle \qquad g_V \in \mathcal{D}(V)$$

thus the coherent states :

$$E_{g_V}(f) = \exp\{-\tfrac{1}{2}(f,f) + 2i\sigma(f,g_V)\} \qquad f \in \mathcal{D}(\mathbb{R}^3)$$

The mean density of particles in the state E_{g_V} is from (29) :

$$\bar{d} = \frac{1}{V} \langle \Omega_{g_V} | N_F | \Omega_{g_V} \rangle = \frac{\|g_V\|^2}{V} \tag{41}$$

We shall look now at the limit of E_{g_V} when $V \to \infty$ and $\bar{d}$ remains constant.

We begin to choose the function of $\mathcal{D}(V)$:

$$g_{V,\vec{k}}(\vec{x}) = (2\pi)^{-3/2} |g_{\vec{k}}| e^{i(\vec{k}\vec{x} + \alpha_{\vec{k}})} \chi_V(\vec{x}) \tag{42}$$

where $\chi_V(\vec{x})$ is some conveniently regularized characteristic function of the volume V. In the limit, this function $g_{V,\vec{k}}$ will describe particles with definite momentum $\vec{k}$.

Then :

$$E_{g_{\vec{k}}}(f) = \lim_{\substack{V\to\infty \\ \bar{d}\ c^t}} E_{g_{V,\vec{k}}}(f) = \exp\{-\tfrac{1}{2}(f,f) + 2i\,\mathrm{Im}(\tilde{f}^*(\vec{k})|g_{\vec{k}}|e^{i\alpha_{\vec{k}}})\} \tag{43}$$

The pure monochromatic coherent state $E_{g_{\vec{k}}}$ is not a Fock one since $f \to \tilde{f}^*(\vec{k})$ is evidently not continuous. As in the previous example, no particle number can be found in $E_{g_{\vec{k}}}$. The situation is different if one restores the gauge invariance by defining the state :

$$E^{inv}_{g_{\vec{k}}}(f) = e^{-\frac{1}{2}(f,f)} \int_0^{2\pi} \frac{d\alpha_{\vec{k}}}{2\pi} \exp\{2i\,\mathrm{Im}(\tilde{f}^*(\vec{k})|g_{\vec{k}}|e^{i\alpha_{\vec{k}}})\} \tag{44}$$

The cyclic Weyl system corresponding to $E^{inv}_{g_{\vec{k}}}$ is $(\mathcal{H}_0 \otimes \mathcal{M}, W, \Omega_0 \otimes \chi_0)$ where $\mathcal{M}$ is the Hilbert space of square integrable functions on the unit circle ,

$$\chi_0(\alpha) = 1 \qquad \forall \alpha \in [0, 2\pi]$$

$$W(f) = W_0(f)\, e^{2i\, \mathrm{Im}\, (\tilde{f}^*(\vec{k})\, |g_{\vec{k}}|\, \Theta)}$$

$$\{\Theta \chi\}(\alpha) = e^{i\alpha} \chi(\alpha) \qquad , \chi \in \mathcal{M}$$

There exists in this system a particle number operator given by :

$$N = N_F \otimes \mathbb{1}_{\mathcal{M}} + \mathbb{1}_{\mathcal{H}_0} \otimes i \frac{d}{d\alpha} \tag{45}$$

and a phase operator Φ defined by

$$e^{i\Phi} = \mathbb{1}_{\mathcal{H}_0} \otimes \Theta \tag{46}$$

with $[N, \Phi] = i$ (47)

We are now in position to show in what sense the coherent state $E_{g_{\vec{k}}}$ is a phase state. Consider the states

$$E_\lambda \longleftrightarrow (\mathcal{H}_0 \otimes \mathcal{M}, W, \Omega_0 \otimes P_\lambda)$$

where the sequence $(P_\lambda)_\lambda$ is cyclic in $\mathcal{M}$ with respect to $e^{i\Phi}$ and such that

$$P_\lambda^2 \xrightarrow[\lambda \to \infty]{} \delta_{\alpha_{\vec{k}}}$$

Then $E_\lambda \longrightarrow E_{g_{\vec{k}}}$ in the weak convergence topology and

$$\langle \Omega_0 \otimes P_\lambda |\, e^{i\Phi} \,| \Omega_0 \otimes P_\lambda \rangle \xrightarrow[\lambda \to \infty]{} e^{i\alpha_{\vec{k}}}$$

i.e. the dispersion in phase around $\alpha_{\vec{k}}$ goes to zero in this limit .

In this sense, the pure monochromatic non-Fock coherent states are actually phase states. Applications of this result are presently used in constructive theories of the laser. Similar results are obtained for n-Fouri component non-Fock coherent states, but there are no phase properties for a continous Fourier spectrum, as is reasonable on physical grounds (11)

REFERENCES

There is evidently a lot of references for such a review talk. One of principal criteria of choice of the following ones is that they present important lists of references...

(1) This definition was first given by
G.ROEPSTORFF, Comm. Math. Phys. 9, 315 (1968).

(2) A complete blibliography on the introduction and properties of coherent states in quantum mechanics can be found in
J.KLAUDER and E.C.G.SUDARSHAN "Fundamentals of Quantum Optics" W.A.Benjamin (1968) Chapter 7.

(3) V.BARGMANN, P.BUTERA, L.GIRARDELLO and J.R.KLAUDER, Reports on Math. Phys. 2, 221 (1971).

(4) V.BARGMANN, Comm. Pure Appl. Math. 14, 185 (1961).

(5) Reference (2), Chapter 8.4.

(6) For example :
F.T.ARECCHI and E.O.SCHULZ-DUBOIS, "Laser Handbook", North-Holland (1972).

(7) R.J.GLAUBER in "Quantum Optics and Electronics" (C. de Witt, A.Blandin and C.Cohen-Tannoudji, Ed.) Gordon and Breach (1964).

(8) This presentation can be found in :
J.P.PROVOST, F.ROCCA and G.VALLEE, Journ. Math. Phys. 16, 832 (1975).

(9) For example :
P.CARRUTHERS and M.M.NIETO, Rev. Mod. Phys. 40, 44 (1968).

(10) F.ROCCA and M.SIRUGUE, Comm. Math. Phys. 34, 111 (1973).
J.P.PROVOST, F.ROCCA, G.VALLEE and M.SIRUGUE, Jour. Math. Phys. 15, 2079 (1974).

(11) More details can be found in
J.P.PROVOST, F.ROCCA and G.VALLEE, "Coherent states, phase states and condensed states" NTH 74/13 (to be published).

COHERENT STATES AND PIPPARD NETWORKS

M. Boon

Institute for Theoretical Physics, Toernooiveld, Nijmegen (NL)

We want to describe the relationship that exists between sets of wave-functions that are used in problems of quantum solid state physics - namely the Dingle functions which are localised eigenfunctions of an electron in an external uniform magnetic field - and coherent state wave functions. We shall see that Dingle functions are unitarily transformed coherent states and therefore that many properties of the latter, notably criteria for completeness, are at the same time properties of the former. This enables us to answer some questions about the completeness or otherwise of the so-called Pippard networks of Dingle functions, as they can be rephrased as questions about the completeness of networks of coherent states. The relevant theorems on coherent state networks have been proved in papers by Bargmann, Butera, Girardello and Klauder [1] and Perelomov [2], and we shall indicate their contents below.

Let us recall the definition of coherent state for a system of one degree of freedom, with space of states $L^2(x)$: for any point $Z = (q + ip)/\sqrt{2\hbar}$ in the complex plane we know the coherent state is the minimal wave-packet of averaged position and momentum as q and p respectively [3]

$$|z> \equiv \frac{1}{(\pi\hbar)} \quad \exp\left[-\frac{1}{2\hbar}(x-q)^2 + \frac{ipx}{\hbar} - \frac{ipq}{\hbar}\right] .$$

These states form a complete and overcomplete set spanning $L^2(x)$, and in references [1] and [2] we find a proof of the following statement: If we restrict ourselves to a subset of points $\vec{z}_{nm} = n\vec{a}_1 + m\vec{a}_2$ (n,m over all integers), where $\vec{a}_1$ and $\vec{a}_2$ are two non-colinear vectors in the complex plane, then for $\mathcal{A}$ the area of the lattice unit cell (in the Z-plane),

$$\mathcal{A} \leqslant \pi \Rightarrow \text{the } |Z_{nm}> \text{ are a complete and overcomplete set in } L^2(x);$$

$$\mathcal{A} > \pi \Rightarrow \text{the } |Z_{nm}> \text{ are incomplete.}$$

Furthermore, Perelomov [2] proves the result that in the limit of $\mathcal{A} = \pi$, the set $\{ |Z_{nm}> \}$ remains complete if any one of the infinite set is removed, but becomes incomplete if two are removed. In other words for $\mathcal{A} = \pi$ we have a basis for $L^2(x)$ if we remove just one member of the set.

The limiting case $\alpha = \pi$ is called the von Neumann lattice (of points in complex space), since von Neumann originally asserted completeness for the corresponding set of wave-packet states, without giving a proof; he was motivated by the fact that the lattice area corresponds to an area of h in the units of the real and imaginary parts q,p of Z, thus giving one state per Planck cell if we identify Z-space with phase space.
Recently, a simplified proof of completeness for the von Neumann lattice has been given by Bacry,Grossmann and Zak [4].(In this reference, the completeness proof applies further to lattices of general states, not only coherent states.)
The coherent state $|Z\rangle$ can be obtained from $|0\rangle$, which is the harmonic oscillator ground state, by using the shift operators

$$D(Z) = \exp\left[\frac{i}{\hbar}(p\hat{x} - q\hat{p})\right] \quad \text{thus}$$

$$D(Z)\,|0\rangle = |Z\rangle \quad ,$$

where $\hat{x},\hat{p}$ are position and momentum operators. It is well known that these operators D(Z), which belong to the Heisenberg-Weyl group, do not commute in general; since

$$D(Z)\,D(Z') = \exp\left[\frac{i}{\hbar}(pq' - p'q)\right] D(Z')\,D(Z)$$

follows from $[\hat{x},\hat{p}] = i\hbar$. However,
in the case of the von Neumann lattice it is easily verified that the phase factor is unity whenever $|Z\rangle$ and $|Z'\rangle$ are both on the lattice: the $|Z_{mn}\rangle$ are therefore generated from $|0\rangle$ by an infinite abelian group $\{D(Z_{mn})\}$ of "translations" in Z-space.

Let us now consider an electron of unit mass in a uniform magnetic field $\vec{B}$; if $\vec{B}$ is along the Z-axis then the Hamiltonian is

$$H = \tfrac{1}{2}\left(\vec{p} + \frac{e}{c}\vec{A}\right)^2 \quad ; \quad \vec{A} = (-\tfrac{1}{2}By, \tfrac{1}{2}Bx, 0)$$

where $\vec{p} = (p_x, p_y, p_z)$ is the momentum operator and $e > 0$ is the electron charge. Ignoring the simple translational motion in the Z-direction, we have

$$H = \tfrac{1}{2}(p_x - \tfrac{1}{2}\beta y)^2 + \tfrac{1}{2}(p_y + \tfrac{1}{2}\beta x)^2 \ ; \ \beta = e\,B/c,$$

and we shall henceforth deal only with the 2-dimensional (x-y) system.

We know the characteristics of the stationary states of this system: the energy levels, called Landau levels, are harmonic oscillator levels of infinite degeneracy. In solid state physics it can often be a good approximation, where we are investigating the energy levels of a metal in a magnetic field, to take the above Hamiltonian as the unperturbed system and add the crystalline field as a perturbation . Where this is valid, the crystalline field leads to a broadening of the Landau levels, the exact details of which are extremely complicated . Among other difficulties, is making a suitable choice of unperturbed wave-functions, since the infinite degeneracy of the unperturbed levels leaves a great deal of freedom. Evidently, the choice of the set of unperturbed levels must accommodate the translational symmetry of the problem in some way. Among the first and most important articles dealing with such a perturbation theory approach is that of Pippard [5], where he introduces the so-called Pippard networks. These are infinite 2-dimensional networks of Dingle functions (localised eigenfunctions of H), one for each Landau level, where for a given level the whole network can be generated from a single member by operating on it with the elements of an infinite discrete 2-dimensional translation group in x-y space. The area of the unit cell of this discrete translation group is determined by the magnetic field strength alone, and is chosen to give the correct density of states.

The analogy with the von Neumann lattice now appears very suggestive: we shall see the analogy is perfect, and that the translation group is abelian, as for the von Neumann lattice; and finally that the unit cell area of h for the von Neumann lattice in phase space is replaced by an area A in the x-y plane where

$$B\,A = hc/e = \text{quantum unit of flux.}$$

One of the unresolved questions about the Pippard networks was the question of their completeness, as they are formed from non-orthogonal functions. This question has, for example, been raised by Capel [6] . We can now answer the question, using the above analogy, by stating that for a Pippard network of area A above, the Dingle functions form a complete set spanning the given Landau level; that if we omit one from the set they remain complete and that if we omit two they form an incomplete set.

We introduce a canonical transformation to put H in a manifestly harmonic oscillator form:

$$Q = (p_x + \tfrac{1}{2}\beta y)/\beta \ ; \ P = (p_y - \tfrac{1}{2}\beta x); \ \bar{Q} = (p_y + \tfrac{1}{2}\beta x)/\beta \ ; \ \bar{P} = (p_x - \tfrac{1}{2}\beta y).$$

It is easily verified that this conserves the canonical commutation relations. Then H transforms into H' with

$$H' = \tfrac{1}{2}\ (\bar{P}^2 + \beta^2\ \bar{Q}^2\ ),$$

where we observe that P and Q do not occur in H'.

The unitary transformation between the state spaces $L^2(x,y)$ and $L^2(Q,\bar{Q})$ in the two representations is known from the theory of linear quantum canonical transformations, and is [7]

$$\psi(x,y) = \frac{\beta}{2\pi\hbar} \iint_{-\infty}^{\infty} dQ d\bar{Q} \exp\left[- \frac{i\beta}{2\hbar} (xy + 2Q\bar{Q} - 2xQ - 2y\bar{Q})\right] \Psi(Q, \bar{Q})$$

Here, $\psi(x,y)$ and $\Psi(Q, \bar{Q})$ are the same state in the x-y and Q-$\bar{Q}$ representations, respectively.

The Q-$\bar{Q}$ representation is very convenient for displaying suitable basis sets of eigenstates for this system. Of course a disadvantage of this representation is that a local potential V(x,y) becomes non-local (when we want to introduce a crystalline field, for example).

It is clear from the form of H' that the Landau levels $\ell\,(=0,1,2,\ldots\infty)$ have harmonic oscillator energies $\beta\ (\ell + \tfrac{1}{2})$, and the ℓth level is spanned by the set of functions

$$U_\ell(\beta^{\frac{1}{2}}\ Q)\ \Psi_\alpha(\ Q),$$

where U_ℓ is the ℓth Hermite function and the $\Psi_\alpha(Q)$ go with α over any set of functions spanning $L^2(Q)$. If we label the transformed functions $\psi_{\ell\alpha}(x,y)$, then evidently for fixed ℓ the functions span the corresponding Landau level as we run over α. With this in mind we can choose our $\Psi_\alpha(Q)$ according to convenience. For example, if we take the set

$$\{\ e^{2\pi i k Q}\ \mid -\infty < k < \infty\ \}$$

for our $\Psi_\alpha(Q)$, the corresponding $\psi_{\ell k}(x,y)$ are the usual Landau functions

$$\psi_{\ell k}(x,y) = e^{\frac{i\beta}{2\hbar}xy}\, U_\ell\left[\beta^{\frac{1}{2}}(y + 2\pi k/\beta)\right] \exp 2\pi ikx \,.$$

If now we choose the ψ_α to be the coherent states $|Z_{nm}>$ in the variable Q where the Z_{mn} form the von Neumann lattice as described above (unit cell area h in phase space), we obtain a set of functions listed $\psi_{\ell;nm}(x,y)$ which, by the theorems quoted earlier and by the above transformations, will

a) span the Landau level ℓ;

b) form a complete set, on omitting any one of them;

c) are incomplete when any two are omitted.

When we calculate the $\psi_{\ell;nm}(x,y)$ explicitly we obtain the following result: Suppose we take the von Neumann lattice defined by

$$Z_{nm} = \left(\frac{\beta}{2\hbar}\right)^{\frac{1}{2}} (na_2 - ima_1) \; ; \; (\beta^{\frac{1}{2}}a_1)(\beta^{\frac{1}{2}}a_2) = h \quad ,$$

then

$$\psi_{\ell;nm}(x,y) = \left(\frac{\beta}{2\pi\hbar\,\ell!}\right)^{\frac{1}{2}} \left(\frac{\beta}{2\hbar}\right)^{\frac{\ell}{2}} (x'+iy')^{\ell} \exp\left[-\tfrac{1}{4}\frac{\beta}{\hbar}(x'^2+y'^2)+\tfrac{1}{2}\frac{i\beta}{\hbar}(x'na_1-y'ma_2)\right] \; ;$$

$$x' = x-ma_1 \, ; \; y' = y-na_2 \, ,$$

which is just the Dingle function **centered at** $x= ma_1 ; y= na_2$. (Note that the Dingle functions are ring-like functions with electron density $\propto \rho^{2\ell} \exp\left[-\frac{1}{2}\frac{\beta}{\hbar}\rho^2\right]$ where ρ is the radius measured from $x= ma_1, y= na_2$.) The whole network is thus just the network of the sort introduced by Pippard [5], with a well defined area for the unit cell of the lattice of centres. This area in real (x-y) space is consequently

$$A= a_1a_2 = h/\beta \, ,$$

so that BA= hc/e is the flux condition through a unit cell, as stated earlier. Thus when the flux condition is precisely satisfied we have the completeness properties (a) - (c) peculiar to the von Neumann lattice. Indeed, the number of states in level ℓ/unit area is $1/A= eB/hc$, agreeing with the standard result for the density of states [8].

As with the von Neumann lattice, the Pippard network is generated from a Dingle function at m=n=0 by a discrete abelian group; the operators are obtained(via the canonical transformation)from

$$D(Z_{nm}) = e^{i(pQ - qP)/\hbar} ; \; q= \beta^{\frac{1}{2}}a_2\, n ; \; p= -\beta^{\frac{1}{2}}a_1\, m \quad ,$$

and they take the form in the (x-y) representation as:

$$T_{nm} = e^{\frac{i\beta}{2h}(xna_2 - yma_1)} e^{-\frac{i}{h}(ma_1p_x + na_2p_y)}$$

Such magnetic translation operators were introduced by Zak [9] .

In conclusion, we make several remarks;

(i) The strength of the field defines a natural Pippard network in x-y space, where the size of the unit cell is fixed but the lattice vectors are otherwise free. In crystals, the commensurability or otherwise of this natural network with the crystal lattice appears to be of great importance in determining the band structure.

(ii) The $Q - \bar{Q}$ representation makes it easy to see that all operators depending on the canonical operators P and Q commute with H; thus the whole Heisenberg- Weyl group in P and Q commutes with H and this is the group theoretical explanation of the infinite degeneracy.

(iii) By taking different basis sets $\Psi_\alpha(Q)$ one can recover familiar basis sets $\psi_{\ell\alpha}(x,y)$. For instance:

$$\Psi_\alpha(Q) \rightarrow U_m(Q) \quad \text{leads to the } \psi_{\ell m}(x,y)$$

defined by Johnson and Lippmann [10] which are simultaneous eigenfunctions of H and of the angular momentum around the Z-axis, of values $\beta(\ell+\frac{1}{2})$ and $\hbar(\ell-m)$, respectively.

Another choice of basis can be made by constructing Fourier sums of the $Z_{nm}(Q)$ for the von Neumann lattice: one defines*

$$\Phi_{\kappa\kappa'}(Q) = \sum_{mn} (-1)^{mn} e^{2\pi i\kappa n} e^{2\pi i\kappa' m} Z_{nm}(Q) \qquad (0 \leq \kappa,\kappa' < 1) ;$$

which carries a one-dimensional representation of the abelian group introduced above. Then

$$\Phi_{\kappa\kappa'}(Q) = A(\kappa,\kappa') \sum_{j=-\infty}^{\infty} e^{2\pi i\kappa j} \delta(\kappa'+j-Q)$$

are essentially functions of the Zak-Cartier [10,11] type. Care must be taken, for it turns out that the normalisation factor $A(\kappa,\kappa')$ is a so-called θ-function with a zero at $\kappa=\kappa'=\frac{1}{2}$ [11]. This is apparently connected with the fact that the original set $Z_{nm}(Q)$ is overcomplete by just one

* For convenience $|Z_{nm}>$ is now denoted $Z_{nm}(Q)$

function [2] . Without the normalisation factor, the $\phi_{\kappa\kappa'}(Q)$ redefined are a basis for $L^2(Q)$, and the corresponding $\psi_{\ell;\kappa\kappa'}(x,y)$ a basis of Bloch functions for Landau level ℓ.

(iv) We know that the coherent states Z(Q) introduced at the beginning are eigenstates of the annihilation operator. From this, one easily shows that

$$a = (p_x + i\, p_y) - \tfrac{1}{2}\, i\beta(x + iy)$$

is an annihilation operator for Dingle functions (as can be seen directly).

(v) One has constructed the Pippard network out of transformed coherent states. The latter are harmonic oscillator ground states. One could easily construct a von Neumann lattice from excited states - with the same lattice area one obtains similar completeness properties as for the coherent states.* The corresponding Pippard network of "excited" Dingle functions - actually displaced functions of the form $\psi_{\ell m}(x,y)$ described above, for a fixed level ℓ and for fixed m, could also provide a starting point for lattice calculations.

REFERENCES

[1] V. Bargmann, P. Butera L. Girardello, J. Klauder, Reps on Math. Phys., 2, 221 (1971).

[2] A. Perelomov, Theor. Math. Phys. 6, 156 (1971).

[3] J. Klauder, E. Sudarshan, "Fundamentals of Quantum Optics" (Benjamin, N.Y. 1968).

[4] H. Bacry, A. Grossmann, J. Zak, Phys. Rev. B (to appear).

[5] A. Pippard, Phil. Trans. Roy. Soc.(London) A256, 317 (1963).

[6] H. Capel, Physica 42, 491 (1969), see p. 504.

[7] M. Boon, T. Seligman, J. Math. Phys. 14, 1224 (1973).

[8] e.g. J. Ziman, "Principles of the Theory of Solids" (Cambridge University Press, 1964).

[9] J. Zak, Phys. Rev. 134A, 1602, 1607 (1964).

[10] J. Zak, Phys. Rev. 168, 686 (1968).

[11] P. Cartier, Am. Math. Soc. Symposia on Pure Mathematics 9, 361 (1966).

---

* This follows from the results of [4]

ATOMIC, MOLECULAR, NUCLEAR, SOLID STATE PHYSICS

The Algebraic Approach to Nuclear Structure Problems

R.M. Dreizler
Institut für Theoretische Physik
Johann Wolfgang Goethe-Universität,
Frankfurt/M.

Conventionally, one uses group theory in nuclear physics for the state labeling problem. I would like to report on some efforts to mobilise group theoretical aspects for a discussion and hopefully a solution (even if approximate) of dynamical aspects in nuclei. The basis of this report is some work I did in collaboration with A. Klein and M. Vallieres and subsequent work of the Philadelphia group.

The Hamiltonian of the many body problem in nuclear physics is usually written down in the form

$$\hat{H} = \Sigma_{\alpha\beta}\ \varepsilon_{\alpha\beta}\ a^{+}_{\alpha}\ a_{\beta} + \frac{1}{4}\ \Sigma_{\alpha\beta\gamma\delta}\ \ V_{\alpha\beta\gamma\delta}\ a^{+}_{\alpha}\ a^{+}_{\beta}\ a_{\delta} a_{\gamma} \tag{1}$$

There seems to be agreement that single particle oscillator states can serve as a suitable basis for the representation. The labels thus stand for

$$\alpha \Rightarrow \{n\ell jm,\ m_{t}\} \tag{2}$$

On the other hand there exists a reasonably confusing selection of matrix elements for the one and two body parts, that are used in actual applications.

I will not attempt to throw any light on this confusion, but rather emphasize the following point: Even if the structure of the two body part is sufficiently simple (and for this reason not to be counted in the class of so-called realistic interactions), the actual determination of eigenstates of the operator or other characteristics of the solution is far from trivial.

This I will take as the motivation to pose the question: Are there, despite the long history of attempts to provide solutions of the problem, some novel methods, which could be turned to some advantage?

The standard methods rely on the actual construction of antisymmetrized wavefunctions. As a point of introduction, it might be worthwhile to have a brief look at the two most reliable approaches.

(1) The shell model or diagonalisation method: The foundation of this method is the Ritz variational principle. After setting up a basis for the states of an A-nucleon system, for instance in the form of Slater determinants

$$|n\rangle = \prod_{i=1}^{A} a^{+}_{\alpha_i} |0\rangle , \qquad (3)$$

the expansion of the eigenstates:

$$|\psi_e\rangle = \Sigma_n \; c_n^{(e)} |n\rangle \qquad (4)$$

is determined via the variational principle

$$\delta_{c_n^*} \langle \psi_e | (\hat{H} - E) | \psi_e \rangle = 0 . \qquad (5)$$

The result are the wellknown eigenvalue equations

$$\Sigma_{n'} \{ \langle n | \hat{H} | n' \rangle - E\, \delta_{nn'} \} c_{n'} = 0. \tag{6}$$

As long as the basis is reasonably complete, this is <u>the</u> method. Technical difficulties, however, prevent the application to nuclei in the intermediate to heavy mass region, despite tremendous advances in the recent years. There is, to my knowledge, no systematic scheme for the selection of a basis on the grounds of a given physical situation except the restriction to major shells. The latter is in general not sufficient to bring the diagonalisation problem (and more so the construction of the matrix to be diagonalised) to manageable proportions. Thus there is no access to aspects summed under the catchword "collective phenomena" via this method.

(2) On the other side of the spectrum of methods are the ground state variational calculations such as Hartree-Fock (HF), the BCS approximation and the Hartree-Fock-Bogoljubov scheme. The common denominator is the reduction of the many body problem to an effective one body problem via the concept of quasiparticles and a special ansatz for the supposed groundstate. For instance for the case of HF one uses for the description of the groundstate

$$| HF \rangle = \Pi_{i=1}^{A} b_{k_i}^{+} | 0 \rangle . \tag{7}$$

The quasiparticles

$$b_k^{+} = \Sigma_\alpha c_k^{(\alpha)} a_\alpha^{+} \tag{8}$$

are determined through the restricted variational principle

$$\delta_{C_\alpha^*} \{ \langle HF | \hat{H} | HF \rangle - \Sigma_k e_k \langle k | k \rangle \} = 0. \tag{9}$$

Although the basic aim is the determination of one state, the groundstate, one obtains, per chance, some information on excited states. The reason is: The HF solution does not necessarily correspond to the symmetries of the original Hamiltonian $\hat{H}$. It thus contains information on excited states, which can be sorted out by projection techniques. As this information is a side product, the question: "In how far is it reliable?" must come to mind.

As the gist of this brief exposition, I would state: On the one hand we have methods, which explicitly try to determine one state, the groundstate. On the other hand we have a method, which would give the complete spectrum, if we could overcome the technical limitations. It seems only fair to raise the original question again in the form: "Is it possible to set up a scheme, which is viable in the large range inbetween these two methods? In more detail I would ask: Is it possible to set up a scheme that is able

(a) to go beyond the groundstate variational approach,

(b) contains the full Ritz variational formulation (in principle)

and (this is the main point)

(c) offers the possibility to incorporate acceptable approximations more easily?

I do not claim to have the answer to this question, at best a first step in one possible direction. The problem encountered in trying to go beyond the relatively compact ansatz for the groundstates, but not use an expansion of large dimensionality in terms of a single basis, is the specification of a suitable form of the wavefunctions to be used. In order to bypass this difficulty one could do away with wavefunctions altogether and attempt a formulation in terms of other quantities.

Today I would like to introduce one option, to which we attached the name algebraic variational method.

We (what I would call the Philadelphia-Orsay-Frankfurt collaboration) also developed one alternative. This is based on a direct determination of the coefficients of fractional parentage (cfp)

$$\langle \psi_e (A-1) | a_\alpha | \psi'_e (A) \rangle = \Sigma_{\nu,n} c^*_\nu c'_n \langle \nu | a_\alpha | n \rangle \tag{10}$$

between eigenstates of the A-1 and A particle systems via equation of motion techniques. It is reasonably obvious, that the parentage coefficients are sufficient for the characterisation of the properties of the A-body system. For a one body operator we have for instance

$$\langle \psi'_e (A) | \hat{O} | \psi_e (A) \rangle = \Sigma_{\alpha\beta} \langle \alpha | \hat{O} | \beta \rangle \tag{11}$$

$$\Sigma_{\ell''} \langle \psi'_e | a^+_\alpha | \psi''_e (A-1) \rangle \langle \psi''_e (A-1) | a_\beta | \psi_e \rangle .$$

Because the coefficients are (in principle) directly observable in pick-up reactions, a close orientation of approximations according to experiment should be possible.

This scheme has met with limited success. Some of the difficulties seem to arise from the simultaneous consideration of even and odd systems. The desire to deal with even systems first, was the starting point for the AVM scheme. The basic idea of this scheme can be described as follows.

We take as the starting point of our theory multipole and two particle transfer operators

$$\hat{O}_\rho \rightarrow \Sigma_{\alpha\beta} \; C_\rho^{\alpha\beta} \; a_\alpha^+ a_\beta$$
$$\Sigma_{\alpha\beta} \; C_\rho^{\alpha\beta} \; a_\alpha^+ a_\beta^+ \qquad (12)$$
$$\Sigma_{\alpha\beta} \; C_\rho^{\alpha\beta} \; a_\alpha a_\beta .$$

The coefficients C can either represent simple angular momentum coupling or some more complicated superposition determined dynamically.

The Hamiltonian can be represented through these operators

$$\hat{H}\,(a^+,a) \Rightarrow \hat{H}\,(\hat{O}_\rho). \qquad (13)$$

The operators O_ρ satisfy a Lie algebra of the form

$$[\,\hat{O}_\rho,\ \hat{O}_\sigma\,] = \Sigma_\tau \; f_{\rho\sigma}^{\tau} \; \hat{O}_\tau . \qquad (14)$$

The structure constants can be determined using the Fermi commutation relations of the creation and destruction operators.

Normally, we are interested in a set of states

$$\{ \ |\psi_n\rangle \ , n = 1, 2, \dots \} , \tag{15}$$

as for instance collective bands etc. If these states belong to a definite representation of the algebra(14) or a subalgebra, then we can evaluate the matrix elements of products in the following fashion

$$\langle \psi_{n'} | \hat{O}_\rho \hat{O}_\tau | \psi_n \rangle = \Sigma_{n''} \langle \psi_{n'} | \hat{O}_\rho | \psi_{n''} \rangle \langle \psi_{n''} | \hat{O}_\tau | \psi_n \rangle . \tag{16}$$

We will assume that all products of interest can be saturated in this fashion by a reasonable number of states of interest. We then have for the energies of the states $|\psi_n\rangle$

$$E_n = \langle \psi_n | \hat{H} | \psi_n \rangle = E_n (\langle \psi_n | \hat{O} | \psi_{n'} \rangle) , \tag{17}$$

a functional of the transition matrix elements within the band (or bands).

If the number of matrix elements under consideration is sufficient to yield an adequate representation of the expectation values, then we can use the matrix elements as variational parameters in the sense of the Ritz variational principle. We try to determine them via

$$\delta_{\langle\rangle} \ E_n (\langle \ \rangle) = 0 . \tag{18}$$

The matrix elements are, however, far from independent. They are to be subjected to a number of constraints.

As constraints we use on the one side a set of kinematical conditions

(1) matrix elements of the Lie algebra, that can be evaluated with the sum rule given above

$$\Sigma_{n''} \{ \langle \psi_{n'} | \hat{O}_\rho | \psi_{n''} \rangle \langle \psi_{n''} | \hat{O}_\sigma | \psi_n \rangle - (\rho \leftrightarrow \sigma) \}$$
$$= \Sigma_\tau f^\tau_{\rho\sigma} \langle \psi_{n'} | \hat{O}_\tau | \psi_n \rangle. \tag{19}$$

(2) Matrix elements of the Casimir invariants of the Lie algebra

$$\langle \psi_{n'} | \hat{C}(\hat{O}_\rho) | \psi_n \rangle = C(\langle \psi_{n''} | \hat{O} | \psi_n \rangle) \tag{20}$$
$$= C\, \delta_{nn'}.$$

The Casimir invariants specify the representation, including its Fermion character. These constraints then are vital to enforce the Pauli principle.

If necessary we also have to consider dynamical constraints, as for instance the requirement that the Hamiltonian be diagonal

$$\langle \psi_n | \hat{H} | \psi_{n'} \rangle = 0 \qquad n \neq n'. \tag{21}$$

This condition has to be considered if it is not automatically satisfied through conservation theorems (e.g. angular momentum).

The application of this scheme to real nuclei is not immediately evident, as the Lie algebras involved are formidable indeed. For instance the representation of $\hat{H}$ as a polynomial in the operators O_ρ can, in general, be effected in many ways.

We do not want to attempt a full solution (the shell model is superior). If one restricts oneself to a subset of states (as indicated above) a selection of suitable kinematical constraints is called for. There are no fast rules for doing this. The selection corresponds roughly to the choice of a wavefunction in standard language.

The only way to develop the "physical intuition" and experience seemed to lie in the investigation of a number of many body models, which have sufficient simplicity.

I would like to stress, however, that in each case the HF or HFB approximations are contained in the scheme as relatively simple limiting situations. They correspond to restriction to a particular subalgebra and the consideration of one state $\{|\psi_n\rangle\} = |\psi_0\rangle$ only. One can also make the connection with the RPA approximation and various quasi-boson methods. As an illustration of the procedure envisaged, I would like to discuss one of the relatively simple many body methods that were investigated.

The model consists of a one shell configuration space $(j)^n$ with a pairing and quadrupole Hamiltonian ($\Omega = j+{}^1/2$)

$$\hat{H} = -G\Omega A^+_{00} A_{00} - \frac{1}{2}\chi \Sigma_q (-)^q B_{2q} B_{2-q} . \tag{22}$$

The operators A^+_{00}, A_{00} and B_{2q} are special cases of the general two-particle transfer and multipole operators that can be constructed from the Fermi operators a^+_{jm} and a_{jm}

$$A^+_{kq} = \frac{1}{\sqrt{2}} \Sigma_m \begin{bmatrix} j & j & k \\ m & q-m & q \end{bmatrix} a^+_{jm} a^+_{j\,q-m}$$

$$B_{kq} = [2k+1]^{-\frac{1}{2}} \Sigma_m \begin{bmatrix} j & j & k \\ m & q-m & q \end{bmatrix} (-)^{j-m+q} a^+_{jm} a_{j\,m-q} . \tag{23}$$

The complete set of operators spans the Lie algebra of the rather formidable group $O(4j+2)$. The structure of the algebra is indicated by the commutation relations

$$[B_{kq}, A^{+}_{k'q'}] = \Sigma_{k''} f^{(AB)}_{kq\,k'q',\,k''} A^{+}_{k'',\,q+q'}$$

$$[A_{kq}, A^{+}_{k'q'}] = \delta_{kk'}\delta_{qq'} - \Sigma_{k''} f^{(AA)}_{kq\,k'q',\,k''} B_{k'',\,q'-q}$$

$$[B_{kq}, B_{k'q'}] = \Sigma_{k''} f^{(BB)}_{kq\,k'q',\,k''} B_{k'',\,q+q'}. \tag{24}$$

The structure constraints can be expressed through Racah and Clebsch-Gordan coefficients.

We first take a brief look at the shell model problem. The solution in the limit $\chi = 0$ is wellknown, it is the seniority scheme. The solution for the case $\chi \neq 0$ is far from trivial if one considers not too small values of j. In order to indicate the labour involved I give a few numbers for the case $(21/2)^6$. For the construction of the Hamiltonian matrix the determination of the basic cfp is required. This requires the diagonalisation of the Casimir operators of SU(2j+1) and Sp(2j+1) with the dimensionalities

36 for J = 0, 173 for J = 2, 302 for J = 4.

The final diagonalisation problem for the Hamiltonian has the dimensionalities

13 for J = 0, 31 for J = 2, 51 for J = 4.

In all: not a trivial problem.

In order to apply the variational method we also have to note the Casimir operator.

It has the form

$$\hat{C}^{(2)} = \Sigma_{kq} \{ A^{+}_{kq}, A_{kq} \}_{(+)} + \Sigma_{\substack{k\neq 0\\ q}} B_{kq} B_{k-q} \tag{25}$$

$$+ \frac{1}{2}\Omega \left(1 - \left(\frac{\hat{N}}{\Omega}\right)^2\right) - \frac{1}{2}\Omega + \Sigma_{k_{even}} (2k+1).$$

The eigenvalue is determined by taking the matrix element with respect to the Fermi vacuum. We now define the reduced matrix elements

$$\langle 0'0\,(N-2)\,|A_{00}|\,00\,(N)\rangle \Rightarrow A_0(00') \tag{26}$$

$$\langle 2'M\,(N)\,|B_{2q}|\,00\,(N)\rangle \Rightarrow B_2(2'0)$$

and assume that the groundstate expectation value of the Hamiltonian is saturated by one intermediate state

$$\langle 0|\hat{H}|0\rangle = -G\Omega\, A_0(00)^2 - \frac{\chi}{2} B_2(20)^2. \tag{27}$$

This is exact in the limit $\chi = 0$.

We then attempt to determine a sufficient set of kinematical constraints. This is only possible if we bring in the additional amplitudes

$$A_2(2(N), 0(N-2)) \text{ and } A_2(0(N), 2(N-2)), \tag{28}$$

matrix elements of quadrupole pairing operators.

The implication of the Casimir constraint takes a bit of arguing involving the statement that correlations are restricted to low multipoles. We finally use the constraints (including the number nonconserving approximation analogous to the BCS ansatz):

$$A_2(20)^2 - A_2(02)^2 = \frac{5(\Omega - N)}{\Omega} \rightarrow [A_2, A_2^+]$$

$$[A_2(20) - A_2(02)]\, B_2(20) = \sqrt{\frac{10}{\Omega}}\, A_0(00) \rightarrow [A_2, B_2]$$

$$A_0(00)^2 = \frac{1}{2}N - \frac{N^2}{4\Omega} + 3\frac{N}{\Omega} - \frac{5}{2}\, B_2(20)^2$$

$$- A_2(02)^2 \rightarrow \text{Casimir Operator.} \tag{29}$$

With this simplest possible approximation (4 amplitudes) we have then the variational equation plus 3 kinematic constraints, which constitute a set of nonlinear equations for the amplitudes, that can be solved numerically (with reasonable ease).

In order to demonstrate the quality of the results we will look at the case $(^{21}/2)^6$.

(1) We first have the excitation energy $\omega_2 = E_2 - E_0$ as a function of the relative coupling strength (Fig.1). This energy can be determined from the results of the variational calculation via the commutation relation

$$<2|\,[\hat{H}, B_{20}]\,|0> \Rightarrow \omega_2\, B_2(20) \tag{30}$$

as

$$\omega_2 = \frac{9}{5j(j+1)(2j+1)}\,\chi + \frac{2}{5}\, G\, \frac{A_0(00)}{B_2(20)}\, [A_2(20) - A_2(02)]. \tag{31}$$

The comparison with the exact result is quite reasonable. The maximum deviation (about 20%) occurs for the intermediate situation, where one expects and sees a relative sharp transition from the seniority scheme to a deformed situation. The

RPA (also shown) fails at exactly this point.

(2) The individual matrix elements $A_0(00)$ and $B_2(20)$ (Fig.2). Here we compare the results of the VAM with the exact values. Further comparison is invited with the results of a HFB calculation with effective angular momentum projection. In real life such a calculation is feasible these days. On the basis of this comparison I would not put too much trust in the HFB method. The quantity $A_0(00)$ (in normal language the gap parameter) is quite acceptable in the seniority limit, the quadrupole matrix element is not. Then we have the often quoted phase transition. It is much too sharp, the gap parameter drops too much, but the quadrupole matrix element becomes quite reasonable.

I also show the same quantities for the rather exotic configurations $(399/2)^{104}$ (Fig.3). An exact calculation is not possible here. We again see the phase transition (spherical to deformed) which is too drastic in the HFB approximation.

The calculation can be extended to include more states. In order to demonstrate one of the dangers we look at the case $(17/2)^4$, three states 0,2,4 (10 amplitudes) and in particular the result for the excitation energy of the J=4 state relative to the J=2 state (Fig.4). In the exact calculation we have two J=4 states, one being characterised by s=2, the other by s=4 in the limit χ=0.The variational calculation with one J=4 state gives reasonable answers in the limiting situations spherical and deformed. In the intermediate region it makes the best of a bad situation by interpolating between the two extremes.

A number of other models have been investigated. The setup and the overall results are comparable (if not better) than the results quoted for the single j-model.

So far the following many body models have been considered in detail

(a) model of Meshkov, Glick and Lipkin with SU(2) symmetry: (14 amplitudes, 3 states).

(b) Several j levels with a pairing interaction, symmetry: Su(2)x Su(2)x... (24 amplitudes).

(c) R(5) model with two orbitals with vibrational and rotational limiting situations,

(d) groundstate correlations for closed shell situations (no applications).

At Frankfurt we are attempting a combination of the shell model methods (the Lanczos basis generating scheme) and the algebraic approach. There are no definite results so far.

In conclusion I would state: we definitely have a novel approach, we have gained considerable experience through the investigation of simple but nontrivial many body models, attempts for a first application to a real life situation are under way. Whether the scheme proves to be successful in the end, remains to be seen.

References

(1) CFP-scheme, general summaries

A. Klein, "Theory of collective motion in nuclei", Lectures in Theoret. Phys., Proc. Boulder Summer Inst. for Theoret. Phys., Vol. XI B (Gordon and Breach, New York, 1968) p.1.

R.M. Dreizler, "Core-Particle Coupling, a General Approach", Proc. of the Topical Conf. on Vibrational Nuclei, Zagreb 1974 (to be published by North Holland).

(2) AVM-Scheme, Summary

A. Klein, "Lie algebras, exactly soluble shell models and theories of collective motion", Rev. Fisica Mexico.

(3) The single j model

R.M. Dreizler and A. Klein, Phys.Lett. 30B (1969), 236.

M. Vallières and R.M. Dreizler, Nucl.Phys. A175, (1971), 272.

M. Vallières, A. Klein and R.M. Dreizler, Phys.Lett. 41B, (1972), 125.

M. Vallières, A. Klein and R.M. Dreizler, Phys.Rev. C7, (1973), 2188.

(4) The MGL Model

G.J. Dreiss, A. Klein and S.C. Pang, Phys.Lett. 29B, (1969), 465.

G.J. Dreiss and A. Klein, Nucl.Phys. A139, (1969), 81.

(5) The Pairing Problem

S.C. Pang and A. Klein, Can.J.Phys. 50 (1972). 655.

C. Dasso, A. Klein, C.Y. Wang-Kaiser, G.J. Dreiss, Nucl.Phys. A205, (1973), 200.

(6) R(5)-Models

P.K. Chattopadhyay, F. Krejs and A. Klein, Phys.Lett. 42B, (1972), 315.

C. Dasso, F. Krejs, A. Klein and P.K. Chattopadhyay, Nucl. Phys. A210, (1973) 429.

C. Dasso and A. Klein, Nucl.Phys. A210, (1973) 443.

C. Dasso and A. Klein, Nucl.Phys. A222, (1974), 445.

(7) Groundstate Correlations

R.M. Dreizler, A. Klein, F.R. Krejs and G.J. Dreiss, Nucl.Phys. A166, (1971), 624.

F. Krejs and A. Klein, J.Math.Phys. 14, (1973), 1155.

Figure Captions

Fig. (1): The excitation energy of the first 2^+ state for the configuration $(21/2)^6$ as a function of the relative coupling strength $\chi/_G$ and $G/_\chi$. Comparison of AVM (———) with the exact shell model calculation (— — —) and RPA (.......).

Fig. (2): The basic quantities $A_0(0,0)$ (the gap parameter Δ) and $B_2(2,0)$ (the reduced quadrupole transition matrix element) are compared to exact results (------) for the configuration $(21/2)^6$. Results of AVM: (— — —) and angular momentum projected (effective) H(F)B: (———). [as a function of the relative coupling strength $\chi/_G$ and $G/_\chi$].

Fig. (3): The same quantities for the configuration $(399/2)^{104}$ (HB (———), AVM (— — —)).

Fig. (4): Excitation energies of the 4^+ states relative to the first 2^+ state for the configuration $(17/2)^4$ as a function of the relative coupling strength $\chi/_G$. The curves are: $\Delta E_4(1)$ (s=2 in the limit χ = 0) and $\Delta E_4(2)$ (s=4 in the limit χ = 0) (full line) and the results of a three state AVM calculation (- ——).

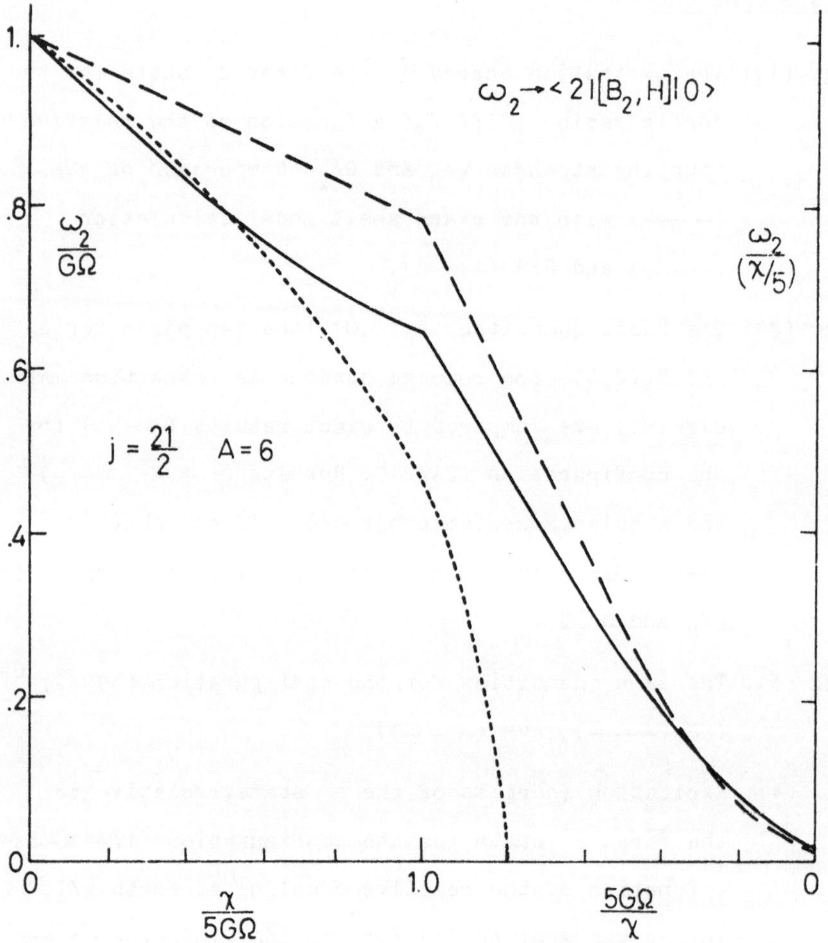

Fig. 1

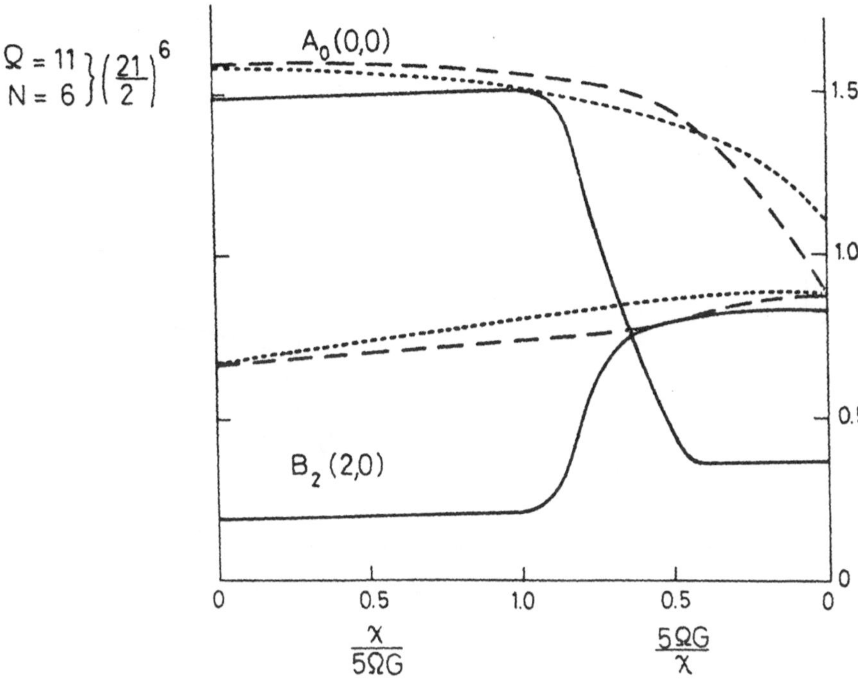

Fig. 2

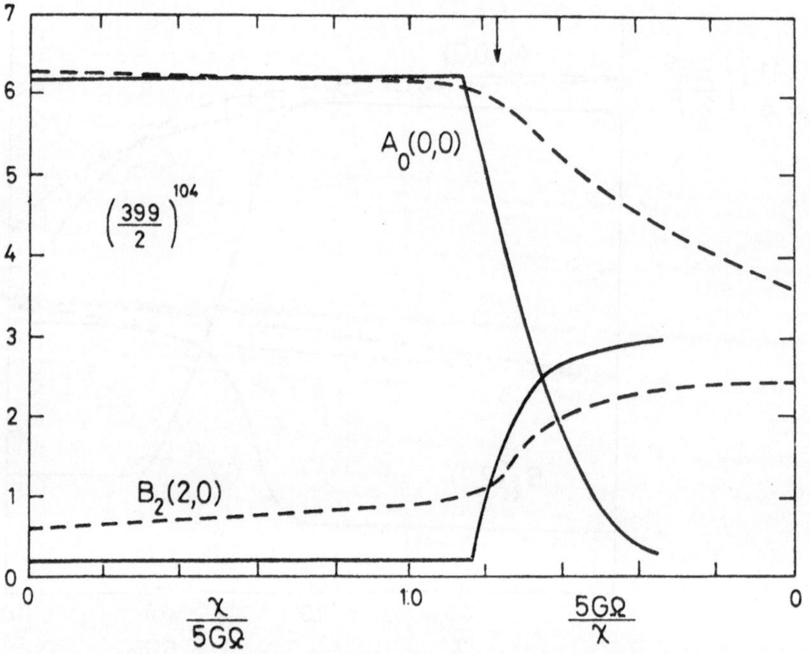
7
6
5
4
3
2
1
0
0
1.0
0
$\left(\frac{399}{2}\right)^{104}$
$A_0(0,0)$
$B_2(2,0)$
$\frac{\chi}{5G\Omega}$
$\frac{5G\Omega}{\chi}$

Fig. 3

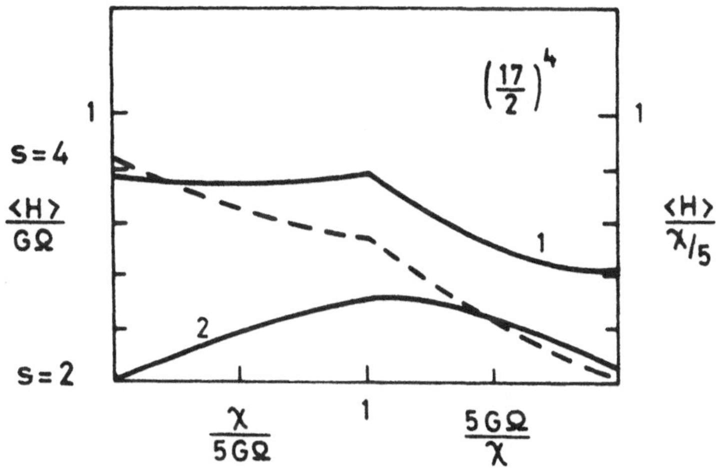

$\left(\frac{17}{2}\right)^4$
1
s=4
$\frac{\langle H \rangle}{G\Omega}$
s=2
1
$\frac{\langle H \rangle}{\chi/5}$
1
2
$\frac{\chi}{5G\Omega}$
1
$\frac{5G\Omega}{\chi}$

Fig. 4

Lie Groups and the Jahn-Teller Effect for a Color Center*

B. R. Judd

The Johns Hopkins University, Baltimore, Maryland

Abstract

The F^+ center in CaO consists of an electron trapped in an oxygen vacancy. The interaction between the electron and the even normal modes of the surrounding octahedron of calcium ions can be represented by an approximate Hamiltonian possessing an oscillator term of U(5) symmetry and an interaction term of O(3) symmetry. To separate repeated angular-momentum quantum numbers L in the symmetric irreducible representations of O(5), four independent methods have been studied. A remarkable coalescence of these approaches in the case of L=6 suggests a natural way to make the separation.

*Work supported in part by the U. S. National Science Foundation

1. INTRODUCTION

The Jahn-Teller (JT) effect refers to the spontaneous distortions that symmetric molecular complexes undergo when the associated electronic state is degenerate.[1] A system of particular interest to us is the F^+ center in CaO, which consists of an electron trapped in an oxygen vacancy. The immediate environment of the electron is an octahedron of calcium ions. The two even modes, which are labelled by the appropriate irreducible representations ϵ and τ_2 of the octahedral group, possess almost exactly coincident angular frequencies ω; moreover, the coupling of both modes to the electronic p state is approximately the same.[2,3] As a consequence of these accidents, the two components of ϵ and the three components of τ_2 can be combined to form a spherical d phonon whose five components are created by the second-rank tensor $a^\dagger$. The Hamiltonian can be written

$$H = \tfrac{1}{2}\hbar\omega(a^\dagger \cdot a + a \cdot a^\dagger) + T^{(2)} \cdot (a^\dagger + a),$$

where the dot means the formation of an O(3) scalar. The first term is an oscillator Hamiltonian for which the symmetry group is U(5): the second term represents phenomenologically the coupling between the phonons and the electron. Its magnitude is determined by the amplitude of the second-rank tensor $T^{(2)}$, which acts only in the space of the p electron.

To solve for the energies and eigenstates of H, we may use a basis determined by either the first or second terms in H. These choices correspond to the weak and strong JT limits respectively. The latter has been recently described in detail.[4]

For present purposes, we take the opposite point of view, and consider the group-theoretical aspects of starting from the weak JT limit. This is the approach of O'Brien,[3] who first showed that the structure of the line $s \rightarrow p$ could be accounted for by a Hamiltonian of the form of H.

2. GROUPS

If we decide to work within the basis provided by the oscillator part of H, then we are led naturally to the scheme

$$U(5) \supset O(5) \supset O(3),$$

in which the five-dimensional irreducible representation [1] of U(5) leads down to (10) of O(5), and thence to (2) of O(3), corresponding to a single d phonon. Since phonons are bosons, we have only to consider the symmetric representations [N] of U(5) and (w0) of O(5). Now, the representations (w0) possess a rather unusual property:[5] under the reduction $O(5) \rightarrow O(3)$, the sequence of L values, taken in order of increasing L, tends to a well-defined pattern as $w \rightarrow \infty$. Thus, if w is exactly divisible by 3, the low-L structure (expressed in the traditional spectroscopic labels) is S, F, G, I, I, K,.... If w is not a multiple of 3, the structure is D, G, H, I, K,.... Most of the previous theoretical work on the F^+ center has concentrated on states whose total angular momentum J is 1, and such states can only arise by coupling the p electron to either S or D. However, as soon as states of higher J are studied, multiplicity difficulties associated with the internal labelling problem arise. The earliest instance of this occurs for L=6, corresponding to the two I states in the first of the two L structures listed above. Because of the comparative simplicity of this case, it was decided to make a detailed study of the various ways in which the two I states can be separated. The aim is not merely to formally define two distinct I states, but rather to find an approach that makes their separation a natural one. Such a result, if it could be obtained, would point the way to a general method for resolving the internal multiplicities.

3. FRACTIONAL PARENTAGE COEFFICIENTS

In the weak JT limit, the oscillator part of H is diagonal with respect to the basis. Our attention is thus directed to a calculation of the matrix elements of $a^\dagger$ and a. The reduced matrix elements of $a^\dagger$ are related to the coefficients of fractional parentage (cfp) by the equation

$$(\psi \| a^\dagger \| \psi') = [(2L+1)N]^{\frac{1}{2}}(\psi\{|\psi'),$$

where ψ is a state of d^N with orbital angular momentum L. The cfp factorize:

$$(\psi\{|\psi') = ([N]W|[N-1]W' + [1](10))(W\beta L | W'\beta' L' + (10)d),$$

where W and W' are irreducible representations of O(5), and repeating values of L and L' are distinguished by β and β'. The two factors on the right-hand side of this equation are isoscalar factors; for example, the second factor is just a Clebsch-Gordan (CG) coefficient for O(5) with the CG coefficient for O(3) extracted. For simplicity, we pick the special case defined by W'=(w-1, 0), W=(w0), and N=w, for which the first isoscalar factor is 1.

Many of the cfp for which a given L and L' occur once in W and W' possess a strikingly simple form. They can be found by applying tensorial techniques to evaluate matrix elements whose values are known from general grounds. For example, all matrix elements of $(a^{\dagger}a^{\dagger})^{(1)}$ and $(a^{\dagger}a^{\dagger})^{(3)}$ are zero, since there are no P or F states in d^2. A number of cfp have been calculated by O'Brien[7] by methods such as these. A typical example of a cfp for which no multiplicity complications arise is

$$(d^{w}(w0)K\{|d^{w-1}(w-1,\ 0)M) \quad = \quad [77(u-2)(u-13)/306u(u-1)]^{\frac{1}{2}}, \qquad (1)$$

where w is a multiple of 3, and u=2w+1. When w=6, for which a K state exists in (60) but there is no M state in (50), the cfp automatically vanishes.

The problem of distinguishing multiply-occurring L values would be ideally solved if cfp of a comparable simplicity to the one above could be introduced to define the states. It seems that this is too much to hope for. The various options open to us will now be considered with particular reference to the repeating I states.

4. GODPARENTS

Perhaps the simplest way of constructing a specific I state is to pick a multiplicity-free state of (w-1, 0) and couple a creation operator to it. For example,

$$(a^{\dagger}|G\rangle)^{(6)}, \qquad (a^{\dagger}|H\rangle)^{(6)}, \qquad (a^{\dagger}|I\rangle)^{(6)}, \qquad (a^{\dagger}|K\rangle)^{(6)}$$

are I states corresponding to the godparents G, H, I, and K. When w is a multiple of 3, the (w0) parts of the four I states above is a linear combination of the two possible I states. It is easy to see that the choice of a particular godparent $|L\rangle$ for one I state, say I_1, implies $(I_2\{|L)=0$ for the orthogonal companion I_2. This condition is enough to determine all remaining cfp for I_1 and I_2.

At first sight, there is not much to choose between the various godparents. All four lead to high primes in the denominators of the cfp. (For w=6, the primes are 251, 2113, 157, and 53 respectively.) This is not suggestive of a simple algebraic structure. However, one godparent turns out to be much more significant than the others. It is the G state, in terms of which we now separate the I_1 state from the I_2 state by means of

$$(I_2\{|G) = 0. \tag{2}$$

We can now show that

$$(I_1\{|G) = [3U/715u(u-1)]^{\frac{1}{2}},$$

$$(I_1\{|H) = 8(u-15)[(u-7)(u-2)/130u(u-1)U]^{\frac{1}{2}},$$

$$(I_2\{|H) = 10[22(u+4)(u+6)(u+11)/91(u-1)U]^{\frac{1}{2}},$$

$$(I_1\{|I) = 8(u+9)[(u+4)(u-7)/u(u-1)U]^{\frac{1}{2}},$$

$$(I_2\{|I) = -22[(u-2)(u+6)(u+11)/35(u-1)U]^{\frac{1}{2}},$$

etc., where $U=61u(u+13)+1470$. Although U does not break up into two linear factors with rational coefficients, its presence in the cfp does not detract too much from our ideal form.

Our definition of I_1 and I_2 coincides with the ostensibly arbitrary separation that Hecht[8] made for the special case for which w=6.

5. INTRINSIC STATES

The problem of defining the angular-momentum states of (w0) has been studied by Williams and Pursey[9] by extending the notion of intrinsic states that Elliott[10] used for SU(3). Although this approach leads to non-orthogonal components, it is of considerable interest to us because it leads to an equivalent definition of I_2.

Consider the (unnormalized) state

$$|\Phi\rangle = (a_1^\dagger)^{2n+2}(a_{-2}^\dagger)^{n-2}|0\rangle$$

of d^w, where $w=3n$. The subscripts to $a^\dagger$ denote magnetic quantum numbers, the total value of which is given by

$$M_L = 2n+2 + (-2)(n-2) = 6.$$

According to the method of Williams and Pursey, $|\Phi\rangle$ is one of the two intrinsic states that separate the I terms. It is only necessary to rotate $|\Phi\rangle$ through some angle defined by the Euler triad Ω, thereby giving $|\Phi\rangle_\Omega$, and then project out an I term by forming the Hill-Wheeler integral

$$|I\rangle = \int D^{(66)}_{\cdot\,-6}(\Omega)\,|\Phi\rangle_\Omega\, d\Omega, \tag{3}$$

where the double tensor $D^{(JJ)}$ is related to the rotation matrices by the equation[11]

$$D^{(JJ)}_{M,-N}(\Omega) = (-1)^{J-N}(2J+1)^{\frac{1}{2}}\mathscr{D}^J_{MN}(\Omega)^*.$$

If, now, the annihilation tensor a is applied to $|I\rangle$, only those components $(D^{(22)}a^{(20)})^{(02)}_{01}$ or $(D^{(22)}a^{(20)})^{(02)}_{0,-2}$ in the rotated frame give a residue when they act on $|\Phi\rangle_\Omega$. In doing so, they introduce as coefficients $D^{(22)}_{\cdot 1}$ or $D^{(22)}_{\cdot -2}$. When these are contracted with $D^{(66)}_{\cdot\,-6}$ in the integrand, the resulting tensors are of the type $D^{(LL)}_{\cdot\,-5}$ or $D^{(LL)}_{\cdot\,-8}$, which implies that $L \geqslant 5$. Thus $\langle G|a|I\rangle = 0$, and so the I state defined by Eq.(3) is identical to I_2.

The second I state that the method of Pursey and Williams provides comes from the intrinsic state $(a_1^\dagger)^{2n+1}(a_{-2}^\dagger)^{n-1}|0\rangle$. There is no point in developing the state in detail, since it is not orthogonal to $|I_2\rangle$. However, it is worth noting that the overlap is very small: for $w=6$ it amounts to only $(6845/1064993)^{\frac{1}{2}}$.

6. GENERALIZED SENIORITY

For spinless bosons, seniority adds no information above that provided by the irreducible representations of O(5). However, the method of generating states of equal seniority by successive application of the scalar operator $(a^\dagger a^\dagger)^{(0)}$ suggests a possible generalization. An obvious candidate for the new scalar operator is $(a^\dagger a^\dagger a^\dagger)^{(0)}$, that is, the operator that creates the S state of d^3. For not only is it the most elementary extension of $(a^\dagger a^\dagger)^{(0)}$, but it connects states of common L in representations (w0) of O(5) for which the low-L structure is identical.

In fact, $(a^\dagger a^\dagger a^\dagger)^{(0)}$ is equivalent to one of the four operators that Sharp and Lam[12] introduced to distinguish multiply-occurring L values in the representations (w0) of O(5). The others are $a_2^\dagger$, $(a^\dagger a^\dagger)_2^{(2)}$, and $(a^\dagger a^\dagger a^\dagger)_3^{(3)}$. Sharp and Lam's idea is to form stretched products of these operators, subject to the condition that $(a^\dagger a^\dagger a^\dagger)_3^{(3)}$ occurs at most once. For example, the $M_L=6$ state of (60) can be formed in two ways, namely

$$(a^\dagger a^\dagger)_2^{(2)}(a^\dagger a^\dagger)_2^{(2)}(a^\dagger a^\dagger)_2^{(2)}|0\rangle, \tag{4}$$

$$(a^\dagger a^\dagger a^\dagger)^{(0)} a_2^\dagger\, a_2^\dagger\, a_2^\dagger\, |0\rangle, \tag{5}$$

and this indicates that an I state occurs twice. If we write (5) in the form

$$(a^\dagger a^\dagger a^\dagger)^{(0)}|d^3\ I\rangle, \tag{6}$$

we see that it represents an I state that has been constructed by precisely the method that we have in mind.

Of course, the actual states (4), (5), and (6) contain not only the stretched components, but others as well. For example,

(6) contains components of (40)I as well as those of (60)I. When the former are projected out, it is found, rather unexpectedly perhaps, that the I state of (60) is identical to I_1. To see why this should be so, the 3-particle cfp $(I_i\{|\, I_j, (30)S)$ are required (where i, j = 1, 2). They can be calculated by combining products of single-particle cfp. A detailed analysis reveals that

$$(I_2\{|\, I_1, (30)S) = 0. \tag{7}$$

There is thus no way to form an I_2 state by adding the phonon triad $(a^\dagger a^\dagger a^\dagger)^{(0)}$ to an I_1 state. This shows that a string of I_1 states can be formed by writing $[(a^\dagger a^\dagger a^\dagger)^{(0)}]^p \,|\, d^3\ I_1\rangle$ and projecting out all states that do not belong to the irreducible representation (3p+3, 0) of O(5).

If $(I_1\{|\, I_2, (30)S)$ were zero, similar statements could be made about the I_2 states. This cfp does not, however, vanish; but a number of remarkable cancellations lead to its being exceptionally small. In fact,

$$\mathop{\mathrm{Lt}}_{u\to\infty} (I_1\{|\, I_2, (30)S)/(I_2\{|\, I_2, (30)S) = 8u^{-3}(385)^{\frac{1}{2}}. \tag{8}$$

So a string of I_2 states could be formed in an analogous way in the limit of large u (or w).

7. DIAGONALIZING A SCALAR OPERATOR

A common method used by physicists to resolve multiplicity difficulties is to separate the states by requiring that they be the eigenfunctions of some convenient operator -- perhaps one of physical interest. The operator must be a scalar in O(3) so as not to mix different L values. To be effective it cannot be scalar in O(5); nor can it transform according to (22) of O(5), since an operator of this type can be constructed from Casimir's operator for O(5) and L^2. The most elementary operator appears to be the three-body operator

$$T = (a^\dagger a^\dagger a^\dagger)^{(0)}(aaa)^{(0)}.$$

By diagonalizing this operator within the I states of (60) we obtain again the orthogonal pair I_1 and I_2; and Eqs.(7) and (8) show that the $(I_1 I_2)$ separation is also obtained when T is diagonalized within the I states of (w0) in the limit $w \to \infty$. In general, however, we must content ourselves with irrational eigenvalues.

Communications from Dr. M. C. M. O'Brien and Professor R. T. Sharp proved very helpful in the work reported here.

REFERENCES

1. H. A. Jahn & E. Teller, Proc. Roy. Soc. (London) A161, 220 (1937).
2. A. E. Hughes, G. P. Pells & E. Sonder, J. Phys. C5, 709 (1972).
3. M. C. M. O'Brien, J. Phys. C4, 2524 (1971).
4. B. R. Judd & E. E. Vogel, Phys. Rev. B11, 2427 (1975).
5. J. Le Tourneux, K. Dan. Vidensk. Selsk. Mat.-Fys. Medd. 34, 11 (1965).
6. G. Racah, Phys. Rev. 76, 1352 (1949).
7. M. C. M. O'Brien, private communication.
8. K. T. Hecht, Nucl. Phys. 63, 177 (1965).
9. S. A. Williams & D. L. Pursey, J. Math. Phys. 9, 1230 (1968).
10. J. P. Elliott, Proc. Roy. Soc. (London) A245, 562 (1958).
11. B. R. Judd, Angular Momentum Theory for Diatomic Molecules, Academic Press, New York (1975).
12. R. T. Sharp & C. S. Lam, J. Math. Phys. 10, 2033 (1969).

SYMMETRIES AND STATISTICS IN NUCLEAR PHYSICS

C. Quesne *

Physique Théorique et Mathématique, Université Libre de Bruxelles, Brussels, Belgium

1. Introduction

In this contribution we are going to consider some relationships between symmetries and statistics in many-particle systems, with particular emphasis on nuclear physics where they have been studied up to now. However the range of application of the methods we are going to review might be larger than that as they could be used in principle whenever the system is described in spectroscopic terms, the states being represented in terms of particles distributed over some finite set $\mathcal{N}$ of single-particle states.

It is well known that the existence of good symmetries, such as angular momentum, isospin and parity, or broken ones, such as SU(4), SU(3), seniority, etc. greatly simplifies the problem of finding the properties of nuclei.

The merit of French and coworkers was to show that there is another general simplifying principle, the existence of a central limit theorem, by virtue of which the distributions of energy, and other additive quantities, are asymptotically normal [1 - 4]. This theorem, which is neither accurately formulated nor rigorously proved, shows up when exact shell model calculations can be carried out and the corresponding distributions are constructed.

Taking into account the two above-mentioned principles - existence of symmetries and central limit theorem - it becomes interesting and feasible to consider subsets of the n-particle space, chosen as representation spaces of some irreducible representations (IR) of a group or a chain of groups, and to study the distribution of the summed intensity of these subsets over the spectrum. This is the aim of the so-called spectral distribution method [1 - 4].

Owing to the central limit theorem, those distributions can be described by a few low order moments. We thus arrive at the problem of calculating operator averages (such as averages of powers of H) over some IR of a group or a chain of groups. The central point of the spectral distribution method is the possibility, that arises in some cases, of propagating operator averages from low to high values of the number of particles. This enables indeed to compute the distributions even in the cases where a complete shell model calculation is not feasible . This opened the way to a variety of applications, such as the calculation of low energy spectra and level densities, and the study of the goodness of symmetries.

----

* Maître de recherches F.N.R.S.

In Secs. 2 and 3 , we are going to discuss two points intimately connected with group theory : the group theoretical formulation of the propagation process, and the study of symmetries through the decomposition of the widths into partial ones.

2. Group Theoretical Formulation of Propagation

Let us consider the average of a k-body operator $\mathcal{O}(k)$ in a subspace of the n-particle space, defined as the representation space of an IR λ of a subgroup G of $U(\mathcal{N})$, and specified by additional quantum numbers φ if necessary,

$$\langle \mathcal{O}(k) \rangle^{n\varphi\lambda} = \left[\dim(\lambda) \right]^{-1} \sum_{\mu} \langle n\varphi\lambda\mu | \mathcal{O}(k) | n\varphi\lambda\mu \rangle . \tag{2.1}$$

Here μ denotes the row of the IR λ and dim (λ) its dimension.

At this stage it is interesting to introduce a new group $\mathcal{G}$ in order to avoid considering chains of groups for defining the subspaces [5] : if G is a subgroup of SU($\mathcal{N}$), then $\mathcal{G}$ = U (1) x G, the single generator of U(1) being the number operator N; if G is not a subgroup of SU($\mathcal{N}$), i.e. if N belongs to its Lie algebra, then $\mathcal{G}$ = G. The IR's of $\mathcal{G}$ are characterized by $\Lambda = (n, \lambda)$, and their dimension is dim (Λ) = dim (λ).

Following French [2], we say that the average of $\mathcal{O}(k)$ can be propagated from its defining subspaces $\varphi'\Lambda'$, $\Lambda' = (k, \lambda')$, if it can be expressed, for any φ and $\Lambda = (n, \lambda)$, as a linear combination of the averages of $\mathcal{O}(k)$ in its defining subspaces. It is straightforward to show [2, 5] that a necessary and sufficient condition for this to happen is that

$$\langle \varphi\Lambda | D^{\varphi''\Lambda'}_{\varphi'\Lambda'} | \varphi\Lambda \rangle = \delta_{\varphi'\varphi''} \langle \varphi\Lambda | D^{\varphi'\Lambda'}_{\varphi'\Lambda'} | \varphi\Lambda \rangle \tag{2.2}$$

for any φ , φ', φ'', Λ and Λ' . Here operator $D^{\varphi''\Lambda'}_{\varphi'\Lambda'}$ is defined by

$$D^{\varphi''\Lambda'}_{\varphi'\Lambda'} = \sum_{\mu'} \mathbb{D}^{\varphi''\Lambda'\mu'}_{\varphi'\Lambda'\mu'} , \tag{2.3}$$

where

$$\mathbb{D}^{\varphi''\Lambda''\mu''}_{\varphi'\Lambda'\mu'} = P^{+}_{\varphi'\Lambda'\mu'} \, P^{\varphi''\Lambda''\mu''} , \tag{2.4}$$

and $P^{+}_{\varphi'\Lambda'\mu'}$ ($P^{\varphi'\Lambda'\mu'}$) is the creation (annihilation) operator of the k-particle state transforming according to the IR λ' of G, belonging to row μ' , and specified by φ' .

When conditions (2.2) are fulfilled, we can define an operator

$$\hat{\mathcal{O}}(k) = \sum_{\varphi'\Lambda'} \langle \mathcal{O}(k) \rangle^{\varphi'\Lambda'} \mathcal{D}_{\varphi'\Lambda'}^{\varphi'\Lambda'} , \qquad (2.5)$$

which is trace-equivalent to $\mathcal{O}(k)$, and is a linear combination of the operators $\mathcal{D}_{\varphi'\Lambda'}^{\varphi'\Lambda'}$, called propagation operators [2, 5]. The calculation of average (2.1) then reduces to the construction of these operators, and the determination of their diagonal matrix elements. Therefore it is worth while to investigate the group theoretical structure to which operators $\mathcal{D}_{\varphi'\Lambda'}^{\varphi''\Lambda'}$ belong.

It is well known [5, 6] that for fermions distributed over some finite set $\mathcal{N}$ of single-particle states, the states can be characterized by the following chain of groups, below each one of which we give the IR to which they correspond :

$$\begin{array}{ccccccc} U(2^{\mathcal{N}}) & \supset & O^{+}(2\mathcal{N}+1) & \supset & U(\mathcal{N}) & \supset & G \\ [1] & & [\tfrac{1}{2}\mathcal{N}] & & [1^{n}] & & \lambda \end{array} . \qquad (2.6)$$

We shall denote by $\mathcal{G}^{C}$ and $\mathcal{G}^{TC}$ the complementary groups [5, 6] of $\mathcal{G}$ within (the IR $[\frac{1}{2}\mathcal{N}]$ of) O^{+} $(2\mathcal{N}+1)$ and (the IR [1] of) U $(2^{\mathcal{N}})$ respectively. $\mathcal{G}^{C}$ does not always exist , while $\mathcal{G}^{TC}$ can be shown to be for any group $\mathcal{G}$

$$\mathcal{G}^{TC} = \sum_{\Lambda} \oplus \; U(d(\Lambda)) , \qquad (2.7)$$

where the sum is taken over all the IR's Λ of $\mathcal{G}$ contained in the IR [1] of U $(2^{\mathcal{N}})$, and $d(\Lambda)$ is their multiplicity. Whenever $\mathcal{G}^{C}$ does exist, the generators of $\mathcal{G}^{TC}$ are polynomials in the generators of $\mathcal{G}^{C}$.

It is straightforward to show that operators $\mathcal{D}_{\varphi'\Lambda'}^{\varphi''\Lambda'}$ are linear combinations of the generators of $\mathcal{G}^{TC}$ [5] . On the other hand the algebra of $\mathcal{G}^{TC}$ is made of all the polynomials in the generators of U $(\mathcal{N})$ which are scalar with respect to G . Therefore the propagation of operator averages is connected with a classical problem of group theory : the construction of subgroup scalars in the universal enveloping algebra of a group.

Let us return now to conditions (2.2). If a state labeling problem appears in the reduction U $(\mathcal{N}) \supset \mathcal{G}$ for the first time for t-particle states, then it is obvious that these conditions are fulfilled for any $k < t$. They are not satisfied in general for $k \geqslant t$ as it can be seen on some simple examples. Therefore there are close connections between the propagation procedure and the state labeling problem.

The main problem that remains now is the construction of the propagation operators when the propagation is possible. This amounts to build the generators of $\mathcal{G}^{TC}$. For this purpose it is interesting to consider a subgroup $\mathcal{H}$ of $\mathcal{G}^{TC}$, whose Lie algebra is made of all the polynomials in the Casimir operators of $\mathcal{G}$. When $\mathcal{H}$ coincides with $\mathcal{G}^{TC}$, then the propagation operators are easily constructed because the Casimir operators of all classical groups are well known, as well as their eigenvalues in all IR's. However when $\mathcal{H}$ is a proper subgroup of $\mathcal{G}^{TC}$, it is necessary to construct an integrity basis of the algebra of $\mathcal{G}^{TC}$. This is a difficult mathematical problem to tackle but some definite progress was made recently in its solution [7] and we may hope that applications to spectral distributions will be made in a near future.

Sufficient conditions for $\mathcal{H}$ and $\mathcal{G}^{TC}$ to coincide can be found easily [5]. In fact the cases of this type which are treated in the physical literature belong essentially to two classes :

(i) $\mathcal{G}$ is the direct product of two complementary groups, $\mathcal{G} = \bar{G} \times \bar{G}^C$. An example of this kind is $\mathcal{G} = U(\mathcal{N}/r) \times U(r)$. Values of r equal to 1, 2 and 4 correspond to scalar, fixed isospin and fixed supermultiplet averaging respectively.

(ii) $\mathcal{G}$ is the direct product of a group $\bar{G}$ and a canonical subgroup $\bar{G}'$ of its complementary group $\bar{G}^C$, $\mathcal{G} = \bar{G} \times \bar{G}'$, $\bar{G}' \subset \bar{G}^C$. An example of this kind is $\mathcal{G} = Sp(\mathcal{N}) \times U(1)$, where U(1) is a canonical subgroup of Sp(2). It corresponds to fixed seniority averaging for identical nucleons.

$\mathcal{H}$ is a proper subgroup of $\mathcal{G}^{TC}$ for some groups $\mathcal{G}$ for which there is no state labeling problem in the reduction $U(\mathcal{N}) \supset \mathcal{G}$, as well as for all groups $\mathcal{G}$ for which there is a labeling problem. In fact in the latter case it is well known that the algebra of $\mathcal{G}^{TC}$ contains the supplementary operators whose eigenvalues define the additional quantum numbers φ. Cases of the first kind are rather pathological and uninteresting, but cases of the second kind are very numerous. Examples are $\mathcal{G} = SU(3) \times U(4)$ (fixed supermultiplet and SU(3) symmetry averaging), $\mathcal{G} = U(\mathcal{N}/4) \times SU(2) \times SU(2)$ (fixed supermultiplet, spin and isospin averaging), $\mathcal{G} = Sp(\mathcal{N}/2) \times U(2)$ (fixed seniority, isospin and reduced isospin averaging), and $\mathcal{G} = U(1) \times O^+(3)$ (fixed angular momentum averaging for high enough $\mathcal{N}$ values). The state labeling problem occurs in the reduction $U(\mathcal{N}/4) \supset SU(3)$, $U(4) \supset SU(2) \times SU(2)$, $Sp(4) \supset U(2)$ and $U(\mathcal{N}) \supset O^+(3)$ respectively.

We proceed now to discuss the use of spectral distributions to study the goodness of symmetries.

3. Study of the Goodness of Symmetries

The first methods used to evaluate the admixings of various IR's λ', λ''.... into the IR λ of G were based on some global properties of distribution functions. One of these procedures is the following : all the exact distribution functions associated with the various IR's of G are replaced by continuous approximations determined from some calculated low-order moments ; then the relative intensities of the various IR's in the ground state region are measured by the relative values of the corresponding distribution functions at the energy of some low-lying excited state [3] . Another procedure is the calculation of the average value of some operator. An example of this kind is the prediction of the fractional occupancies of single-particle orbits in the ground state [4] .

Although the methods sketched above can give valuable information about symmetry breaking, it is possible to devise more elaborate methods based on group theory. They rest on the decomposition of the widths of the distributions of the various subsets λ ,

$$\sigma^2(n,\lambda) = [\dim(\lambda)]^{-1} \sum_{\mu} \langle n\lambda\mu | [H - \mathcal{E}(n,\lambda)]^2 | n\lambda\mu\rangle = \sum_{\lambda'} \sigma^2(n, \lambda \to \lambda'), \tag{3.1}$$

into partial widths

$$\sigma^2(n,\lambda \to \lambda') = [\dim(\lambda)]^{-1} \sum_{\mu\mu'} |\langle n\lambda'\mu' | H | n\lambda\mu\rangle|^2 - \delta_{\lambda\lambda'} [\mathcal{E}(n,\lambda)]^2 . \tag{3.2}$$

Here

$$\mathcal{E}(n,\lambda) = [\dim(\lambda)]^{-1} \sum_{\mu} \langle n\lambda\mu | H | n\lambda\mu\rangle \tag{3.4}$$

is the centroid energy of the states belonging to λ and we have dropped for commodity all additional quantum numbers as they play no essential part. The partial width $\sigma^2(n,\lambda\to\lambda')$, for $\lambda' \neq \lambda$, is the average over the states of λ of the squares of the matrix elements connecting subspaces λ and λ' . It gives of course a very useful information over the admixing of λ' into λ .

The starting point for the calculation of the partial widths is the expansion of the Hamiltonian in terms of irreducible tensors with respect to G. When the latter has been obtained two possibilities may arise.

The first one is that there is a 1-to 1 correspondence between the irreducible parts of H and the partial widths. Then each partial width is nothing less than the total width for the corresponding irreducible tensor, and the technique developed for calculating the total widths can be used for the partial ones too. Examples of this kind correspond to configuration [4] and quasi-particle [8] distributions ; the subspaces are defined by the number of particles in every orbit or the number of quasi-particles respectively.

The second possibility is that there is no 1-to-1 correspondence between the irreducible parts of H and the partial widths. Special techniques have now to be found. Two cases are completely solved. The first one is associated with $\mathcal{G} = U(\mathcal{N}/r) \times U(r)$. Values of r equal to 2 and 4 correspond to fixed isospin and fixed supermultiplet distributions respectively. The partial widths are shown to be proportional to some Racah coefficients of the unitary group in the chain $U(\mathcal{N}) \supset \mathcal{G}$, and the latter can be calculated [9].

The second case is associated with $\mathcal{G}$ = Sp $(\mathcal{N})$ x U (1), and corresponds to fixed seniority distributions for identical nucleons. The seniority is either that in a single subshell [10] or the multi-shell generalized seniority [1, 11]. It is convenient to use double tensors with respect to the symplectic group Sp$(\mathcal{N})$ and the quasi-spin group Sp(2). It can be shown that the partial widths satisfy some recursion relations, which can be solved for each irreducible tensor separately. The results of ref. 11 show the importance of the single-particle Hamiltonian in giving rise to the seniority breaking and that of the transitions with $|\Delta v| = 2$ in comparison with those with $|\Delta v| = 4$.

By a straightforward generalization of the preceding case, it is possible to get the partial widths of the configuration-seniority distributions, associated with $\mathcal{G} = \sum_a \oplus [Sp(\mathcal{N}_a) \times U(1)]$, where $\mathcal{N}_a$ is the degeneracy of orbit a [12]. The subspaces are now defined by $[\vec{n}\,\vec{v}] = [n_a v_a n_b v_b \ldots]$, where n_a and v_a are the number of particles and the seniority in orbit a respectively. If one is interested in the seniority quantum numbers, it is possible to sum over all the states with the same total number of particles $n = \sum_a n_a$, and the same seniority in every orbit, thus getting in this way seniority distributions. Table 1 contains the partial widths of these distributions, $\sigma^2([\vec{v}] \to [\vec{v}'])$, for the lowest lying $[\vec{v}]$ subspaces, calculated in Ni^{62} with the Auerbach interaction [12]. It is clear that only one subspace is separated from the other ones, namely [000].

Table 1

Partial widths $\sigma^2([\vec{v}] \to [\vec{v}'])$ in Ni^{62}. The ordering of $[\vec{v}]$ subspaces is according to increasing centroid energy.

$[\vec{v}]$ \ $[\vec{v}']$	[000]	[200]	[101]	[110]	[011]	[020]	[211]	[031]	[220]	[121]	[130]	[231]
[000]	3.647	0	0	0	0	0	0.090	0	0.224	0.263	0.003	0
[200]	0	0.442	0.689	0.014	0.037	0.109	0.005	0	0	0.209	0.003	0
[101]	0	0.431	1.845	0.064	0.045	0.068	0.125	0.001	0.131	0.146	0	0.004
[110]	0	0.002	0.011	1.436	0.026	0.019	0.456	0.023	0.066	0.109	0.063	0.117
[011]	0	0.010	0.020	0.069	1.939	0.040	0.014	0	0.087	0.275	0.083	0.168
[020]	0	0.019	0.020	0.032	0.026	2.310	0.034	0.043	0.096	0.150	0.094	0.102

Once the partial widths have been obtained, there still remains a long way before getting an idea of the magnitude of the admixtures in the wave functions. The admixture coefficient [13]

$$I^2(n, \lambda \to \lambda') = \sigma^2(n, \lambda \to \lambda') \Big/ \left[\mathcal{E}(n,\lambda) - \mathcal{E}(n,\lambda') \right]^2 , \tag{3.4}$$

used by many authors, is not a very reliable measure when the internal widths of the distributions are not small in comparison with the distance between the centroids [12]. We think that careful studies of the range of usefulness of this coefficient remain to be made, as well as a search for other ways of estimating the admixtures in the wave functions.

References

1. J.B. French, in Nuclear Structure, A. Hossain, Harun-ar-Rachid, and M. Islam, Eds. (North-Holland, Amsterdam, 1967).
2. J.B. French and K.F. Ratcliff, Phys. Rev. C3 (1971) 94.
3. K.F. Ratcliff, Phys. Rev. C3 (1971) 117.
4. F.S. Chang, J.B. French, and T.H. Thio, Ann. Phys. (N.Y.) 66 (1971) 137.
5. C. Quesne, to be published.
6. M. Moshinsky and C. Quesne, J. Math. Phys. 11 (1970) 1631.
7. B.R. Judd, W. Miller Jr., J. Patera, and P. Winternitz, J. Math. Phys. 15 (1974) 1787.
8. H. Nissimov, R. Arvieu, and O. Bohigas, Nucl. Phys. A190 (1972) 514.
9. K.T. Hecht and J.P. Draayer, Nucl. Phys. A223 (1974) 285.
10. M. Nomura, Progr. Theoret. Phys. 47 (1972) 1858.

11. C. Quesne and S. Spitz, Ann. Phys. (N.Y.) 85 (1974) 115 ;
C. Quesne, to be published.

12. S. Spitz and C. Quesne, to be published.

13. J.C. Parikh and S.S.M. Wong, Nucl. Phys. A182 (1972) 593.

GROUP THEORY IN POLYMER PHYSICS*

I.B. Božović,+ M. Vujičić* and F. Herbut*

+Dept. of Physics, Faculty of Science, University of Belgrade and Institute of Physics, Belgrade

*Institute "Boris Kidrich", Vinča, Belgrade

Polymer materials possess a characteristic chemical structure: their basic building units contain a very large number of atoms (in some cases exceeding 10^5) covalently bound into a molecule. Such a macromolecule in its turn consists of equal smaller monomer units each containing up to a few dozens of atoms. In stereoregular polymers these subunits form geometrically regular patterns, possessing certain symmetry elements.[1] This class of polymers is a significant one also because only stereoregular forms of more complex polymers can crystallize.[2] The stereoregular macromolecules are found in chainlike forms periodical along a line, i.e., they remain invariant under the translation determined by a vector $\underline{c}$ (of the length of about 3-50 A). Further, they may have rotational and screw axes of the order $n=2,3,\ldots$; rotational axes of the order two perpendicular to the chain axis are also possible. So are mirror and glide planes containing the chain axis, as well as mirror planes perpendicular to it, and finally the inversion through the origin.

The type of the screw axis has been experimentally determined for a few hundreds of polymers.(An extensive list of such data published before 1962 can be found in Ref. 1; there are also numerous more recent references dealing with other polymers.) The appearance of other symmetry elements is illustrated in Ref. 3.

The above symmetry elements can be combined into line groups[4] (called rod groups in Ref. 5). They are defined as those subgroups L of the three-dimensional Euclidean group E all the pure translations of which act along a line and form a discrete group T. (We are not interested in L's with T continuous.) It is known that E is a semidirect product of the group of all translations by the group of all rotations round a fixed point. Consequently, T is an invariant subgroup of L, and $L/T \cong P$, where P is the isogonal point group of L, consisting of the rotational parts of the elements of L. Therefore, each line group is an extension of a point group which leaves a line invariant by T. It can be shown that every such P is a subgroup of some $D_{nh} = (C_n \wedge C_2') \times C_{1h}$, $n=1,2,\ldots$, where C_n is the cycle of the rotation through $2\pi/n$ around the z-axis; in C_2' the prime denotes that that the axis of rotation is orthogonal to the z-axis; C_{1h} is the group of the reflection in the xy- plane. The above factorization of D_{nh} is inherited by its subgroups, so that each P can be decomposed into an analogous product of no more than three cyclic factors. This fact can be made use of in the construction of the line groups isogonal to P due to the following theorem:

Let $P = P_1 \wedge P_2$ and let L_1 and L_2 be extensions of P_1 and P_2 respectively by T. Let further s_1 and s_2 be the homomorphisms mapping L_1 onto P_1 and L_2 onto P_2 respectively. Finally, let $L_1 \cap L_2 = T$ and $x_2 L_1 x_2^{-1} = L_1$, for every $x_2 \in L_2$. Then:

> $L = L_1 L_2$ is an extension of P by T, and the corresponding homomorphism s of L onto P is given by $s(x) = s_1(x_1)s_2(x_2)$, where $x = x_1 x_2$, $x_1 \in L_1$, $x_2 \in L_2$,

if and only if

$$s_2^{-1}(x_2)s_1^{-1}(x_1)(s_2^{-1}(x_2))^{-1} = s_1^{-1}(x_2 x_1 x_2^{-1}) \text{, for every } x_1 \in L_1,\ x_2 \in L_2 . \tag{1}$$

What is more, it can be shown that it is sufficient to test (1) for the generators of P_1 and P_2 only. Thus one can construct L by deriving extensions of cyclic point groups only, and subsequently forming their generalized semidirect product (GSP).[6]

Since $T \triangleleft L$, one can write $L = \sum_{i=1}^{|P|} (R_i|\underline{v}_i)T$, where $R_i \in P$, $\underline{v}_i$ is a translation not necessarily belonging to T. The coset representatives satisfy:

$$(R_i|\underline{v}_i)(R_j|\underline{v}_j) = (R_i R_j|\underline{v}_i + R_i \underline{v}_j) = (\mathcal{E}|\underline{w}_{ij})(R_i R_j|\underline{v}_{ij}) , \tag{2}$$

where $\mathcal{E}$ is the identity transformation, $(\mathcal{E}|\underline{w}_{ij}) \in T$, and $(R_i R_j|\underline{v}_{ij})$ is also a coset representative. In fact, (2) is sufficient for $\sum_{i=1}^{P} (R_i|\underline{v}_i)T$ to be a line group.

There is a nondenumerable infinity of line groups in E ; but for application to physical systems some of them are not essentially distinct. Namely, two line groups L_1 and L_2 are considered equivalent if one can be taken into the other either changing the length unit or conjugating one of them by an element of E . Our procedure[7] of construction of the line groups consists in obtaining a representative out of each equivalence class. This task is simplified by the following result:

> In each equivalence class of line groups there is one the translational parts of whose elements are all collinear.

Now, choosing all the translations $\underline{v}_i$ to be parallel to the axis of rotation, (2) becomes easy to solve directly for any cyclic P , and thus derive the corresponding line groups. These are in turn multiplied into GSP's to obtain the rest of the line groups. However, not every pair of line groups can be combined in this way. Those (and only those) can that pass the compatibility test (1), which boils down to

$$\underline{v}_2 + R_2\underline{v}_1 - R'\underline{v}_2 = \underline{v}' + \underline{t} , \qquad \underline{t} \in T , \tag{3}$$

where $R' = R_2R_1R_2^{-1}$ and $\underline{v}'$ is the corresponding translational part. Actually, it suffices to apply (3) to the generators of P_1 and P_2 only, so that in practice we deal with no more than two such equations.

For applications in physics the irreducible representations (IR's) of the line groups are also required. These are obtained easily owing to the fact that in our construction of the line groups the second factor in the GSP form of the noncyclic L's is always cyclic, so that their IR's are direct-product-like.

The vibrational spectroscopy of polymers is a rather developed field of research nowadays, and numerous theoretical studies are published.[8] Since polymer molecules contain a very large number of atoms, the calculations are usually complex and tedious, even with the help of a powerful computer. Therefore, making maximal use of the symmetry of the problem is highly desirable. In the case of the infra-red absorption and Raman spectra the approximate point-group symmetry has been used for one-phonon processes; however, for two-phonon processes the IR's of the line groups are indispensable (for derivation of selection rules etc.).

A similar situation is encountered in the determination of the electronic band structure of chain molecules. Several large computational systems have been developed to this purpose in recent years.[9] We expect that the systematic development of the theory of the line groups and their representations will give some contribution to the solution of such problems.

REFERENCES

1. Geil Ph., Polymer Single Crystals, Interscience, N.Y., 1963
2. Allen G., The Dynamics of Polymer Chains, in "Neutron Inelastic Scattering", IAEA, Vienna, 1972
3. Veinshtein B.K., Kristallografiya 4 (1959) 842
4. Alexander E., Z. Kristallogr. 70 (1929) 367

5. Shubnikov A.V., Belov N.V., Coloured Symmetry, Pergamon Press, Oxford, 1964

6. Šijački Đ., Vujičić M., Herbut F., J. Math. Phys. 13 (1972) 1755

7. Vujičić M., Božović I., Herbut F., Proc. 3rd Int. Coll. on Group Theor. Methods in Physics, Marseille, 1974; an extensive treatment should appear soon.

8. Oleinik E.F., Kompaniets V.Z., Achievements in the Vibrational Spectroscopy of Polymers, in "New Developments in Polymer Research Methods", Eds. Rogovin Z.A., Zubov V.P., Mir, Moscow, 1968 (in Russian); Zbinden R., Infrared Spectroscopy of High Polymers, Mir, Moscow, 1966 (in Russian)

9. Shipman L.L., Christoffersen R.E., J.Am.Chem.Soc. 95 (1973) 1408; Duke B.J., O'Leary B., Chem.Phys.Lett. 20 (1973) 459; Suhai S., Ladik J., Int.J.Quantum Chem. 7 (1973) 547.

GROUP THEORETICAL APPROACH TO BLOCH ELECTRONS IN ANTIFERROMAGNETS

H.D. BUTZAL
Institut für Physik III
Universität Regensburg
D-8400 Regensburg, Germany

To appear in J. of Magnetism.

For the electron in the spinor representation, the symmetry of the time-independent Schrödinger equation is investigated in antiferromagnetic single crystals. The one-particle Hamiltonian

$$H = \frac{1}{2m} (\vec{p}-e\vec{A})^2 + e\phi - \frac{e\hbar}{2m} (\vec{\sigma}.\vec{B}) - \frac{e\hbar}{4m^2c^2} (\vec{\sigma}.[\vec{E}x(\vec{p}-e\vec{A})]),$$

where A and B depend on an underlying rigid antiferromagnetic structure, is expected to give a good description of electronic states not contributing to the magnetic structure.

A double-valued representation to the relevant Shubnikov space group gives the symmetry group of the Hamiltonian. The lattice periodicity of the Hamiltonian leads to the Bloch theorem. The symmetry properties of the energy bands are examined systematically for each type of Shubnikov space groups. In some groups, the energy bands are allowed by symmetry to have terms in $\vec{k}$; this holds even if spin is neglected. Such a behaviour can have measurable consequences.

Finally it is discussed, whether the extension of Shubnikov groups to Kitz groups ("spin groups") is useful or not for the problem under discussion.

$U(5) \supset O(5) \supset O(3)$ AND THE EXACT SOLUTION FOR THE PROBLEM OF QUADRUPOLE VIBRATIONS OF THE NUCLEUS

E. Chacón*, M. Moshinsky*,**

Instituto de Física, Universidad de México (UNAM) México, D.F.

and

R.T. Sharp***

Centre de Recherches Mathématiques,

Université de Montreal, Canada

Over twenty years ago A. Bohr discussed the quantum mechanical problem of the quadrupole vibrations in the liquid drop model of the nucleus. States of definite angular momentum could not be obtained exactly except when $L = 0, 3$. In the present paper we indicate how we can determine states for arbitary angular momentum L and definite number of quanta ν in terms of polynomials of the creation operators characterized by irreducible representation (IR) of the chain of groups $U(5) \supset O(3)$. We furthermore characterize the states by a definite IR λ of $O(5)$ by replacing the creation operators by traceless ones. These states are fully determined by an extra label μ that gives the number of triplets of traceless creation operators coupled to angular momentum zero. We show then how all the wave functions of the problem discussed by A. Bohr can be obtained in a recursive fashion and briefly discuss some of their applications.

---

*Member of the Instituto Nacional de Energía Nuclear.

**Member of El Colegio Nacional.

***On leave of absence from the University of McGill, Montreal, Canada.

WAVE VECTOR SELECTION RULES FOR SPACE GROUPS.

A.P. Cracknell

Carnegie Laboratory of Physics, University of Dundee, DUNDEE DD1 4HN

and

B.L. Davies

School of Mathematics and Computer Science, University College of North Wales, BANGOR LL57 2UW.

Abstract.

A programme of work is in progress for the construction of tables of wave vector selection rules (WVSRs) for all the classical space groups. The general theory is briefly reviewed and some practical considerations which have arisen in the construction of these tables are also discussed. The use of induced compatibility tables (ICTs) is described and this is illustrated with an example taken from the space group $F\bar{4}3m$ (T_d^2).

1. Introduction.

For many years a considerable amount of energy was expended on the determination of the irreducible representations of the classical space groups. The principal motivation for studying the irreducible representations of the space groups is the fact that they are relevant to the quantum-mechanical treatment of particles or quasiparticles in a crystalline solid. Suppose that a certain particle or quasiparticle belongs to an irreducible representation Γ_i of a space group $\underset{\sim}{G}$. Then the transformation properties of the wave function ψ_i of the particle, under the symmetry operations of $\underset{\sim}{G}$, will be those of one component of a basis $\langle\phi_i|$ of Γ_i, while the degeneracy of the corresponding energy eigenvalue, E_i, for the particle will be equal to the degeneracy of Γ_i. These facts have been widely exploited to simplify the calculation of energy eigenvalues, E_i, for electrons, phonons, and magnons in crystalline solids. The labels Γ_i form the basis of a convenient scheme for labelling the eigenvalues E_i.

For our present purposes we shall assume that the irreducible representations of all the 230 classical space groups are readily available (for references see, for example, Bradley and Cracknell (1972)). We shall be concerned with the problem of the reduction of the Kronecker products of these representations, that is, with determining the coefficients $C^{\underset{\sim}{k}_i\underset{\sim}{k}_j,\underset{\sim}{k}_l}_{pq,r}$ in the decomposition

$$(\Gamma_p^{\underset{\sim}{k}_i}\uparrow \underset{\sim}{G})\boxtimes(\Gamma_q^{\underset{\sim}{k}_j}\uparrow \underset{\sim}{G}) \equiv \sum_r \sum_l C^{\underset{\sim}{k}_i\underset{\sim}{k}_j,\underset{\sim}{k}_l}_{pq,r}(\Gamma_r^{\underset{\sim}{k}_l}\uparrow \underset{\sim}{G}). \qquad (1)$$

$(\Gamma_p^{\underset{\sim}{k}_i}\uparrow \underset{\sim}{G})$ is an irreducible representation of the space group $\underset{\sim}{G}$ induced from the irreducible representation $\Gamma_p^{\underset{\sim}{k}_i}$ of the little group $\underset{\sim}{G}^{\underset{\sim}{k}_i}$. The reason for constructing these tables is that the reductions of these products enable one to determine selection rules for various physical processes involving scattering between quantum-mechanical states of particles or quasiparticles in crystalline solids. These processes include infra-red absorption, Raman scattering, magnon sidebands on optical spectral lines in magnetically-ordered crystals, solid-state phase transitions, electron scattering and neutron scattering (for further details see Birman and Berenson (1974) and section 7 of Cracknell (1975)). We do not have sufficient space to discuss these processes in detail here. The formal theory of the reduction of products of the form given in equation (1) is already available and it has been applied to the determination of complete tables of the reductions for a few special space groups (references are quoted in section 7.1 of Cracknell (1975)). We have been turning our attention to the problem of producing complete tables of these reductions for all the space groups. We have found it convenient to divide our work into two stages. First, there is the determination of wave vector selection rules, that is identifying the (relatively few) values of $\underset{\sim}{k}_l$ that arise for each given pair of $\underset{\sim}{k}_i$ and $\underset{\sim}{k}_j$. Secondly, once this is done, it remains to find the actual coefficients $C_{pq,r}^{\underset{\sim}{k}_i\underset{\sim}{k}_j,\underset{\sim}{k}_l}$.

2. Theory.

We shall give a very brief résumé of the necessary theory; further details may be found, for example, in section 4.7 of Bradley and Cracknell (1972).

2.1 Wave vector selection rules (WVSRs).

The values of $\underset{\sim}{k}_l$ which may appear on the right-hand side of equation (1) are restricted by the condition

$$R_\beta \underset{\sim}{k}_i + R_\alpha \underset{\sim}{k}_j \equiv \underset{\sim}{k}_l \tag{2}$$

where $\{R_\beta | \underset{\sim}{v}_\beta\}$ and $\{R_\alpha | \underset{\sim}{v}_\alpha\}$ are elements of $\underset{\sim}{G}$. The values of R_α and R_β are restricted so that $\{R_\alpha | \underset{\sim}{v}_\alpha\}$ and $\{R_\beta | \underset{\sim}{v}_\beta\}$ form quite a small subset of all the elements of the group $\underset{\sim}{G}$. The allowed values of R_α and R_β have to be determined from a detailed examination of certain double-coset decompositions of $\underset{\sim}{G}$ (see Bradley and Cracknell (1972)). Thus the allowed values of R_α are found from writing

$$\underset{\sim}{G} = \sum_\alpha \underset{\sim}{G}^{\underset{\sim}{k}_l} \{R_\alpha | \underset{\sim}{v}_\alpha\} \underset{\sim}{G}^{\underset{\sim}{k}_j} \tag{3}$$

and, for each allowed R_α, the corresponding allowed values of R_β are found from the double-coset decomposition

$$\underset{\sim}{G} = \sum_\beta (\underset{\sim}{G}^{\underset{\sim}{k}_l} \cap \underset{\sim}{G}^{R_\alpha \underset{\sim}{k}_j}) \{R_\beta | \underset{\sim}{v}_\beta\} \underset{\sim}{G}^{\underset{\sim}{k}_i}. \tag{4}$$

With the allowed values of R_α and R_β the wave vector selection rules, that is the identification of the allowed $\underset{\sim}{k}_l$ for a given pair of $\underset{\sim}{k}_i$ and $\underset{\sim}{k}_j$, can be determined by the use of equation (2). In practice the restrictions on R_α and R_β are so severe that equation (2) frequently leads to only one value of $\underset{\sim}{k}_l$ for which the coefficients $C_{pq,r}^{\underset{\sim}{k}_i\underset{\sim}{k}_j,\underset{\sim}{k}_l}$ do not automatically vanish.

2.2 Determination of coefficients $C_{pq,r}^{\underset{\sim}{k}_i\underset{\sim}{k}_j,\underset{\sim}{k}_l}$.

Assuming that the wave vector selection rule for a given $\underset{\sim}{k}_i$ and $\underset{\sim}{k}_j$ has been determined, the coefficient $C_{pq,r}^{\underset{\sim}{k}_i\underset{\sim}{k}_j,\underset{\sim}{k}_l}$ can be obtained by

using the formula which is given, for example, on p. 211 of Bradley and Cracknell (1972)

$$C_{pq,r}^{\underset{\sim}{k}_i \underset{\sim}{k}_j, \underset{\sim}{k}_l} = \sum_{\alpha} \sum_{\beta} (|\underset{\sim}{T}|/|N_{\alpha\beta}|) \sum_{\{R_\gamma|\underset{\sim}{v}_\gamma\} \in \underset{\sim}{N}_{\alpha\beta}/\underset{\sim}{T}} \chi_p^{\underset{\sim}{k}_i}(\{R_\beta|\underset{\sim}{v}_\beta\}^{-1}\{R_\gamma|\underset{\sim}{v}_\gamma\}\{R_\beta|\underset{\sim}{v}_\beta\})$$
$$\chi_q^{\underset{\sim}{k}_j}(\{R_\alpha|\underset{\sim}{v}_\alpha\}^{-1}\{R_\gamma|\underset{\sim}{v}_\gamma\}\{R_\alpha|\underset{\sim}{v}_\alpha\}) \chi_r^{\underset{\sim}{k}_l}(\{R_\gamma|\underset{\sim}{v}_\gamma\})^* \qquad (5)$$

where $\{R_\gamma|\underset{\sim}{v}_\gamma\}$ are elements of $\underset{\sim}{G}$, $\chi_p^{\underset{\sim}{k}_i}(\{R|\underset{\sim}{v}\})$ is the character of the element $\{R|\underset{\sim}{v}\}$ in $\Gamma_p^{\underset{\sim}{k}_i}$, and $\underset{\sim}{N}_{\alpha\beta}$ is $\underset{\sim}{G}^{R_\beta \underset{\sim}{k}_i} \cap \underset{\sim}{G}^{R_\alpha \underset{\sim}{k}_j} \cap G^{\underset{\sim}{k}_l}$.

Our task is to determine the WVSRs and the coefficients on the right-hand side of equation (1) for all possible sets of p, q, $\underset{\sim}{k}_i$, and $\underset{\sim}{k}_j$ in the decomposition of the left-hand side of equation (1) for each of the 230 classical space groups. This involves a very large amount of tabulation and all that we can hope to do here is to describe some of the difficulties that we have encountered along the way and to indicate the form of the results for a few examples.

2.3 Symmetrized and antisymmetrized powers.

In addition to the reductions of the ordinary Kronecker products, there is also the special case when $\underset{\sim}{k}_i = \underset{\sim}{k}_j$ and p = q. The product $(\Gamma_p^{\underset{\sim}{k}_i} \uparrow \underset{\sim}{G}) \boxtimes (\Gamma_p^{\underset{\sim}{k}_i} \uparrow \underset{\sim}{G})$, or the square of $(\Gamma_p^{\underset{\sim}{k}_i} \uparrow \underset{\sim}{G})$, can be separated into symmetrized and antisymmetrized parts and these symmetrized and antisymmetrized squares of space-group representations are of considerable importance in a number of applications. Symmetrized and antisymmetrized cubes, and higher powers, of space-group representations can also be considered by an adaptation of the theory already outlined (for details see Bradley and Davies (1970), Lewis (1973), Gard (1973a, 1973b)). We are also planning to tabulate the results of the reductions of symmetrized and antisymmetrized powers of the irreducible representations of all the 230 space groups; the extra tabulation involved in doing this is quite small since $\underset{\sim}{k}_i = \underset{\sim}{k}_j$ and p = q.

3. Identification of space-group representations.

For our purposes it is necessary to have available a set of tables of the space-group representations themselves, such tables being complete, correct, and in a notation that is unambiguously defined. In a paper which is being submitted elsewhere (Davies and Cracknell 1975) we have examined the problem of establishing such a definitive set of tables. This has involved some synthesis, and also some extension, of the work of Miller and Love (1967) and of Bradley and Cracknell (1972). To construct unambiguous tables of

WVSRs we have found it necessary to include all the special planes of symmetry; these were not included in any of the published sets of tables of space-group representations.

4. Determination of wave vector selection rules.

If both $\underset{\sim}{k}_i$ and $\underset{\sim}{k}_j$ are wave vectors corresponding to points of symmetry, the determination of WVSRs is quite easy and the results for any given space group can be presented quite concisely. On the other hand, if either $\underset{\sim}{k}_i$ or $\underset{\sim}{k}_j$ is a wave vector corresponding to a line or plane of symmetry then the allowed vectors $\underset{\sim}{k}_l$ in equation (2) will be linearly dependent on one or more parameters. If now the parameters of either $\underset{\sim}{k}_i$ or $\underset{\sim}{k}_j$ take on special values or are related to each other, then the symmetry of $\underset{\sim}{k}_l$ may be increased from a lower to a higher member of the sequence:

general $\underset{\sim}{k}$, plane of symmetry, line of symmetry, point of symmetry.

Thus for completeness we need to identify the special values of $\underset{\sim}{k}_l$ that arise from the use of special values of the parameters. For these special values of $\underset{\sim}{k}_l$ it will be necessary to re-label the representations of $\underset{\sim}{G}$ associated with $\underset{\sim}{k}_l$ in terms of this higher symmetry. We illustrate what is involved by considering an example.

We consider F$\bar{4}$3m (T_d^2), which is the space group of the zinc blende structure, since this is a space group for which a lot of previous relevant work already exists (Birman 1962, 1963, Bradley and Davies 1970). We shall follow the notation used by Miller and Love (1967) in labelling the space-group representations, although we shall follow the notation of Bradley and Davies (1970) in labelling the symmetry operations. Suppose we consider the reduction of Kronecker products of representations belonging to two different DT (Δ) wave vectors so that

$$\underset{\sim}{k}_i = (0,\alpha,0); \quad \underset{\sim}{k}_j = (0,\alpha',0). \tag{6}$$

In general we can assume no special relationship between the values of α and α'. By performing the appropriate analysis (see equations (3) and (4)) it is straightforward to show that the allowed vectors $\underset{\sim}{k}_l$ may be chosen to be

(i) $(0,\alpha+\alpha',0)$ where $R_\alpha = E = R_\beta$, and $\underset{\sim}{k}_l$ is a DT wave vector which is different from both $\underset{\sim}{k}_i$ and $\underset{\sim}{k}_j$;

(ii) $(0,-\alpha+\alpha',0)$ where $R_\alpha = E$, $R_\beta = C_{2x}$, and $\underset{\sim}{k}_l$ is a DT wave vector which is different from both $\underset{\sim}{k}_i$ and $\underset{\sim}{k}_j$;

(iii) $(\alpha,\alpha',0)$ where $R_\alpha = E$, $R_\beta = C_{31}^-$, and $\underset{\sim}{k}_l$ is an A wave vector, which corresponds to a plane of symmetry.

The symmetry labels DT and A apply when no restriction is placed on the values of α and α'. However, it is also necessary to consider the possibility of a special relationship between α and α' or that α and α' may take special values. The possibilities are indicated in table 1 in which the appropriate symmetry labels for the corresponding wave vectors $\underset{\sim}{k}_1$ are also given. One of these special cases, namely $\alpha' = \alpha$, was included in the tables given by Bradley and Davies (1970).

Table 1.

α, α' values	Stars of possible $\underset{\sim}{k}_1$		
α, α' unrelated and unrestricted	$(0,\alpha+\alpha',0)$ DT	$(0,-\alpha+\alpha',0)$ DT	$(\alpha,\alpha',0)$ A
$\alpha'=\alpha$	$(0,2\alpha,0)$ DT	$(0,0,0)$ GM	$(\alpha,\alpha,0)$ SM
$\alpha'=-\alpha$	$(0,0,0)$ GM	$(0,-2\alpha,0)$ DT	$(\alpha,-\alpha,0)$ SM
$\alpha'=\alpha=\frac{1}{4}$	$(0,\frac{1}{2},0)$ X	$(0,0,0)$ GM	$(\frac{1}{4},\frac{1}{4},0)$ SM
$\alpha'=-\alpha=\frac{1}{4}$	$(0,0,0)$ GM	$(0,\frac{1}{2},0)$ X	$(-\frac{1}{4},\frac{1}{4},0)$ SM

By using the formula in equation (5) one can determine the reduction of DTI $\boxtimes$ DTJ where I and J may each take any value from 1 to 5. Let us consider DT1 $\boxtimes$ DT1 for example; we obtain

$$\underset{(0,\alpha,0)}{\underset{(6)}{\mathrm{DT1}}} \boxtimes \underset{(0,\alpha',0)}{\underset{(6)}{\mathrm{DT1}}} = \underset{(0,\alpha+\alpha',0)}{\underset{(6)}{\mathrm{DT1}}} + \underset{(0,-\alpha+\alpha',0)}{\underset{(6)}{\mathrm{DT1}}} + \underset{(\alpha,\alpha',0)}{\underset{(24)}{\mathrm{A1}}} \qquad (7)$$

where the appropriate wave vector is given below each representation and no special assumption is made about the values of α and α'. The numbers in brackets indicate the degeneracies of the induced representations $(\Gamma_p^{\underset{\sim}{k}_i} \uparrow \underset{\sim}{G})$, $(\Gamma_q^{\underset{\sim}{k}_j} \uparrow \underset{\sim}{G})$, and $(\Gamma_r^{\underset{\sim}{k}_1} \uparrow \underset{\sim}{G})$. The corresponding reductions for the special values of α and α' can also be determined by using equation (5) directly:

$$\underset{(0,\alpha,0)}{\underset{(6)}{\mathrm{DT1}}} \boxtimes \underset{(0,\alpha,0)}{\underset{(6)}{\mathrm{DT1}}} = \underset{(0,2\alpha,0)}{\underset{(6)}{\mathrm{DT1}}} + \underbrace{\underset{(1)}{\mathrm{GM1}} + \underset{(2)}{\mathrm{GM3}} + \underset{(3)}{\mathrm{GM4}}}_{(0,0,0)} + \underbrace{\underset{(12)}{\mathrm{SM1}} + \underset{(12)}{\mathrm{SM2}}}_{(\alpha,\alpha,0)} \qquad (8)$$

$$\underset{(0,\alpha,0)}{\underset{(6)}{\mathrm{DT1}}} \boxtimes \underset{(0,-\alpha,0)}{\underset{(6)}{\mathrm{DT1}}} = \underbrace{\underset{(1)}{\mathrm{GM1}} + \underset{(2)}{\mathrm{GM3}} + \underset{(3)}{\mathrm{GM4}}}_{(0,0,0)} + \underset{(0,-2\alpha,0)}{\underset{(6)}{\mathrm{DT1}}} + \underbrace{\underset{(12)}{\mathrm{SM1}} + \underset{(12)}{\mathrm{SM2}}}_{(\alpha,-\alpha,0)} \tag{9}$$

$$\underset{(0,\frac{1}{4},0)}{\underset{(6)}{\mathrm{DT1}}} \boxtimes \underset{(0,\frac{1}{4},0)}{\underset{(6)}{\mathrm{DT1}}} = \underbrace{\underset{(3)}{\mathrm{X1}} + \underset{(3)}{\mathrm{X3}}}_{(0,\frac{1}{2},0)} + \underbrace{\underset{(1)}{\mathrm{GM1}} + \underset{(2)}{\mathrm{GM3}} + \underset{(3)}{\mathrm{GM4}}}_{(0,0,0)} + \underbrace{\underset{(12)}{\mathrm{SM1}} + \underset{(12)}{\mathrm{SM2}}}_{(\frac{1}{4},\frac{1}{4},0)} \tag{10}$$

$$\underset{(0,-\frac{1}{4},0)}{\underset{(6)}{\mathrm{DT1}}} \boxtimes \underset{(0,\frac{1}{4},0)}{\underset{(6)}{\mathrm{DT1}}} = \underbrace{\underset{(1)}{\mathrm{GM1}} + \underset{(2)}{\mathrm{GM3}} + \underset{(3)}{\mathrm{GM4}}}_{(0,0,0)} + \underbrace{\underset{(3)}{\mathrm{X1}} + \underset{(3)}{\mathrm{X3}}}_{(0,\frac{1}{2},0)} + \underbrace{\underset{(12)}{\mathrm{SM1}} + \underset{(12)}{\mathrm{SM2}}}_{(-\frac{1}{4},\frac{1}{4},0)} . \tag{11}$$

5. Induced compatibility tables.

If one is to construct complete tables of the reductions of all the Kronecker products for a space group, one needs to include all the equations like equations (8) to (11) giving the reductions for all the special α, α' values as well as equations like equation (7) for the general case. The question which we then had to consider was to see whether it is necessary to tabulate explicitly the reduction of DT1 $\boxtimes$ DT1 for all the special cases of α and α' or whether these reductions could be deduced in a simple manner from equation (7). It happens in fact that it is not necessary to tabulate separately the reductions for all the special cases of α and α' because they can be obtained quite easily from equation (7) using what we may describe as "induced compatibility tables", which we call ICTs for short. These are not identical with the compatibility tables which one normally encounters. However, Raghavacharyulu and Shrestha (1966) demonstrated a very useful result. Suppose that $G^{\underset{\sim}{k}}$ is a subgroup of $G^{\underset{\sim}{k}_o}$, that

$$\Gamma_\lambda^{\underset{\sim}{k}_o} \downarrow G^{\underset{\sim}{k}} = \sum_\mu c_{\lambda\mu} \Gamma_\mu^{\underset{\sim}{k}} \tag{12}$$

and that

$$(\Gamma_\mu^{\underset{\sim}{k}} \uparrow \underset{\sim}{G}) = \sum_\lambda c'_{\mu\lambda} (\Gamma_\lambda^{\underset{\sim}{k}_o} \uparrow \underset{\sim}{G}). \tag{13}$$

The Frobenius reciprocity theorem then enables one to show that

$$c'_{\mu\lambda} = c_{\lambda\mu} . \tag{14}$$

The question of degeneracies may be a little puzzling at first sight in the understanding of the construction of ICTs using equation (14). Suppose that $\underset{\sim}{k}_o$ and $\underset{\sim}{k}$ differ by a small vector $\underset{\sim}{\kappa}$, so that

$$\underset{\sim}{k} = \underset{\sim}{k}_o + \underset{\sim}{\kappa} \tag{15}$$

where $\underset{\sim}{\kappa}$ may be arbitrarily small. The conventional compatibility tables can be regarded as "subduced compatibility tables"; that is, they identify $C_{\lambda\mu}$ in equation (12) in the reduction of the subduced representation $(\Gamma_{\lambda}^{\underset{\sim}{k}_o} \downarrow \underset{\sim}{G}^{\underset{\sim}{k}})$. Since it is the small representations, i.e. representations of $\underset{\sim}{G}^{\underset{\sim}{k}_o}$ or $\underset{\sim}{G}^{\underset{\sim}{k}}$, which are commonly used in the group-theoretical labelling of electronic band structures, and of phonon dispersion relations etc. it is the subduced compatibility tables that are used in this connection. But when we consider the reductions of Kronecker products of space-group representations it is really the product of the induced representations $(\Gamma_p^{\underset{\sim}{k}_i} \uparrow \underset{\sim}{G})$ and $(\Gamma_q^{\underset{\sim}{k}_j} \uparrow \underset{\sim}{G})$ which we are considering (see equation (1)); that is, we require to know $C'_{\mu\lambda}$ in equation (13). Recalling that $\underset{\sim}{G}^{\underset{\sim}{k}}$ is a subgroup of $\underset{\sim}{G}^{\underset{\sim}{k}_o}$, we see that the subduced representation $(\Gamma_{\lambda}^{\underset{\sim}{k}_o} \downarrow \underset{\sim}{G}^{\underset{\sim}{k}})$ may be reducible. Also if $(\Gamma_{\mu}^{\underset{\sim}{k}} \uparrow \underset{\sim}{G})$ is regarded as $((\Gamma_{\mu}^{\underset{\sim}{k}} \uparrow \underset{\sim}{G}^{\underset{\sim}{k}_o}) \uparrow \underset{\sim}{G})$ then $(\Gamma_{\mu}^{\underset{\sim}{k}} \uparrow \underset{\sim}{G}^{\underset{\sim}{k}_o})$ may be a reducible representation of $\underset{\sim}{G}^{\underset{\sim}{k}_o}$. That is, a reduction in degeneracies of the irreducible (small) representations occurs for $\Gamma_{\nu}^{\underset{\sim}{k}}$ as one <u>decreases</u> the symmetry of $\underset{\sim}{k}$, but a reduction of degeneracies also occurs for $(\Gamma_{\nu}^{\underset{\sim}{k}} \uparrow \underset{\sim}{G})$ as one <u>increases</u> the symmetry of $\underset{\sim}{k}$. Whereas the former is widely appreciated, the latter is much less widely appreciated.

One can illustrate the use of induced compatibility tables very easily by showing how they can be used in the case of DT1 ⊠ DT1 for F$\bar{4}$3m. The conventional compatibility tables for this group are given, for example, on page 387 of Miller and Love (1967); we have used those tables to construct that part of the induced compatibility tables for this group that is relevant to equation (7), see table 2. The degeneracies of the induced representations have also been included in table 2 for reference. If $\alpha = \alpha'$ equation (7) becomes

$$\begin{array}{ccccccccc} DT1 & \boxtimes & DT1 & = & DT1 & + & (DT1\uparrow GM) & + & (A1\uparrow SM) \\ (6) & & (6) & & (6) & & (6) & & (24) \\ (0,\alpha,0) & & (0,\alpha,0) & & (0,2\alpha,0) & & (0,0,0) & & (\alpha,\alpha,0) \end{array} \qquad (16)$$

and by using table 2 we see that this leads immediately to equation (8). One can obtain equations (9), (10), and (11) very easily in a similar manner. The important point for our purpose is that although one needs to use equation (5) to determine equation (7), one does not then need to use equation (5) again to obtain equations (8) - (11). This simplifies our task of constructing tables of reductions of Kronecker products; it means that

(i) We only need to tabulate the reductions of $(\Gamma_p^{\underset{\sim}{k}_i} \uparrow \underset{\sim}{G}) \boxtimes (\Gamma_q^{\underset{\sim}{k}_j} \uparrow \underset{\sim}{G})$

for general values of α and α'.

(ii) Reductions for special values of α and α' can be obtained by using ICTs.

(iii) We do not need to tabulate the ICTs explicitly because they can be obtained in a trivial manner from the compatibility tables in Miller and Love (1967).

Table 2. Part of the induced compatibility tables for F$\bar{4}$3m.

	DT1 (6)	DT2 (6)	DT3 (6)	DT4 (6)	DT5 (12)
GM	GM1 (1) GM3 (2) GM4 (3)	GM2 (1) GM3 (2) GM5 (3)	GM4 (3) GM5 (3)	GM4 (3) GM5 (3)	GM6 (2) GM7 (2) 2GM(8) (2(4))
X	X1 (3) X3 (3)	X2 (3) X4 (3)	X5 (6)	X5 (6)	X6 (6) X7 (6)

	A1 (24)
SM	SM1 (12) SM2 (12)

The arguments illustrated above can be extended to products involving planes of symmetry. All we need to do is to make some additions to the compatibility tables of Miller and Love (1967) to cover the additional points, lines, and planes of symmetry that they did not include.

6. Conclusion.

We are now well advanced in determining WVSRs for the orthorhombic and cubic space groups and we hope to complete the work for the other space groups too within the next few months.

References.

J.L. Birman, 1962, Phys. Rev. 127, 1093.

J.L. Birman, 1963, Phys. Rev. 131, 1489.

J.L. Birman and R. Berenson, 1974, Phys. Rev. B 9, 4512.

C.J. Bradley and A.P. Cracknell, 1972, The mathematical theory of symmetry in solids: representation theory for point groups and space groups (Clarendon Press, Oxford).

C.J. Bradley and B.L. Davies, 1970, J. math. Phys. 11, 1536.

A.P. Cracknell, 1975, Group theory in solid-state physics (Taylor and Francis, London).

B.L. Davies and A.P. Cracknell, 1975, in preparation.

P. Gard, 1973a, J. Phys. A: Math. Nucl. Gen. 6, 1807.
P. Gard, 1973b, J. Phys. A: Math. Nucl. Gen. 6, 1829.
D.H. Lewis, 1973, J. Phys. A: Math. Nucl. Gen. 6, 125.
S.C. Miller and W.F. Love, 1967, Tables of irreducible representations of space groups and co-representations of magnetic space groups (Pruett Press, Boulder, Col.).
I.V.V. Raghavacharyulu and C.B. Shrestha, 1966, Can. J. Phys. 44, 444.

A CHEMIST LOOKS AT THE STRUCTURE OF SYMMETRY GROUPS

On a generic scheme of important point groups for rigid molecular frames

H. P. Fritzer

Institut für Physikalische und Theoretische Chemie,
Technical University, A-8010 Graz, Austria

Abstract

A practical method for generating larger symmetry groups from smaller ones is presented. It is based upon the construction of the abstract-unique group H(G) called the holomorph of a given starting group G. The extension of G by its full automorphism group A(G) is given in great detail as permutation realizations for both the cyclic and the abelian group of order 4. A selection of point groups generated by this method of the holomorph is given for some important symmetries.

Contribution for the Proceedings of the 4th International Colloquium on "Group Theoretical Methods In Physics", University of Nijmegen, The Netherlands, 1975.

1. Introduction

Symmetry considerations have always been important in various branches of chemistry from both qualitative (or geometrical) and quantitative (i.e. group theoretical) points of view. It is the concept of structural symmetry or (mentioning L. Pauling's words about)the "architecture of molecules", resp., that is relating experimental observations like optical spectra, dipole moments, electric and magnetic susceptibilities, optical activity and chirality, etc., to theoretical calculations based upon quantum mechanics. Therefore, group theory is connecting very efficiently the world of problems in chemical statics and dynamics to the world of abstractly operating computational machinery supplied by physics and mathematics.

The traditional interest of chemists in group theory stems from areas as classification of molecules by means of point groups, normal vibrations analysis, MO theory, crystal and/or ligand field theory (a terrible semantics since these topics have nothing to do with a physical "field theory"), selection rules, and so forth. The tools needed are: character systems of the various groups involved in the chemical problem, classification of wave functions and state vectors (i.e. the orbital business), and some capability in understanding energy level diagrams. It is probably true that character tables will maintain their leading position in practical applications also in the near future. The same is true for the use of the 32 crystallographic point groups in solid state chemistry and some other special groups like $O^+(3)$, $SU(2)$, and some members of the family of the symmetric group $G = S_m$ on m symbols.

In the last five years, however, there is an increasing interest in the chemical literature for more abstract concepts from group theory. Among others are topics like symmetry adaption[1], the role of the group S_m in the m-electron problem[2], and, recently, the molecular dynamics of the so-called non-rigid molecules[3], respectively. The last example forces chemists to the idea of feasible symmetry operations[4] which are combinations of permutations of identical entities within the nuclear framework and traditional operations (rotations, inversions, etc.). Because of these features new symmetry operations together with unconventional symmetry groups appear whose character tables are not known beforehand.

The work of the present author is following along these lines having objectives from both scientific and educational view points simultaneously. In the science area, we are looking at simple methods to generate symmetrized states for the many-body problem(important for optical and magnetic spectral data). On the other hand for teaching purposes, we can control the proper level of abstractness in courses of both theoretical chemistry and chemical physics by the feedback of the audience being bombarded by group theoretical "food"[5].

In this first paper out of a planned series a technique for the construction of chemical relevant point groups from the simplest possible subgroups is presented. We use the method of the holomorph that is relating in an abstractly unique manner a given group G (in principle arbitrary concerning both order and structure) to a strongly related group H(G) via the full automorphism group A(G) of G. This construct stems from the early days of group theory[6] and has not been used consciously in theoretical chemistry. In the following sections we are exploiting the method with typical examples from the area of finite point groups that are of importance for rigid molecular frameworks. The treatment of the non-rigid molecules is postponed to a forthcoming paper. Furthermore, after having looked at the chosen examples some general remarks are given. In conclusion of this section, the present author must mention that he is addressing mainly those readers using group theory for chemical applications.

2. Method of the Holomorph

As examples are taken the two structural different groups of order 4, i.e. in the chemical notation the cyclic group $G = C_4$ and the Abelian group $G = D_2$ (= F. Klein's "Vierergruppe"), resp. As point groups, we have the realizations

$$
\begin{aligned}
C_4 &= \mathrm{gp}\,\{C_4 / C_4^4 = E\} = \{E, C_4, C_2, C_4^-\} , \\
D_2 &= \mathrm{gp}\,\{2_z, 2_y / 2_z^2 = 2_y^2 = E\} = \{E, 2_z, 2_y, 2_x\} ;
\end{aligned}
\qquad (2.1)
$$

here, E is the identity operation; x,y,z refer to the orientation of the Cartesian axes for the symmetry elements in question. For the following arguments it is convenient to represent both groups as regular permutation groups (due to the Cayley-theorem[7]) defined by the multiplication tables, that is

$$
\begin{aligned}
\underline{C_4}:\ E &\leftrightarrow (E)(C_4)(C_2)(C_4^-) = (E) = P(E) , \\
C_4 &\leftrightarrow (EC_4C_2C_4^-) = P(C_4) = P(C_4^-)^{-1}, \qquad (2.2) \\
C_2 &\leftrightarrow (EC_2)(C_4C_4^-) = P(C_2) ; \quad \text{and} \\
\underline{D_2}:\ E &\leftrightarrow (E) = P(E) , \\
2_z &\leftrightarrow (E2_z)(2_y2_x) = P(2_z), \text{ and so on.}
\end{aligned}
$$

In the next stage we construct the automorphism groups A(G) of G where the image group is G itself. Due to the properties of automorphisms and bijective mappings of groups onto itself one has to consider that two corresponding elements preserve their order under any mapping, and identity operations are always mapped onto itself[7]. Therefore, it is clear that all allowed automorphisms ϕ are permuting the group elements in a definite way. Again, the multiplication table, i.e. the structure of G, generates the complete set $\{\phi_i\}$ having group properties. The result is the group A(G) whose binary operation is the succession of the isomorphic mappings. In general the set consists of both outer and inner automorphisms (the latter being induced by conjugation, i.e. equivalence relations on the set). The subset of the inner automorphisms forms the normal subgroup I(G) of A(G), i.e. $I(G) \triangleleft A(G)$. This is all standard abstract group theory. To see how it works we return to our examples.

Instead of the abstract group A(G) it is more appropriate to work with the isomorphic permutation group $PA(G) = \{P(\phi_i)/i = 1 \text{ to } /A(G)/\}$, /A(G)/ being the order of the group. It can easily be shown that for $G = C_4$ there are only two allowed automorphic mappings, given by

$$\begin{aligned} \underline{PA(C_4)}:\ \phi_1 &\longleftrightarrow (E)(C_4)(C_2)(C_4^-) = (E) = P(\phi_1) \quad , \\ \phi_2 &\longleftrightarrow (C_4C_4^-) \qquad\qquad\qquad\;\, = P(\phi_2) \ . \end{aligned} \tag{2.3}$$

On the other hand, $G = D_2$ has three elements of order two, cf. Eq.(2.1), and we are dealing with six different permutations as representations of the allowed automorphisms, given by

$$\begin{aligned} \underline{PA(D_2)}:\ &\phi_1 \longleftrightarrow (E) = P(\phi_1), \phi_2 \longleftrightarrow (2_z2_x2_y) = P(\phi_2) = P(\phi_3)^{-1}, \\ &\phi_4 \longleftrightarrow (2_z2_y) = P(\phi_4), \text{ and so on,} \end{aligned} \tag{2.4}$$

which is isomorphic to the symmetric group S_3 on three symbols.

Looking at the cycle structures of the groups $P(C_4) =$

$= \{P(R_k)/k = 1 \text{ to } /C_4/\} \longleftrightarrow C_4$ and $PA(C_4)$ from Eqs.(2.2) and (2.3) on the one hand, and analogously for $P(D_2) \longleftrightarrow D_2$ and $PA(D_2)$ in Eqs.(2.2) and (2.4) on the other hand, we can extract the following important informations:

a) $P(C_4) \cap PA(C_4) = \{(E)\}$; $P(D_2) \cap PA(D_2) = \{(E)\}$,

b) $P(C_4)$ is invariant under the action of $PA(C_4)$; the same is due for $P(D_2)$ under the action of $PA(D_2)$,

c) the reverse statement of item b is not true.

As a reminder, the reader should have no difficulties to identify point groups that are isomorphic to the groups PA(G) and A(G), respectively. For instance, $PA(C_4) \longleftrightarrow C_2, C_s, C_i$ whereas $PA(D_2) \longleftrightarrow D_3, C_{3v}$.

Now we are interested to form a larger group from the subgroups P(G) and PA(G) encouraged by items a to c (see above). That means, one has to make the set of ordered pairs $\{P(R_k), P(\phi_i);$ all $R_k \in G$ and all $\phi_i \in A(G)\}$ into a group. Looking at our examples we are indeed successful; for convenience we list all elements of the new group $PH(C_4)$ that has been formed from $P(C_4)$, the elements of which are always taken first in the ordered pairs, and $PA(C_4)$, i.e. using Eqs.(2.2) and (2.3) one obtains:

$$\begin{aligned}
&P(E\phi_1) = (P(E), P(\phi_1)) = (E),\ P(E\phi_2) = (C_4C_4^-),\\
&P(C_4\phi_1) = (EC_4C_2C_4^-),\ P(C_4\phi_2) = (EC_4)(C_2C_4^-),\\
&P(C_2\phi_1) = (EC_2)(C_4C_4^-),\ P(C_2\phi_2) = (EC_2),\\
&P(C_4^-\phi_1) = (EC_4^-C_2C_4^-),\ P(C_4^-\phi_2) = (EC_4^-)(C_4C_2).
\end{aligned} \qquad (2.5)$$

This group $PH(C_4)$ is nothing else but the permutation realization of the isomorphic group $H(C_4)$ called the holomorph of $G = C_4$. The experienced reader will agree that our construction is identical with the formation of both a semi-direct product and/or a split extension of C_4 by $A(C_4)$. Reaching the aim of our efforts is, ultimately, equivalent to the finding of a point group isomorphic to the holomorph. By inspection, we find from Eq.(2.5) that $PH(C_4)$ is isomorphic to the <u>dihedral</u> group $D_4 = C_4 \circledS C_2'$.(The prime in the subgroup C_2 of the semi-direct product is for the relative orientation of the symmetry elements.) Furthermore, we learn from Eq.(2.5) the following properties:

$$/PH(C_4)/ = /P(C_4)//PA(C_4)/ \quad ; \quad P(C_4) \triangleleft PH(C_4) \; ; \qquad (2.6)$$
$$PH(C_4)/P(C_4) \longleftrightarrow PA(C_4).$$

Lastly, we formulate the product of elements as ordered pairs for $PH(C_4)$, i.e. for all $R_k \in C_4$ and all $\phi_i \in A(C_4)$:

$$(P(R_k),P(\phi_i))(P(R_l),P(\phi_j)) = \qquad (2.7)$$
$$= (P(R_k)P(\phi_i)P(R_l)P(\phi_i)^{-1},P(\phi_i)P(\phi_j)) = (P(R_p),P(\phi_t)) \in PH(C_4).$$

The other example for $G = D_2$ can be treated in the same manner resulting in the permutation realization $PH(D_2)$ of the holomorph $H(D_2)$ that-with proper choice of axes for the subgroups-is found to be isomorphic to the <u>octahedral</u> rotation group O, that is:$O = D_2 \circledS D_3'$. Here too, it is found that $A(D_2) \longleftrightarrow D_3$ normalizes D_2 in O, and the factor group O/D_2 is isomorphic to the subgroup D_3, cf. Eq.(2.6).

3. <u>Generalization of the concept</u>

From the group theoretical point of view, a holomorph belongs to the category of extensions of a group G by other groups F having definite relations with G via special homomorphism conditions which are treated elsewhere[7-9]. In the special case of the holomorph these relations are trivial[6]. We have only to generalize Eq.(2.6) and find, that

$$H(G) = G \circledS A(G); \; /H(G)/ = /G//A(G)/; \qquad (3.1)$$
$$G \triangleleft H(G); \; H(G)/G \longleftrightarrow F; \; F \longleftrightarrow A(G).$$

The composition law for this split extension H(G) is given by

$$(R_k,\phi_i)(R_l,\phi_j) = (R_k\phi_i R_l \phi_i{}^{-1},\phi_i\phi_j) \; , \qquad (3.2)$$

fulfilling the closure condition, cf. Eq.(2.7). A general inverse element of H(G) is therefore

$$(R_k,\phi_i)^{-1} = (R_k',\phi_i') \text{ , with } R_k' = \phi_i^{-1}R_k^{-1}\phi_i \text{ and } \phi_i' = \phi_i^{-1} . \tag{3.3}$$

For our problem finding suitable pointgroups as extensions of certain given point groups using the method of the holomorph, one has to analyse both the group and class structure of G and H(G), respectively. The results are used to find new point or symmetry groups in general that are isomorphic to H(G), i.e. $H(G) \longleftrightarrow K = G \circledS F$, with $F \longleftrightarrow A(G)$. Some examples are given in Table 1 where the point groups K have been constructed along the lines of section 2.

Table 1. Generic scheme for some important point groups K ($K = G \circledS F$)

G	/A(G)/	F	/H(G)/	K	Remarks about K
C_2	1	C_1	2	C_2	trivial extension
C_3	2	C_2	6	D_3	
C_3	2	C_s	6	C_{3v}	$C_{3v} \longleftrightarrow D_3$
C_4	2	C_2	8	D_4	
C_4	2	C_s	8	C_{4v}	$C_{4v} \longleftrightarrow D_4$
C_5	4	C_4	20	?	
C_6	2	C_2	12	D_6	
C_6	2	C_s	12	C_{6v}	$C_{6v} \longleftrightarrow D_6$
C_∞	2	C_2	∞	D_∞	infinite dihedral group
C_∞	2	C_s	∞	$C_{\infty v}$	$C_{\infty v} \longleftrightarrow D_\infty$
D_2	6	D_3	24	O	octahedral group

The chemically interested reader will probably miss two interesting point groups, viz. the tetrahedral group T and the icosahedral group I with /T/ = 12 and /I/ = 60. Indeed, the group T that is very important in ligand field theory and chemistry in general cannot be generated by our method directly. If however one is considering that $T \triangleleft O$, it turns out: $T = D_2 \circledS C_3'$, with

$C_3 \triangleleft D_3$; cf. last line in Table 1. This semi-direct product (together with others not mentioned here) using certain isomorphic subgroups of A(G) for a given starting group G is important for extensions in the theory of space groups[9,10]. Finally, the group I is not tractable with our proposed method since it is simple[11].

In conclusion, we like to mention that this method of the holomorph is useful to generate the character systems of symmetry groups important in chemistry by "inducing" of representations from the starting group G that is normal in the holomorph.

Acknowledgement. Partial support by the Austrian Science Foundation (project 1489) is very much appreciated.

References.

1. D.J.Klein, C.H.Carlisle, and F.A.Matsen, Adv.Quant.Chem. (P.-O.Löwdin, ed.) 5,219 (1970).
2. G.A.Gallup, ibid. 7,113 (1973).
3. J.Serre, ibid. 8,1 (1974).
4. H.C.Longuet-Higgins, Molec.Phys. 6,445 (1963).
5. H.P.Fritzer, Lectures on "The Symmetric Group S_m in Modern Chemistry", Univ.East Anglia (England), Aug./Sept. 1974.
6. W.Burnside, Theory of Groups of finite order, Dover, 1955.
7. See any standard text on group theory.
8. O.Schreier, Mh.Math.Phys. 34,165 (1926); M.Hall, Ann.Math.39, 220 (1938).
9. F.Herbut,M.Vujičić,I.Božović,and Dj.Šijački, Proc.3rd Intern. Coll.Group theoretical methods in physics, Marseille 1974, (A.Janner and T.Jannssen, eds.), Nijmegen University Press, p.441,447 and 577 in Vol.2 , and refs. therein.
10. S.L.Altmann, Phil.Trans.Roy.Soc., A255, 216 (1963); Proc. Roy.Soc.,A298, 184 (1967).
11. N.B.Backhouse, and P.Gard, J.Phys.A: Math.,Nucl.Gen., 7, 2101 (1974).

CACNONICAL TRANSFORMATIONS AND GAUSSIAN INTEGRAL KERNELS IN NUCLEAR PHYSICS[x)]

G. John and P. Kramer

Institute for Theoretical Physics, University of Tübingen

Tübingen, German Federal Republic

1. Canonical transformations and integral kernels

In 2n-dimensional phase space a linear canonical transformation has the form

$$\begin{pmatrix} x' \\ p' \end{pmatrix} = g \begin{pmatrix} x \\ p \end{pmatrix} = \begin{pmatrix} A & B \\ C & D \end{pmatrix} \begin{pmatrix} x \\ p \end{pmatrix} \tag{1.1}$$

where g is an element of the symplectic group Sp(2n,R):

$$g^t \begin{pmatrix} 0 & -I_n \\ I_n & 0 \end{pmatrix} g = \begin{pmatrix} 0 & -I_n \\ I_n & 0 \end{pmatrix} \tag{1.2}$$

In quantum mechanics this transformation implies a unitary transformation of states

$$|x'\rangle = \mathcal{U}\, |x\rangle \tag{1.3}$$

or

$$\varphi'(x') = \int K_g(x',x)\, \varphi(x)\, dx$$

where the kernel is given by

$$K_g(x',x) = [(2\pi)^n \det B]^{-\frac{1}{2}} \exp \tfrac{i}{2} \{ x'^t (B^t)^{-1} D^t x' - 2x'^t (B^t)^{-1} x + x^t A^t (B^t)^{-1} x \} \tag{1.4}$$

if det B ≠ 0 and

$$K_g(x',x) = [\det A]^{-\frac{1}{2}} \exp \tfrac{i}{2}\, x'^t (CA^{-1}) x'\, \delta(A^{-1}x' - x)$$

if B = 0 [1] . If det B = 0 and B ≠ 0 no simple expression exists and a coordinate transformation must be performed, which allows to

x) Supported by Deutsche Forschungsgemeinschaft

split the kernel into a product of either type. Fortunately this case does not occur in applications made so far.

The mapping

$$g \rightsquigarrow K_g$$

is a unitary projective representation of the group of linear canonical transformations in the Hilbert space of square integrable functions:

$$K_{gg'}(x,x') = \int K_g(x,x'')\, K_{g'}(x'',x')\, dx'' \tag{1.5}$$

This relation may be checked by performing the integration.

The integration in (1.5) is possible, even if the matrix elements are complex, at least as long as certain restrictions are observed. This suggests the continuation of the representation to a subset of the group Sp(2n,C) of complex symplectic transformations. It has been shown [1, 2] that this can be done for a semigroup which in one dimension contains matrices of the form

$$\omega = \begin{pmatrix} 1 & 0 \\ ic & 1 \end{pmatrix} \tag{1.6}$$

They are represented by kernels

$$K_\omega(x';x) = \exp - \frac{c}{2} x'^2 \, \delta(x'-x) \tag{1.7}$$

Matrix elements of these kernels between square integrable functions coincide with matrix elements of Gaussian potentials

$$V(x) = V_0 \exp - \frac{c}{2} x^2$$

which turns out to be very useful for numerical calculations in many-body systems.

The eigenstates of the Hamiltonian

$$H = \frac{1}{2}(p^2 + x^2)$$

provide a convenient basis of Hilbert space. Using such functions the calculation of matrix elements is considerably smplified by introducing the representations of canonical transformations in the Bargmann-Hilbert space [3] of entire complex analytic functions, which are square integrable in C with the measure

$$d\mu_z = \frac{1}{\pi} e^{-\frac{1}{2}|z|^2} \, d\,Re(z) \, d\,Im(z)$$

The unitary isomorphism between ordinary Hilbert space and Bargmann space is given by the integral transform

$$f(z) = (A\varphi)(z) = \int \pi^{-\frac{1}{4}} \exp\{-\tfrac{1}{2}x^2 - \tfrac{1}{2}z^2 + \sqrt{2}\, z x\} \varphi(x)\, dx \tag{1.8}$$

and its inverse

$$\varphi(x) = (A^{-1}f)(z) = \int \pi^{-\frac{1}{4}} \exp\{-\tfrac{1}{2}x^2 - \tfrac{1}{2}(z^*)^2 + \sqrt{2}\, z^* x\} f(z)\, d\mu \tag{1.9}$$

This transformation maps oscillator states onto polynomials

$$h_n(x) \rightsquigarrow \frac{z^n}{\sqrt{n!}}$$

and therefore matrix elements of kernels onto Taylor coefficients of the transformed kernel.

That means that the kernel may be interpreted as a generating function of oscillator matrix elements. Especially a kernel of type (1.4) is mapped onto a kernel

$$K_h(z',z) = [\det\beta]^{\frac{1}{2}} \exp(-z'^t \alpha z' + z'^t \beta z^* + \tfrac{1}{2} z^{*t} \gamma z^*) \tag{1.10}$$

where the matrices α, β, γ are determined by the complex symplectic matrix

$$h = \begin{pmatrix} T & -iR \\ iV & S \end{pmatrix} = \begin{pmatrix} \sqrt{\tfrac{1}{2}}\, I & -i\sqrt{\tfrac{1}{2}}\, I \\ -i\sqrt{\tfrac{1}{2}}\, I & \sqrt{\tfrac{1}{2}}\, I \end{pmatrix} \begin{pmatrix} A & B \\ C & D \end{pmatrix} \begin{pmatrix} \sqrt{\tfrac{1}{2}}\, I & i\sqrt{\tfrac{1}{2}}\, I \\ i\sqrt{\tfrac{1}{2}}\, I & \sqrt{\tfrac{1}{2}}\, I \end{pmatrix} \tag{1.11}$$

with $\alpha = V T^{-1}$; $\beta = (T^t)^{-1}$; $\gamma = T^{-1} R$

K_h represents Sp(2n,R) in the same way in Bargmann space as relation (1.4) in ordinary Hilbert space.

2. Matrix elements of operators between cluster states of atomic nuclei

The relations of the first section are used to calculate interaction matrix elements in the nuclear cluster model. This shall be demonstrated by a solution of the integrals which appear in the calculation of the norm of antisymmetrized wave functions.

In the integral

$$\int \psi^*(\xi_1,\dots,\xi_n)\, P\, \psi(\xi_1,\dots,\xi_n)\, d\xi_1 \dots d\xi_n \tag{2.1}$$

P permutes the coordinates of the right hand function which has the form of a cluster state of the atomic nucleus. In such a state the system is split into a set of k subsystems (clusters) of $n_1,\dots n_k$ particles.

$$\psi \sim \psi_1(\eta_1^1,\dots,\eta_{n_1-1}^1)\dots\ \psi_k(\eta_1^k,\dots\eta_{n_k-1}^k)\,\chi(\sigma_1,\dots,\sigma_k) \tag{2.2}$$

where the functions ψ_i have the simple structure

$$\psi_i(\eta_1^i,\dots \eta_{n_i-1}^i) = \exp - \frac{1}{2d_i^2} \sum_{\nu=1}^{n_i-1} (\eta_\nu^i)^2$$

and

$$\chi(\sigma_1,\dots,\sigma_k) = h_{d_{\sigma_1}}^{\nu_1}(\sigma_1)\dots\ h_{d_{\sigma_k}}^{\nu_k}(\sigma_k)$$

The $h_{d_{r_i}}^{\nu_i}$ are eigenstates of harmonic oscillators with frequencies $(d_{\sigma_\nu})^{-2}$. The internal Jacobi coordinates of the clusters depend on the differences of the single particle coordinates only while $\sigma_1,\dots,\sigma_{k-1}$ are relative coordinates between the centres of clusters and σ_k is the c.m. coordinate of the system. This choice of coordinates allows a separation of the c.m. motion.

To give an algebraic interpretation of the integral defined in (2.1) we adjust all different frequencies to a single unit frequency by applying the dilatation matrix

$$D = \begin{pmatrix} d_{\sigma_1} & & & & & & & & & 0 \\ & \ddots & & & & & & & & \\ & & d_{\sigma_k} & & & & & & & \\ & & & d_1 & & & & & & \\ & & & & \ddots & & & & & \\ & & & & & d_1 & & & & \\ & & & & & & \ddots & & & \\ & & & & & & & d_k & & \\ & & & & & & & & \ddots & \\ 0 & & & & & & & & & d_k \end{pmatrix} \quad \text{(entries } d_1 \text{ repeated: } n_1 - 1 \text{)} \tag{2.3}$$

Then (2.1) becomes a kernel corresponding to the coordinate transformation

$$L = \mathcal{D}^{-1}\, \mathcal{U}^{-1}\, P\, \mathcal{U}\, \mathcal{D} \tag{2.4}$$

The matrix $\mathcal{U}$ is unitary and transforms from single particle coordinates to relative coordinates.

According to (1.11) the construction of the kernel in Bargmann space involves the calculation of the matrix

$$(T^t)^{-1} = \frac{1}{2}\,(L^t + L^{-1})^{-1}$$

which is rather troublesome in systems of many particles. But the direct calculation of T^{-1} can be avoided. For this purpose the transformation   has to be split into the product $\mathcal{U}_1 \cdot \mathcal{U}_2$ [5]

The transformation $\mathcal{U}_2$ changes only the coordinates $\sigma_1, \ldots \sigma_k$ and replaces them by the c.m. vectors of the k clusters. We call the resulting set of coordinates center coordinates.

Introducing a new diagonal matrix

$$\Delta = \begin{pmatrix} d_1 & & & & & & 0 \\ & \ddots & \updownarrow n_1 & & & & \\ & & d_1 & & & & \\ & & & \ddots & & & \\ & & & & d_k & & \\ & & & & & \ddots \;\updownarrow n_1 & \\ 0 & & & & & & d_k \end{pmatrix} \tag{2.5}$$

and writing

$$L = D^{-1}\, \mathcal{U}_2^{-1}\, \Delta\, \Delta^{-1}\, \mathcal{U}_1^{-1}\, P\, \mathcal{U}_1\, \Delta\, \Delta^{-1}\, \mathcal{U}_2\, D \tag{2.6}$$

we get

$$L = (\Delta^{-1}\mathcal{U}_2 D)^{-1}\, \mathcal{U}_1^{-1}\, (\Delta^{-1} P \Delta)\, \mathcal{U}_1\, (\Delta^{-1}\mathcal{U}_2 D)$$

since Δ commutes with $\mathcal{U}_1$. The matrix $\Delta^{-1}\mathcal{U}_2 D$ splits into a direct sum of a unit matrix with respect to the internal coordinates and a matrix which connects center coordinates and relative coordinates. Then the representation property is used

$$K_L = K_{L_1^{-1}} \circ K_{\mathcal{U}_1^{-1}} \circ K_{\Delta^{-1}P\Delta} \circ K_{\mathcal{U}_1} \circ K_{L_1} \tag{2.7}$$

(with $L_1 = \Delta^{-1}\mathcal{U}_2 D$) , and the kernels are transformed separately into Bargmann space. Here $K_{\mathcal{U}_1}$ implies a coordinate transformation, just as in ordinary space since $\mathcal{U}_1$ is orthogonal, while K_{L_1} acts trivially with respect to the internal coordinates.

Integrating over these coordinates means setting them zero in the kernel [1] and that implies that the single particle coordinates become $z_i/\sqrt{n_i}$. Denoting complex coordinates in Bargmann space by those Latin letters which correspond to the Greek letters used in ordinary space we finally obtain:

$$k_L(s_1', \dots, s_k', s_1, \dots s_k) = K(s_1', \dots, s_k', 0 \dots 0, s_1, \dots, s_k, 0 \dots 0) =$$

$$= \int K_{L_1^{-1}}(s_1' \ldots s_k', 0, \ldots, 0, z_1', \ldots z_k', 0, \ldots, 0) \cdot K_{\Delta^{-1}P\Delta}\left(\frac{z_1'}{\sqrt{n_1}}, \ldots \frac{z_1'}{\sqrt{n_1}}, \ldots \frac{z_k'}{\sqrt{n_k}} \ldots \frac{z_k'}{\sqrt{n_k}}, \frac{z_1}{\sqrt{n_1}} \cdots \frac{z_k}{\sqrt{n_k}}\right) \tag{2.8}$$

$$\cdot K_{L_1}(z_1, \ldots, z_k, 0, \ldots, 0, s_1, \ldots, s_k, 0, \ldots, 0)\, d\mu_{z_1'} \ldots d\mu_{z_k'}\, d\mu_{z_1} \ldots d\mu_{z_k}$$

The integral transforms of (2.8) can be performed by the methods used for integration of Gauss-functions of quadratic forms.

It remains to calculate the kernel in the middle. Permuting Δ and P we get

$$\Delta^{-1} P \Delta = \Delta^{-1} P \Delta P^{-1} P = \Delta^{-1} \Delta_P P$$

where Δ_P is a diagonal matrix obtained from Δ by permuting the elements of the diagonal.

The matrix

$$T_{\Delta^{-1} P \Delta} = \frac{1}{2}\left(\Delta^{-1} \Delta_P P + \Delta \Delta_P^{-1} P\right)$$

can be inverted explicitly. Moreover we make use of the special block form of the permutation matrix [4]

$$P = \begin{pmatrix} E_{11} & \cdots & E_{1k} \\ \vdots & & \\ E_{k1} & \cdots & E_{kk} \end{pmatrix} \tag{2.9}$$

where each submatrix E_{rs} has n_r rows and n_s columns and processes ε_{rs} elements different from zero. Then the kernel splits into a product

$$K_{\Delta^{-1}P\Delta}\left(\frac{z_1'}{\sqrt{n_1}}, \ldots, \frac{z_k'}{\sqrt{n_k}}, \frac{z_1}{\sqrt{n_1}} \cdots \frac{z_k}{\sqrt{n_k}}\right) = \prod_{r,s} \left(k_{rs}(z_r', z_s)\right)^{\varepsilon_{rs}} \tag{2.10}$$

with

$$k_{rs}(z_r', z_s) = f_{rs}^{+\frac{1}{2}} \exp\left(-\frac{1}{2} \frac{g_{rs}}{n_r} z_r'^2 + \frac{f_{rs}}{\sqrt{n_r n_s}} z_r' z_s^* + \frac{1}{2} \frac{g_{rs}}{n_s} (z_s^*)^2\right)$$

$$f_{rs} = \frac{1}{2}\left(d_r'^{-1} d_s + d_r' d_s^{-1}\right)$$

$$g_{rs} = \left(d_r' d_s^{-1} - d_r'^{-1} d_s\right)\left(d_r' d_s^{-1} + d_r'^{-1} d_s\right)^{-1}$$

The cluster states as given in eq. (2.2) are characterized by the distribution of nucleons corresponding to $n_1, \dots n_k$ and by the internal frequencies $\alpha_1^{-2} \dots \alpha_k^{-2}$ and relative frequencies $\alpha_{r_1}^{-2} \dots \alpha_{r_k}^{-2}$. The exchange type of the integral between the states is determined by the number ε_{rs} characterizing the decomposition of the permutation P with respect to the splitting into clusters. After performing the integration over the internal coordinates we are left with the kernel eq. (2.10) which clearly shows the dependence on the parameters. A similar algebraic treatment can be applied to the interaction integrals and leads to a corresponding kernel which then depends also on the parameter c appearing in eq. (1.7).

The present example shows the advantage of analysing the cluster ansatz in terms of canonical transformations. By relating all essential parameters of the manybody system to symplectic matrices and using the representation condition we are in a position to perform the necessary integrations algebraically and calculate the kernels which determine the dynamical equations for the relative motion of the clusters.

REFERENCES

[1] P. Kramer, M. Moshinsky, T. H. Seligman in:Group Theory and its Applications, Vol. 3, Ed.: E. M. Loebl, Acad. Press 1975

[2] J. M. Brunet, P. Kramer,Contribution to this Colloquium

[3] V. Bargmann, Com. Pure and Appl. Math. 14, 187 (1961)

[4] P. Kramer, T. H. Seligman, Nucl. Phys. A 186, 49 (1972)

[5] W. Zahn, to be published

CRYSTALS AS DYNAMICAL SYSTEMS : A NEW CLASS OF MODELS

by

Peter Kasperkovitz

1.Institut für theoretische Physik, Technische Hochschule

Vienna, Austria

1. Introduction

Dynamical systems that can be interpreted as models of crystals are introduced and their symmetry properties are studied. The new models take an intermediate position between realistic models (the best to our present knowledge but, due to their complexity, unsuited for straightforward calculations) and the usual ones (based on the adiabatic and the harmonic approximations [1]). The new models clarify the relations between rigid motions, permutations and the so-called 'crystal symmetries'. Their essential features and their relation to the realistic and the usual models can be clearly exhibited even for 1-dimensional systems. This is done in the following; the generalization to more dimensions is given elsewhere [2],[3].

The systems considered here consist of a finite number of point particles and evolve in time according to a non-relativistic theory (classical or quantum mechanics). The models discussed below differ only in the form and the symmetry properties of the interaction between the particles and in the constraints imposed on their motion.

2. Realistic models

In a realistic description the motion of the particles is subject to no constraints and governed solely by the 'true' interaction, represented e.g. by a sum of suitable 2-body potentials.

$$V(x_{11},..,x_{nN_n}) = (1/2) \sum_{hi \neq h'i'} v_{hh'}(x_{hi}-x_{h'i'})$$

$$v_{hh'}(y) = v_{h'h}(y) = v_{hh'}(-y) \qquad (1)$$

In (1) the index $h=1,..,n$ is used for the different kinds of particles and the index $i(=i_h)=1,..,N_h$ for the different particles of one kind. If there are quasi-free electrons in the crystal they are labelled by $h=n$; in this case we put $n'=n-1$, otherwise $n'=n$.

Each particle is allowed to move throughout the whole space. However in states of low energy the particles are expected to cluster together within a finite region. Moreover there should be a state of lowest energy where the structure of a finite but otherwise perfect crystal is clearly perceivable. In the 1-dimensional case considered here the finite structure $s_{\bar{N}}$ is given by the following equations :

$$s_{\bar{N}} = s \cap (0,\bar{N}] \quad , \quad s = \{ m-b_{hk} : m \text{ integer}; h=1,..,n'; k=0,..,n_h-1 \}$$
$$0 \leq b_{hk} < 1 \ (=\text{lattice distance}) \quad , \quad b_{hk}=b_{h'k'} \iff h=h', k=k'$$
$$\{ m-b_{hk}+c : m \text{ integer}; h=1,..,n'; k=0,..,n_h-1 \} = s \implies c \text{ integer} \qquad (2)$$

This implies that the numbers of the different particles are chosen in such a way that

$$i = 1+\bar{N}k \ ; \ 1 = 1,..,\bar{N} \ ; \ k = 0,..,n_h-1 \ ; \ N_h/n_h = \bar{N} \ (=\text{crystal volume}) \qquad (3)$$

holds. A state where such a structure is realized is a static equilibrium state if a classical description is allowed and the potential (1) is bounded from below. If the potential is not bounded from below a semi-classical (Bohr-Sommerfeld) or completely quantum mechanical description is needed to get at all a lower bound for the energy. Due to the uncertainity principle a precise localization of the particles is then impossible but all information about the structure should be obtainable from the density function corresponding to the ground state.

In any case the only justification for calling such a dynamical system a model of a crystal is the existence of these special solutions of the equation(s) of motion. Their regularities are by no means straight consequences of the symmetries of the Hamiltonian since the latter is only invariant under some 'natural' transformations, namely

$$\begin{array}{lll} \text{rigid translations} : & x_{hi} \longrightarrow x_{hi} + c & \\ \text{rigid reflection} : & x_{hi} \longrightarrow -x_{hi} & \\ \text{permutations} : & x_{hi} \longrightarrow x_{h,r^{-1}i} & (4) \end{array}$$

($i \rightarrow ri$ is a permutation of the numbers $1,..,N_h$). The transformations (4) close into a group

$$G_{realistic} \cong T(1) \circledS (S_{N_1} \times .. \times S_{N_n} \times C_2) \qquad (5)$$

where T(1) is the additive group of the real numbers, $C_2 \cong S_2$ and S_M the symmetric group of order M! .

3. Usual models

In the usual description the structure of the crystal is put into the model from the very beginning although, if quasi-free electrons are present, in a rather different way for these particles than for the lattice constituents. The motion of these two groups of particles is essentially decoupled by the adiabatic approximation : the electrons are assumed to move as if the lattice constituents formed a static structure; on the other hand it is assumed that an effective potential depending on the motion of the electrons and the position of the lattice constituents forces these particles to arrange themselves into the desired structure.

This means that the substitution

$$v_{nh'}(x_{ni}-x_{h'i'}) \longrightarrow v_{nh'}(x_{ni}-l'+b_{h'k'}) \qquad \text{for } h'=1,..,n' \tag{6}$$

is made in the equation(s) governing the motion of the electrons (h=n). Thereby the symmetry properties of this part of the problem are reduced to the permutations of the electrons. To improve this situation and to express the regularities of the structure (2) in a purely group-theoretical language the (external) potentials (6) are substituted by potentials which are periodic functions with a period $\bar{N}$.

$$v_{nh}(y) \longrightarrow w_{nh}(y) \;, \qquad w_{nh}(y) = w_{hn}(y) = w_{nh}(-y) = w_{nh}(y+\bar{N}) \tag{7}$$

The shorter the range of the potentials v_{nh} the less is the motion of the electrons influenced by the substitution (7) if w_{nh} is chosen to be equal to v_{nh} for $|y|<\bar{N}/2$ and if all electrons are just moving within the crystal ($(c-\bar{N})/2 < x_{ni} < (c+\bar{N})/2$ for $i=1,..,N_n$; c see equ.(10) below). However to garantuee the last condition constraints have to be imposed onto the motion of the electrons. The simplest (but somewhat obscure) choice are periodic boundary conditions demanding the solutions of the equation(s) of motion (wave functions or orbits in phase space) to be invariant under all

<u>periodicity translations +)</u> : $\quad x_{ni} \longrightarrow x_{ni} + m_{ni}\bar{N}$, m_{ni} integer
+) of the electrons (8).

To avoid inconsistencies the substitution (7) has then also to be made for the electron-electron interactions. This leads to a Hamiltonian which is invariant under

lattice translations +) : $\quad x_{ni} \longrightarrow x_{ni} + m$, m integer
<u>permutations +)</u> : $\quad x_{ni} \longrightarrow x_{n,r^{-1}i}$
+) of the electrons (9).

If the following condition is satisfied,

there exists a $c \in [0,1)$, so that for $h=1,..,n'$

$$\left\{ m+b_{hk}+c \,:\, m \text{ integer};\ k=0,..,n_h-1 \right\} = \left\{ m-b_{hk} \,:\, m \text{ integer};\ k=0,..,n_h-1 \right\} \quad (10),$$

then the Hamiltonian is also invariant under the

lattice reflection +) : $x_{ni} \longrightarrow -x_{ni} + c$

+) of the electrons (11).

The symmetry group is therefore isomorphic to $Z^+ \circledS C_2 \times S_{N_n}$ or merely to $Z^+ \times S_{N_n}$ depending on whether condition (10) is satisfied or not (Z^+ = additive group of the integers). Due to the periodic boundary conditions however the group of 'visible' symmetry transformations is only a finite homomorphic image of this group, namely

$$G^{\text{electrons}}_{\text{usual}} \cong C_{\bar{N}} \left[\circledS\, C_2 \right] \times S_{N_n} \quad (12)$$

(C_M is the cyclic group of order M).

The adiabatic approximation of the interaction of the lattice constituents is a function which differs from (1) in two respects : the sum is restricted to $h,h'=1,..,n'$ and the direct 2-body interactions $v_{hh'}$ are supplemented by indirect (lattice-electron-lattice) contributions $v^{el}_{hh'}$. These depend on the density of the electrons (corresponding to the ground-state of the many-electron system) and on the special form of the functions v_{hn} ($h<n$). When considering the electrons then all these functions were assumed to be periodic ones; thus the same periodicity might be expected for the functions $v^{el}_{hh'}$. This is however not evident from the expressions given in literature [4] since these interactions are introduced there together with and in terms of the harmonic approximation. This consists in expanding the interaction of the lattice constituents around the equilibrium configuration $x_{h,l+\bar{N}k} = l-b_{hk}$ and in retaining only those terms which are at most quadratic in the displacements. The effective interaction $\bar{V}$ approximated this way coincides with the function V of equ.(1) if there are no quasi-free electrons.

$$\bar{V}(x_{11},..,x_{n'N_{n'}}) = (1/2) \sum_{hlk \neq h'l'k'}{}' \bar{v}_{hh'}(x_{h,l+\bar{N}k} - x_{h',l'+\bar{N}k'}) \longrightarrow$$

$$\bar{V}^{harm}(x_{11},..,x_{n'N_{n'}}) = A + \sum_{hlk}{}' (x_{h,l+\bar{N}k} - l + b_{hk})\, B(hlk) + \sum_{\substack{hlk\\h'l'k'}}{}' (x_{h,l+\bar{N}k} - l + b_{hk})(x_{h',l'+\bar{N}k'} - l' + b_{h'k'})\, C(hlk;h'l'k') \quad (13)$$

The symbol $\sum'$ indicates that the range of h,h' is 1,..,n' (n' see above). The coefficients of the linear terms

$$B(hlk) = \sum_{h'l'k'(\neq hlk)} \bar{v}'_{hh'}(l-b_{hk}-l'+b_{h'k'}) \qquad (14)$$

are assumed to vanish for the given structure ore are forced to do so by fitting parameters in the potentials $\bar{v}_{hh'}$. The coefficients

$$C(hlk;h'l'k') = C(h'l'k';hlk) = (1/2)\Big[\delta_{hh'}\delta_{ll'}\delta_{kk'} \sum_{\substack{h''l''k'' \\ (\neq hlk)}} \bar{v}''_{hh'}(l-b_{hk}-l''+b_{h''k''}) - (1-\delta_{hh'}\delta_{ll'}\delta_{kk'})\, \bar{v}''_{hh'}(l-b_{hk}-l'+b_{h'k'})\Big] \qquad (15)$$

are collected into a ('dynamical') matrix C. They satisfy

$$\sum_{hlk} C(hlk;h'l'k') = 0 \qquad (16)$$

which shows that $\bar{V}^{harm}$ is invariant under

rigid translations +) : $x_{h,l+\bar{N}k} \longrightarrow x_{h,l+\bar{N}k} + c$

+) of the lattice constituents (17).

These are however all the symmetries common to $\bar{V}$ and $\bar{V}^{harm}$. Other symmetry relations of the dynamical matrix such as

$$C(hlk;h'l'k') = C(h,l+m+M\{l+m\},k;h',l'+m+M\{l'+m\},k') \qquad (18)$$

$$C(hlk;h'l'k') = C(hl^*k^*;h'l'^*k'^*) \qquad (19)$$

where the function $M : \mathbb{R} \rightarrow \{ m\bar{N} : m \text{ integer} \}$ is defined by

$$x + M\{x\} \in (0,\bar{N}] \ ; \quad M\{x\} \equiv 0 \ (\text{modulo } \bar{N}) \qquad (20)$$

and the numbers l^*,k^* by

$$l^* - b_{hk^*} = -l + b_{hk} + c + M\{-l+b_{hk}+c\} \qquad (21)$$

are by no means obvious. They imply that $\bar{V}^{harm}$ is invariant under

lattice translations +) : $x_{h,l+\bar{N}k} \longrightarrow x_{h,l+m+M\{l+m\}+\bar{N}k} - m - M\{l+m\}$, m integer

+) of the lattice constituents (22)

and, if condition (10) is satisfied, also under the

lattice reflection +) : $x_{h,l+\bar{N}k} \longrightarrow - x_{h,l^*+\bar{N}k^*} + c + M\{-l+b_{hk}+c\}$ (23).

+) of the lattice constituents

These were straight consequences of (10) and the invariance properties of $\bar{V}$ if these transformations were composed only of the 'natural' symmetries (4). It is however easily seen that each of the transformations (22),(23) does not only contain a permutation and a rigid motion but also one of the periodicity translations +) : $x_{hi} \longrightarrow x_{hi} + m_{hi}\bar{N}$, m_{hi} integer (24).

+) of the lattice constituents

Therefore the symmetries (18),(19) of the dynamical matrix C could be traced back to symmetry properties of the effective potential if, as it was done for the electrons, all 2-body interactions $\bar{v}_{hh'}$ were replaced by periodic functions,

$$\bar{v}_{hh'}(y) \longrightarrow w_{hh'}(y), \quad w_{hh'}(y) = w_{h'h}(y) = w_{hh'}(-y) = w_{hh'}(y+\bar{N}) \tag{25}$$

before the harmonic approximation is performed.

No matter for what reasons the substitutions (22) (and (23) if (10) holds) are accepted it is easy to show that they combine with the rigid translations into

$$G^{lattice}_{usual} \cong (T(1) \times C_{\bar{N}}) \left[Ⓢ \; C_2 \right] \tag{26}$$

The symmetry group of the total system is then $G^{electrons}_{usual} \times G^{lattice}_{usual}$; if further interactions between the electrons and the lattice constituents are taken into account to improve the adiabatic approximation this group reduces to

$$G_{usual} \cong (T(1) \times C_{\bar{N}}) \left[Ⓢ \; C_2 \right] \times S_{N_n} \tag{27}$$

4. Periodic models

The models to be introduced here are based on the substitutions (7),(25), i.e. all 2-body potentials are assumed to be periodic functions the periodicity being given by the size of the crystal. The recourse to 2-body potentials can be avoided by simply postulating the translations (8),(24) to be symmetry transformations. As in the case of the external potential of the electrons this modification of the realistic potential (which is here the interaction of all crystal constituents) can be expected to be an approximation only for certain situations. These have to be fixed by constraints which are again periodic boundary conditions adapted to the imposed periodicities. The present situation is slightly more complicated since the potential is not an external one and has therefore to be invariant under the rigid translations (4) meaning that its proper domain is the set of fibres $(\sqrt{\mu_1}[x_{11}+c],\ldots,\sqrt{\mu_n}[x_{nN_n}+c])$, $-\infty < c < \infty$. (The masses μ_h are introduced here only for easier splitting of the uniform motion of the center of mass). The N-dimensional lattice $L = \{ (\sqrt{\mu_1}\, m_{11}\bar{N},\ldots,\sqrt{\mu_n}\, m_{nN_n}\bar{N}) : m_{hi} \text{ integer} \}$ corresponding to the new translational symmetries is therefore 'smeared out' into the direction $(\sqrt{\mu_1},\ldots,\sqrt{\mu_n})$ thereby defining in the orthogonal subspace an (N-1)-dimensional lattice

$$L[\text{int}] = \left\{ (\sqrt{\mu_1}[m_{11}-\bar{m}]\bar{N},\ldots,\sqrt{\mu_n}[m_{nN_n}-\bar{m}]\bar{N}) : m_{hi} \text{ integer}; \; \bar{m} = (\sum_h \mu_h N_h)^{-1} \sum_{hi} \mu_h m_{hi} \right\} \tag{28}$$

The periodic boundary conditions introduced here are given by the sublattice

$$\tilde{L}[\mathrm{int}] = \left\{ (y_{11},\dots,y_{nN_n}) : (y_{11},\dots) \in L[\mathrm{int}];\ y_{hi} - N_h^{-1}\sum_{i'} y_{hi'} = \sqrt{\mu_h}\, m_{hi}\bar{N},\ m_{hi} \text{ integer} \right\} \quad (29).$$

In case of the electrons the special choice of the sublattice characterizing the boundary conditions was motivated by the fact that there was a symmetry-adapted cell of this lattice (namely the N_n-cube $(c-\bar{N}/2) < x_{ni} < (c+\bar{N}/2)$) where the new (periodic) potential approximated the original (aperiodic) one. The choice (29) is made for similar reasons. It can be shown [3] that the lattice (29) posesses a cell so that for all configurations $(\sqrt{\mu_1}\, x_{11},\dots,\sqrt{\mu_h}\, x_{nN_n})$ belonging to it the inequalities

$$(x_{hi} - x_{hi'}) \leqslant \bar{N} \qquad h=1,\dots,n;\quad i,i'=1,\dots,N_h \tag{30}$$

$$\left[x_{(h)} - x_{(h')}\right] < \bar{N}\ ,\qquad x_{(h)} = N_h^{-1}\sum_i x_{hi} \qquad h,h'=1,\dots,n \tag{31}$$

are satisfied. Equs.(30) correspond to all situations where the particles can be enclosed into a a freely movable volume of the size of the crystal, each kind into its own volume. Equs.(31) show that their centers of mass must not seperate by more than the diameter of the volume. The region characterized by (30),(31) is the largest one where a periodic potential can be expected to approximate the rapid oscillations of the realistic potential suppressing however at the same time its mean variation within this region.

The cell of the lattice $\tilde{L}[\mathrm{int}]$ can be chosen to be invariant under all the transformations (4). It is therefore natural to base the new models onto the following assumption : the potential is not only invariant under rigid motions and permutations (as is the realistic one) but (contrary to it) also under the periodicity translations (8),(24). Its symmetry group is therefore

$$G_{\mathrm{periodic}} \cong (\ T(1) \times T[\mathrm{int}]\ )\ \circledS\ (\ C_2 \times S_{N_1} \times \dots \times S_{N_n}) \tag{32}$$

where $T[\mathrm{int}]$ is the group of

$$\text{internal periodicity translations :}\quad x_{hi} \longrightarrow x_{hi} + y_{hi}\ ,\ (\sqrt{\mu_1}\, y_{11},\dots,\sqrt{\mu_n}\, y_{nN_n}) \in L[\mathrm{int}] \quad (33).$$

As outlined above the choice of such a periodic interaction is (hopefully) justified by introducing also constraints : periodic boundary conditions demanding certain periodicity translations, namely those corresponding to the sublattice (29), to be 'invisible' (trivially represented). This reduces the symmetry group of the present models to the finite group $G_{\mathrm{periodic}}/\tilde{T}[\mathrm{int}]$ where $\tilde{T}[\mathrm{int}]$ is the translation group corresponding to $\tilde{L}[\mathrm{int}]$. It should be noted that both groups G_{periodic} and $\tilde{T}[\mathrm{int}]$ depend only on the lattice but not on the basis of the structure.

5. Reasons for investigating the new models

There are mainly two questions arising from the models defined in the preceding section :

(1) Are the new models extensions of the usual ones (in the sense that the results of the latter approximate the results of the former) ?

(2) Does the symmetry of the new models suffice to simplify the quantum mechanical eigenvalue problems to such an extent that calculations of practical interest can be performed ?

The answer to the first question is 'yes' since as pointed out before the harmonic approximation of a periodic potential has just the symmetry properties commonly used. This result does not rely on the use of 2-body potentials. It can be proven quite generally by determining those transformations of $G_{periodic}$ which either leave invariant the (N_n+1)-dimensional set

$$\left\{ \begin{array}{l} (\sqrt{\mu_1}[1-b_{11}+c],\ldots,\sqrt{\mu_{n'}}[\bar{N}-b_{n',n'-1}+c],\sqrt{\mu_n}[x_{n1}+c],\ldots,\sqrt{\mu_n}[x_{nN_n}+c]) : \\ -\infty<c<\infty \;\; ; \; -\infty < x_{ni} < \infty \quad \text{for } i=1,..,N_n \end{array} \right\} \tag{34}$$

or the 1-dimensional set

$$\left\{ (\sqrt{\mu_1}[1-b_{11}+c],\ldots,\sqrt{\mu_n}[\bar{N}-b_{n,n-1}+c]) \; : \; -\infty<c<\infty \right\} \tag{35}$$

depending on whether the system contains quasi-free electrons or not. In the second case one finds [3] that the group of these transformations coincides with the usual symmetry group (26) generated by the transformations (22) (and (23) if (10) holds). In the first case the allowed transformations are composed of rigid translations of all particles (equ.(4)), of the lattice translations of the lattice constituents (equ.(22)), of periodicity translations and permutations of the electrons (equs.(8), (4)), and provided that (10) holds also of a lattice reflection given by equs.(11), (23), the parameter c being uniquely determined by the basis $\{b_{hk}\}$ .If the usual periodic boundary conditions are chosen for the elelctrons the symmetry group reduces again to the usual one (equ.(27)). These results show that the models introduced here are at least suited to elucidate the relations between the general symmetries common to all systems (rigid motions, permutations) and the special ones typical for systems representing crystalline solids (crystal symmetries).

It remains to be seen whether the new models can also be applied to physical situations not covered by the usual concepts (e.g. non-vibrational motions of some lattice constituents). It is rather obvious that the multi-dimensional lattices and the corresponding translation groups lead directly to a generalization of crystallography and band theory to more than three dimensions. This poses some algebraic problems; but guessing from the solutions obtained so far [2] it are not these problems that make the applicability of the new models dubious. The decision on this point depends rather on the success or failure of further assumptions resulting in a drastic reduction of either the state space ('tight-binding or plane-wave approximations') or the number of considered particles ('long-wavelength limit') without affecting the convenient symmetry properties of the models.

References

[1] These approximations are treated in most textbooks on solid state theory. For a detailed discussion see e.g. A.Haug, Theoretische Festkörperphysik I (Wien: Deuticke 1964) 15-22 ; G.V.Chester and A.Houghton, Proc.Phys.Soc.73(1959)609--22 ; G.Weinreich, Solids : Elementary theory for advanced students (New York: Wiley 1965)50-4

[2] P.Kasperkovitz, New models of crystals. I. Basic ideas and simple examples. II. General formulation of the problems and first solutions (submitted for publication)

[3] P.Kasperkovitz (in preparation)

[4] W.A.Harrison, Solid state theory (New York: McGraw-Hill 1970)427-33 ; W.Jones and N.H.March, Theoretical solid state physics, vol.1 (London: Wiley 1973) 250-9

NON LINEAR CANONICAL TRANSFORMATIONS AND THEIR REPRESENTATIONS IN QUANTUM MECHANICS

M. Moshinsky and P.A. Mello

Instituto de Física, Universidad de México (UNAM), México 20, D.F.

In the last few years an extensive literature has developed on linear canonical transformations and their representation in quantum mechanics. Applications of these results have been made to clustering theory in nuclei, problems of accidental degeneracy, etc. In the present paper we wish to turn our attention to non-linear canonical transformations. We show that by dealing with appropriate functions f_α $(\alpha = 1, \cdots 2n)$ of x_i, p_i $(i = 1, \cdots n)$ rather than with these variables themselves, we can in principle set unambiguously the equations that determine the representation in quantum mechanics of the canonical transformation under study. This result holds when the old and new functions f_α have the same spectrum. We discuss specific examples when this last condition is satisfied: Non-linear canonical transformations in the radial variable that were obtained from projection of linear ones in higher dimensional spaces; canonical transformations that take us from one

Hamiltonian to another with the same spectrum, be this one continuous or discrete; canonical transformations that relate two sets of integrals of motion (which include the Hamiltonians) when we are dealing with phase spaces of dimensions higher than 2, etc. We discuss briefly, the possibility of extending our analysis to canonical transformations that do not conserve the spectrum of the relevant operators.

The present paper, will be published in Journal of Mathematical Physics (September or October 1975).

INVARIANCE GROUPS OF YOUNG OPERATORS; PAULING NUMBERS

R.W.J. ROËL

Institute of Theoretical Chemistry, University of Amsterdam, Amsterdam, Holland

I *DEFINITION AND NOTATION*

1 *YOUNG OPERATOR*

Consider any positive integer n and its partitions $(\lambda) = \ldots j^{\lambda_j}\ldots$, $j=1,2,\ldots,n$, defined by $\sum_{j=1}^{n} j\lambda_j = n$. Let $\sum_{j=1}^{n} \lambda_j = n_\lambda$. (1.1.1)

Corresponding to any (λ) a *Young diagram*, YD, D^λ is defined as an arrangement of n_λ rows ordered according to length, λ_j rows of length j for each j, from the top to the bottom.

Corresponding to any D^λ a set of *standard Young tableaux*, YT, T_a^λ is defined by filling the YD with the integers 1,2,...,n so that they increase from left to right *and* from top to bottom. Their number, $a=1,2,\ldots,|\lambda|$, equals the dimension of the irreducible representation, IR, labelled $[\lambda]$ of S_n.

Corresponding to any T_a^λ a *row group* R_a^λ is defined as the direct product

$$R_a^\lambda = \mathop{X}_{j=1}^{n} \{ \mathop{X}_{\alpha=1}^{\lambda_j} S_j(\alpha)\} \qquad (1.1.2)$$

in which the factor $S_j(\alpha)$ is the symmetric group defined on the indices in the α-th row of length j in T_a^λ. Analogously, a *column group* C_a^λ is defined. Let $P_a^\lambda = \sum_p p$, $p \in R_a^\lambda$ and $N_a^\lambda = \sum_q q\ \mathrm{sgn}(q)$, $q \in C_a^\lambda$ denote the row symmetrizer and column antisymmetrizer, respectively.

Given a "first" YT, T_1^λ, any T_a^λ can be written $f_a T_1^\lambda$ in which $f_a \in S_n$ transforms the first YT into the a-th. Correspondingly, row symmetrizer and column antisymmetrizer pertaining to T_a^λ are given by $N_a^\lambda = f_a N_1^\lambda \bar{f}_a$ and $P_a^\lambda = f_a P_1^\lambda \bar{f}_a$, respectively. Let (1.1.4)
$Y_{11}^\lambda = N_1^\lambda P_1^\lambda$. For any pair of YT, T_a^λ and T_b^λ a *Young operator* can be defined [1] by

$$Y_{ab}^\lambda \equiv f_a Y_{11}^\lambda \bar{f}_b = f_a N_1^\lambda P_1^\lambda \bar{f}_b = N_a^\lambda f_a \bar{f}_b P_b^\lambda = Y_{aa}^\lambda f_a \bar{f}_b = f_a \bar{f}_b Y_{bb}^\lambda \qquad (1\ 1.5)$$

and because of this series of equivalent expressions we will concentrate on Y_{11}^λ, abbreviated Y=NP. Obviously, $qYp = qNPp = NP\ \mathrm{sgn}(q) = Y\ \mathrm{sgn}(q)\ \forall q \in C, \forall p \in R$ (1.1.6)
i.e., row- and column group leave Y invariant up to a sign if applied from the left and right respectively. They are left- and right-invariance groups of the Young operator.

2 *NORMALIZER*

Consider any subgroup $H \subset G$. The *normalizer* of H, N(H), is defined [2]

$$N(H) = \{g \in G \ /\ gH\bar{g} = H\}, \text{ i.e.,} \qquad (1.2.1)$$

N(H) is the largest subgroup of G in which H is normal, $H < N(H) \subset G$.

With $G=S_n$, $H=R^\lambda$ (1.1.2) one has, [3]:

$$(a)\quad N(R^\lambda) \cong \mathop{X}_{j=1}^{n} [\{ \mathop{X}_{\alpha=1}^{\lambda_j} S_j(\alpha)\} \mathbin{s} \{ \mathop{\boxtimes}_{k=1}^{j} S_{\lambda_j}(k \bmod j)] \qquad (1.2.2)$$

in which s means semi-direct and $\boxtimes$ means inner-direct. The meaning of $S_j(\alpha)$ and $S_{\lambda_j}(k \bmod j)$ is explained on the next page.

(b) $R^{\lambda}<N(R^{\lambda})$ by definition and (c) $N(R^{\lambda})/R^{\lambda}\sqsubset C^{\lambda}$, evidently. (1.2.3)

With $R^{\lambda} \cong C^{\tilde{\lambda}}$, $\tilde{\lambda}$ the associated partition, the same holds for the column group as well, i.e., (a) $N(C^{\lambda}) \cong N(R^{\lambda})$, (b) $C^{\lambda}<N(C^{\lambda})$, and (c) $N(C^{\lambda})/C^{\lambda}\sqsubset R^{\lambda}$ (1.2.4)

In expression (1.2.2), $S_j(\alpha)$ is the symmetric group defined on the numbers in the α-th row of the j-th part of the YT (horizontal arrow), and $S_{\lambda_j}(k \bmod j)$ is the symmetric group defined on the numbers in the k-th column of the j-th part of the YT (vertical arrow).

	1	2	k	j
	j+1	j+2	j+k	2j
	2j+1	2j+2	... 2j+k ...	3j
→	αj-j+1		... αj-j+k ...	αj
			↑	

j-th part of YT : λ_j rows of length j

3 *DOUBLE COSET*

Consider any pair of subgroups $H,K\subset G$. A *double coset*, DC, in G wrt H on the lhs and K on the rhs is defined [2] as the set of elements

$HfK = \{g\in G \;/\; g=hfk,\ h\in H,\ k\in K\}$ with fixed generator $f\in G$. (1.3.1)

The main properties of DC are given by:

(a) $Hf'K \cap Hf''K = \emptyset$ if $f''\notin Hf'K$

$= Hf'K$ if $f''\in Hf'K$ so that $G = \bigcup_{\{f\}} HfK$ (1.3.2)

in which {f} is the largest subset of G the elements of which generate disjoint DC.

(b) The *length* $l_f = |HfK|$ is the number of *different* elements in the DC.

The *frequency* $d_f = |H\cap fK\bar{f}|$ is the number of times each element occurs in HfK, if all elements are written out. Obviously, $l_f d_f = |H||K|$. (1.3.3)

(c) The number of DC is given by [3,4] $\dfrac{|G|}{|H||K|}\sum_C \dfrac{|C\cap H||C\cap K|}{|C|}$ (1.3.4)

in which the summation extends over all classes C of the group G.

An alternative expression [3,5] is given by

$\langle t_H\uparrow G|t_K\uparrow G\rangle = \langle t_H|(t_K\uparrow G)\downarrow H\rangle = \langle (t_H\uparrow G)\downarrow K|t_K\rangle$ (1.3.5)

in which $\langle\Gamma|\Lambda\rangle$ denotes the frequency with which the representation Γ occurs in Λ, t_H and t_K denote the trivial IR of H and K respectively and "↑", "↓" denote induction and subduction.

4 *PAULING NUMBER*

Pauling numbers are defined wrt products of Young operators with their hermitian adjoints, [6]. Such products are elements of the group algebra of S_n and they can be written as $Y^{\dagger}Y = (NP)^{\dagger}NP = PNNP = |C|PNP = |C|\sum_f \sigma_f f$, $f\in S_n$ (1.4.1)

in which σ_f is the PNP Pauling number and C (order |C|) denotes the column group.

Also $YY^{\dagger} = NP(NP)^{\dagger} = NPPN = |R|NPN = |R|\sum_f \tau_f f$, $f\in S_n$ (1.4.2)

in which τ_f is the NPN Pauling number and R (order |R|) denotes the row group.

Thus by definition,

the $\genfrac{}{}{0pt}{}{PNP}{NPN}$ number $\genfrac{}{}{0pt}{}{\sigma_f}{\tau_f}$ is the coefficient with which f occurs in $\genfrac{}{}{0pt}{}{PNP}{NPN}$ (1.4.3)

II *FRAME WORK*

1 *INTRODUCTION*

The hamiltonian H which characterizes a system of n identical particles commutes with the permutations on particle coordinates: $[H,f]_-=0 \forall f \in S_n$. Consequently its eigenstates are labelled by IR of S_n i.e., by partitions of n. n-Fermion eigenstates transform according to $[\lambda]=[1^n]$ as required by the Pauli-principle. (2.1.1)

Orbital n-particle-product space is most commonly employed in quantummechanical applications. This space is spanned by product functions of the type $|t_\lambda\rangle = \ldots \cdot |j\rangle|j\rangle \ldots |j\rangle \cdot \ldots$, where the orbital $|j\rangle$ occurs λ_j times $\forall j$, and where t_λ denotes the trivial IR of R, the invariance group of this function. Induction yields an n-particle space $V(t_\lambda \uparrow)$ which upon symmetry adaptation, SA, decomposes into irreducible subspaces according to

$V(t_\lambda\uparrow) = \sum_\Lambda \sum_k V_k(t_\lambda\uparrow,\Lambda)$ in which Λ ranges over the IR of S_n and k (2.1.2)

ranges over the frequency with which Λ occurs in $t_\lambda\uparrow$, $k=1,2,\ldots,\langle\Lambda|t_\lambda\uparrow\rangle$.

Note, [7]: $\langle\Lambda|t_\lambda\uparrow\rangle=0$ if $(\lambda)>(\Lambda)$ in which the ordering between the two partitions is defined as follows:

$(\lambda) = \ldots j^{\lambda_j} \ldots > (\Lambda) = \ldots j^{\Lambda_j} \ldots$, j decreasing from the left to the right if the first non-vanishing difference $\lambda_j - \Lambda_j$ is positive. (2.1.3)

Thus e.g., with $(\Lambda) = 2^p 1^{n-2p}$, $\langle\Lambda|t_\lambda\uparrow\rangle = 0 \;\; \forall(\lambda)$ for which $\lambda_j \neq 0$ for any $j>2$.

In *electron* theory the symmetric group on particle coordinates is $S_n=S_n^\rho \otimes S_n^\sigma$, S_n^ρ acting on spatial-, S_n^σ on spin-coordinates. On account of the two-ness property of spin, $[\sigma]=[n-p,p]$ and consequently $[\rho]=[\tilde{\sigma}]=[2^p 1^{n-2p}]$ in order to satisfy Pauli's principle. Thus, according to (2.1.3) orbital n-electron-product space is restricted to functions in which *at most* two electrons are in the same orbital. This property characterizes therefore the allowed symmetries of n-electron spin-free eigenstates. (2.1.4)

In *nucleon* theory the symmetric group on particle coordinates is $S_n=S_n^\rho \otimes S_n^\sigma \otimes S_n^\tau$ in which the separate factors act on spatial-, spin- and isospin coordinates respectively. Because of the two-ness property of both spin and isospin, the isospin-free n-nucleon eigenstates are allowed the same symmetries as spin-free n-nucleon eigenstates whereas the allowed symmetries of spin- and isospin-free n-nuleon eigenstates are given by the associates of quaternary partitions i.e., $(\lambda)=\ldots j^{\lambda_j}\ldots, j\leq 4$, which in view of (2.1.3) restricts n-nucleon-product space to orbitals in which *at most* four nucleons can take place. (2.1.5)

Hence, s-ary partitions, $s\leq 4$, and their associates are of prime importance. Induction of orbital n-particle-product space and subsequent SA are prerequisites in order to describe an n-particle system appropriately.These two aspects are combined in e.g.:

(*a*) Wigner operator SA: (2.1.6)

$$Q^\Lambda_{ts} = |\Lambda||G|^{-1} \sum_{f\in S_n} \Lambda(\bar{f})_{st} f, \text{ fixed } s,\; t=1,2,\ldots,|\Lambda|,$$

$\Lambda(f)$ a unitary matrix representation, yield an orthogonal basis for the IR Λ of S_n if applied to any appropriate primitive product function. The "diagonal" operators

Q^{Λ}_{rr} are idempotent, which property they share with the "diagonal" Young operators Y^{Λ}_{rr}.

(b) Young operator SA: $Y^{\Lambda}_{ts} = f_t Y^{\Lambda}_{11}\bar{f}_s$, fixed t, $s=1,2,\ldots,|\Lambda|$, (2.1.7)

yield a non-orthogonal basis for the IR Λ if applied to any appropriate primitive product function.

The eigenvalue problem in terms of such SA bases involve bra-ket bracketed operators

$$Q^{\dagger}_{t's}Q_{t''s} = Q_{st'}Q_{t''s} = \delta(t',t'')Q_{ss} \quad \text{and} \tag{2.1.8}$$

$$Y^{\dagger}_{ts'}Y_{ts''} = (f_t NP\bar{f}_{s'})^{\dagger}(f_t NP\bar{f}_{s''}) = f_{s'}Y^{\dagger}_{11}Y_{11}f_{s''} \sim f_{s'}PNP\bar{f}_{s''} \tag{2.1.9}$$

respectively. In the latter case we see that the determination of the Pauling numbers is essentially the algebraic part of the eigenvalue problem.

Operators $Y^{\dagger}=PN$ [8,9,10,11,12,13] are the spin-free analogues of the antisymmetrized Löwdin spin-projection-operator [14]. The operators Y=NP have also been discussed extensively in the literature [6,12,13,15,16]. They correspond to the valence-bond structure or bond function projection operator.

Formulaes for NPN have been derived previously in the spin formulation [8,9,10,11] as well as in a spin-free formulation [12,13].

Formulaes for PNP have been derived previously for

a special cases [6,12,15,16]

b the general case by making use of the invariance with respect to the row group [13].

The treatment of PNP in ref [13] as well as the results given there are incorrect. The prescription by means of which the Pauling numbers are to be calculated is not generally applicable if the number of electrons $n\geq 8$ and gives wrong results for $n\geq 6$.

Here we present the treatment of PNP for the general case on the basis of the maximal invariance group we have at our disposal, the normalizer of the row group.

2 *INVARIANCE GROUPS FOR Y=NP, $Y^{\dagger}Y$=PNP AND $YY^{\dagger}$=NPN.*

The normalizers N(C) and N(R) are right- and left-invariance groups for Y. (2.2.1)

* The elements of N(C) can be written $n=qt$, $q\in C$, $t\in N(C)/C\subset R$

N(R) $m=ps$, $p\in R$, $s\in N(R)/R\subset C$. Then,

$nYm=qtNPps=t\bar{t}qtNPps$. With $\bar{t}qt=q'$, $q'N=N\,\mathrm{sgn}(q')=N\,\mathrm{sgn}(q)$, and $Pp=P$,

$tNPs\,\mathrm{sgn}(q)=tN\bar{t}s\bar{s}tPs\,\mathrm{sgn}(q)$. With $tN\bar{t}=N$, $Ns=N\,\mathrm{sgn}(s)$, $tP=P$, and $\bar{s}Ps=P$

$NP\,\mathrm{sgn}(sq)=Y\,\mathrm{sgn}(sq)$.**

By definition, the groups N(C) and N(R) are the largest such invariance groups. Thus the normalizers N(C) and N(R) are *the* invariance groups for $YY^{\dagger}$=NPN and $Y^{\dagger}Y$=PNP respectively (2.2.2)

* $n'NPNn''=NPN\,\mathrm{sgn}(q'q'')$ with $n'=q't'$ and $n''=q''t''$,

$m'PNPm''=PNP\,\mathrm{sgn}(s's'')$ with $m'=p's'$ and $m''=p''s''$ are easily derived.**

3 *DOUBLE COSET DECOMPOSITIONS*

Consider $PNP = \sum_f \sigma_f f$, f ranging over $S_n=G$.

Let $\{\zeta\}$ be a set of DC generators pertaining to the decomposition of G wrt N(R). The generator of a DC which has a non-empty intersection with C is chosen from C. This choice is not unique in general.

Let $\{\gamma\}$ be a set of DC generators pertaining to the decomposition of G wrt R. Since $N(R)\ \zeta\ N(R)=R\{N(R)/R\ \zeta\ N(R)/R\}R$ we choose $\{\gamma\} \subseteq N(R)/R\ \{\zeta\}\ N(R)/R$, which choice is not unique in general. According to (1.1.6) and (1.4.3),

$$\sigma_{p'\gamma p''}=\sigma_\gamma \ \forall p',p''\in R \quad \text{so that} \quad PNP=\sum_f \sigma_f f=\sum_\gamma \sigma_\gamma \bar{d}_\gamma \sum_{p',p''} p'\gamma p''=\sum_\gamma \sigma_\gamma \sum_{f\sim\gamma} f \qquad (2.3.1)$$

in which the second summation is over the different elements $f\in R\gamma R$ only.

Hence, the number of Pauling numbers to be calculated equals the number of DC in the decomposition of S_n wrt R.

With $\gamma=s'\zeta s''$, $s',s''\in N(R)/R$ one has $\sigma_\gamma=\sigma_\zeta\ sgn(s's'')=\sigma_\zeta\ sgn(\zeta\gamma)$ on account of (1.1.6, 1.2.3, 1.4.3). Thus,

$$PNP=\sum_\zeta \sigma_\zeta\ sgn(\zeta) \sum_\gamma sgn(\gamma) \sum_{f\sim\gamma} f \qquad (2.3.2)$$

in which the second summation is over all $\gamma\in\{\gamma\} \cap \{N(R)/R\ \zeta\ N(R)/R\}$.

The Pauling number σ_ζ is the coefficient of ζ in PNP so that with

$$PNP= \sum_{p',p''} \sum_q p'qp''\ sgn(q)= \sum_q d_q\ sgn(q) \sum_{f\sim q} f$$

$$\text{and (2.3.1)} \quad = \sum_\gamma \sigma_\gamma \sum_{f\sim\gamma} f \qquad \text{one arrives at}$$

$$\sigma_\gamma = \sum_{q\sim\gamma} d_q\ sgn(q)=d_\gamma \sum_{q\sim\gamma} sgn(q), \text{ i.e., } \sigma_\zeta=d_\zeta \sum_{q\sim\zeta} sgn(q) \quad \text{This yields:} \qquad (2.3.3)$$

$$\sigma_f=d_\zeta\ sgn(\zeta\gamma) \sum_{q\sim\zeta} sgn(q)=d_\zeta \sum_{q\sim\gamma} sgn(q) \quad \forall f\in R\gamma R \subset N(R)\ \zeta\ N(R).$$

This expression is invariant for all possible choices of ζ and γ. Hence,

$$PNP= \sum_\zeta d_\zeta \Big(\sum_{q\sim\zeta} sgn(q\zeta)\Big) \sum_\gamma \sum_{f\sim\gamma} f\ sgn(\gamma) \qquad (2.3.4)$$

The same line of argument in which P is replaced by N, N by P, $\{\zeta\}$ by $\{\xi\}$, $\{\gamma\}$ by $\{\delta\}$ and q by p yields:

$$NPN= \sum_\xi d_\xi \Big(\sum_{p\sim\xi} sgn(p\xi)\Big) \sum_\delta \sum_{f\sim\delta} f\ sgn(f\delta) \qquad (2.3.5)$$

Hence:

(*a*) The number of PNP Pauling numbers to be calculated equals the number of DC in the decomposition of S_n wrt N(R).

(*b*) Each Pauling number is purely group theoretical and can be calculated once a set of generators $\{\zeta\}$ has been chosen.

(*c*) For 2-columned Young tableaux a simplification occurs, namely,

$$\sum_{q\sim\zeta} sgn(q\zeta)=|R\zeta R\cap C| \ \text{ if } \zeta\in C \quad \text{and} \quad \sum_{p\sim\xi} sgn(p\xi)=|C\xi C\cap R| \ \text{ if } \xi\in R \qquad (2.3.6)$$

$$= 0 \qquad \zeta\notin C \qquad\qquad = 0 \qquad \xi\notin R \qquad (2.3.7)$$

which renders the determination of Pauling numbers for these cases more simple

The reduction in the number of Pauling numbers to be calculated is seen in the examples on the next page:

(λ)	$\lvert\{\gamma\}\rvert$	$\lvert\{\zeta\}\rvert$	(λ)	$\lvert\{\gamma\}\rvert$	$\lvert\{\zeta\}\rvert$
4^4	10.147	43	2^8	545.007.960	22
4^5	2.224.955	264	$2^4 1^8$	93.176.758.080	69

Here (λ) defines the row-group R, $\{\gamma\}$ (order $\lvert\{\gamma\}\rvert$) denotes a set of DC generators for the decomposition of $G=S_n$ wrt R and $\{\zeta\}$ (order $\lvert\{\zeta\}\rvert$) denotes a set of DC generators for the decomposition of $G=S_n$ wrt $N(R)$.

III *ELECTRON SYSTEMS AND PAULING NUMBERS*

1 *INTRODUCTION AND NOTATION*

Let

$$R = \mathop{X}_{j=1}^{p} S_2(2j-1,2j) \quad \text{then } C = S_{n-p}(1,3,,,2p-1,2p+1\ldots,n) \times S_p(2,4\ldots,2p)$$

and

$$N(R) = [\{\mathop{X}_{j=1}^{p} S_2(2j-1,2j)\} \wr \{S_p(1,3\ldots,2p-1) \boxtimes S_p(2,4\ldots,2p)\}] \times S_{n-2p}(2p+1\ldots,n) \tag{3.1.1}$$

and

$$N(C) = [\{S_p(1,3\ldots,2p-1) \times S_p(2,4\ldots,2p)\} \wr \{\mathop{\boxtimes}_{j=1}^{p} S_2(2j-1,2j)\}] \text{ if } n=2p$$
$$= C \text{ if } n>2p. \tag{3.1.2}$$

Note: $\mathop{\boxtimes}_{j=1}^{p} S_2(2j-1,2j) = Z(N(R))$, the centre of $N(R)$, consists of two elements, e and $c=(1,2)(3,4)\ldots(2p-1,2p)$. (3.1.3)

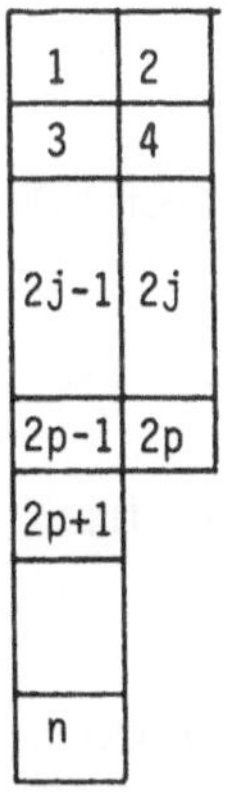

The numbers $1,3..,2p-1;2,4\ldots,2p;2p+1,2p+2\ldots,n$ will be denoted by $\alpha';\alpha'';\beta$, respectively. Numbers $2j-1$ and $2j$ from the same row of the YT will be called *corresponding* indices. The YT is filled as shown in the figure on the right. We will use the following abbreviations: $S_p(1,3\ldots,2p-1)=S'_p$ with elements u; $S_p(2,4\ldots,2p)=S''_p$ with elements $v=cu\bar{c}$; $S_{n-2p}(2p+1,2p+2\ldots,n)=S_\beta$ with elements t and $S_{n-p}(1,3\ldots2p-1,2p+1\ldots,n)$ $=S'_{n-p}$ with elements w. The group $S'_p \boxtimes S''_p$ contains elements $uv=ucu\bar{c}$ which will be denoted by s.

The sign $\approx$ between two elements means that the generate the same DC wrt a normalizer whereas the sign $\sim$ will be reserved to denote that they belong to the same DC wrt the subgroup on which the normalizer is defined.

2 *NPN, $n>2p$*

The generators ξ in (2.3.5) may be chosen from R. A canonical choice is

$$\{\xi\}=\{\xi_j \;/\; \xi_j=(1,2)(3,4)\ldots(2j-1,2j);0\leq j\leq p,\xi_0\equiv e\} \tag{3.2.1}$$

Since $n>2p$, one has $N(C)=C$ so that $\{\delta\}=\{\xi\}$ and the sum over δ drops out.

One easily derives $d_j= \lvert C\cap\xi_j C\bar{\xi}_j\rvert=(n-p-j)!(j!)^2(p-j)!$ (3.2.2)

and $\lvert C\xi_j C\cap R \rvert= p!/j!(p-j)!$ so that

$$\text{NPN} = p! \sum_{j=0}^{p} (n-p-j)!j!(-1)^j \sum_{f\sim\xi_j} f \,\text{sgn}(f)$$

confirming previous results, [8,9,10,11,12].

3 *NPN, n=2p*

From (3.1.2) it is seen that the set (3.2.1) of canonical generators reduces to $\{\xi\}=\{\xi_j \,/\, \xi_j=(1,2)(3,4)\ldots(2j-1,2j);0\le j\le q \text{ for } p=2q,2q+1\}$

and that the sum over δ contains two terms (except for ξ_q,p=2q) which have the same sgn for even p and opposite for odd p. Apparently the use of the normalizer does not improve the treatment dramatically, unlike the next two cases to be investigated.

4 *PNP, n=2p*

The generators ζ in (2.3.4) may be chosen from C, say $\zeta=u_jv_k$. With $u_jv_k=u_j\bar{u}_ku_kv_k=$ $=u_j\bar{u}_ks_k\approx u_j\bar{u}_k=u\approx su\bar{s}\,\forall s\in N(R)/R$, the generators ζ may be chosen as class representatives of S'_p(or S''_p). From [3]:

Any set of class representatives, one element from each class of S'_p, is a complete set for the decomposition of S_{2p} wrt $N(S_2^p)=N(R)$. Thus, the DC $N(R)\zeta N(R)$ are characterized by partitions (λ) of p. (3.4.1)

A canonical set $\{\zeta_{(\lambda)} \,/\, \forall(\lambda)\}$ is obtained by the prescription: (3.4.2)

(*a*) Order the cycle structure (λ) according to cycle lengths, and

(*b*) fill in the odd indices in natural order,

A particular $\zeta_{(\lambda)}$ can be represented by an *ordered-1-row symbol*

$$(\ldots k_\lambda \ldots) \quad \left.\begin{matrix} k=p,p-1\ldots,1 \\ \lambda=1,2\ldots,\lambda_k\forall k \end{matrix}\right\} \tag{3.4.3}$$

From [3]:

$$|R\cap\zeta_{(\lambda)}R\bar{\zeta}_{(\lambda)}|=2^{\lambda_1},|R\zeta_{(\lambda)}R\cap C|=2^{n_\lambda-\lambda_1}. \tag{3.4.4}$$

Moreover:

$$\text{sgn}\,s=+1\forall s\in N(R)/R \quad \text{implying } \text{sgn}(\gamma)=\text{sgn}(\zeta)\forall\gamma\approx\zeta$$

so that

$$PNP = \sum_{(\lambda)\in p} 2^{n_\lambda}\text{sgn}(\zeta_{(\lambda)}) \sum_{f\approx\zeta_{(\lambda)}} f \tag{3.4.5}$$

in which the second summation is over all different $f\approx\zeta_{(\lambda)}$.

5 *PNP, n>2p*

Again, the generators ζ in (2.3.4) are chosen from C, say $\zeta=w_jv_k$. Since $w_jv_k=$ $=w_j\bar{u}_ku_kv_k=w_j\bar{u}_ks_k\approx w_j\bar{u}_k=w'\in S'_{n-p}$, the generators may be chosen from the first column. If w' contains cycles with two or more β-indices there always exist $t\in S_\beta$ such that any cycle in w't=w contains at most one β-index. The cycles of w can be ordered according to the number of α'-indices they contain, and within a set of cycles having the same number of α'-indices they can be ordered according to the presence of a β-index. As a generalization of the previously introduced (trivial) 1-row symbols we define a set of canonical generators by means of *ordered 2-row symbols*, [3]:

$$\begin{pmatrix} \ldots k_\lambda & \\ & k'_\lambda\ldots \end{pmatrix} \quad \begin{matrix} k=p,p-1\ldots,1 \;;\; \lambda=1,2\ldots\lambda_k\forall k \\ 1\ge k'_1\ge k'_2\ge\ldots\ge k'_{\lambda_k}\ge 0\forall k, \end{matrix} \tag{3.5.1}$$

λ ranges over the partitions of p. The associated canonical generator is obtained as follows: (3.5.2)

(*a*) The column $\begin{pmatrix} k_\lambda \\ k'_\lambda \end{pmatrix}$ represents $\begin{matrix} (\alpha'_{1\lambda},\alpha'_{2\lambda}\ldots,\alpha'_{k\lambda}) & \text{if } k'_\lambda=0 \\ (\beta,\alpha'_{1\lambda})(\alpha'_{1\lambda},\alpha'_{2\lambda}\ldots,\alpha'_{k\lambda}) & \text{if } k'_\lambda=1 \end{matrix}$

(*b*) The α'- as well as the β-indices are naturally ordered starting with the first column of the symbol.

Thus, a canonical generator is defined by a partition (λ) of p and a set of binary partitions (λ'_k,λ''_k) of λ_k $\forall k$ in which λ'_k is the number of times 1 appears in the k-th part of the second row. We denote a particular generator by $\zeta_{(\lambda,\vec{\lambda}'')}$ and assert, [3]:

The set of canonical generators $\{\zeta_{(\lambda,\vec{\lambda}'')}$ / $\forall\lambda$, $\vec{\lambda}''$ is complete for the DC decomposition of PNP wrt N(R) (3.5.3)

From [3]: $|R\cap\zeta_{(\lambda,\vec{\lambda}'')}R\bar{\zeta}_{(\lambda,\vec{\lambda}'')}| = 2^{\lambda''_1}$, $|R\zeta_{(\lambda,\vec{\lambda}'')}R\cap C| = 2^{n_{\lambda''}-\lambda''_1}$; (3.5.4)

one has $\sigma_f = 2^{n_{\lambda''}}$ $\forall f\approx\zeta_{(\lambda,\vec{\lambda}'')}$, up to sign. (3.5.5)

Note: $n_{\lambda''}$ is the number of zeroes in the second row.

Since, $S_\beta\subset N(R)/R$ and sgnt=±1 some phase is involved namely α_f=sgn(γ) if f$\sim\gamma$ so that

$$PNP = \sum_{(\lambda)\in p}\ \sum_{\vec{\lambda}''} 2^{n_{\lambda''}} \sum_{f\approx\zeta_{(\lambda,\vec{\lambda}'')}} \alpha_f f \tag{3.5.6}$$

6 *DC-DECOMPOSITION OF S_n WRT N(R), 3(4)-ROW SYMBOLS.* [3].

The decomposition of S_n wrt N(R) might also be of interest. In a Wigner SA procedure e.g. (2.1.6) all elements of S_n are used, as opposed to a Young SA procedure. We present only the main results.

A set of canonical generators for the decomposition of S_n wrt N(R) can be defined by means of *ordered 3-row symbols* which appear as straightforward generalizations of the above-defined 2-row symbols, (3.5.2). (3.6.1)

$$\begin{pmatrix} \ldots k_\lambda & \\ k'_\lambda & \\ k''_\lambda & \ldots \end{pmatrix} \quad \begin{matrix} k=p,p-1\ldots,1\ ;\ \lambda=1,2\ldots,\lambda_k\forall k \\ 1\geq k'_1\geq k'_2\geq\ldots\geq k'_{\lambda_k}\geq 0\ \forall k \\ k\geq k'_1\geq k'_2\geq\ldots\geq k''_{\lambda_k}\geq 0\ \forall k\ ;\ k''_\lambda=0 \text{ if } k'_\lambda=0 \end{matrix}$$

Again, λ ranges over the partitions of p. This ordered 3-row symbol represents a canonical generator, obtained as follows:

a The column $\begin{pmatrix} k_\lambda \\ k'_\lambda \\ k''_\lambda \end{pmatrix}$ represents $\begin{matrix} (\alpha'_{1\lambda},\alpha'_{2\lambda}\ldots,\alpha'_{k\lambda}) & \text{if } k'_\lambda=k''_\lambda=0 \\ (\beta',\alpha'_{1\lambda})(\alpha'_{1\lambda},\alpha'_{2\lambda}\ldots,\alpha'_{k\lambda}) & \text{if } k'_\lambda=1,\ k''_\lambda=0 \\ (\beta',\alpha'_{1\lambda})(\alpha'_{1\lambda},\alpha'_{2\lambda}\ldots,\alpha'_{k\lambda})(\alpha''_{k''\lambda},\beta'') & \text{if } k'_\lambda=1,\ k''_\lambda\neq 0 \end{matrix}$

in which $\alpha'_{k''\lambda}$ and $\alpha''_{k''\lambda}$ are corresponding indices, and

b the α'- as well as the β-indices are filled in in natural order from the left to the right starting with the first column.

We assert: (3.6.3)

The set of canonical generators is complete for the decomposition of S_n wrt N(R) if p<4 and all n≥2p, *or* n<2p+4 and all p≥1. In all other cases the set is overcomplete.

The redundancies are spurious however and can easily be eliminated. Extend any 3-row symbol by an auxiliary 4-th row the entries of which are given by $k_\lambda - k''_\lambda$ if $k''_\lambda \neq 0$, by 0 if $k''_\lambda = 0$ and apply the rule, [3]:

4-Row symbols which have equal third rows (up to the order of the entries) *and* equal fourth rows (again up to order) represent one and the same DC wrt N(R). From a set of such symbols select the one which has an ordered fourth row to arrive at a set of 3(4)-row symbols which is precisely complete.

In any application which makes use of DC the frequencies of the DC always enter the description. If ζ is a canonical generator associated with a 4-row symbol which has M columns, $N_k^{(1)}$ columns {k,0,0}, $N_k^{(2)}$ columns {k,1,0}, $M_k^{(a)}$ entries k in the a-th row and $M_\phi^{(a)}$ a-th row entries unequal to 0, then $d_\zeta = |N(R) \cap \zeta\, N(R)\, \bar{\zeta}\,|$ is given by

$$2^{M_0^{(2)}+2M_\phi^{(3)}} (M_0^{(4)}-M_0^{(3)})!\,(n-2p-M_\phi^{(2)}-M_\phi^{(3)})!\,\prod_{k=1}^{p} k^{N_k^{(1)}}\, N_k^{(1)}!\,N_k^{(2)}!\,M_k^{(3)}!\,M_k^{(4)}! \tag{3.6.4}$$

The associated Pauling number is $2^{M_0^{(2)}}\,\delta(M_0^{(3)},M)$, up to sign. (3.6.5)

Note: From section (3.5) and expression (3.6.5) it follows that only canonical generators represented by 3-row symbols with a vanishing third row enter the equation derived for PNP. (3.5.6). Thus the reduction in the number of Pauling numbers to be calculated will be even more pronounced than the examples at the end of section (2.3) show. Indeed, for the case $(2^4 1^8)$ one has $|\{\zeta_{(\lambda,\vec{\lambda}'')}\}| = 20$ as compared with $|\{\zeta\}| = 69$.

7 *PNP PAULING NUMBER FOR ARBITRARY* $f \in S_n$.

Depending on the specific property of the primitive product function it may be convenient to determine σ_f for each f separately, e.g. the use of orthogonal orbitals leads to the survival of transpositions only. With $f \in N(R)\ \zeta\ N(R)$ *if and only if* $fc\bar{f}\bar{c} = n\zeta c\bar{\zeta}\bar{c}\bar{n}$ for some $n \in N(R)$, $c \in Z(N(R))$ and the uniqueness of the cycle structure of the product $\zeta c\bar{\zeta}\bar{c}$, it is easy to read $n_1\zeta\bar{n}_1$ with $f = n_1\zeta n_2$ directly from $fc\bar{f}\bar{c}$. From the definition of ζ (3.6.2) and the associated 3-row symbol (3.6.1) one can infer that the product $\zeta c\bar{\zeta}\bar{c}$ contains $M_\phi^{(3)}$ cycles in which two β-indices occur. Also, if and only if $M_\phi^{(3)} = 0$, this product contains $M_1^{(2)}$ cycles with 1β- and $2M_0^{(2)}$ cycles, including cycles of length 1, with no β-index. The same holds for $fc\bar{f}\bar{c}$, any $f \in S_n$, from which:

$$|\sigma_f| = 2^{M_0^{(2)}}\,\delta(M_\phi^{(3)},0) \tag{3.7.1}$$

in which $M_\phi^{(3)}$ and $2M_0^{(2)}$ are the numbers of cycles in $fc\bar{f}\bar{c}$ which contain 2 β-indices and no β-indices respectively.

The sign of $|\sigma_f|$ can be determined as follows: (3.7.2)

Calculate $n_1\bar{\zeta}\bar{n}_1 f = n_1 n_2$; this product contains cycles in which no β-indices occur, and cycles in which only β-indices occur, the latter constituting $t \in S_\beta$. The sign of $|\sigma_f|$ is given by sgn (ζt).

These rules do *not* agree with the general results of ref.[13].

IV *CONCLUSION AND DISCUSSION*

The DC decomposition of the symmetric group S_n wrt the normalizer of subgroups defined by partitions $(2^p 1^{n-2p})$ is presented, (3.6).
A set of canonical generators is defined by means of ordered 3(4)-row symbols which carry all the information about the associated DC. They are very convenient in actual applications.
For the application considered in this paper, i.e., matrix representations on Young-SA-bases, only those symbols with a zero third row i.e., 2-row symbols, are needed.
The algebraic problem arising in the description of an n-electron system i.e. the calculation of Pauling numbers, is solved. The solutions are expressed explicitly in terms of the entries of the 2-row symbols, (3.5).
For applications which involve a limited number of group elements an efficient method is given for determining the Pauling number for any given individual element, (3.7).
Partitions $(\lambda)=\ldots j^{\lambda_j}\ldots, j>2$ present additional intricacies:
(*a*) The choice of a set of canonical generators is less obvious on account of the fact that $Z(N(R))=e$ in general. However, there always exists a non trivial $R'\subset R$ the elements of which commute elementwise with the elements of $N(R)/R$, ($R'=\{e,c\}$ for the case of section III), which may be of help in establishing a canonical set of elements.
(*b*) The simplification (2.3.6) i.e., all vertical permutations $q'\in RqR$ have equal sgn, is not applicable in general. The DC-symbol or arrow-diagram technique may be of help to overcome this inconvenience.

References:

[1] See e.g. Salmon, W.I., "Advances in Quantum Chemistry, Vol. 8", 1974. Academic Press, New York, London, and refs. therein. [2] See e.g. Huppert, B., "Endliche Gruppen, I", 1967. Springer-Verlag, Berlin, Heidelberg, New York. [3] Roël, R.W.J., "Thesis", 1975, Amsterdam. [4] Ruch, N., Theor. Chim. Acta 19, 288 (1970). [5] Seligman, H., "Hab. Schrift", 1974, Tübingen. [6] Matsen, F.A., Cantu, A.A., Poshusta, R.D., J. Phys. Chem. 70, 1558 (1966). [7] Murnaghan, F.D., "The Theory of Group Representations", 1963. Dover Publications, Inc., New York. [8] Mc.Intosh, H.V., J. Math. Phys. 1, 453 (1960). [9] Goddard, W.A., Phys. Rev. 157, 73 (1967). [10] Harris, R.E., J. Chem. Phys. 46, 2769 (1967). [11] Heldman, G., Int. J. Quantum Chem. 2, 785 (1968). [12] Gallup, G.A., J. Chem. Phys. 50, 1206 (1969). [13] Klein, D.J., Junker, B.R., J. Chem. Phys. 54, 4290 (1971); 55, 5532 (1971). [14] Löwdin, P.O., Phys. Rev. 97, 1509 (1955). [15] Pauling, L., J. Chem. Phys. 1, 280 (1933). [16] Shull, H., J. Chem. Phys. 3, 523 (1969).

Applications of Group Theory to Nuclear Reactions : A Critical Survey [+]

T.H. Seligman

Institut für Theoretische Physik
der Universität zu Köln
5 Köln 41, Germany

I. Introduction

In recent years the group theoretical structure of the resonating group method (1) and its extensions (2) has been analyzed in detail (3 - 11) . We recall that these methods (2) decompose a system into substructures often called clusters for which a simple ansatz is made. The Schroedinger equation for n particles is reduced to an approximative equation for the relative motion modes of the clusters by integration over the internal variables . Some nuclear structure calculations were carried out in this way but the main field of applications is that of nuclear reactions of complex systems where this approach is essential . The range of application reaches from polarization studies of reactions in light systems (12) to heavy ion reactions (13) .

Yet these applications do not make use of any of the group theoretical techniques described in (3 - 11) and it seems therefore necessary to ask three critical questions which I believe should always be asked when a mathematical formalism is introduced into a theory in physics :

1) What physical insight into the problem does the formalism yield ?
2) What advantages does the formalism yield for computations of results that compare to experiment ?
3) To what extent have these advantages been obtained without the formalism by other ('pedestrian') techniques ?

Under this point of view we shall consider the applications of the group theoretical methods developed in (3 - 11) , while we shall not consider the obvious applications of Racah algebra . We devide these methods into three categories : First, the simplification of antisymmetrization by double coset decompositions of S (n) (3 - 5) , second, the expansion of

[+] Work supported by the Bundesministerium für Forschung und Technologie, Bonn-Bad Godesberg reaction channels into supermultiplets (4,6) and finally the use of complex canonical transforms for the evaluation of exchange integrals (7 - 11) . Upon this last point we shall dwell somewhat longer to present some new results about the interpretation of the Hill-Wheeler and similar transforms useful in cluster theory (14 - 16) in terms of complex canonical transforms (10,11) .

II. Antisymmetrization and double cosets

As this is the oldest of the methods discussed we shall not describe it in detail but only outline the main idea : The cluster ansatz splits the compound system into fragments which are taken to have good permutational symmetry (antisymmetric if the entire function is considered and a definite supermultiplet if as usual the orbital functions are discussed) . The ansatz is thus symmetry adapted to a subgroup of S (n) and the elements of this subgroup may be applied readily to bra and ket whenever matrix elements of permutations are formed . Thus only permutations that are in different double cosets of S(n) with respect to the symmetry groups for the ansatz in bra and ket will yield different integrals . The double cosets characterize the exchange-type and are in turn characterized by certain matrices of positive integers called DC - symbols .

The i, j - th matrix element of a DC - symbol indicates the number of particles transformed from cluster j to cluster i by a given permutation ; DC - symbols thus simplifying the physical interpretation of exchange terms . Also they simplify the task of antisymmetrization considerably. So questions 1 and 2 are positively answered . As far as 3 is concerned, we must mention that DC - symbols were used (17) under the name of coarse permutation matrices before the corresponding group theoretical work (3 - 5) was done and indeed were taken over from (17) in (3,4) . Nevertheless the full use of group theory including fractional parentage coefficients (18) for the spin isospin part and the normalizer of the symmetry group of the ansatz for the orbital part (5) yield improvements over the method of (17) in complicated cases .

III. Supermultiplet expansions of reaction channels

The supermultiplet expansion of reaction channels (4,6) assumes given supermultiplets or orbital permutational symmetry for the fragments . In the case of two particle channels the asymptotic wave functions are

$$\Psi_A = \left[\frac{n!}{n_a!\,n_b!}\right]^{1/2} \mathcal{A}\, |\alpha^{n_a}\{1^{n_a}\}\, a\, S_a\, M_{S_a}\, T_a\, M_{T_a}\rangle\, |\alpha^{n_b}\{1^{n_b}\}\, b\, S_b\, M_{S_b}\, T_b\, M_{T_b}\rangle\, X(R_a - R_b)$$

where α^{n_λ} stands for configuration of n_λ particles with orbital symmetry λ, spin S_λ, spin projection M_{S_λ}, Isospin T_λ, Isospin projection M_{T_λ}; $\lambda = a, b$ and a relative motion function $X(R_a - R_b)$. $\mathcal{A}$ is the antisymmetrizer.

Ψ_A may be rewritten as

$$\Psi_A = \left[\frac{n!\,|a|\,|b|}{n_a!\,n_b!}\right]^{1/2} \sum_{STF\varphi} |f| \langle \varphi \hat{F} S T \{| \hat{a}\, S_a T_a\, \hat{b}\, \hat{S}_b\, T_b \rangle$$

$$\langle S\, M_{S_a} + M_{S_b} \,|\, S_a\, M_{S_a}\, S_b\, M_{S_b} \rangle \langle T\, M_{T_a} + M_{T_b} \,|\, T_a\, M_{T_a}\, T_b\, M_{T_b} \rangle$$

$$|\alpha^n \{1^n\}\, (a\, b\, \varphi)\, f\, S\, M_{S_a} + M_{S_b}\, T\, M_{T_a} + M_{T_b} \rangle$$

where f characterizes an IR of S(n) and is the channel symmetry as S is the channel spin and T the channel isospin, and φ a multiplicity label. $\langle \{| \rangle$ indicates a spin-isospin fractional parentage coefficient and $\hat{}$ denotes an associate representation for the permutation group (18). The state on the right hand side finally is an antisymmetric two-cluster state with good orbital symmetry, whose matrix elements may be calculated by methods described in (3-5) as mentioned in Section II.

We note that the expansion in supermultiplets is quite analogous to that in channel spin and isospin. Indeed this analogy is further emphasized if we consider the SU (4) interpretation where the above expansion is nothing but a Clebsch-Gordan series for SU (4) in a chain SU (4) ⊃ SU (2) x SU (2). The fractional parentage coefficients are essentially reduced Wigner-coefficients of SU (4) (19).

It turns out that as far as computations go, this method is not useful because the cases considered are too simple. On the other hand, the concept of supermultiplet channels has been of considerable importance for the study of certain nuclear reactions. Their reaction mechanisms are understood by assuming dominance of one of such channels. This occurs for both resonances (20,21) and doorway states (22,10). As these applications are quite new one may expect other cases to be found, but already several are understood that way and e.g. the prediction (21) of high spin threshold states in the ^9B, ^9Be, ^{13}C and ^{13}N compound systems have been partially verified by experiment since their publication (23).

IV. Integral transforms, canonical transforms and exchange integrals

If we use in the cluster ansatz harmonic oscillator functions (possibly of different width) and expand the interaction in terms of Gaussians, derivatives, powers and spherical harmonics, the exchange integrals may be evaluated analytically (17). For larger systems it proved convenient to use integral transforms in this process, for oscillator functions of equal frequencies (7,8,13-16). Recently, this restriction could be lifted (9-11, 24).

Complex canonical transforms enter the picture in two ways. First the operators occurring may be interpreted as such. This is seen from the example of a Gaussian interaction corresponding to a Moshinsky-Quesne (10,25) representation of a complex transform as (8)

$$e^{-\alpha x^2} \Leftrightarrow \begin{pmatrix} 1 & 0 \\ 2i\alpha & 1 \end{pmatrix}$$

But even point transforms turn complex after integration over the internal variables (7,8). The results of (8) and (9) concerning this point were reported at this conference (26), in a formulation that uses the Bargmann - Segal transform.

We shall therefore focus our attention to more recent developments that include the integral transforms into the picture. Recently Zahn (10) has shown, that the Hackenbroich transform (15) can be interpreted as a complex canonical transform in the Moshinsky-Quesne representation. Indeed this result can be generalized and the correspondences are listed in table I for three useful transforms (14-16) including the Hill-Wheeler transform.

The main idea is to drop the former restriction (8) to bounded operators, which turns out to be unnecessary. Indeed in the cases examined either the transform, its 'inverse' or both are unbounded on $L^2(R)$, but may be defined (27) on a dense subspace of appropriate exponential growth. The inverse is given by the Moshinsky-Quesne representation of the inverse complex canonical transformation.

A similar interpretation can be given to the Bargmann transform according to Wolf (28), but the connection with the complex transforms resulting from the potentials is somewhat artificial due to the changes of measure. This correspondence is also included in Table I.

	integral transform	compl. can. transformation	type of operator		change of measure
			transform	inverse	
Hill - Wheeler (14)	$\left(\frac{\alpha}{2\pi}\right)^{1/2} \exp\left\{\frac{-\alpha}{2}(x-x')^2\right\}$	$\begin{pmatrix} 1 & \frac{1}{i\alpha} \\ 0 & 1 \end{pmatrix}$	bounded	unbounded	no
Hackenbroich - Wiedmann (15)	$(\pi)^{-1/2} \exp\left\{\frac{-\alpha}{2}x^2 + 2ixx' + \frac{1}{\alpha}x'^2\right\}$	$\begin{pmatrix} \frac{i\alpha}{2} & \frac{1}{2} \\ -1 & \frac{-i}{\alpha} \end{pmatrix}$	unbounded	unbounded	no
Sünkel - Wildermuth (16)	$\left(\frac{\alpha}{2\pi}\right)^{1/2} \exp\left\{\frac{-\alpha}{2}(x-ix')^2\right\}$	$\begin{pmatrix} i & \frac{-1}{\alpha} \\ 0 & -i \end{pmatrix}$	unbounded	unbounded	no
Bargmann - Segal (Wolf 28)	$(\pi)^{-1/4} \exp\left\{\frac{-1}{2}x^2 - \frac{1}{2}z^2 + \sqrt{2}\,xz\right\}$	$\frac{1}{\sqrt{2}}\begin{pmatrix} 1 & -i \\ -i & 1 \end{pmatrix}$	unitary	unitary	yes

Table I The complex canonical transformations corresponding to four integral transforms as well as the operator properties of the transforms are listed

Using these results it is therefore possible to describe an even larger portion of the problem in terms of complex canonical transforms as the entire kernel including integral transforms appears in that way . Yet we must keep in mind that only integrals over Gaussians or operators and functions generated by Gaussians are carried out . These integrals can also be obtained by the standard method of completing squares and indeed in certain special cases (9) only this method works . Also for the kinetic energy and even more so for spin-orbit coupling the detour over the corresponding generating transforms (8,9) is inadequate for computational purposes . Points 2 and 3 thus find a negative answer and it remains to be asked whether we gain any physical insight by the use of complex canonical transforms . To this question a definite answer does not yet exist . But one should be sceptical due to the necessity of differentiations that do not fit into the framework . Also we actually need linear combinations of many terms .

Much more important is the following consideration : The main point of the resonating group method is actually the reduction of an n-particle to a $\varkappa$ - cluster problem, i.e. the elimination of dynamical variables . Formally this may be viewed as integration or projection ; in practice it amounts to the use of the generating function properties of the integral transform .

To allow important physical insight, the method of complex canonical transforms would have to include this point, but up to now it doesn't . Yet there is hope to include it if we consider diastrophic canonical transformation introduced by Boiteux (29) .

These are defined roughly as follows : Consider a set of canonical variables $q_i', p_i' ; i = 1 \ldots n$ that fulfill

$$[q_i', p_j'] f = i \delta_{ij} f \quad ; \quad f \in D^n$$

where D^n is a dense subspace of a Hilbert space over n variables, and another set $q_\alpha, p_\alpha, \alpha = 1 \ldots n+1$ with

$$[q_\alpha, p_\beta] F = i \delta_{\alpha\beta} F \quad ; \quad F \in D^n$$

Then the transformation

$$(q_i', p_i' ; i = 1 \ldots n) \longrightarrow (q_\alpha, p_\alpha ; \alpha = 1 \ldots n+1)$$

is called a diastrophic canonical transformation . Boiteux points out that a problem in D^n may be transformed into a problem in D^{n+1} with a superselection rule .

Our problem is in a sense inverse as we wish to drop rather than to add dynamical variables . But we could view the $\mathcal{R}$-cluster problem as related to the n-particle problem by a diastrophic canonical transformation plus a superselection rule consisting in the restriction of the internal cluster functions . This superselection rule would only hold approximatively in the real n-particle problem .

The proposed procedure appears straight forward and we hope to supply a convenient formalism soon . The question whether this interpretation is useful must be postponed until the formal work is completed .

I wish to thank M. Boiteux, H.H. Hackenbroich and W. Zahn for stimulating discussions and advice .

1) J.A. Wheeler , Phys. Rev. 52 (1937) 1083

2) Clustering Phenomena in Nuclei, Bochum, 1969 (IAEA, Vienna , 1969)

3) P. Kramer, T.H. Seligman , Nucl. Phys. A 136 (1969) 545 , A 186 (1972) 49

4) T.H. Seligman BMBW FB K-69 42 , Karlsruhe , 1969

5) T.H. Seligman , Rev. Méx. Fiz 22 (1973) 151

6) G. John, T.H. Seligman , Nucl. Phys. A 236 (1974) 397

7) P. Kramer, D. Schenzle , Rev. Méx. Fiz 22 (1973) 25

8) P. Kramer, M. Moshinsky, T.H. Seligman in Group Theory and its Applications III, Ed. E.M. Loebl, Academic Press , New York , 1975

9) T.H. Seligman and W. Zahn (submitted for publication)

10) W. Zahn "Kollektive Moden und Integraltransformationen in der Clustertheorie" Burg Verlag, Basel , 1975

11) T.H. Seligman, W. Zahn (to be published)

12) H.H. Hackenbroich in Clustering Phenomena in Nuclei, Maryland , 1975

13) H. Friedrich , Nucl. Phys. A 224 (1974) 537

14) D.L. Hill , J.A. Wheeler , Phys. Rev. 89 (1953) 1102

15) B. Wiedmann, Diplomarbeit Köln , 1972

16) W. Sünkel, K. Wildermuth , Phys. Lett. 41 B (1972) 439

17) H.H. Hackenbroich , Z. Physik 231 (1970) 266

18) J.P. Elliot, J. Hope and H.A. Jahn , Phil. Trans. Roy. Soc. A 246 (1953) 241

19) P. Kramer and M. Moshinsky in Group Theory and its Applications Ed. E.M. Loebl , Academic Press, New York , 1968

20) H.H. Hackenbroich , T.H. Seligman , Phys. Letts. 41 B (1972) 102

21) J. Niewisch, D. Fick (to be published)

22) W. Gruhle, T. Bauer, T.H. Seligman, H.H. Hackenbroich , Z. Physik 262 (1973) 271

23) F. Seiler , Nucl. Phys. A 224 (1975) 236

24) B. Giraud, L. Le Tourneux , Nucl. Phys. A 240 (1975) 365

25) M. Moshinsky and C. Quesne , J. Math. Phys. 12 (1971) 1772

26) G. John, P. Kramer , These Proceedings

27) L. Martignon and T.H. Seligman (to be published)

28) K.B. Wolf, J. Math. Phys. 15 (1974) 1295

29) M. Boiteux , Physics 75 (1974) 603

MATHEMATICAL PHYSICS

THE CANONICAL RESOLUTION OF THE MULTIPLICITY PROBLEM FOR U(3): AN EXPLICIT AND COMPLETE CONSTRUCTIVE SOLUTION

L. C. Biedenharn*, M. A. Lohe†, and J. D. Louck‡

ABSTRACT. The multiplicity problem for tensor operators in U(3) has a unique (canonical) resolution which we utilize to effect the explicit construction of all U(3) Wigner and Racah coefficients. We employ methods which elucidate the structure of the results; in particular, we describe the significance of the denominator functions entering the structure of these coefficients, and the relation of these denominator functions to the null space of the canonical tensor operators. An interesting feature of the denominator functions is the appearance of new, group theoretical, polynomials exhibiting several remarkable and quite unexpected properties.

INTRODUCTION. At a conference on group theoretical methods, it is quite unnecessary for us to emphasize the basic importance of tensor operator constructions in the unitary groups: this is the well-known problem of generalizing to U(n) the famous Racah-Wigner calculus.[1] The basic difficulty in this program is equally well-known: to find the appropriate concepts for a canonical resolution of the multiplicity problem. (Here canonical means: free of all arbitrary choices to within suitably defined equivalence.)

The purpose of the present discussion is to announce the explicit constructive solution to this problem for U(3). We shall present results tantamount to the algebraic determination of all Wigner and Racah function. (The fact that this canonical resolution existed was demonstrated several years ago.[2]) To comply with the conference format, we shall limit our discussion to a single, but fundamental, element in the construction: the denominator function. This structural element is, as we shall discuss, not only an invariant characterization of the operator, but also defines the class of the boundary Racah functions completely.[3]

Although it is clearly useful to have explicit algebraic results available, in our view the more important problem lies in interpreting the structure characteristic of the canonical multiplicity splitting. It is precisely here that our results are most interesting, perhaps even surprising. We shall show, in fact, that the canonical resolution re-introduces the concept of a weight space, but in an unexpected context.

NULL SPACE. The existence proof for the canonical splitting of the multiplicity in U(3) was based on the concepts of the pattern calculus[4] and the geometrical properties of arrow patterns[5] ("flow patterns"). It was found later[6] that precisely the same canonical splitting could be achieved by employing a more general concept, that of the null space of a tensor operator. We define: the null space of an operator is the set of vectors annihilated by the operator.

Consider the action of an operator T characterized by the tensor operator labels $[M] = [M_{13}M_{23}M_{33}]$. That is, $T: \mathcal{H} \to \mathcal{H}$, $T \varepsilon [M]$. From the Kronecker product structure, one knows that the action of T on vectors of $\mathcal{H}$ belonging to the irrep space having U(3) labels $[m] = [m_{13}m_{23}m_{33}]$ may be characterized by the intertwining number $I([m^f];[M],[m^i])$. For SU(2) the intertwining number is either one or zero, but for SU(3) the intertwining number may be any non-negative integer. The 'multiplicity problem' is then to find a canonical partitioning of the set of operators $\{T:T\varepsilon[M]\}$ into subsets having the intertwining number one or zero.

It is useful at this point to introduce some notations to facilitate the discussion. We first introduce the partial hooks, $p_{i3} = m_{i3} + 3 - i$ $(i = 1,2,3)$, in place of the labels m_{i3} of an irrep space of $\mathcal{H}$. With each triple (p_{13},p_{23},p_{33}) we next associate the triple of symmetric variables (x_1,x_2,x_3), where $x_i = p_{j3} - p_{k3}$ and (ijk) is a positive permutation of 1,2,3.

In view of the relation $\sum_i x_i = 0$, the variables $(x) = (x_1,x_2,x_3)$ are coordinates in a two-dimensional plane (Möbius plane, related to barycentric coordinates).

In view of our association $[m] \to (x)$, one sees that each U(3) irrep is mapped onto one SU(3) irrep which, in turn, is mapped to one lattice point in the Möbius plane. Choosing the canonical ordering $m_{13} \geqslant m_{23} \geqslant m_{33}$, determines the lexical region of the Möbius plane (one-sixth of the plane).

Consider now the set of irreducible tensor operators $T_{[M]}$ transforming in the same manner as the irrep space having labels [M]. This set of operators is first partitioned by the shift labels $(\Delta_1\ \Delta_2\ \Delta_3)$, where $\Delta_i = m_{i3}^{\text{final}} - m_{i3}^{\text{initial}}$. That is, $T_M^{(\Delta)}$: $[m^i] \to [m^f]$.

<u>The multiplicity problem may now be stated in the form</u>: <u>How do we distinguish between operators $\{T_{[M]}^{(\Delta)}\}$ having the same Δ-pattern</u>, $[\Delta_1\ \Delta_2\ \Delta_3]$? The key to the answer lies in the properties of the intertwining number $I([m^f];[M],[m^i]$ written in the form $I([m] + [\Delta];[M],[m])$, where we now write $[m^i] = [m]$ and $[m'] = [m^f] = [m] + [\Delta]$, and $[\Delta]$ may be any specified weight of irrep [M]. The idea is to determine the values of $I([m] + [\Delta];[M],[m])$, where $[\Delta]$ and [M] are specified (since they are associated with the set of operators $T_{[M]}^{(\Delta)}$), and the labels [m] of an irrep space of $\mathcal{H}$ can assume all integral values consistent with $m_{13} \geqslant m_{23} \geqslant m_{33}$. Since the intertwining number considered in this form depends, in fact, only on the coordinates (x), the set of values $\{I([m] + [\Delta];[M],[m])\}$ is most appropriately displayed by exhibiting the intertwining number at each point of the Möbius plane.

The calculation of the intertwining number (in the form described above) as a function of (x) is a nontrivial matter, although techniques for this calculation are well-known.[8] We have found the following method of procedure to be best suited for our form of the intertwining number. We had earlier found[6] the result: the intertwining number I belonging to our triple m'Mm, i.e., the multiplicity of $[m] + [\Delta]$ in $[M] \times [m]$, where $[\Delta]$ is a Δ-pattern belonging to [M] is given by $\mathcal{I} = \mathfrak{M}_{123} + \mathfrak{M}_{231} + \mathfrak{M}_{312} - \mathfrak{M}_{213} - \mathfrak{M}_{132} - \mathfrak{M}_{321}$, where $\mathfrak{M}_{ijk}$ is the multiplicity of the Δ-pattern $[p_{i3} + \Delta_i, p_{j3} + \Delta_j, p_{k3} + \Delta_k] - [p_{13}p_{23}p_{33}]$ belonging to [M]. The calculation of I now becomes a fairly straightforward procedure when supplemented with the following closed formula giving the multiplicity $\mathfrak{M}$ of any Δ-pattern (weight), $(\Delta_1\ \Delta_2\ \Delta_3)$, of [M]:

$$\mathcal{M} = \begin{cases} (M_{23} - M_{33} + 1) - (\lambda_1 + \lambda_2 + \lambda_3) \\ 0 \text{ unless } \Delta_1 + \Delta_2 + \Delta_3 = \sum_i M_{i3} \text{ and } M_{13} \geqslant \Delta_i \geqslant M_{33} \end{cases},$$

where λ_i is the "step function" defined by $\lambda_i = \max(0, M_{23} - \Delta_i)$.

The results of this calculation are displayed in Fig. 1 which we call the Intertwining Number Null Space Diagram. The <u>level lines</u> of I appear as the bent lines designated by $I = 1, \cdots$, $I = k, \cdots$, $I = \mathcal{M} - 1$. The boundary lines are the bent lines designated by $I = 0$ and $I = \mathcal{M}$. I is zero at all lexical lattice points in the region II, and it is $\mathcal{M}$ at all lattice points in region I.

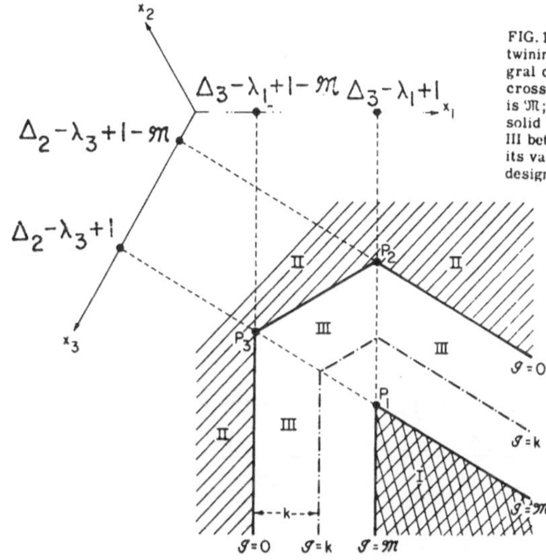

FIG. 1. The intertwining number-null space diagram. The intertwining number $\mathcal{I}$ is defined at each lattice point (points having integral coordinates) of the Möbius plane. At each lattice point in the cross-hatched region I, including the bent solid line, the value of $\mathcal{I}$ is $\mathfrak{M}$; at each lattice point in the shaded region II, including the bent solid line, the value of $\mathcal{I}$ is zero; at each lattice point in the region III between the two bent solid lines, the value of $\mathcal{I}$ is $1, 2, \ldots, \mathfrak{M} - 1$, its value being k at the lattice points on the bent dash-dot line designated by $\mathcal{I} = k$.

The first bit of information which may be read off from Fig. 1 is that there exist precisely $\mathcal{M}$ independent tensor operators in the set $\{T_M^{(\Delta)}\}$. Since $\mathcal{M}$ is the multiplicity of a weight, it follows that the independent tensor operators in this set may be enumerated by patterns (called operator patterns to distinguish them from the similar Gel'fand patterns) of the form

$$\begin{pmatrix} M_{13} & & M_{23} & & M_{33} \\ & \Gamma_{12} & & \Gamma_{22} & \\ & & \Gamma_{11} & & \end{pmatrix},$$

where the entries in this pattern obey the familiar "betweenness relations", $M_{13} \geqslant \Gamma_{12} \geqslant M_{23} \geqslant \Gamma_{22} \geqslant M_{33}$, $\Gamma_{12} \geqslant \Gamma_{11} \geqslant \Gamma_{22}$, and where the Δ-pattern of the triangular array is given by $\Delta_1 = \Gamma_{11}$, $\Delta_2 = \Gamma_{12} + \Gamma_{22} - \Gamma_{11}$, $\Delta_3 = \Sigma_i M_{i3} - \Gamma_{12} - \Gamma_{22}$. (Since $[\Delta]$ and $[M]$ are specified, Γ_{12} and Γ_{22} may vary over all values which satisfy the betweenness conditions and $\Gamma_{12} + \Gamma_{22} = \Delta_1 + \Delta_2$, thus enumerating the independent operators in $\{T^{(\Delta)}_{[M]}\}$.)

The preceding considerations leads one to introduce (following (Racah[1]) the set of unit tensor operators [7] (tensor operators which are orthogonal and normalized:

$$\left\{ \left\langle \begin{matrix} (\Gamma_t) \\ [M] \\ \cdot \end{matrix} \right\rangle : [\Delta(\Gamma_t)] = [\Delta] \right\} ,$$

where the operator pattern is inverted (a notational convenience), and (M) is a Gel'fand pattern specifying the transformation properties of one of these tensor operators. The operators in this set are, accordingly, enumerated by $\mathcal{M}$ distinct operator patterns $\Gamma_1, \Gamma_2, \cdots, \Gamma_{\mathcal{M}}$ of the form described above.

The second bit of information which may be read off from Fig. 1 is the null space of each of the unit tensor operators. Observe that the level line $I = \mathcal{M} - t$ partitions the lexical region of the plane into two parts (see Fig. 2): those points $\{(x)\}$ corresponding to those irrep spaces $\{[m]\}$ which are annihilated by the unit tensor operator labelled (Γ_t)--this set of irrep spaces defines the null space $\mathcal{N}_t$; and the set of points $\{(x)\}$ in the shaded region corresponding to those irrep spaces which are not annihilated by the unit tensor operator labelled (Γ_t). (These properties are discussed more completely in Ref. 6.) The principal result obtained from this analysis is: <u>The null spaces of the tensor operators labelled by</u> (Γ_t), $t = 1,2,\cdots,\mathcal{M}$ <u>are simply ordered by inclusion</u>. That is, $\mathcal{N}_1 \subset \mathcal{N}_2 \subset \cdots \subset \mathcal{N}_{\mathcal{M}}$.

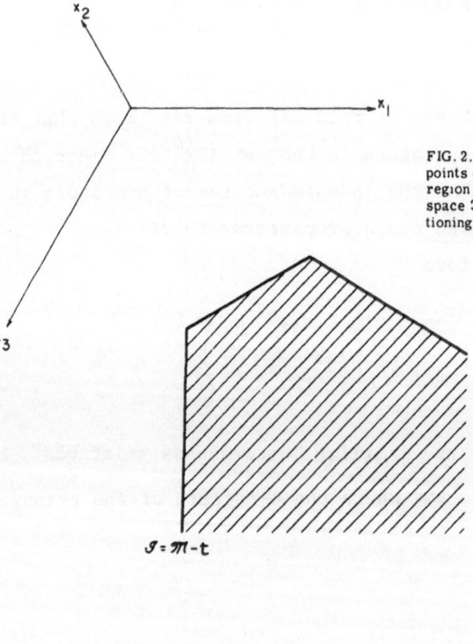

FIG. 2. The null space of the $U(3)$ Wigner operator (Γ_t). Lattice points on the bent solid line $\mathcal{I} = \mathcal{M} - t$ and exterior to the shaded region define the set of irrep labels $\{[m]\}$ which belong to the null space $\mathcal{N}_t$ of the Wigner operator designated by (Γ_t). The exact positioning of the solid line is determined from Fig. 1.

Remarks:

(1) The concept of a tensor operator inherently assigns a different rôle to each of the three representations associated with a given intertwining number.

(2) The canonical splitting of the multiplicity space is characterized by the entire null space associated with a given tensor operator; to examine the behavior at a given point in the Möbius plane is insufficient.

(3) The null space inclusion property does not assign the specific numerical operator pattern (Γ_t)--this is accomplished by yet another property (limits) discussed elsewhere.[3,6,9]

THE IMPORTANCE OF THE DENOMINATOR FUNCTION. Let us consider now the canonical tensor operators in U(3); we have mentioned in the previous section that such an operator carries the labelling

$$\left\langle \begin{matrix} (\Gamma) \\ [M] \\ (M) \end{matrix} \right\rangle \quad \begin{matrix} \leftarrow \text{operator pattern} \\ \leftarrow \text{irrep label (Young frame)} \\ \leftarrow \text{Gel'fand labels} \end{matrix} \quad .$$

The first simplification comes from splitting off the Gel'fand (state) labels (M) using the U(3) ⊃ U(2) subgroup decomposition. This defines the "isoscalar" factors,[10] or the U(3): U(2) projective operators:

$$\left\langle \begin{matrix} (\Gamma) \\ [M] \\ (M) \end{matrix} \right\rangle = \sum_{\gamma} \begin{bmatrix} (\Gamma) \\ [M] \\ (\gamma) \end{bmatrix} \left\langle \begin{matrix} & \gamma & \\ M_{12} & & M_{22} \\ & M_{11} & \end{matrix} \right\rangle .$$

Next one establishes a structural form for these projective operators:

$$\begin{bmatrix} (\Gamma) \\ [M] \\ (\gamma) \end{bmatrix} = \frac{(\mathrm{NPCF})}{D_3 \cdot D_2} \cdot f$$

where: (a) The Numerator Pattern Calculus Factor is a square root of linear factors known completely from the pattern calculus for all U(3):U(2) projective operators; (b) D_2 is a known denominator function involving the U(2) labels only; (c) D_3 is a denominator function involving the U(3) labels $(x_1 x_2 x_3)$ only; and (d) f is a function taking the value 1 for suitable parameter choices.

The significance of this structural form is that: (1) the function D_3 is well-defined--it is, in fact, the norm of the tensor operator (Γ), and (2) the denominator function D_3 vanishes at points determined by the null space of the operator.

This structural form is very complicated! How much of this complication is really essential, and how much is simply "group theoretic bookkeeping"? Put differently, what do we really need to know to determine the operator completely?

For SU(2), this question has been examined in detail[11,12] and the answer is surprisingly simple: knowledge of the denominator function alone (the numerator pattern calculus factors are always known) suffices to construct all Wigner functions and Racah operators! (This happens because the denominator functions suffice to define the fundamental Racah operators and these in turn suffice to generate all the remaining results.)

Although the situation for SU(3) is more complicated (the denominator functions of themselves do not suffice to determine all fundamental Racah functions), nevertheless we shall concentrate on the U(3) denominator functions, since they constitute definitive progress toward the complete answer.

SYSTEMATICS OF THE DENOMINATOR FUNCTION; THE WEIGHT SPACE PROPERTY: The zeroes of the denominator function are completely specified by the null space that defines the operator. In this section we will show that these zeroes have quite remarkable properties, and are, in fact, determined by SU(3) weight space patterns.

We begin by observing (cf. Fig. 1) that each line of zeroes lying in the lexical region determines a unique linear factor. For example, the denominator function for the $\langle 630\rangle$, $\Delta = (333)$ operator having maximum null space has six lines of zeroes: $\prod_{i=1}^{3} (x_1 - i)(x_3 - i)$.

Next one notes that relaxing the lexicality conditions allows one to define the denominator function over the entire Möbius plane, and that this extended function is symmetric under the group $S_3 \times Z_2$ (permutations of the $\{x_i\}$ and reflections: $x_i \to - x_i$.)

It follows that in this example, $\langle 630\rangle$, there are <u>18</u> linear factors in all.

Let us now remove these (known) linear factors and consider: $\mathcal{D} \equiv$ [Denominator]^2 $\times$ [product of linear factors].

From the fact that the null space contains, besides these 'null lines', a finite number of zeroes one concludes: the function $\mathcal{D}$ is, in the general case, <u>a ratio of polynomials</u>.

We have investigated the properties of $\mathcal{D}$ numerically using a computer and discovered to our surprise that the $\mathcal{D}$'s are ratios of a single family of polynomials G_q^t (discussed in the following section), each polynomial being characterized by an SU(3) weight space pattern of zeroes.

This result is most easily understood from examples. In Fig. 3 we illustrate the reduced denominator functions $\mathcal{D}$ for the four operators $\langle 6\overset{\Gamma_t}{3}0\rangle$, $\Delta = (333)$, $t = 1, 2,3,4$. Note that the <u>poles</u> of $\mathcal{D}$ (squares in the figure) are known (finite) zeroes in the null space, while the <u>zeroes</u> of $\mathcal{D}$ (circles) are at intersections of null lines. By putting in <u>cancelling zeroes and poles</u> one obtains the ratio of weight space patterns shown. This implies that $\mathcal{D}$ is a <u>ratio</u> of polynomials G_q^t: For the example shown in Fig. 3 one has

$$\mathcal{D}(\Gamma_1) = G_3^1 \Big/ (G_3^0 = 1) \quad ; \quad \mathcal{D}(\Gamma_2) = G_3^2 \Big/ G_3^1 \quad ;$$

$$\mathcal{D}(\Gamma_3) = G_3^3 \Big/ G_3^2 \quad ; \quad \mathcal{D}(\Gamma_4) = (G_3^4 = 1) \Big/ G_3^3 \; .$$

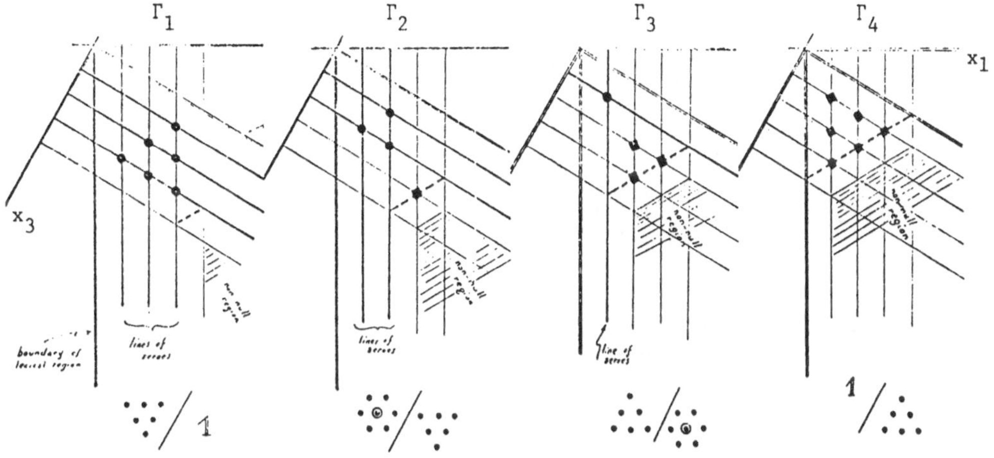

Fig. 3. The family of ⟨630⟩ Δ = (333) denominator functions, illustrating the weight space property.

We have proven that these properties are valid for the general case (the G_q^t being irreducible in general) and constitute a canonical resolution of the multiplicity.

It is of interest to note that the weight space patterns characterizing a given tensor operator multiplicity set $\{(\Gamma_t): t = 1,2,\cdots,\mathcal{M}\}$ are precisely that set of patterns having <u>diameter</u> $\mathcal{M} - 1$.

FUNCTIONS DEFINED BY WEIGHT SPACE ZEROES. The net result of the previous section has been to show that the denominator function defined by the canonical splitting involves functions defined by weight space zeroes. It is very natural to turn the question around and ask: can one construct polynomials in (x), of the proper symmetry and degree, which vanish precisely in the weight space pattern, and are positive, non-zero in the lexical region? A second question would be: are these functions unique?

We have been able to construct explicitly a set of polynomial functions $G_q^t(\Xi;x)$ $(t = 1,2,\cdots,q; q = 0,1,\cdots)$ which indeed show precisely the correct degree and weight space zeroes; we have verified that these functions possess many of the desired symmetry properties. These functions are:

$$G_q^t\begin{pmatrix} \xi_{11} & \xi_{12}+x_1 & \xi_{13}-x_1 \\ \xi_{21} & \xi_{22}+x_2 & \xi_{23}-x_2 \\ \xi_{31} & \xi_{32}+x_3 & \xi_{33}-x_3 \end{pmatrix} = G_q^t(\Xi;x) = \prod_{k=1}^{t}\frac{(q-k+1)!}{(k-1)!}\sum_{\lambda\mu\nu\rho} h(\lambda\mu\nu\rho)\,\frac{\prod_{s=1}^{t}(\Sigma+2t-q-s)_{\rho_s}}{M([\rho])}$$
$$\cdot F_{q-t+1,[\lambda]}(\xi_{11},\ \xi_{12}+x_1,\ \xi_{13}-x_1)$$
$$\cdot F_{q-t+1,[\mu]}(\xi_{21},\ \xi_{22}+x_2,\ \xi_{23}-x_2)$$
$$\cdot F_{q-t+1,[\nu]}(\xi_{31},\ \xi_{32}+x_3,\ \xi_{33}-x_3)\ .$$

In this result we have used the definitions:

$$\Sigma = \Sigma_i \xi_{ij} = \Sigma_j \xi_{ij} \text{ (magic square condition)} \quad ;$$

$h(\lambda\mu\nu\rho)$ = number of times irrep $[q-t+1\ q-t+1 \cdots q-t+1]$ of $U(t)$ is contained in the direct product representation $\lambda \times \mu \times \nu \times \rho$, where $\lambda = [\lambda_1 \lambda_2 \ldots \lambda_t]$, etc.;

$$M([\rho]) = \prod_{r<s=1}^{t} (\rho_r - \rho_s + s - r) \Big/ \prod_{s=1}^{t} (\rho_s + t - s)! \quad ;$$

$$F_{q-t+1,[\lambda]}(x,y,z) = (1/M([\lambda])) \prod_{s=1}^{t} [x+t-s]_{q-t+1-\lambda_s} [y+s-1]_{\lambda_s} [z+s-1]_{\lambda_s} \quad ;$$

$$(x)_a = x(x+1) \cdots (x+a-1)\ ,\ [x]_a = x(x-1) \cdots (x-a+1)\ .$$

We have been able to prove that the function $G_q^t(\Xi; x)$ vanishes on the set of points $\{(x_1,x_2,x_3)\}$, where $x_1 = \xi_{13} - i$, $x_2 = -\xi_{13} - \xi_{33} + q + 1 - 2t - j$, $x_3 = \xi_{33} - k$ in which (i,j,k) is any weight of the irrep $[q-t\ \ 0-t+1]$. Furthermore, the order of each zero equals the multiplicity of the corresponding weight.

In the $\langle 530 \rangle$, $\Delta = (333)$ example discussed earlier the polynomials in question are $G_3^t(\Xi;x)$ in which $\xi_{ij} = 4 - t$ and $t = 1,2,3$. $G_3^t(\Xi;x)$ then vanishes on the set of points $\{(4-t-i,\ -4-j,\ 4-t-k):\ \ (i,j,k)$ is a weight of $[3-t\ \ 0-t+1]\}$.

The polynomials $G_q^t(\Xi;x)$ are quite remarkable in their properties. We will confine ourselves to mentioning only two results related to these properties.

(1) The proof that the functions vanish on a weight space of zeroes hinges upon demonstrating a far-reaching generalization of the Saalschütz identity.[13] Let us define a generalization of the Gauss (hypergeometric) function:

$$F(abc;x) = \sum_{[\mu]} \langle F(abc) | [\mu] \rangle \langle [\mu] | x \rangle \quad ,$$

where $(x) = (x_1 x_2 x_3 \cdots x_t)$, $[\mu] = [\mu_1 \mu_2 \cdots \mu_t]$, the notation $\langle [\mu] | x \rangle$ denotes the classical Schur function,[8] and

$$\langle F(abc) | [\mu] \rangle = \frac{1}{M([\mu])} \prod_{s=1}^{t} \frac{(a-s+1)_{\mu_s} (b-s+1)_{\mu_s}}{(c-s+1)_{\mu_s}} \ .$$

The proof that the weight space zeroes actually occur is equivalent to showing that:

$$F(abc;x)\ F(c-a-b,\ b,\ b;\ x) = F(c-a,\ c-b,\ c;\ x) \quad ,$$

and this result has been demonstrated to be valid. One sees that this relation is a generalized version of the famous Kummer identity for the hypergeometric function.

Combining these results and using the properties of the Schur functions it is now straightforward to prove:

$$\sum_{[\mu],[\nu]} g(\mu\nu\lambda)\,\langle F(abc)|[\mu]\rangle\,\langle F(c-a-b,b,b)|[\nu]\rangle = \langle F(c-a,\ c-b,c)|[\lambda]\rangle \quad .$$

In this result $g(\mu\nu\lambda)$ is the multiplicity of λ in $\mu \times \nu$.

This relationship generalizes the Saalschütz identity.

(2) The second remarkable feature of the G_q^t is this. Observe that the definition of this function involves a highly non-trivial group-theoretic intertwining number $h(\lambda\mu\nu\rho)$. This intertwining number has a known computational algorithm, but this is not our point. The surprising result is that determining an SU(3) group-theoretic function G_q^t--defined canonically to split a U(3) operator multiplicity--requires of itself intertwining numbers from <u>arbitrarily large</u> U(t)! [Conversely, one may deduce U(t) information from the SU(3) results.] This linking together of all the U(n) groups was, to us, unexpected.

We must admit that, as of now, we have not completed the proof that the function G_q^t and the SU(3) group-theoretic functions are identical. But we are encouraged by the remark of G.-C. Rota that the functions G_q^t are apparently quite new to combinatorists and may be of considerable interest in their own right.

REFERENCES

1. G. Racah, Ergeb. Exakt. Naturw. <u>37</u>, 28 (1965).

2. L. C. Biedenharn, A. Giovannini, and J. D. Louck, J. Math. Phys. <u>8</u>, 691 (1967).

3. J. D. Louck, M. A. Lohe, and L. C. Biedenharn (submitted for publication in J. Math. Phys.).

4. L. C. Biedenharn and J. D. Louck, Commun. Math. Phys. <u>8</u>, 89 (1968).

5. L. C. Biedenharn, J. D. Louck, E. Chacón, and M. Ciftan, J. Math. Phys. <u>13</u>,1957 (1972).

6. L. C. Biedenharn and J. D. Louck, J. Math. Phys. <u>13</u>, 1985 (1972).

7. G. E. Baird and L. C. Biedenharn. J. Math. Phys. <u>5</u>, 1730 (1964).

8. D. E. Littlewood, <u>The Theory of Group Characters and Matrix Representations of Groups</u> (Oxford University Press, London, 1950), 2nd ed.

9. J. D. Louck and L. C. Biedenharn, J. Math. Phys. <u>11</u>, 2368 (1970).

10. J. J. DeSwart, Rev. Mod. Phys. <u>35</u>, 916 (1963).

11. L. C. Biedenharn, in <u>Spectroscopic and Group Theoretical Methods in Physics</u>, F. Bloch <u>et al</u>., eds.)North-Holland, Amsterdam, 1968), p. 59.

12. J. D. Louck and L. C. Biedenharn, Revista Mexicana Física <u>23</u>, 221 (1974).

13. A. Erdélyi, <u>Higher Transcendental Functions</u>, Vol. 1, McGraw-Hill (1953).

*†Department of Physics, Duke University, NC27706: Supported in part by NSF grant GP-14116.

†Supported in part by an Overseas Scholarship from the Royal Commission for the Exhibition of 1851.

‡Theoretical Division, Los Alamos Scientific Laboratory, University of California, Los Alamos, NM 87545: Work performed under the auspices of the USERDA.

On Space-Time Groups

by

Hans Zassenhaus and Wilhelm Plesken
California Institute of Technology

About the non finite discrete subgroups of the Poincaré group very little is known. We call the discrete subgroups with compact left coset space space-time groups in analogy to the space groups which are the discrete subgroups of the isometry group of euclidean spaces with compact left coset space.

THEOREM 1 : (a) Any space-time group contains four independent translations.

(b) The translations of G form a free abelian group of rank four.

(c) The group G is similar to a group $\tilde{G}$ of matrices

$$g = \begin{pmatrix} 1 & t(g) \\ 0^{4\times 1} & H(g) \end{pmatrix}$$

with $t(g)$ a 4-row and $H(g)$ a rational integral matrix of degree 4 such that

1) H is a homomorphism of $\tilde{G}$ into the orthogonal group

$$Aut(S/\mathbb{Z}) = \{X \mid X \in \mathbb{Z}^{4\times 4} \ \& \ XSX^T = S\}$$

relative to the symmetric rational integral matrix S of signature 3, 1,

2) the index of $H(\tilde{G})$ in $Aut(S/\mathbb{Z})$ is finite,

3) the kernel of H is the full group of all rational integral translation matrices:

(1) $\ker H = T(4,\mathbb{Z})$

$$= \left\{ \begin{pmatrix} 1 & \bar{x} \\ 0 & I_4 \end{pmatrix} \middle| \bar{x} \in \mathbb{Z}^{1\times 4} \right\},$$

(d) conversely

for any rational integral symmetric matrix S of signature 3, 1 and for any subgroup $\hat{H}$ of Aut(S($\mathbb{Z}$))) of finite index there are only finitely many conjugacy classes of groups $\tilde{G}$ with $H(\tilde{G}) = \hat{H}$ under transformation by real translations

$$\begin{pmatrix} 1 & \bar{x} \\ 0 & I_4 \end{pmatrix} \qquad (\bar{x} \in \mathbb{R}^{1\times 4})$$

and the left coset space of Aut(S/$\mathbb{R}$) over $\hat{H}$ is compact.

Part (a) of theorem 1 is a simple application of

Theorem 2: Let G a discrete reducible linear group of finite degree n over a field F such that for some non singular matrix X of degree n over F

$$\tilde{G} = X^{-1} G X = \left\{ \tilde{g} = \begin{pmatrix} \Delta_1(\tilde{g}) & \Lambda(\tilde{g}) \\ 0^{n_2\times n_2} & \Delta_2(\tilde{g}) \end{pmatrix} \right\}$$

where $\Delta_j(\tilde{g}) \in F^{n_j \times n_j}$, $\Lambda(\tilde{g}) \in F^{n_1\times n_2}$ then the elements g of G for which the diagonal component

$$(\Delta_1 \dot{+} \Delta_2)(\tilde{g}) = \begin{pmatrix} \Delta_1(\tilde{g}) & \\ & \Delta_2(\tilde{g}) \end{pmatrix}$$

$$(\tilde{g} = X^{-1} g X)$$

belongs to the 1-component of the diagonal group $(\Delta_1 \dot{+} \Delta_2)\tilde{G}$ form a nilpotent normal subgroup G_1 of G with discrete factor group and abelian diagonal group $\Delta G_1 = (\Delta_1 \dot{+} \Delta_2)(X^{-1}G_1X)$.

Here the topologization of the rectangular matrices over F is based on a Kuerszak valuation Φ

$$Y \to \mathbb{R}^{\geq 0} : \Phi \qquad (Y \in F^{p\times q})$$

such that

(2a) $$Y\Phi = 0 \Leftrightarrow Y = 0^{p\times q}$$

(2b) $(Y_1 + Y_2)\Phi \leq Y_1\Phi + Y_2\Phi \qquad (Y_1, Y_2 \in F^{p\times q})$

(2c) $(Y_1 Y_2)\Phi \leq Y_1\Phi\, Y_2\Phi \qquad (Y_1 \in F^{p\times q}, Y_2 \in F^{q\times r})$

(2d) $I_p\Phi = 1$

$(p, q = 1, 2, \ldots)$.

The best known way of constructing Kuerszak valuations Φ is based on the knowledge of mappings

$$F^{1\times p} \to \mathbb{R}^{\geq 0} : \Phi_p$$

subject to the conditions

(2e) $y\Phi_p = 0 \Leftrightarrow y = 0^{1\times p} \qquad (y \in F^{1\times p})$

(2f) $(y_1 + y_2)\Phi_p \leq y_1\Phi_p + y_2\Phi_p \quad (y_1, y_2 \in F^{1\times p})$

(2g) $(\lambda y)\Phi_p = (\lambda\Phi_1)y\Phi_p \qquad (\lambda \in F, y \in F^{1\times p})$

by means of the Banach construction

(2h) $$Y\Phi = \operatorname{lub}_{0\neq y\in F^{1\times p}} (yY)\Phi_q / y\Phi_p$$

$$(Y \in F^{p\times q}).$$

This Kuerszak valuation restricts to Φ_p on $F^{1\times p}$ and Φ_1 is a multiplicative valuation of F. Among the better known methods of constructing the Φ_p's for given multiplicative valuation Φ_1 of F are the following

I cartesian valuation : $(y_1, \ldots, y_p)\Phi_p = \left(\sum_{i=1}^{p} (y_i\Phi_1)^2\right)^{\frac{1}{2}}$,

II hypercubic valuation : $(y_1, \ldots, y_p)\Phi_p = \max y_i\Phi_1,\ 1 \leq i \leq p$,

III hyperoctahedral valuation: $(y_1 \ldots, y_p)\Phi_p = \sum_{i=1}^{p} y_i\Phi_1$.

Using Kuerszak valutation Φ the pq-dimensional linear space of the p × q - matrices over F turns into a metric space with distance function given by

(2i) $d_\Phi(Y_1, Y_2) = (Y_2 - Y_1)\Phi \ (Y_1, Y_2 \in F^{p\times q})$.

Thus every linear group G of degree n over F is a topological group. It is discrete precisely if

(2j) $\underset{I_n \neq g \in G}{\mathrm{glb}} \quad (g - I_n)\Phi > 0.$

The proof of theorem 2 was given in Zassenhaus [1], see also Auslander, Wang.

Part (b) of theorem 1 follows from a well known theorem of the theory of geometric lattices, s. Cassels.

Part (c) of theorem 1 follows from (a),(b) for some real symmetric matrix S of degree 4 with signature 3, 1 which is uniquely determined by $H(\tilde{G})$ up to non zero real factor as will be shown later on. In other words the solutions of the system of linear homogeneous equations

$$XS_1X^T = S_1 \qquad (X \in H(\tilde{G}))$$

for the coefficients of the real symmetric matrix S_1 form a one-dimensional linear space over $\mathbb{R}$. Since the coefficients of the system are rational integers it follows that there is precisely one basic primitive rational integral solution up to $\pm$. Let that be S.

Part (d) of theorem 1 is a consequence of Theorem 3: Let f be a homogeneous polynomial of degree d on the $\mathbb{R}$-linear space $\mathbb{R}^{1\times n}$. Let $\mathrm{Aut}(f/\mathbb{R})$ be the multiplicative group of all non singular linear matrices of $\mathbb{R}^{n\times n}$ preserving f. Then the subgroup $\mathrm{Aut}(f/\mathbb{Z})$ formed by the rational integral matrices of $\mathrm{Aut}(f/\mathbb{R})$) can be finitely presented and has compact left coset factor space. Proof see Zassenhaus [2],[3].

In order to prove part (d) of theorem 1 let us form the module $X(\hat{H})$ of all mappings

$$\hat{H} \to \mathbb{R}^{1\times 4} : f$$

subject to the conditions

$$(h_1h_2)f \equiv (h_1f)h_2 + h_2f \pmod{\mathbb{Z}^{1\times 4}} \quad (h_1, h_2 \in H)$$

suitable to form a representative set of matrices

$$\begin{pmatrix} 1 & hf \\ & h \end{pmatrix} \quad (h \in \hat{H})$$

of $\tilde{G}$ over $T(4,\mathbb{Z})$).

Furthermore let us form the submodule $X_0(\hat{H})$ of the mappings of $\hat{H}$ in $\mathbb{Z}^{1\times 4}$ which may be added to any element f of $X(\hat{H})$ without changing $\tilde{G}$.

Finally let us form the submodule $X_1(\hat{H})$ of $X(\hat{H})$ consisting of the mappings

$$\hat{H} \to \mathbb{R}^{1\times 4} : f_y \quad (y \in \mathbb{R}^{1\times 4})$$

$$hf_y = y(I_4 - h).$$

If one of these mappings is added to an element f of $X(\hat{H})$ then $\tilde{G}$ is transformed by the element

$$\begin{pmatrix} 1 & y \\ & I_4 \end{pmatrix}$$

of $T(4,\mathbb{R})$.

The finiteness statement (d) of theorem 1 is implied by the finiteness of the module theoretic index of $X(\hat{H})$ over $X_0(\hat{H}) + X_1(\hat{H})$.

According to theorem 3 there is a finite generator set

$H_1 = \{z_1, \ldots, z_\nu\}$ of $\hat{H}$.

Lemma 1: The restriction of the mappings belonging to $X(\hat{H})$ to H_1 establishes an isomorphism res_{H_1} of $X(\hat{H})$ on a module $X(H_1)$ formed by certain mappings of H_1 in $\mathbb{R}^{1\times 4}$.

Proof: See Zassenhaus [4].

We have to prove the finiteness of the module theoretic index of $X(H_1)$ over module theoretic index of $X(H_1)$ over $X_0(H_1)+X_1(H_1)$ where $X_0(H_1) = X_0(\hat{H})\, \mathrm{res}_{H_1}$ is the module of the mappings of H_1 into $\mathbb{Z}^{1\times 4}$ and $X_1(H_1) = X_1(\hat{H})\, \mathrm{res}_{H_1}$ is the module of the mappings f_h restricted to H_1.

The module $X(H_1)$ is a submodule of the $\mathbb{R}$-linear space with finite R-basis β_{hi} $(h \in H_1,\ 1 \leq i \leq 4)$ where β_{hi} is the mapping of H_1 in $\mathbb{R}^{1\times 4}$ that maps h on the i-th unit row and h' on 0 if $h' \in H$, $h' \neq h$.

Lemma 2: There is a rectangular matrix $\mathbb{R} = (\rho_{ik})$ of $\mathbb{Z}^{\delta\times 4\nu}$ such that $X(H_1)$ consists of all linear combinations $\sum_{h\in H_1} \sum_{i=1}^{4} \xi_{hi}\, \beta_{hi}$ with real coefficients ξ_{hi} subject to the congruence conditions

$$\sum_{i=1}^{4} \sum_{k=0}^{\nu-1} \xi_{ki}\ \rho_{j,i+4k} \equiv 0 \pmod{\mathbb{Z}}. \quad (1 \leq j \leq p)$$

Proof: see Zassenhaus [4].

Lemma 3: Any submodule $\mathfrak{R}/\mathbb{R}$ of $\mathbb{R}^{1\times q}$ defined as the submodule of the real solutions of some set $\mathfrak{R}$ of congruences of the form

$$(*) \qquad \sum_{i=1}^{q} \lambda_i\, \xi_i \equiv 0 \pmod{\mathbb{Z}} \qquad (\lambda_i \in \mathbb{Z},\quad 1 \leq i \leq q)$$

is the direct sum of finitely many submodules $M_0, \mathbb{Z}u_1, \ldots, \mathbb{Z}u_\mu$ of $\mathbb{R}^{1\times q}$ such that $0 \leq \mu \leq q$, M_0 is an $\mathbb{R}$-linear subspace of $\mathbb{R}^{1\times q}$ of dimension $q-\mu$,

$$\mathbb{Z}^{1\times q} = M_0 \cap \mathbb{Z}^{1\times q} \dotplus \sum_{i=1}^{\mu} \mathbb{Z} e_i u_i$$

where $e_1, \ldots, e_\mu$ are natural numbers satisfying the conditions $e_i | e_{i+1}$ $(1 \leq i < \mu)$.

The numbers $\mu, e_1, \ldots, e_\mu$ and the $\mathbb{R}$-linear space M_0 are uniquely determined by $\mathfrak{R}/\mathbb{R}$.

Proof: see Zassenhaus [4].

Corollary 1 : The submodule $\mathcal{R}/\mathcal{Z}$ of $\mathbb{R}^{1\times q}$ formed by the rational solutions $(\xi_1, \ldots, \xi_q)$ of the congruences (*) is the direct sum $(M_0 \cap \mathbb{Z}^{1\times q})\mathcal{Z} \dot{+} \dot{\sum}_{i=1}^{\mu} \mathbb{Z}e_i u_i$. Again the $\mathcal{Z}$-linear space $(M_0 \cap \mathbb{Z}^{1\times q})\mathcal{Z}$ of $\mathcal{Z}$-dimension $q-\mu$ is uniquely determined by $\mathcal{R}/\mathcal{Z}$.

Corollary 2 : Every $\widetilde{G}$ can be transformed by a suitable translation into a new group which has only rational translative parts.

Consequently it suffices to study the submodule $Y(H_1)$ of all mappings of H_1 into $\mathcal{Z}^{1\times 4}$ belonging to $X(H_1)$ and to show the finiteness of the module theoretical index of $X(H_1)$ over $X_0(H_1)+Y_1(H_1)$ where $Y_1(H_1) = X_1(H_1)\cap Y(H_1)$ consists of the restrictions of f_y to H_1 for y of $\mathcal{Z}^{1\times 4}$. The $\mathcal{Z}$-dimension of $Y_1(H_1)$ is 4.

Let $e_1, e_2, \ldots, e_\mu$ be the natural numbers determining the structure of the module $Y(H_1)$. It fallows that the factor module of $Y(H_1)$ over $X_0(H_1) + Y_1(H_1)$ is the direct sum of an $\big(4(\nu - 1) - \mu\big)$ – dimensional linear $\mathcal{Z}$-space and of μ cyclic modules of orders $e_1, e_2, \ldots, e_\mu$.

We want to show that the factor module is finite which is tantamount to the statement that

$$4\nu = 4 + \mu .$$

For this purpose we consider a $\mathcal{Z}$-linear subspace $Y_2(H_1)$ of $Y(H_1)$ with zero intersections with $X_0(H_1)$. For any element f of $Y_2(H_1)$ the congruences are satisfied as equations.

Lemma 4: Let $\hat{H}$ be a group acting on a module M and $\hat{H}_1$ a subgroup of $\hat{H}$ of finite index. Then there is the sequence

$$H^n(\hat{H},M) \underset{\mathrm{res}_{H_1}}{\rightarrow} H^n(\hat{H}_1,M) \underset{I}{\rightarrow} H^n(\hat{H},M)$$

such that the product of the restriction map and the inductions map is equal to $(H:\hat{H}_1)\underline{1}$.

This is a well known statement of cohomology theory (s. McLane) which implies the

Corollary: If M is finitely generated and if $H^n(\hat{H}_1,M)$ is finite then $H^n(\hat{H},M)$ is finite.

As a consequence of the corollary for n=1 we are permitted to replace the group $\hat{H}$ of theorem 1(d) by a subgroup of finite index,

for example by the

subgroup of $\hat{H}$ formed by all matrices of $\hat{H}$ that are congruent to I_4 modulo some prime number p. Without loss of generality we can assume that already every matrix of $\hat{H}$ is congruent to I_4 modulo p. This reduction makes it possible to use a p-adic method for the proof of Theorem 1(d).

Lemma 5: The limits of the sequences of matrices of elements of $\hat{H}$ such that the 4^2 coefficient sequences are p-adically convergent form a p-adically complete subgroup H_p of $Aut(S/\mathbb{Z}_p)$.

The p-adic logarithms of the elements of $\hat{H}_p$ over the ring $\mathbb{Z}_p$ of the p-adic integers.

Proof: See Loonstra.

Corollary: If the index of $\hat{H}$ over a subgroup $\hat{H}_1$ is finite then it is equal to the module theoretic index of log $\hat{H}_p$ over $\log(\hat{H}_1)_p$ and we have $\hat{H}(\hat{H}_1)_p = \hat{H}_p$, $\hat{H}\cap(\hat{H}_1)_p = \hat{H}_1$.

The proof of the corollary is reduced to the case of a maximal subgroup H_1 of $\hat{H}$ in which case $\hat{H}_1$ is a normal subgroup of $\hat{H}$ of index p.

Whenever $\hat{H}_1$ is a normal subgroup of $\hat{H}$ with abelian p-factor group one shows by the use of the Hausdorff-Baker formula that the factor-group of $\hat{H}/\hat{H}_1$ is multiplicative-additively isomorphic to the factormodule of $\log \hat{H}_p/\log(\hat{H}_1)_p$.

Lemma 6: The p-adic completion $\tilde{G}_p$ of $\tilde{G}$ is a subgroup of $GL(5,\mathbb{Z}_p)$ with an epimorphism H_p on the p-adic completion $\hat{H}_p$ of $\hat{H}$ such that $\mathrm{res}_{\tilde{G}} H_p = H$, $\ker H_p = T(4,\mathbb{Z}_p)$.

Proof: There is a natural number λ such that $\lambda z_i f$ is integral for $i = 1,2,\ldots,\nu$. In view of the formulae $t(gg') = t(g)H(g') + t(g')$ $t(g^{-1}) = -t(g)H(g^{-1})$ $(g, g' \in \tilde{G})$ with integral matrices $H(g), H(g')$ it follows that $t(\tilde{G}) = t(\langle T(4,\mathbb{Z}), \tilde{z}_1,\ldots,\tilde{z}_\nu\rangle$ is contained in the module $\lambda^{-1}\mathbb{Z}_p^{1\times 4}$ which is compact in the p-adic topology. Using also the corollary of Lemma 5 the Lemma 6 is demonstrated.

Corollary 1: The p-adic limits of the elements of $\tilde{G}_p$ form a complete linear group for the p-adic topologization of $\mathbb{Z}_p^{5\times 5}$ such that the p-adic logarithms of the elements of $\tilde{G}_p$ form a Lie ring $\log \tilde{G}_p$ over $\mathbb{Z}_p$. (See Loonstra.)

Corollary 2: Every element g of $\tilde{G}_p$ is of the form $\begin{pmatrix} 1 & t_p(g) \\ & H_p(g) \end{pmatrix}$

where t_p is a continous mapping of $\tilde{G}_p$ in $Q_p^{1\times 4}$ in the p-adic topologization.

Proof: Use the fact that the congruences α are satisfied as equations and use also Corollary 1 of Lemma 5.

Corollary 3:

$$\log \tilde{G}_p = \left\{ \begin{pmatrix} 0 & t \\ & 0^{4\times 4} \end{pmatrix} \middle| \, t \in \mathbb{Z}_p^{1\times 4} \right\}$$

$$\dot{+}\left\{\begin{pmatrix} 0 & hf(\log h/(h-1)) \\ & \log h \end{pmatrix} \middle| h \in \tilde{H}_p \right\} .$$

where the function hf(log h/(h-1)) defined on $\log \tilde{H}_p$ is a Lie 1-cocycle that is a 1-coboundary only if f = 0. From the assumption made above it follows that the 1-cohomology of $\log \hat{H}_p$ acting on $\mathbb{Q}_p^{1\times\varphi}$ is not trivial. Or, equivalently, the 1-cohomology of the Lie algebra $\mathcal{Q}_p \log \hat{H}_p$ acting on $\mathbb{Q}_p^{1\times 4}$ is not trivial. On the other hand we know from the Whitehead lemma that the 1-cohomology of a finite dimensional semisimple Lie algebra of zero characteristic acting on a finite dimensional representation space always is trivial. The conclusion is that the Lie algebra $\mathcal{Q}_p \log \hat{H}_p$ over $\mathcal{Q}_p$ is not semisimple. But we have

Lemma 7: The linear Lie algebra $\mathcal{Q}_p \log \hat{H}_p$ is absolutely irreducible.

Proof: If this linear Lie algebra $\mathcal{Q}_p \log \hat{H}_p$ permits a reduction after suitable field extension of $\mathcal{Q}_p$ then the same applies to $\hat{H}_p = \exp(\log \hat{H}_p)$ and therefore also $\hat{H}$ can be reduced after suitable extension of the real number field. But $\hat{H}$ is a subgroup of the Lorentz group Aut(X/ℝ) with compact left coset space. It follows that a maximal reducible subgroup V of Aut(X/ℝ) containg $\hat{H}$ as subgroup also has compact left coset space. Because of its maximal property V is a closed linear subgroup of Aut(X/ℝ).

It is well known that the left coset space of the maximal closed linear subgroups of the Lorentz group is non compact (s.e.g. Winternitz; Friš).

Hence $\mathcal{Q}_p \log \hat{H}_p$ is absolutely irreducible. Since its members have zero trace it follows that $\mathcal{Q}_p \log \hat{H}_p$ is semi-simple.

Thus, by Whitehead's lemma, Theorem 1 part (d) is demonstrated.

A. Schild dealt with the discrete subgroup Aut(S/Z) formed by the integral Lorentz matrices relative to

$$S = \begin{pmatrix} -1 & & & \\ & 1 & & \\ & & 1 & \\ & & & 1 \end{pmatrix} .$$

They form a discrete subgroup with compact left cosetspace. The group is generated by the reflection matrices

$$A = \begin{pmatrix} 2 & -1 & -1 & -1 \\ 1 & 0 & -1 & -1 \\ 1 & -1 & 0 & -1 \\ 1 & -1 & -1 & 0 \end{pmatrix}, \quad B = \begin{pmatrix} 1 & 0 & 0 & 0 \\ 0 & -1 & 0 & 0 \\ 0 & 0 & 1 & 0 \\ 0 & 0 & 0 & 1 \end{pmatrix},$$

$$C = \begin{pmatrix} 1 & 0 & 0 & 0 \\ 0 & 1 & 0 & 0 \\ 0 & 0 & 0 & 1 \\ 0 & 0 & 1 & 0 \end{pmatrix}, \quad D = \begin{pmatrix} 1 & 0 & 0 & 0 \\ 0 & 0 & 1 & 0 \\ 0 & 1 & 0 & 0 \\ 0 & 0 & 0 & 1 \end{pmatrix}$$

with defining relations

$$A^2 = B^2 = C^2 = D^2 = (AB)^4 = (AC)^2 = (AD)^2 = (BC)^2 = (BD)^4 = (CD)^3 = 1$$

We want to determine all possible vector systems $\mathrm{Aut}(S/Z) \to \mathbb{R}^{1\times 4}$: f modulo the trivial vector-systems $g \to y(I_4 - g)$: f_y. As described earlier it suffices to compute AG, Bf, Cf, and Df which are determined as solutions of the following system of defining relations of the A,B,C,D as described in Zassenhaus [4]:

$$Af(I_4 + A) \equiv Bf(I_4 + B) \equiv Cf(I_4 + C) \equiv Df(I_4 + D) \equiv$$

$$\equiv Af(B((AB)^3 + (AB)^2 + AB + I_4)) + Bf((AB)^3 + (AB)^2 + AB + I_4 \equiv$$

$$\equiv Af(C(AC + I_4)) + Cf(AC + I_4) \equiv Af(D(AD + I_4)) + Df(AD + I_4) \equiv$$

$$\equiv Bf(C(BC + I_4)) + Cf(BC + I_4) \equiv$$

$$\equiv Bf(D((BD)^3 + (BD)^2 + BD + I_4)) + Df((BD)^3 + (BD)^2 + BD + I_4) \equiv$$

$$\equiv Cf(D((CD)^2 + CD + I_4)) + Df((CD)^2 + CD + I_4) \equiv 0 \bmod Z$$

By treating this congruence system in the usual way as described in Zassenhaus [4] one gets:

$$Af \begin{pmatrix} 1 & 0 & 0 \\ 0 & 1 & 0 \\ 0 & 0 & 1 \\ 1 & -1 & -1 \end{pmatrix} \equiv Bf \begin{pmatrix} 2 & 0 & 0 \\ 0 & 0 & 0 \\ 0 & 1 & 0 \\ 0 & 1 & 2 \end{pmatrix} \equiv Cf \begin{pmatrix} 1 & 0 & 0 \\ 0 & 1 & 0 \\ 0 & 0 & 1 \\ 0 & 0 & 1 \end{pmatrix} \equiv Df \begin{pmatrix} 1 & 0 & 0 \\ 0 & 1 & 0 \\ 0 & 1 & 0 \\ 0 & 0 & 1 \end{pmatrix} \equiv 0 (\bmod Z)$$

So all possible vector systems f depend on 4 continuous parameters $a,b,c,d \in R$:

$$Af = (-a, a, a, a), \quad Bf = (\frac{\beta_1}{2}, b, \frac{-\beta_2}{2}, \frac{\beta_2}{2}),$$

$$Cf = (0, 0, -c, c), \quad Df = (0, -d, d, 0)$$

with $\beta_1, \beta_2 \in \{0,1\}$.

But the trivial vector system

$g \to y(I_4 - g) : f_y$ with

$y = (a + \frac{3}{2}b + c + 2d, \frac{b}{2}, \frac{b}{2} + d, \frac{b}{2} + c + d)$

is equal to the vector system above with $\beta_1 = \beta_2 = 0$. So we get four essentially different space groups. They are generated by

$$\bar{A} = \begin{pmatrix} 1 & 0^{1\times 4} \\ 0^{4\times 1} & A \end{pmatrix}, \quad \bar{C} = \begin{pmatrix} 1 & 0^{1\times 4} \\ 0^{4\times 1} & C \end{pmatrix}, \quad \bar{D} = \begin{pmatrix} 1 & 0^{1\times 4} \\ 0^{4\times 1} & D \end{pmatrix}, \quad E(y) = \begin{pmatrix} 1 & y \\ 0^{4\times 1} & I_4 \end{pmatrix}$$

with y = (1, 0, 0, 0), (0, 1, 0, 0), (0, 0, 1, 0), (0, 0, 0, 1) and

$$\bar{B}(Z) = \begin{pmatrix} 1 & z \\ 0^{4\times 1} & B \end{pmatrix}$$ where z is equal to (0,0,0,0) for the first group,

$(\frac{1}{2},0,0,0)$ for the second, $(0,0,-\frac{1}{2},\frac{1}{2})$ for the third and $(\frac{1}{2},0,-\frac{1}{2},\frac{1}{2})$ for the last. Finally we have to show that no two of these groups are conjugated under the affine group. By Zassenhaus [4] we have to determine the orbits of the vector systems under the action of the normalizer of Aut(S/$\mathbb{Z}$) in G(4,$\mathbb{Z}$). But his normalizer is equal to Aut(S,$\mathbb{Z}$) by the following lemma.

<u>Lemma</u>: Let G be a absolutely irreducible subgroup of Aut(S/$\mathbb{Z}$) where S is a symmetric non singular matrix in $\mathbb{Z}^{n\times n}$. Then the normalizer of G in GL(n,$\mathbb{Z}$) is contained in Aut(S/$\mathbb{Z}$).

Proof: Let Δ denote the natural representation of G defined by $G \rightarrow GL(n,\mathbb{Z}) : \Delta$ with $g\Delta = g$.

For all $g \in G$ we have $g\, S\, g^T = S$ implying $g\Delta S = S(g\Delta^{-T})$, where Δ^{-T} is the inverse transposed representation of Δ given by $g\Delta^{-T} = (g^{-1}\Delta)^T$. The irreducibility of Δ now implies by Schur's Lemma that S is uniquely determined up to scalar multiples, i.e. S is essentially the only matrix fixed by G. Now let h be an element of the normalizer of G in GL(n,$\mathbb{Z}$). We have $h^{-1}Gh = G$ and for all $g \in G$ the equation h $h^{-1}ghS(h^{-1}gh)^T = S$ holds which implies : $g(hSh^T)g^T = hSh^T$ for all $g \in G$.

So hSh^T has to be a scalar multiple of S. But det h = $\pm$ 1 and hence $hSh^T = S$, proving $h \in Aut(S/\mathbb{Z})$, q.e.d.

Theorem 1 is capable of generalization to discrete subgroups with compact left coset space in Lie groups of type B,C,D. Jointly with T. Janssen, A. Janner and E. Ascher the question is raised whether there exist observable structures in Einstein-Minkowski spaces formed by a finite number of substantial orbit subsets of a space-time group (space-time crystal)?

References

Auslander, Louis: Bieberbach's Theorem on Space Groups and discrete uniform Subgroups of Lie Groups. I Ann. of Math. (2) 71 (1960), 579-590, II Amer. J. Math. 83 (1961), 276-280.

Janssen, T.; Janner, A.; Ascher, E.: Crystallographic Groups in Space and Time. I. General definition and basic properties. Physica 41 (1969), 541-565. II. Central Extensions. Physica 42 (1969), 41-70.

Loonstra, Frans: Die Lieschen Fundamentalsätze in bewerteten Körpern. Nederl. Akad. Wetensch., Proc. 44 (1941), 568-576.

MacLane, Saunders: Homology. Berlin, Springer 1963.

Schild, Alfred E.:

[1] Discrete space-time and integral Lorentz transformations. Physical Review (2) 73 (1948), 414-415.

[2] Discrete space-time and integral Lorentz transformations. Canadian J. Math. 1 (1949), 29-47.

Wang, Hsien-Chung: On the deformations of lattice in a Lie group. American Journal Of Mathematics 85 (1963), 189-212.

Whitehead, J.H.C.: Certain equations in the algebra of a semisimple infinitesimal group. Quart. J. Math. (Oxford Ser.) 8 (1973), 220-237.

Winternitz, P.; Friš, I.: Invariant expansions of relativistic amplitudes and subgroups of the proper Lorentz group. Jadernaja Fiz. 1 (1965), 889-901; Translated As Soviet J. Nuclear Phys. 1 (1965), 636-643.

Zassenhaus, Hans:

[1] Beweis eines Satzes über diskrete Gruppen. Abh. Math. Sem. Hansische Univ. 12 (1938), 289-312

[2] On the units of orders. Journal of Algebra 20 (1972), 368-395.

Zassenhaus, Hans:

[3] Gauss theory of ternary quadratic forms, an example of the theory of homogeneous forms in many variables, with applications. California Institute of Technology Lecture Notes October 1975.

[4] Über Einen Algorithmus Zur Bestimmung Der Raumgruppen. Math. Helv. 21 (1948), 117-141.

FRAME'S CONJUGATING REPRESENTATION AND GROUP EXTENSIONS

Nigel Backhouse

Department of Applied Mathematics and Theoretical Physics, University of Liverpool, Liverpool, L69 3BX, England.

Introduction

One of the first things the beginner learns of in representation theory is that every group has two natural representations, namely the trivial representation and the regular representation. There is, however, another natural representation, Frame's Conjugating Representation, which deserves attention for the following reasons: it is easily defined, providing useful examples for students; it presents some unsolved problems and conjectures; it has aroused independent interest in the Pure Mathematics and Mathematical Physics literature in recent years (see ref. 1 for a bibliography).

Let G be a finite group with group algebra A(G) = {complex linear sums of group elements, with product the linear extension of the multiplication in G}. The conjugating representation F is defined on A(G) (or any faithful representation of A(G)) by $g \to F_g$, where $F_g a = gag^{-1}$, for $g \in G$, $a \in A(G)$. By interpreting A(G) as an algebra of group functions, the definition of F can be carried over to the compact group case - see ref. 2 for the proper definition.

F is completely reducible so one can ask for its irreducible constituents. We first observe that F_g is the identity if $g \in Z$, the centre of G, hence F can only contain irreducibles which arise from irreducibles of G/Z. Even so, there are examples which show that F need not contain all irreducibles of G/Z. However, since F is a faithful (and real) representation of G/Z, we know from a theorem of Burnside that $\exists$ an integer N such that $\otimes^n F$ contains all irreducibles of G/Z $\forall$ $n \geq N$.

If G acts faithfully on something, for example a space of wave functions, then A(G) becomes an algebra of operators which transform tensorially among themselves under F. On reducing F we see that we have a source of irreducible tensor operators for G. F is sometimes called the tensor representation.

Having noted above that F gives rise only to tensor operators which correspond to irreducibles of G/Z, and even then not necessarily all of them directly, we investigate in this paper generalisations of F which can be used to associate tensor operators with all irreducibles of G, hence answering in the affirmative a question of de Vries. Finally, we show that all irreducibles of G/Z appear in F for $G = SU(3)$.

<u>Generalised conjugating representation</u>

Two obvious generalisations of F are obtained by (1) letting G act by conjugation on the group algebra of a group G′ containing G; (2) letting G act on its twisted group algebra. Putting these together, suppose $G \leq G'$ and let $A(G', \omega)$ be the twisted group algebra of G′ corresponding to the factor system ω (recall from ref. 1 that $A(G', \omega)$ is a module over the complex field, with basis the set of objects $\{\nu(g): g \in G'\}$, which has as a multiplication the linear extension of the law $\nu(g_1)\nu(g_2) = \omega(g_1, g_2)\nu(g_1 g_2)$). Now define the representation F^{ω} of G by $g \to F_g^{\omega}$, where $F_g^{\omega}a = \nu(g)a\nu(g)^{-1}$ for $g \in G$, $a \in A(G', \omega)$. Using the properties of factor systems and $\nu(g)^{-1} = \nu(g^{-1})/\omega(g, g^{-1})$, we can check that F^{ω} is indeed a representation of G and that $F_g^{\omega}\nu(g') = \nu(gg'g^{-1})\omega(g, g')/\omega(gg'g^{-1}, g)$ for $g \in G$, $g' \in G'$. So F_g^{ω} is the identity operation iff $g \in Z'$, the centre of G′, and $\omega(g, g') = \omega(g', g)$ for all $g' \in N'(g)$, the centraliser of g in G′ (which for central g is G′ itself). But the latter is precisely the condition that g be ω-regular in G′ - see ref. 1. If R^{ω} is the set (not in general a group) of ω-regular elements in G′ then it is clear that the kernel of F^{ω} is $G \cap Z' \cap R^{\omega} = K^{\omega}$, which of course must be a subgroup of G. Evidently $A(G', \omega)$ has become a faithful G/K^{ω}-module, hence is a faithful G-module iff K^{ω} is trivial.

To calculate the irreducible constituents of F^{ω} we first compute its character, which is the restriction to G of the character χ^{ω} of the conjugating representation of G′ on $A(G', \omega)$. Using the fact that, for fixed $g \in G'$, $g' \to \omega(g, g')/\omega(g', g)$ is a linear character on $N'(g)$, we find that $\chi^{\omega}(g) = |N'(g)|$ if $g \in N'(g) \cap R^{\omega}$, but zero otherwise. Then if $\chi^{\omega} = \sum_{\mu} c_{\mu}\chi^{(\mu)}$, we find that $c_{\mu} = \sum_i{}' \chi_i^{(\mu)}$, where $\chi_i^{(\mu)}$ is the value of the μth irreducible character of G′ on the ith conjugacy class, and where the prime restricts the summation to ω-regular classes only. Another expression is $\chi^{\omega} = \sum_{\lambda} \theta^{\omega,\lambda}\theta^{*\omega,\lambda}$, where the $\theta^{\omega,\lambda}$ are the inequivalent irreducible ω-characters of G′. Thus we have two ways of computing the irreducibles of F^{ω}: either summing entries in the rows of the ordinary character table of G′ or using the Clebsch-Gordon series for projective representations of G′, and then restricting to G. Which method one chooses of course depends on context and available information, but we must remark that it is only the second method which makes proper sense and is indeed valid in the case of a non-finite compact group.

Examples

E. de Vries has posed the problem: given a finite group G, find some way of associating tensor operators to all irreducibles of G. I hope that the following is the best possible general solution in the context of this paper.

G acts on itself by left translations, thus for $g \in G$, define λ_g by $\lambda_g g' = gg'$. The set $\{\lambda_g : g \in G\}$ forms a group of permutations isomorphic to G itself on the set of elements of G, and hence embeds G in the full permutation group $S_{|G|}$ of order $|G|!$ Now $S_{|G|}$ is centreless for $|G| > 2$, hence $A(S_{|G|})$ provides a faithful G-module. In ref. 1 we show that $A(S_{|G|})$ in fact carries all irreducibles of G. Of course we knew in advance that $\otimes^N A(S_{|G|})$ carries all irreducibles of G for some suitable integer N, but it requires a calculation to prove that N can be taken as unity. I know of no sufficient condition which, in the general case, allows one to take N = 1. Kasperkovitz and Dirl, ref. 2, have suggested that a sufficient condition might be the existence of a faithful irreducible representation, but as yet the conjecture is unproved. If the conjecture is true then it implies in particular my own conjecture that Frame's conjugating representation for $G = S_n$ contains all irreducibles of S_n for $n > 2$ - it is strongly verified by looking at character tables for n = 3, 4,...10.

Let me now look at the compact groups SO(3), SU(2), SU(3).

(a) SO(3) is centreless, and has irreducibles D^j, j = 0, 1, 2... We compute the conjugating representation $\bigoplus_{j=0}^{\infty} (D^{2j} \oplus D^{2j-1} \ldots \oplus D^0)$, hence contains all D^{J}'s infinitely many times.

(b) The projective representations of SO(3) are D^j, $j = 1/2, 3/2, 5/2, \ldots$, hence the twisted conjugating representation of SO(3) is $\bigoplus_{j=1/2,3/2} (D^j \otimes D^{j*}) = \bigoplus_{j=1/2,..} (D^{2j} \oplus D^{2j-1} \ldots \oplus D^0)$, so again all irreducibles of SO(3) appear infinitely many times.

(c) SU(2) is the covering group of SO(3), so its representations are the ordinary and projective representations lifted from SO(3). The ordinary conjugating representation (there is no non-trivial conjugating representation) thus contains all D^{J}'s, J an integer, infinitely many times. But these are precisely the ones trivial on the centre of SU(2).

(d) The case of SU(3) requires a little more setting up. I

begin by reminding ourselves of some aspects of the representation theory of SU(n). Irreducibles of SU(n) are labelled by Young tableaux $\underline{\lambda} = (\lambda_1, \lambda_2, \ldots \lambda_{n-1})$ containing at most n - 1 rows of square boxes. Actually Young tableaux with at most n rows will do, but it then turns out that $(\lambda_1, \lambda_2, \ldots \lambda_n) \equiv (\lambda_1 - \lambda_n, \lambda_2 - \lambda_n, \ldots, \lambda_{n-1} - \lambda_n)$. Now if $D^{\underline{\lambda}}$ has tableau $\underline{\lambda}$ then the tableaux of $D^{\underline{\lambda}^*}$ is given by rotating through 180° the shaded region in the diagram below.

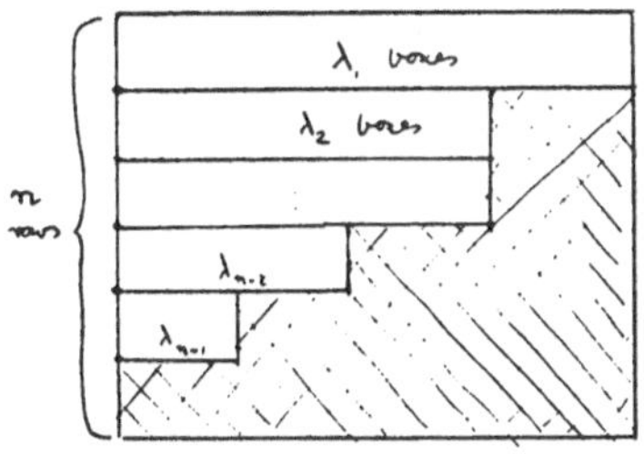

Also $D^{\underline{\lambda}}$ is trivial on the centre of SU(n) iff $\sum_{i=1}^{n-1} \lambda_i$ is a multiple of n.

Now in order to reduce the product $D^{\underline{\lambda}} \otimes D^{\underline{\mu}}$ to irreducibles, we set up the tableaux $\underline{\lambda}$, $\underline{\mu}$ side by side, writing in a fixed symbol, a_1 say, in the μ_1 boxes in the first row of $\underline{\mu}$, a symbol, a_2 say, in the μ_2 boxes of the second row of $\underline{\mu}$, etc. Then we consider all Young tableaux obtainable from $\underline{\lambda}$ by the adjunction one by one of the labelled boxes of $\underline{\mu}$ consistent with the following restrictions:

(1) at each stage in the process the augmented diagrams must be Young tableaux with at most n rows;

(2) adjoin all boxes from the ith row of $\underline{\mu}$ before adjoining any from the $i + 1^{th}$;

(3) no two boxes with the same label can be in the same column;

(4) each final tableau must be such that if one records the occurrence of the symbols a_1, a_2, etc., reading the rows as one would read lines of mirror English, then at each stage in the count the number of a_1's $\geq$ number of a_2's ... $\geq$ number of a_{n-1}'s.

Finally all tableaux with n rows can be reduced to n - 1 rows. This procedure only tells one whether or not a given irreducible occurs in a Kronecker product, but not its multiplicity. Now let me apply the above to show that the conjugating representation of SU(3) contains all irreducibles of SU(3)/Z. Given $\underline{\lambda} = (\lambda_1, \lambda_2)$ with $\lambda_1 + \lambda_2 = 3r$, r an integer, I will find $\underline{\mu} = (\mu_1, \mu_2)$ such that $D^{\underline{\lambda}}$ occurs in $D^{\underline{\mu}} \otimes D^{\underline{\mu}^*}$. Now always we have $\lambda_1 \geq 3r/2$, but either $\lambda_1 \leq 2r$ or $\lambda_1 > 2r$.

<u>Case $\lambda_1 \leq 2r$</u>

Write $\lambda_1 = 2r - k$, $\lambda_2 = r + k$, where $k \leq r/2$. I claim that I can take $\underline{\mu} = (r, k)$ (then $\underline{\mu}^* = (r, r - k)$). The proof is implicit in the following tableau multiplication

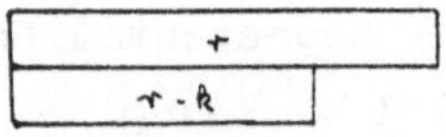

⊗

contains
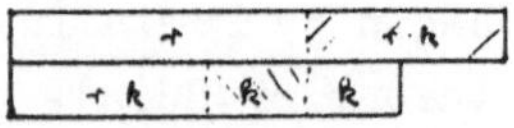

which is precisely $\underline{\lambda}$.

Case $\lambda_1 > 2r$

Write $\lambda_1 = 2r + k$, $\lambda_2 = r - k$, where $0 < k \leqslant r$. I take $\underline{\mu} = (2r, r)$, in which case $\underline{\mu}^* = \underline{\mu}$. Then

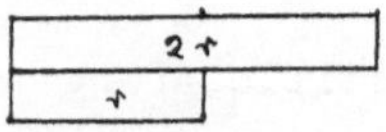

⊗
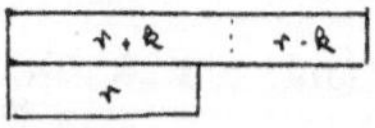

contains
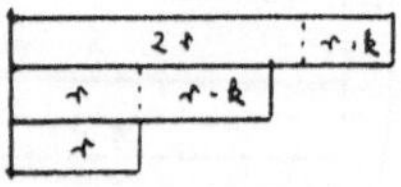

which can be reduced to $(2r + k, r - k) = \underline{\lambda}$. This concludes the proof.

References

1. N. B. Backhouse, J. Math. Phys. 16 (1975), 443-7.
2. P. Kasperkovitz and R. Dirl, J. Math. Phys. 15 (1974), 1203-10.

Symmetries of Differential Equations in Mathematical Physics

Charles P. Boyer
CIMAS, Universidad Nacional de México

The study of continuous symmetry groups of differential equations dates back almost a century to the oringinal work of Sophus Lie[1]. Since that time most of the work on Lie's theory has turned away from the connection with differential equations probably because it is not general enough; not every differential equation admits a nontrivial Lie group. However, it is precisely the differential equations which are of interest to physicists and applied mathematicians that do admit symmetries[2]. The recent books of Miller[3] and Vilenkin[4] have demonstrated the close connection between Lie theory and the special functions arising from certain differential equations. The symmetry group provides a degree of order and understanding to the conglomerate of special funtions identities. However, one can gain further geometrical insight by starting from certain fundamental differential equations, like the Laplace-Beltrami equation on a Riemannian space of constant curvature. The symmetries of such an equation can be linked to the coordinate curves in which the equations admits a separation of variables. The symmetry group can then be used to derive identities between the various separable solutions. I believe that the major part if not all of special function theory can be understood in this light. This approach has now been developed to the stage where one is gaining information, much of which is new, about not only the more common special functions, but also about Lamé and Ince functions, spheroidal, ellipsoidal, and anharmonic oscillator wave functions and others. I would like to mention at this point that the original idea of relating symmetries of differential equations to separable coordinate systems was formulated about ten years ago by P. Winternitz and his collaborators[5] in the Soviet Union and much of the development since then is due to E. G. Kalnins and W. Miller Jr.[6].

Up until now almost all that I have said pertains to linear differential equations and it is this case which will be discussed in the sequel. However a few words are in order about the more difficult and less extensively developed nonlinear equations. It was noticed some time ago[7] how symmetries could be used to generate similarity solutions of nonlinear differential equations where more general techniques are lacking. These similarity methods are especially applicable to boundary value problems which occur in thermo and hydrodynamics. However, it has only been recently that a systematic approach to these problems has been given[2,8,9].

As far as the separation of variables is concerned, the one nonlinear case which is fairly well known is the spatial separation of the Hamilton-Jacobi equation[10,11]

$$\frac{1}{2m} S_{\underline{x}} \cdot S_{\underline{x}} + S_t = 0$$

The separation of this equation in mixed space and time coordinates has recently been studied[12] and it appears that it can be related to only a subgroup of the full symmetry group[13]. This subgroup is the subgroup of linear transformations and is related to the time dependent Schrödinger equation, and hence to a quantization prescription[13]. We will have more to say about the separation of variables and symmetries of the Schrödinger equation later. Suffice it now to say that the general connection between Lie theory and separation of variables that will be described breaks down (although not entirely) for nonlinear equations. Indeed the concept of separability is, in general, not well defined.

A. General prescription

Given a general at most second order linear differential equation in n variables x_i (assumed to be real)

$$Qu = 0 \tag{1}$$

with $Q = \alpha_{ij}(\underline{x})\partial_{x_i x_j} + \beta_i(\underline{x})\partial_{x_i} + \gamma(\underline{x})$ where α_{ij}, β_i, γ are locally C^∞ functions. We will only consider second order differential equations although the generalization to higher order can be made.

1) Determine the symmetry group for (1), i.e. the group of local transformations of the form

$$[T(g)u](\underline{x}) = \mu(\underline{x};g)\, u(\underline{x}\cdot g) \tag{2}$$

where μ is a C^∞ function and $x' = (\underline{x}\cdot g)$ denotes the group action over R^n, such that $Q'[T(g)u] = 0$, whenever u satisfies (1), where Q' is Q written in the primed variables. It follows from Lie theory that the set of such transformations forms a local Lie group G. For practical purposes it is more convenient to work infinitesimally. Then writing $T(g) \simeq 1 + \epsilon L(\underline{x}, \partial_{\underline{x}}) + \cdots$, we have the existence of a C^∞ function $\lambda(x)$ such that

$$[Q, L] = \lambda Q \tag{3}$$

Writing $L = a_i(\underline{x})\partial x_i + b(\underline{x})$ for a_i, $b \in C^\infty$, Eq. (3) determines the functions a_i and b and thus the Lie algebra $\mathcal{g}$ of G. As mentioned previously there may be no symmetries at all. Of course, we are interested in the case where these is a nontrivial symmetry group.

2) Determine the second order symmetries of (1), i.e. the set $\mathcal{S}$ of differential operators (modulo Q) $S = c_{ij}(\underline{x})\partial_{x_i x_j} + a_i(\underline{x})\partial_{x_i} + b(\underline{x})$ such that (3) is satisfied upon replacing L by S and where now $\lambda(\underline{x})$ is a first order differential operator with C^∞ coefficients. It is emphasized that $\mathcal{S}$ does not necessarily form a Lie algebra; however, it does form a vector space which carries a representation of G. It is important to notice that G acts on $\mathcal{S}$ and splits $\mathcal{S}$ into G-orbits. There are two relevant types[14] of second order symmetries

Type 1: the elements of $\mathcal{S}$ are all second order members of the universal enveloping algebra $\mathcal{U}$ of $\mathcal{G}$.

Type 2: At least one element of $\mathcal{S}$ is not a member of $\mathcal{U}$. In general we are interested in classifying all orbits of (n-1) commuting operators of $\mathcal{S}$.

3) Find the coordinate systems such that (1) separates variables, i.e. introduce new variables v_i with $x_i = X_i(v_j)$ such that $u = R(v_j) \prod_{i=1}^{n} U_i(v_i)$ reduces (1) to a set of ordinary differential equations for the functions $U_i(v_i)$. The function $R(v_i)$ is called a multiplier or modulation factor[15] and is determined from the analysis. When Eq. (1) takes the form

$$\left[\Delta + E - V(x_i)\right]u = 0 \tag{4}$$

where $E \neq 0$ and Δ is the Laplace-Beltrami operator for a space of constant Riemannian curvature, the method of Stäckel[10,16,17] can be used to find the separable coordinates. When (1) does not have the form (4) we need recourse to other methods. For example for the wave or Laplace equation (i.e. $E = V = 0$) one can find orthogonal coordinates by the method of obtaining confocal cyclides from hyperspherical coordinates[15,18,6j]. This seems to be related to conformal symmetries. When none of these methods work, one simply uses brute force to find the coordinates. This usually applies to nonorthogonal coordinates[6k]. It should be mentioned that the classification of separable coordinate systems is really a classification of equivalence classes (i.e. orbits) of separable systems defined by some reasonable group of geometric symmetries H which may or may not be the symmetry group G. In general we are interested in both H-and G-inequivalent separable systems.

4) Associate with the (n-1) separation constants λ_i for a given separable system of (1) an H or G orbit of (n-1) commuting members $\{S_i\}$ of $\mathcal{S}$, such that

$$S_i u = \lambda_i u \tag{5}$$

For equations of the form (4), the existence of such a set S_i has been shown[19]. For all other equations treated so far it has always turned out to be the case although no general theorem as yet exists. Let us also mention that if a variable appears in (1) only to first order in its derivative, the corresponding operators will also be of first order and thus a member of $\mathcal{g} \subset \mathcal{L}$.

5) Find a simpler model for the Lie algebra $\mathcal{g}$ and Lie group G in the sense that it acts over a space of lower dimension. If possible construct a Hilbert space and a unitary representation of G for both models and the associated unitary transformation between the two models. Do all calculations in the simple model such as spectral analysis, computation of overlap functions, etc. This provides the derivation of many expansion theorems between the generalized eigenbases for each separable system. If the construction of a Hilbert space is not possible, use Weisner's method[3,20] to obtain generating functions relating various bases.

Steps 1)-4) provide the basic procedure for obtaining and relating the symmetries of a differential equation to the separable coordinate systems for that equation. Although step 5) is somewhat extra its importance cannot be overestimated. It is precisely this step which enables one to derive the kind of information about the solutions of a given differential equation which physicists and applied mathematicians are interested in. It provides deep insight into integral relations and expansion formalae between the various special functions which occur as eigenbases for unitary representations of Lie groups. Some of these eigenbases can be related to subgroup reductions $G_n \subset \cdots \subset G_1 \subset G$ where $G_0, \cdots, G_n$ denote continuous subgroups of G. The remaining eigenbases have been called nonsubgroup reductions although it appears possible to relate these to subgroup reduction where the subgroups may now be discrete[21].

In order to illustrate the above procedure in more concrete terms we discuss some examples.

B. Helmholtz Eq. in Euclidean 2-space

We consider the equation

$$u_{x_1 x_1} + u_{x_2 x_2} + \lambda u = 0 \tag{6}$$

1) It is straight forward to calculate the symmetry group for (6) and the results give the well known group E(2) with its Lie algebra e(2). A basis for e(2) is given by

$$M = x_1 \partial_{x_2} - x_2 \partial_{x_1} \qquad P_1 = \partial_{x_1} \qquad P_2 = \partial_{x_2} \tag{7}$$

whose integrated group action is well known and will not be given.

2) The second order symmetries $\mathcal{S}/\mathcal{G}$ are of type 1 which are spanned by the second order members of the enveloping algebra M^2, P^2_1, P_2, P_1P_2, $MP_1 + P_1M$, $MP_2 + P_2M$. The action of E(2) splits $\mathcal{S}/\mathcal{G}$ into the orbits $M^2 + \ell P_1^2$, $MP_1 + P_1M$, M^2, P_1^2.

3) The separation of variables for (6) is well known[17,22] and there are four E(2)-inequivalent systems given by i) Cartesian, ii) polar, iii) parabolic, iv) elliptic.

4) The correspondence between orbit representatives and separable systems is one-one[5c]. Respectively we have

i) $x_1, x_2 \longleftrightarrow P_1^2$

ii) $x_1 = r\cos\theta$, $x_2 = r\sin\theta \longleftrightarrow M^2$

iii) $x_1 = \frac{1}{2}(\xi_1^2 - \xi_2^2)$, $x_2 = \xi_1\xi_2 \longleftrightarrow MP_1 + P_1M$

iv) $x_1 = \ell \operatorname{ch}\rho \cos\sigma$, $x_2 = \ell \operatorname{sh}\rho \sin\sigma \longleftrightarrow M^2 + lP_1^2$

5) A simpler model for e(2) is

$$\mathcal{M} = \partial_\theta \qquad \mathcal{P}_1 = i\lambda\cos\theta \qquad \mathcal{P}_2 = i\lambda\sin\theta \tag{8}$$

We can set up Hilbert spaces $L^2(R_2)$ such that (7) are skew-adjoint operators and $L^2(S^1)$ where S^1 is the unit circle such that (8) are skew-adjoint operators. A unitary transform between $L^2(S^1)$ and $L^2(R_2)$ is

$$F(\underset{\sim}{x}) = \int_0^{2\pi} d\theta\, e^{i\underset{\sim}{k}\cdot\underset{\sim}{x}\lambda} f(\theta) \tag{9}$$

where $\underset{\sim}{k} = (\cos\theta, \sin\theta)$ and $f \in L^2(S^1)$. We then find the following basis functions for the four systems in the two models.

Type		$L^2(S^1)$	$L^2(R_2)$
i)	Cartesian	delta	Product of exponentials
ii)	Polar	exponential	Product of Bessel and exponential
iii)	Parabolic	Powers of trig. functions	Products of Parabolic Cylinder
iv)	Elliptic	Mathieu	Product of Mathieu

Writing (9) explicitly for each of the above basis functions gives us integral identities. Computing overlap functions in the $L^2(S^1)$ model allows us to derive expansion formulas for the $L^2(R_2)$ model.

C. <u>Free particle Schrödinger equation in two space and one time dimensions</u>[6g,h]

We consider the equation

$$u_{x_1x_1} + u_{x_2x_2} + iu_t = 0 \tag{10}$$

It should also be mentioned here that most of the following remains unchanged if instead of the Schrödinger equation we remove the i and consider the heat equation[8].

1) The symmetry group for (10) can be calculated[23] and we find the structure $G = [SL(2,R) \otimes O(2)] \circledS W_2$ the semi-direct product of the special linear group $SL(2,R)$ with the 6 dimensional Weyl group W_2. A basis for its Lie algebra is

$$\begin{aligned} K_{-2} &= \partial_t \qquad K_2 = t^2\partial_t + t\,\underline{x}\cdot\partial_{\underline{x}} + \tfrac{i\underline{x}^2}{4} + \tfrac{t}{2} \\ D &= 2t\partial_t + \underline{x}\cdot\partial_{\underline{x}} + 1 \qquad M = x_1\partial_{x_2} - x_2\partial_{x_1} \\ P_j &= \partial_{x_j} \qquad B_j = -t\partial_{x_j} + \tfrac{i}{2}x_j \qquad E = i \end{aligned} \tag{11}$$

Again (11) can be integrated to a local Lie group whose details are omitted.

2) The second order symmetries are of type 1 and turn out to be all symmetric quadratic forms of B_j, P_j, E and M, plus the elements of $\mathcal{G}$. Here we are interested in G-orbits of both $\mathcal{G}$ and the factor space $\mathcal{S}/\mathcal{G}$.

3) We can first separate off the t variable[6f,g,24]. This can be done in 4 different G-inequivalent ways. This reduces the problem to the Helmholtz equation with the addition of one of four different potentials which correspond to the four different t separations, i.e.

$$u_{v_1 v_1} + u_{v_2 v_2} + [E - V(v_1, v_2)]u = 0 \tag{12}$$

where V is one of the following four potentials $V = 0, v_1, \pm(v_1^2 + v_2^2)$; i. e. a free particle, linear potential, attractive or repulsive harmonic oscillator. Then the problem reduces to finding the separable systems of the Helmholtz equation which are compatible with the added potentials. In all there are 15 G-inequivalent systems listed as follows: free particle, systems i)-iv); linear potential, system i) and iii); attractive and repulsive oscillators with systems i), ii) and iv) each; three systems obtained by separating the x_1 variable and solving the corresponding one-space dimensional version[6f] of (10). However, the whole group G is not so easy to visualize geometrically so it is of interest to classify the separable systems up to equivalence of the more geometrically meaningful subgroup $D \circledS G_2$ of the two-dimensional Galilei group extended by dilatations. With this definition of equivalence, there are 26 separable coordinate system. A doubling occurs for all but the attractive harmonic oscillator types.

4) To each separable coordinate system there corresponds a commuting orbit pair (K,S) where $K \in \mathcal{G}/\{E\}$ and $S \in \mathcal{S}/\mathcal{G}$. The correspondence, however, is not one-one. The following orbits in $\mathcal{G}/\{E\}$ correspond to t-separation: K_{-2} or $K_2 \longleftrightarrow$ free particle; $K_{-2} + B_1$ or $K_2 + P_1 \longleftrightarrow$ linear potential; $K_{-2} + K_2$ or $D \longleftrightarrow$ repulsive harmonic oscillator; $K_{-2} - K_2 \longleftrightarrow$ attractive harmonic oscillator. The choice for S then is similar to what we had

previously for the Helmholtz equation with an additional term corresponding to the potential.

5) A simpler two variable model for $\mathcal{G}$ is obtained by putting $t = 0$ in (11) and making the replacement $\partial_\lambda \rightarrow i(\partial_{x_1x_1} + \partial_{x_2x_2}) = i\Delta$. We will denote the generators in this model by script letters corresponding to the generators (11). The group action for this model then gives rise to a group representation by integral transforms[25]. A one-parameter subgroup of this group representation is the transform which connects the two models

$$F(\underline{x}, t) = e^{t\mathcal{K}_2} f(x) = \frac{1}{4\pi i t} \iint_{-\infty}^{\infty} d^2y \exp[\bar{i}(\underline{x} - \underline{y})^2/4t] \, f(\underline{y}) \tag{13}$$

This transform is a unitary transform from $L^2(R_2)$ to $L^2(R_2)$ solutions of (10). We provide an explicit example by applying (13) to the case of the linear potential in parabolic coordinates. In this case the separable coordinates are $x_1 = \tau(\xi_1^2 - \xi_2^2)/2 + \frac{1}{\tau}, x_2 = \xi_1\xi_2\tau, t = \tau$ with the commuting orbit pair $(K_2 + P_1, B_2M + MB_2 + P_2^2)$ and the corresponding script generators for the simple model. Without going into further details[24] we simple remark that the script generators give rise to an exponential times an anharmonic oscillator wave function, i. e. an $L^2(R_1)$ solution of

$$h'' - (\mu + \lambda x^2 + \tfrac{1}{4}x^4) = 0$$

denoted by $h_n(x; \lambda, \tfrac{1}{4})$. Then in the three variable model we have the separable solutions $h_n(\xi_1; \lambda, \tfrac{1}{2})\, h_n(i\xi_2; \lambda, -\tfrac{1}{2})$ times a modular factor (exponential here). The transform (13) then gives rise to an integral identity

$$h_n(\xi_1; \lambda, \tfrac{1}{2})\, h_n(i\xi_2; \lambda, -\tfrac{1}{2}) = \tau_n(\lambda) \int_{-\infty}^{\infty} dy \, e^{i\xi_1\xi_2 y} Ai\left[2^{-1/3}\left(\frac{\xi_1^2 - \xi_2^2}{2} - 2\lambda + \frac{y^2}{2}\right)\right] h_n(y; \lambda, \tfrac{1}{4})$$

where $Ai(z)$ is an Airy function. It is remarkable that Lie theory can provide information about problems as asymmetrical as the cuartic anharmonic oscillator. Many other identities can be found by similar techniques. An addition theorem[26] can be derived for the functions h_n. This method has also been applied to the wave functions for the Stark effect[26] of the H-atom, the sextic anharmonic oscillator[27] as well as many other functions.

C. <u>Conclusion</u>

As a conclusion we present a table listing the differential equations which have been studied from the point of view of symmetry groups and separation of variables along with the relevant references to the literature. It should also be mentioned that there is a forthcoming book on the subject of symmetries and separation of variables by W. Miller, Jr.

Equation	Group	Separable System	Reference
$\Delta u + \lambda u = 0$			
i) 2-dimensions			
Euclidean			
$\lambda \neq 0$	E(2)	4, in text	5c,6a
$\lambda = 0$	infinite	infinite, conform.	
Pseudo-Euclidean			
$\lambda \neq 0$	E(1,1)	11	6c,d
$\lambda = 0$	infinite	infinite, conform.	
Sphere	O(3)	2, sph., ellpt.	5d,6e,22,28a
2-sheet Hyperboloid	O(2,1)	9	5a,b,d,6e,22,28b
ii) 3-dimensions			
Euclidean			
$\lambda \neq 0$	E(3)	11	6b,17,22,29,30
$\lambda = 0$	O(4,1)	17	18,30
Pseudo-Euclidean			
$\lambda \neq 0$	E(2,1)	53 or more	6c,j,22
$\lambda = 0$	O(3,2)	over 90	6i-k
Sphere	O(4)	6	17,21,22
2-sheet Hyperboloid	O(3,1)	34	19,22
$\Delta u + iu_t = 0$			
i) 1-dimension	$SL(2,R) \rtimes W_1$	7	6f
ii) 2-dimensions			
Euclidean	$[SL(2,R) \otimes O(2)] \rtimes W_2$	26	6g,h
Pseudo-Euclidean	$[SL(2,R) \otimes O(1,1)] \rtimes W_2$	58	31
$\Delta u + iu_t - \frac{\alpha}{x_1^2} - \frac{\beta}{x_2^2}$			
Euclidean $\alpha, \beta \neq 0$	SL(2,R)	15	26
$\alpha = 0$	$SL(2,R) \rtimes W_1$	25	26
$\Delta u + \frac{\alpha}{x_1^2} u = 0$			
Euclidean	SL(2,R)	9	6ℓ,30
Pseudo-Euclidean	SL(2,R)	9	6ℓ

1. L. Lie, Theorie der Transformationsgruppen. Vol. 1, 2, 3, Leipzig (reprinted by Chelsea, New York, 1970).

2. L. Ovsjannikov, Group Properties of Differential Equations (Acad. Sci. USSR, Novasibirsh, 1962) (In Russian) (Translated by G. Bluman, 1967); W. Miller, Jr., SIAM J. Math. Anal. 4, 314 (1973); G. W. Bluman and J. D. Cole, Similarity Methods for Differential Equations (Springer-Verlag, New York, 1974).

3. W. Miller, Jr., Lie Theory and Special Functions, (Acad., New York, 1968).

4. N. Ja. Vilenkin, Special Functions and the Theory of Group Representations (Amer. Math. Soc., Providence, 1968).

5. a) P. Winternitz and I. Fris, Sov. Phys., JNP 1, 636 (1965); b) P. Winternitz, Ja. A. Smorodinskiǐ, and M. Uhlir, ibid 1, 113 (1965); c) P. Winternitz, Ja. A. Smorodinskiǐ, M. Uhlir, and I. Fris, ibid 4, 444 (1967); d) P. Winternitz, I. Lukac, and Ja. A. Smorodinskiǐ, ibid 7, 139 (1968).

6. a) W. Miller, Jr. SIAM J. Math. Anal, 5, 626 (1974); b) ibid 822 (1974); c) E. G. Kalnins, ibid 6, 340 (1975); d) E. G. Kalnins and W. Miller, Jr., J. Math. Phys. 15, 1025 (1974); e) ibid 1263 (1974); f) ibid 1728 (1974); g) C. P. Boyer, E. G. Kalnins, and W. Miller, Jr., ibid, 16, 499 (1975); h) ibid 512, (1975); i) E. G. Kalnins and W. Miller, Jr., Univ. of Minnesota preprint (to appear in J. Math. Phys.); j) E. G. Kalnins and W. Miller, Jr., Univ. de Montreal preprint CRM 467; k) ibid CRM 489; l) E. G. Kalnins and W. Miller, Jr., Univ. of Minnesota preprint.

7. G. Birkhoff, Hydrodynamics (Princeton U. P. Princeton, 1950).

8. G. W. Bluman and J. D. Cole, J. Math. Mech. 18, 1025 (1969).

9. B. K. Harrison and F. B. Estabrook, J. Math. phys. 12, 653 (1971); H. D. Wahlquist and F. B. Estabrook, ibid 16, 1 (1975).

10. P. Stackel, Habil-Schr. Halle 1891. Math. Ann. 49, 145 (1897). T. Levi-Civita, Math. Ann. 59, 383 (1904).

11. P. Havas, J. Math. Phys. 16, 1461 (1975).

12. C. P. Boyer and E. G. Kalnins, to be published.

13. C. P. Boyer and M. Penafiel N. preprint UNAM, Comun. Tech. 6-94 (1975).

14. W. Miller, Jr., Proceedings of the Seminar on Special Functions, Madison, Wis. (1975).

15. P. M. Morse and H. Feshbach, Methods of Theoretical Physics (McGraw-Hill, New York, 1953) Vol. sec. 5.1.

16. H. P. Robertson, Math. Ann. 98, 749 (1928).

17. L. P. Eisenhart, Ann. Math. 35, 284 (1934)

18. M. Bôcher, Uber die Reihenentwickelungen der Potentialtheorie (Druck und Verlag, Leipzig, 1894).

19. Ja. A. Smorodinskiǐ and I. I. Tugov, Sov. Phys. JETP, 23, 434 (1966).

20. L. Weisner, Pac. J. Math. 5, 1033 (1955).

21. E. G. Kalnins, W. Miller, Jr., and P. Winternitz, Univ. de Montreal preprint CRM 416 to appear in SIAM J. of Appl. Math.

22. M. P. Olevskiǐ, Mat. Sb. 27, 379 (1950).

23. U. Niederer, Helv. Phys. Acta, 46, 191 (1973); C. P. Boyer, ibid 47, 589 (1974).

24. E. G. Kalnins, Proceedings of the 3rd International Conference on Group Theoretical Methods in Physics, Marseille, 1974.

25. K. B. Wolf, J. Math. Phys. 15, 1295 (1974); ibid 15, 2101 (1974) C. P. Boyer and K. B. Wolf, ibid, 16, 1493 (1975).

26. C. P. Boyer, UNAM preprint Comun. Tech. to appear in SIAM J. Math. Anal.

27. C. P. Boyer and K. B. Wolf, UNAM preprint Comun. Tech. to appear in J. Math. Phys.

28. a) J. Patera and P. Winternitz, J. Math. Phys. 14, 1130 (1973); b) N. MacFadyen and P. Winternitz, J. Math. Phys. 12, 281 (1971).

29. A. A. Makarov, Ja. A. Smorodinskiǐ, Kh. Valiev, and P. Winternitz, Nuovo Cimento 52A, 1061 (1967).

30. C. P. Boyer, E. G. Kalnins, and W. Miller, Jr., Nagoya Math. J. to appear.

31. C. P. Boyer and E. G. Kalnins, to appear.

ON THE DETERMINATION OF FACTOR SYSTEMS OF PUA - REPRESENTATIONS

P.M. van den Broek

Institute for theoretical physics, University of Nijmegen, the Netherlands

Abstract: A method is developed to obtain a complete set of inequivalent factor systems of PUA - representations of a group with a subgroup of index two from the factor systems of this subgroup.

1. Introduction

Let G be a group which has a subgroup H of index two. A projective unitary-antiunitary (PUA-) representation of G is a mapping D from G into the operators on some Hilbert space $\mathcal{H}$ such that

i) the operator D(g) is unitary if $g \in H$ and antiunitary if $g \notin H$

ii) $D(g)\, D(g') = \sigma(g,g')\, D(gg')$ $\forall g,g' \in G$ for some mapping

$$\sigma: G \times G \rightarrow U(1).$$

It is customary to choose D(e) = I, where e is the identity of G and I is the identity operator on $\mathcal{H}$. Then σ satisfies

$$\sigma(g,e) = \sigma(e,g) = 1 \qquad \forall g \in G \tag{1.1}$$

and

$$\sigma(g_1,g_2)\, \sigma(g_1g_2,g_3) = \sigma(g_1,g_2g_3)\, \sigma^{g_1}(g_2,g_3) \;\forall g_1,g_2,g_3 \in G \tag{1.2}$$

where λ^g is defined by

$$\lambda^g = \begin{cases} \lambda & \text{if } g \in H \\ \lambda^* & \text{if } g \notin H \end{cases},$$

the asterisk denoting complex conjugation.

A mapping $\sigma: G \times G \rightarrow U(1)$ which satisfies (1.1) and (1.2) is called a factor system of G with respect to H. In the following a factor system of G shall always mean a factor system of G with respect to H. If D is a PUA-representation with factor system σ and c is a mapping from G into U(1) with c(e) = 1 then $D'(g) = c(g)\, D(g)$ is a PUA-representation of G with factor system

$$\sigma'(g_1,g_2) = \frac{c(g_1)\, c^{g_1}(g_2)}{c(g_1g_2)}\, \sigma(g_1,g_2) \tag{1.3}$$

Two factor systems σ and σ' are called equivalent if a mapping $c: G \rightarrow U(1)$ with c(e) = 1 exists such that (1.3) holds. Factor systems of H are defined in an analogous way, the only difference being the absence of the complex conjugation in

(1.2) and (1.3). The theory of PUA-representations and its use in physics is described by Murthy [6] , Parthasarathy [7] , Janssen [4] and Shaw & Lever [9] , [10] . Factor systems of PU-representations are studied quite extensively [1], [2] ,[5] ,[8] . This however is not the case for factor systems of PUA-representations.

It is the aim of this paper to determine a complete set of inequivalent factor systems of G when the factor systems of H are known. This problem has already been attacked by Bradley & Wallis [3] , but they have not obtained the general solution.

Janssen [4] has given a method to obtain the factor systems of G in the case where G is finite, without using the factor systems of the subgroup.

2. Reduction of the problem

First we choose an element a_o from $G \setminus H$ which remains fixed during the following. We can write all elements of $G \setminus H$ as a_oh or $h'a_o$ for some $h,h' \in H$. Suppose σ is a factor system of G.

The restriction σ_H of σ to $H \times H$ is then a factor system of H. If σ_H is a factor system of H we may ask the question whether or not there exists a factor system σ of G such that its restriction to $H \times H$ is σ_H. If such σ exists it is called an extension of σ_H.

Lemma 1 An extension of a factor system σ_H of H to a factor system σ of G is completely determined by the elements $\sigma(a_o,a_o)$, $\sigma(h,a_o)$ and $\sigma(a_o,h)$ for all $h \in H$.

Proof The following relations follow immediately from (1.2):

$$\sigma(a_oh,h') = \frac{\sigma(a_o,hh')\ \sigma^*(h,h')}{\sigma(a_o,h)} \qquad (2.1)$$

$$\sigma(h,h'a_o) = \frac{\sigma(h,h')\ \sigma(hh',a_o)}{\sigma(h',a_o)} \qquad (2.2)$$

$$\sigma(ha_o,a_oh') = \frac{\sigma(h,a_o^2)\,\sigma(a_o,a_o)\,\sigma(ha_o^2,h')}{\sigma(h,a_o)\sigma^*(a_o,h')} \qquad (2.3)$$

This proves the lemma.

If these relations are substituted in (1.2) we obtain after laborious manipulation the following three equations for $\sigma(a_o,a_o)$, $\sigma(a_o,h)$ and $\sigma(h,a_o)$:

$$\sigma^*(h,a_o)\sigma^*(a_o,a_o^{-1}hh'a_o)\,\sigma(a_o^{-1}ha_o,a_o^{-1}h'a_o)\sigma^*(h',a_o)\,\sigma(a_o,a_o^{-1}ha_o)$$

$$\sigma(hh',a_o)\;\;\sigma(h,h')\,\sigma(a_o,a_o^{-1}h'a_o) = 1 \quad \forall h,h' \in H \qquad (2.4)$$

$$\sigma^*(a_oha_o^{-1},a_o^2)\,\sigma(a_o^2,a_o^{-1}\;ha_o)\,\sigma^*(a_o,h)\,\sigma(a_o,a_o^{-1}\;ha_o)$$

$$\sigma(a_oha_o^{-1},a_o)\,\sigma^*(h,a_o) = 1 \qquad \forall h \in H \qquad (2.5)$$

$$\sigma(a_o,a_o)\;\;\sigma(a_o,a_o)\;\;\sigma(a_o^2,a_o)\,\sigma^*(a_o,a_o^2) = 1 \qquad (2.6)$$

Note that for each solution $\sigma(h,a_o)$ and $\sigma(a_o,h)$ of (2.4) and (2.5) we obtain two values of $\sigma(a_o,a_o)$ from (2.6).

The following theorem has now been derived:

__Theorem 1__ All extensions of a given factor system σ of H to factor systems of G are obtained from the solutions $\sigma(h,a_o)$ and $\sigma(a_o,h)$ of the equations (2.4) and (2.5).
For each solution of (2.4) and (2.5) there are two extensions which are given by the equations (2.6), (2.1), (2.2) and (2.3).

To obtain a complete set of inequivalent factor systems of G it is only necessary to consider extensions of inequivalent factor systems of H. This follows from the fact that if σ_H and σ'_H are two equivalent factor systems of H and σ is an extension of σ_H then σ'_H has an extension σ' which is equivalent with σ . On the other hand inequivalent factor systems of H have inequivalent extensions. So in order to obtain a complete set of inequivalent factor systems of G we have to find all inequivalent extensions of one representative of each class of equivalent factor systems of H.

3. Solution of the problem

In this section we present without proof a method to obtain a complete set of inequivalent extensions of a factor system of H. First we give a criterion to decide whether there exist extensions or not. Then we give for the case where extensions do exist a set of extensions which contains a complete set of inequivalent ones. Finally we obtain this complete set.

Define an equivalence relation in H: h and h' are called equivalent if there is a $n \in \mathbb{Z}$ such that $a_o^n h a_o^{-n} = h'$. In this way H is divided into classes. Let H_o be a set of elements of H which contains exactly one element from each class with an even number of elements and none from each other class. Let σ be a factor system of H and define the mapping D_σ from H into the complex numbers of modulus unity by

$$D_\sigma(h) \begin{cases} \prod_{\substack{n=0 \\ n \text{ even}}}^{p-2} \left[\sigma^*(a_o^{n+1} h\, a_o^{-(n+1)}, a_o^2)\, \sigma(a_o^2, a_o^{n-1} h\, a_o^{-(n-1)})\right] & \text{if } h \in H_o \\ 1 & \text{if } h \notin H_o \end{cases}$$

where p is the number of elements in the class containing h. If σ' is a factor system of H' which is equivalent with σ then $D_\sigma = D_{\sigma'}$.

<u>Theorem 2</u> σ can be extended to a factor system of G if and only if there is a factor system σ' of H which is equivalent with σ and obeys

$$\sigma'(h,h')\, \sigma'(a_o^{-1} h a_o, a_o^{-1} h' a_o) = D_\sigma(h)\, D_\sigma(h')\, D_\sigma^*(hh') \tag{3.1}$$

and

$$\sigma'^*(a_o h a_o^{-1}, a_o^2)\sigma'(a_o^2, a_o^{-1} h a_o) = D_\sigma^*(a_o h a_o^{-1})\, D_\sigma(h) \quad . \tag{3.2}$$

Let R(H) be the set of all unitary one-dimensional representations Δ of H with the property $\Delta(h) = \Delta(a_o h a_o^{-1})$

Theorem 3 If σ satisfies (3.1) and (3.2) then a complete set of inequivalent extensions of σ is contained in the set extensions given by $\sigma(a_o,h) = 1$ and $\sigma(h,a_o) = \Delta(h)\, D_\sigma(h)$ where $\Delta \in R(H)$.

The only thing we still need is a criterion to decide when two extensions of this set are equivalent.

Theorem 4 Let σ_1 and σ_2 be two extensions of the set defined above. Then there exists a $\Delta \in R(H)$ with $\dfrac{\sigma_1(h,a_o)}{\sigma_2(h,a_o)} = \Delta(h)$.

σ_1 and σ_2 are equivalent if and only if there exists a one-dimensional unitary representation Δ_o of H with the properties

$$\Delta_o(ha_o^{-1}ha_o) = \Delta(h) \quad \forall\, h \in H \quad \text{and} \quad \Delta_o^*(a_o^2) = \frac{\sigma_1(a_o,a_o)}{\sigma_2(a_o,a_o)} .$$

References

[1] Backhouse, N.B., Quart.J.Math. Oxford (2), 21 (1970) 277.

[2] - , and C.J. Bradley, Quart.J.Math. Oxford (2), 23 (1972) 225.

[3] Bradley, C.J., and Wallis, D.E., Quart.J.Math. Oxford (2), 25 (1974) 85.

[4] Janssen, T., J.Math.Phys. 13 (1972) 342.

[5] Mackey, G.W., Acta Math. 99 (1958) 265.

[6] Murthy, M.V., J.Math.Phys. 7 (1966) 853.

[7] Parthasarathy, K.R., Commun.Math.Phys. 15 (1969) 305.

[8] - , Multipliers on locally compact groups, Springer Verlag, Berlin, Heidelberg, New York 1969.

[9] Shaw, R., and J. Lever, Commun.Math.Phys. 38 (1974) 257.

[10] - , ibid. 38 (1974) 278.

COMPLEX EXTENSION OF THE REPRESENTATION OF THE SYMPLECTIC GROUP ASSOCIATED WITH THE CANONICAL COMMUTATION RELATIONS x)

M. Brunet and P. Kramer

Institute for Theoretical Physics, University of Tübingen,

Tübingen, German Federal Republic

I. Introduction

As is well knwon, the Weyl form of the canonical commutation relations (CCR's) together with the Neumann's theorem leads to a strongly continuous representation of the real symplectic group. This representation has been studied by Bargmann [1] with the carrier space taken as Hilbert space with a reproducing kernel. In this case the operators onto which the group elements are mapped are describable as kernel operators, the kernel being a continuous function of the group elements. By considering the real symplectic group as a subgroup of the complex symplectic group it is then possible to formally extend the kernel function to a function of the elements of the complex symplectic group. The question then is what are the properties of the operators which these extended kernel functions define, in particular when are the operators bounded. It has been shown by Kramer, Moshinsky and Seligman [2] that in the case of the group Sp(2,$\mathbb{C}$) not all of these operators are bounded but those that are define a subsemigroup of Sp(2,$\mathbb{C}$). We shall show that this is also the case for Sp(2n,$\mathbb{C}$) for arbitrary n.

To discuss more explicitly the representation which we wish to extend we require a few facts about the Hilbert space employed by Bargmann. Let μ be the measure on $\mathbb{C}^n$ defined by the weighting function $z \mapsto \pi^{-n} \exp(-\|z\|^2)$ and for fixed n let

x) Supported by Deutsche Forschungsgemeinschaft

$\mathcal{Y} = \mathcal{L}^2(\mu, \mathbb{C})$ be the Hilbert space of complex valued functions square integrable with respect to this measure. With $\mathcal{E}$ denoting the space of all entire functions on C^n the Hilbert space $\mathcal{F} = \mathcal{E} \cap \mathcal{Y}$ has a reproducing kernel $K^r(z,w) = e^{\bar{z}\cdot w}$ where $\bar{z}\cdot w = \bar{z}_1 w_1 + \cdots + \bar{z}_n w_n$ i.e. for all $f \in \mathcal{F}$ the relation $f(z) = \int_{C^n} K^r(z,w) f(w)\, d\mu(w)$ obtains. The Hilbert space $\mathcal{F}$ has the remarkable property that each bounded operator on $\mathcal{F}$ has a kernel, i.e. if T is bounded on $\mathcal{F}$. then $(Tf)(z) = \int_{C^n} K(z,w)\, f(w)\, d\mu(w)$. Related to the existence of a reproducing kernel is the existence of a family $e_a,\ a \in \mathbb{C}^n$ of elements of $\mathcal{F}$ with the property that $\langle e_a, f \rangle = f(a)$ for all f in $\mathcal{F}$. These vectors, the principal vectors, are given explicitly by $e_a(z) = \exp(\bar{a}\cdot z)$. They are complete and the subspace $\mathcal{D}$ generated by the principal vectors is dense in $\mathcal{F}$. This property is extremely useful in checking the continuity of mappings into $\mathcal{F}$.

Recall that the complex symplectic group $H = Sp(2n, \mathbb{C})$ is the group of invertible isometries of the antisymmetric form $\{\ ,\ \}$ on C^{2n}, $\{z, w\} = \sum_{i=1}^{n} (z_i w_{i+n} - z_{i+n} w_i)$. Denoting an element of H by the matrix $h = \begin{pmatrix} \lambda & \mu \\ \nu & \rho \end{pmatrix}$ the H adjoint of h is given by $h^{\#} = \begin{pmatrix} {}^t\rho & -{}^t\mu \\ -{}^t\nu & {}^t\lambda \end{pmatrix}$. The symplectic property of h is then $h^{-1} = h^{\#}$, i.e. $h^{\#}h = hh^{\#} = 1$. The real symplectic group G_o is defined similarly by restricting the form $\{\ ,\ \}$ to $\mathbb{R}^{2n} \times \mathbb{R}^{2n}$. Rather than working directly with G_o it is convenient to work with the isomorphic group $G = WG_oW^{-1}$ where $W = \frac{1}{\sqrt{2}}\begin{pmatrix} 1 & i \\ 1 & -i \end{pmatrix}$. The group G is sometimes called the complex form of the real symplectic group (not to be confused with H) and is equal to the intersection of H with U(n,n). The condition that $g \in$ G is that g have the form $g = \begin{pmatrix} \lambda & \mu \\ \bar{\mu} & \bar{\lambda} \end{pmatrix}$ where λ and μ satisfy $\lambda^*\lambda - \mu^*\mu = 1$ and ${}^t\lambda\bar{\mu} = \mu^*\lambda$ where ${}^t\lambda$ denotes the transpose of λ. With these definitions the representation of G (and therefore also of G_0) studied by Bargmann is defined as follows. To each element $g \in G$ we assign the kernel K_g,

$$K_g(z,w) = \exp\left\{ \tfrac{1}{2} z \cdot \bar{\mu} \lambda^{-1} z - \tfrac{1}{2} \bar{w} \cdot \lambda^{-1} \mu \bar{w} + z \cdot {}^t\lambda^{-1} \bar{w} \right\} \tag{1}$$

and a complex number $\sigma_g = \det^{-1/2} \lambda$ (We adopt the convention that $z^{1/2} = |z| \exp(\tfrac{1}{2} i \operatorname{Arg} z)$. Bargmann has shown that K_g defines a bounded oprator S_g and that if we define $T_g = \sigma_g S_g$ the double valued mapping $g \mapsto (T_g, -T_g)$ is a double valued unitary representation of G. Of course this double valued representation can and should be viewed as a representation of the universal covering group $\tilde{G}$ (G is not simply connected).

To define an extension of the above representation to H it is necessary to extend both the mappings $g \mapsto K_g$ and $g \mapsto \sigma_g$. We are immediately forced by the appearance of λ^{-1} in (1) to restrict our attention to the set $H_0 \subset H$ consisting of those $h \in H$ such that λ is invertible. Then the formal replacement $\begin{pmatrix} \lambda & \mu \\ \bar{\mu} & \bar{\lambda} \end{pmatrix} \to \begin{pmatrix} \lambda & \mu \\ \nu & \rho \end{pmatrix}$ defines a mapping $h \mapsto K_g$ where K_h is given by

$$K_h(z,w) = \exp\left\{ \tfrac{1}{2} z \cdot \nu \lambda^{-1} z - \tfrac{1}{2} \bar{w} \cdot \lambda^{-1} \mu \bar{w} + z \cdot {}^t\lambda^{-1} \bar{w} \right\} \tag{2}$$

Some comments concerning the mapping $h \mapsto K_h$ are in order. Observe first that $K_h(\ ,w)$, $K_h(\ ,w)(z) = K_h(z,w)$ is an entire function. Suppose that it should happen that K_h defines a bounded operator S_h on $\mathcal{G}$. It is readily computed that for a principal vector e_a, $S_h e_a = K_h(\ ,a)$ whence $S_h(\mathcal{D}) \subset \mathcal{F}$. From the assumed continuity of S_h and the fact that $\mathcal{F}$ is closed in $\mathcal{G}$ it follows that $S_h(\mathcal{F}) \subset \mathcal{F}$. This is highly fortuitous since it is somewhat easier to check boundedness on $\mathcal{G}$ (which is an $\mathcal{L}^2$ space) than on $\mathcal{F}$.

It is instructive to consider K_g in the case where h is the diagonal matrix $\begin{pmatrix} \delta & 0 \\ 0 & \delta^{-1} \end{pmatrix}$, δ being diagonal. One then finds that S_h is bounded on $\mathcal{Y}$ (and therefore on $\mathcal{F}$)if and only if $\|\delta\| \geqslant 1$, where $\| \ \|$ is the operator norm. This example shows that while non-trivial extensions exist they are not guaranteed.

II. Extending the Representation to a Subsemigroup of Sp(2n,$\mathbb{C}$)

In what follows it will be necessary to have conditions for the absolute convergence of integrals of the type

$$I = \int_{\mathbb{C}^n} \exp\left\{\tfrac{1}{2} z\cdot \gamma z + \tfrac{1}{2}\bar{z}\cdot\bar{\delta}\bar{z} + z\cdot \varepsilon \bar{z} + a\cdot z + b\cdot \bar{z}\right\} d\mu(z) \tag{3}$$

where γ and δ are $n\times n$ complex symmetric matrices, ε is an $n\times n$ selfadjoint matrix, and $a, b \in \mathbb{C}^n$. Using the standard technique of reducing convergence to that of a Laplace integral we have found that the integral in (3) is absolutely convergent if and only if

$$1 - \bar{\varepsilon} > 0 \tag{4a}$$

and

$$1 - \bar{\varepsilon} - \tfrac{1}{4}(\gamma + \delta)(1-\varepsilon)^{-1}(\bar{\gamma} + \bar{\delta}) > 0 \tag{4b}$$

where >0 (resp. $\geqslant 0$) denotes positive definiteness (resp. positiveness). We are interested in the value of the integral only in the case where $\varepsilon = 0$. This integral has been computed by Itzykson [3] and is given by

$$\begin{aligned} I = {} & \det{}^{-1/2}(1-\gamma\bar{\delta}) \times \\ & \times \exp\left\{\tfrac{1}{2} a\cdot\bar{\delta}(1-\gamma\bar{\delta})^{-1} a + \tfrac{1}{2} \bar{b}\cdot(1-\gamma\bar{\delta})^{-1}\gamma b + \right. \\ & \left. \qquad + b\cdot(1-\gamma\bar{\delta})^{-1} a \right\} \end{aligned} \tag{5}$$

where the sign of $\det(1-\gamma\bar{\delta})^{-1/2}$ is obtained by analytic continuation.

The possibility of extending the representation to a semigroup rests on

Lemma 1. Suppose that for $h_1, h_2 \in H$ the kernels K_{h_1} and K_{h_2} define operators S_{h_1} and S_{h_2} which are bounded on $\mathcal{G}$ and therefore on $\mathcal{F}$. Then $K_{h_1 h_2}$ defines a bounded operator $S_{h_1 h_2}$ on $\mathcal{F}$ and

$$S_{h_1 h_2} = \tau_{h_1, h_2} S_{h_1} S_{h_2} \tag{6}$$

where

$$\tau_{h_1, h_2} = \det{}^{-1/2}(1 - \lambda_1^{-1} \mu_1 \nu_2 \lambda_2^{-1}) \tag{7}$$

The proof of the above lemma is carried out by direct computation using the properties of the integral given by (3) discussed above and the symplectic property of h_1 and h_2. It follows immediately from this lemma that if $\mathcal{B}$ denotes the set of all $h \in H$ such that K_h defines a bounded operator on $\mathcal{F}$ then $\mathcal{B}$ is a semigroup.

Our starting point in determining the semigroup $\mathcal{B}$ is the determination of those $h \in H_0$ such that K_h defines a Hilbert Schmidt (HS) operator on $\mathcal{G}$. It follows easily that in this case K_h also defines an HS operator on $\mathcal{F}$. Since the product of two HS operators is HS the set $\mathcal{S}$ of all such h is also a semigroup.

In the $\mathcal{L}^2$ space $\mathcal{G}$ the condition that a kernel K defines an HS operator is equivalent to

$$\int_{\mathbb{C}^n} \int_{\mathbb{C}^n} |K(z,w)|^2 d\mu(z)\, d\mu(w) < \infty \tag{8}$$

This condition is fairly easy to compute. Using the symplectic nature of h we find that $\bar{K}_h(w,z) = K_{h^\#}(z,w)$ where $h^\# = \begin{pmatrix} \lambda^* & -\nu^* \\ -\mu^* & \rho^* \end{pmatrix}$. By Lemma 1 and Fubini's theorem we find that the condition expressed by (8) is equivalent to

$$\int_{\mathbb{C}^n} K_{h^\#}(w,z)\, K_h(z,w)\, d\mu(z) < \infty \tag{9a}$$

and

$$\int_{\mathbb{C}^n} K_{h^*h}(w,w)\, d\mu(w) < \infty \tag{9b}$$

The conditions expressed by (9) may now be explicitly computed using (4). Let $u: H_0 \to \mathfrak{S}$, where $\mathfrak{S}$ denotes the set of all $2n \times 2n$ selfadjoint matrices, be defined by

$$u(h) = \begin{pmatrix} \alpha(h)-1 & \gamma(h) \\ \gamma^*(h) & 1-\beta(h) \end{pmatrix} \tag{10}$$

and where

$$\alpha(h) = \lambda^*\lambda - \nu^*\nu \quad \text{(11a)} \qquad \beta(h) = \rho^*\rho - \mu^*\mu \quad \text{(11b)}$$

and

$$\gamma(h) = \lambda^*\mu - \nu^*\rho \tag{11c}$$

The conditions (9) and therefore the condition that $h \in \mathcal{S}$ may then be stated as $u(h) > 0$

Since no HS operator is unitary we see that $\mathcal{S} \cap G = \emptyset$. Hence a semigroup larger than $\mathcal{S}$ is required. Recall that whenever a topological semigroup is embedded in a Lie group the closure of that semigroup is again a semigroup. It turns out that $\mathcal{S}^-$, the closure of $\mathcal{S}$, contains G and is in fact equal to the semigroup $\mathcal{B}$.

It is possible here only briefly to indicate how the above assertions may be verified. First, using the fact that the set of all positive $2n \times 2n$ matrices is closed in $\mathfrak{S}$ and the continuity of u one shows that if $h \in \mathcal{S}^-$ then $u(h) \geq 0$. The reverse inequality is then established by showing that every neighborhood of the identity of H contains a point h_1 such that $u(h_1) > 0$ and then using the equation $u(h_1 h) = h_1^* u(h) h_1 + u(h_1)$ and the fact that H is a homogeneous space. One has then established that $\mathcal{S}^-$ is the set of all those $h \in H_0$ such that $u(h) \geq 0$. Since $u(g) = 0$ for all $g \in G$ it follows that $G \subset \mathcal{S}^-$. It remains to show that for all $h \in \mathcal{S}^-$, K_h defines a bounded operator on $\mathcal{F}$. To do this one proceeds as follows. Denoting the operator defined by K_h as S_h

we compute, using (4), that for any principal vector e_a , $S_h e_a$ is defined and has finite norm. Then for an element $v = \sum_i \xi_i e_{a_i}$ of $\mathcal{D}$ we find that $\| S_h v \|$ is finite if and only if $h \in \mathcal{S}^-$ and in that case $\| S_h v \|^2 \leqslant \tau_{h\#h} \| v \|$.

The above results permit us to define a representation of the semigroup $\mathcal{B}$ which maps semigroup elements onto contraction operators on $\mathcal{F}$. Let $\sigma_g = \det^{-1/2} \lambda$ and define $T_h = \sigma_h S_h$. Since $\tau_{h\#h} = |\det \lambda|^2 \det^{-1/2} \alpha(h)$ we find that $\| T_h \|^2 \leqslant |\sigma_h|^2 \tau_{h\#h} = \det^{-1/2} \alpha(h)$. But $h \in \mathcal{S}^-$ implies that $u(h) \geqslant 0$ and hence by (10) $h \in \mathcal{S}^-$ implies $\alpha(h) \geqslant 1$ and hence that $\det \alpha(h) > 1$. Thus T_h is a contraction operator for each $h \in \mathcal{S}^-$. With this normalization we find that $T_{hh'} = (\sigma_h \sigma_{h'} \tau_{hh'} / \sigma_{hh'}) T_{hh'} = \pm T_{hh'}$ by virtue of the symplectic nature of h and h' . Thus if we assign to each $h \in \mathcal{B} = \mathcal{S}^-$ the pair $(T_h, -T_h)$ we obtain a double valued representation of $\mathcal{B}$. Since our normalization agrees with the original normalization of Bargmann it follows that when restricted to G this representation becomes unitary and is in fact the representation given by Bargmann. Hence the original representation has been extended.

Lastly, we point out that the above representation of $\mathcal{B}$ is strongly continuous. Owing to the existence of the principal vectors this is not too difficult to show. It follows from the fact that the representation is contractive that it is strongly continuous if and only if for all principal vectors the mapping $h \mapsto T_h e_a$ is continuous. This is not difficult to check.

In way of summary we can say the following. We began with a double valued, strongly continuous representation of $G \subset H$ and showed that there exists a semigroup $\mathcal{B}$, $G \subset \mathcal{B} \subset H$ with its interior $\mathcal{S}$ also a semigroup. The original representation of G extends to a double valued, strongly continuous representation of $\mathcal{B}$ by contraction operators on $\mathcal{F}$, $h \mapsto T_h$ and for $h \in \mathcal{S}$, T_h

is a Hilbert Schmidt operator.

III. The Algebraic Structure of the Semigroup

We begin by pointing out that the semigroup $\mathcal{B}$ has a simple geometric interpretation. If (,) denotes the U(n,n) inner product on $\mathbb{C}^{2n}$ then the set of all invertible mappings A such that $(Az, Az) \geqslant (z,z)$ for all $z \in \mathbb{C}^{2n}$ is easily seen to define a semigroup, the semigroup of all U(n,n) expansion operators. Using the condition $h \in \mathcal{B}$ if and only if $u(h) \geqslant 0$ together with (10) it may be shown that the intersection of this semigroup with H is precisely $\mathcal{B}$. Similarly the intersection of the semigroup obtained from the condition $(Az, Az) > (z,z)$ with H is $\mathcal{S}$. Recalling that $G = U(n,n) \cap H$ this is not unreasonable.

Using the above model and the Jordan decomposition we are then able to show that every $h \in \mathcal{B}$ may be decomposed as $h = g_1 b g_2$ where $g_1, g_2 \in G$ and b is a direct sum of matrices of the form $\begin{pmatrix} 2 & 1 \\ -1 & 0 \end{pmatrix}$ and $\lambda \cdot \mathbf{1}$ with $\lambda \geqslant 1$ and $\begin{pmatrix} 2 & 1 \\ -1 & 0 \end{pmatrix} \in \mathcal{B}$. It then follows that $\mathcal{B}$ is the semigroup generated by G and those matrices having the form of b.

IV. Relation to the Canonical Commutation Relations

Define the unbounded operators Z_k and Y_k in $\mathcal{F}$ by

$(Z_k f)(z) = z_k f(z)$ and $(Y_k f)(z) = \dfrac{\partial f}{\partial z_k}$. If

$z'_k = \Sigma_\ell g^{-1}_{k\ell} z_\ell$ and $Y'_k = \Sigma_\ell g^{-1}_{k\ell} Y_\ell$

for $g \in G$ then the fact that the original representation of G was associated with a representation of the CCR's may be expressed by

$$Y'_k = T_{h^{-1}} Y_k T_h \quad (12a) \qquad \text{and} \qquad Z'_k = T_{h^{-1}} Z_k T_h \quad (12b)$$

this having meaning only when the operators are applied to functions in dom (Z_k) $\cap$ dom (Y_k) . In attempting to extend (12) to the representation of $\mathcal{B}$ we encounter the problem that if $h \in \mathcal{B}$ then $T_{h^{-1}}$ is in general unbounded. Nevertheless, we are able to show the following. If T_h has the operator defined by $\sigma_{h^{-1}} K_{h^{-1}}$ as its left inverse then $T_{h^{-1}}$ has a dense domain which includes the domains of Z_k and Y_k and (12) holds for functions in this domain.

REFERENCES

[1] V. Bargmann, "Group Representations on Hilbert Spaces of Analytic Functions" in Analytic Methods in Mathematical Physics, (Gilbert and Newton, eds), Gordon and Breach, New York (1968)

[2] P. Kramer, M. Moshinsky and T. H. Seligman "Complex Extensions of Canonical Transformations in Physics" in Group Theory and its Applications (E.M. Loebl ed.) Academic Press, New York (1975)

[3] C. Itzykson, Commun. Math. Phys. (N.Y.) 6, 301 (1959).

CONTINUOUS UNITARY PROJECTIVE REPRESENTATIONS OF POLISH GROUPS: THE BMS-GROUP *)

U. CATTANEO
Fachbereich Physik
Universität Kaiserslautern
D-6750 Kaiserslautern (Germany)

ABSTRACT

It is shown that every continuous unitary projective representation of a Polish group can be lifted to a Borel multiplier representation (i.e., to a representation "up to a Borel factor") and that this, in turn, can be derived from a continuous (ordinary) representation of a Polish group obtained from a central topological extension of the group considered by the multiplicative group of all complex numbers of absolute value 1. One determines the factors of all Borel multiplier representations of the Bondi-Metzner-Sachs group when the subgroup of "supertranslations" is the additive group of a separable real Hilbert space.

*) Supported by the Deutsche Forschungsgemeinschaft

1. Introduction

The typical projective structure of conventional quantum mechanics requires the study of a kind of group representations which are different in many aspects from the ordinary ones, namely of unitary projective representations. These are representations on the projective space deduced from a separable complex Hilbert space, the restriction to "unitary" ones being motivated by the physical assumption of conservation of transition probabilities. Moreover, one studies only continuous unitary projective representations (CUP-reps) on the basis of an assumption suggested by the observation that transition probabilities vary continuously with the group element of a topological symmetry group.

The theory of CUP-reps of second countable locally compact groups [1,2] is well suited to the majority of groups which are met in quantum mechanics. However, in recent years, topological groups not locally compact have been considered by physicists. One of these is the Bondi-Metzner-Sachs group (BMS-group), the symmetry group of asymptotically flat (four-dimensional) space-times, that is receiving particular attention as a possible substitute for the neutral component of the Poincaré group (see [3] and references therein). At the present time, the name "BMS-group" is given to many different topological groups, because a definitive choice of the subgroup of the so-called "supertranslations" has not yet been made [3]. Following McCarthy [3], we shall assume in this paper that the group of "supertranslations" is the additive group of a separable real Hilbert space (cf. Sec.3 for details). The BMS-group belongs to a class of topological groups (the Polish groups, i.e., second countable metrizable complete groups) which can be considered the most natural generalizations of second countable locally compact groups.

Our goal being the study of CUP-reps of the BMS-group, we prepare the field in Sec.2 by extending the theory of Mackey from second countable locally compact to Polish groups. We show that all CUP-reps of a Polish group can be derived from (ordinary) continuous unitary representations of Polish groups obtained from central topological extensions of the group considered by the multiplicative group $\underset{\sim}{U}(1)$ of all complex numbers of absolute value 1. In Sec.3, this result is applied to the BMS-group B. We determine the relevant extensions and show that all CUP-reps of B can be obtained from continuous unitary representations of its universal covering group. This problem was partially solved by McCarthy [3]. We give the complete solution.

Throughout this paper, $\mathfrak{h}$ will stand for a <u>separable complex Hilbert space</u> with inner multiplication $(.|.)$. We shall use the additive notation (with the neutral element denoted by 0) for the additive groups of vector spaces and the multiplicative notation (with the neutral element denoted by 1) for any other group considered. The symbol e_G (neutral element of the group G) stands for 0 or 1.

2. CUP-reps and their Borel liftings [4]

To begin with, let us recall some preliminary definitions.

(a) The <u>projective space</u> $\underset{\sim}{P}(\mathfrak{h})$ is the set of all rays of $\mathfrak{h}$: if $\varphi \in \mathfrak{h}-\{0\}$, we denote by $\tilde{\varphi}$ the ray of $\mathfrak{h}$ generated by φ, i.e., the subset $\tilde{\varphi}=\{\xi\varphi \mid \xi \in \underset{\sim}{C} \text{ and } \xi \neq 0\}$ of $\mathfrak{h}$.

(b) Let $\underset{\sim}{U}(\mathfrak{h})$ stand for the <u>unitary group of</u> $\mathfrak{h}$ equipped with the weak topology (which coincides with the strong one). It is a Polish group and a sequence (U_n) in $\underset{\sim}{U}(\mathfrak{h})$ converges to U if and only if $\lim_{n\to\infty}(U_n\varphi|\psi) = (U\varphi|\psi)$ for all φ,ψ in $\mathfrak{h}$.

(c) The <u>projective unitary group of $\mathfrak{h}$</u> (denoted by $\underset{\sim}{PU}(\mathfrak{h})$) is the set of all mappings $\tilde{U}:\underset{\sim}{P}(\mathfrak{h}) \to \underset{\sim}{P}(\mathfrak{h})$ such that $U \in \underset{\sim}{U}(\mathfrak{h})$ and

$$\tilde{U}\tilde{\varphi} = \widetilde{U\varphi} \ , \tag{2.1}$$

with the composition of mappings as multiplication and topologized as follows. Let Ω be the mapping of $\underset{\sim}{U}(\mathfrak{h})$ onto $\underset{\sim}{PU}(\mathfrak{h})$ defined by $\Omega(U) = \tilde{U}$, where $\tilde{U}$ is given by (2.1). A subset A of $\underset{\sim}{PU}(\mathfrak{h})$ is an open set provided $\Omega^{-1}(A)$ is open in $\underset{\sim}{U}(\mathfrak{h})$. Then $\underset{\sim}{PU}(\mathfrak{h})$ is a Polish group and Ω is a continuous and open group homomorphism with $\mathrm{Ker}\Omega = \{\xi \mathrm{Id}_{\mathfrak{h}} \mid \xi \in \underset{\sim}{U}(1)\}$.

A <u>CUP-rep</u> $\tilde{u}$ of a Polish group G on $\underset{\sim}{P}(\mathfrak{h})$ is a continuous group homomorphism of G into $\underset{\sim}{PU}(\mathfrak{h})$. It can be shown that this definition agrees with the one of Wigner [5] and Bargmann [6] which was based on the assumption that all transition probabilities vary continuously with the group element. Consider the mapping $u:g \mapsto \Sigma(\tilde{u}(g))$ of G into $\underset{\sim}{U}(\mathfrak{h})$, where Σ is chosen according to the following

<u>Lemma 1.</u> [7] There exists a Borel mapping $\Sigma:\underset{\sim}{PU}(\mathfrak{h}) \to \underset{\sim}{U}(\mathfrak{h})$ such that $\Omega(\Sigma(\tilde{U})) = \tilde{U}$ for all $\tilde{U} \in \underset{\sim}{PU}(\mathfrak{h})$ and $\Sigma(\Omega(\mathrm{Id}_{\mathfrak{h}})) = \mathrm{Id}_{\mathfrak{h}}$. ∎

It follows that u is Borel and that $u(g)u(g')u(gg')^{-1} \in \mathrm{Ker}\Omega$ for all g,g' in G. Therefore

$$u(g)u(g') = \mu(g,g')u(gg') \tag{2.2}$$

for all g,g' in G, where μ is a mapping of $G \times G$ into $\underset{\sim}{U}(1)$ (endowed with its canonical topology, which makes it into a Polish group). A Borel mapping u of a Polish group G into $\underset{\sim}{U}(\mathfrak{h})$ is said to be a <u>Borel unitary multiplier representation (BUM-rep)</u> of G on $\mathfrak{h}$ if $u(e_G) = \mathrm{Id}_{\mathfrak{h}}$

and if there exists a mapping $\mu : G \times G \rightarrow \underline{U}(1)$ (the <u>multiplier</u> of u) such that (2.2) is satisfied. More precisely: u is a <u>BUμ-rep</u> of G on $\mathfrak{h}$. One also says that u is a <u>Borel lifting</u> of $\tilde{u} = \Omega \circ u$ because the following diagram is commutative

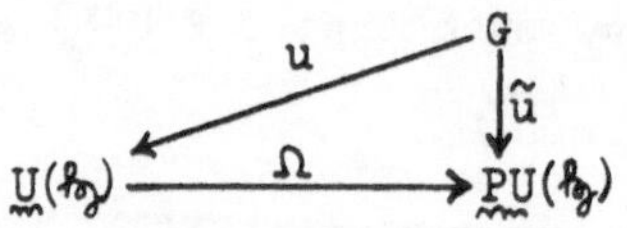

If $\mu(g,g') = 1$ for all g,g' in G, the multiplier μ is said to be <u>trivial</u>. The significance of non-trivial multipliers for quantum mechanics arises from the fact that they are associated with superselection rules: for instance, the univalence if G is the Poincaré or the Galilei group and the superselection rule of the non-relativistic mass if G is the Galilei group.

We make a brief digression, in order to explain some terminology and notation of the cohomology theory of Polish group. Let G and A be Polish groups with A Abelian. A topological operation Ψ of G on A is a group homomorphism of G into the group of all automorphisms of A such that the mapping $(g,a) \mapsto \Psi(g)a$ of $G \times A$ into A is (jointly) continuous. Then A, equipped with Ψ, is said to be a Polish G-module that we denote by A_Ψ. For each integer $n \geqslant 0$, we define $C^n_b(G,A_\Psi)$, $Z^n_b(G,A_\Psi)$, $B^n_b(G,A_\Psi)$ to be, respectively, the subgroups of all Borel mappings of the Eilenberg-MacLane groups of normalized n-cochains, n-cocycles, n-coboundaries of G with values in A [8]. The quotient group $H^n_b(G,A_\Psi) = Z^n_b(G,A_\Psi)/B^n_b(G,A_\Psi)$ is the Mackey-Moore cohomology group of degree n of G with values in A. If $f \in Z^n_b(G,A_\Psi)$, we shall denote by [f] the cohomology class of f.

Returning now to our BUM-rep u of G on $\mathfrak{h}$, we check easily that its multiplier μ is a Borel mapping and satisfies

$$\mu(g,e_G) = \mu(e_G,g) = 1$$

and

$$\mu(g,g')^{-1}\mu(g',g'')\mu(gg',g'')^{-1}\mu(g,g'g'') = 1$$

for all g,g',g" in G. Therefore $\mu \in Z_b^2(G,\underset{\sim}{U}(1)_I)$, where I stands for the trivial operation of G on $\underset{\sim}{U}(1)$. Using a well-known theorem of Banach, we can see that every BUM-rep u of the Polish group G on $\mathfrak{h}$ determines a CUP-rep $\tilde{u} = \Omega \circ u$ of G on $\underset{\sim}{P}(\mathfrak{h})$. However, notice that u is not uniquely fixed by $\tilde{u}$: it depends on the choice of Σ. Two Borel liftings u and u' of the same CUP-rep $\tilde{u}$ of G on $\underset{\sim}{P}(\mathfrak{h})$ are said to be <u>similar</u>. They satisfy the relation $u'(g) = \nu(g)u(g)$ for all $g \in G$, where ν is a Borel mapping of G into $\underset{\sim}{U}(1)$. Moreover, if u and u' are, respectively, a BUμ- and a BUμ'-rep, then

$$\mu'(g,g') = \mu(g,g')\nu(g)\nu(g')\nu(gg')^{-1}$$

for all g,g' in G. In other words, μ' is in the same cohomology class of μ, i.e., $[\mu']=[\mu] \in H_b^2(G,\underset{\sim}{U}(1)_I)$. For this reason, $\tilde{u}(=\Omega\circ u'=\Omega\circ u)$ is also said to be a <u>CU[μ]-rep.</u> The elements of $H_b^2(G,\underset{\sim}{U}(1)_I)$ which are cohomology classes of multipliers of BUM-reps of G constitute a subgroup that we denote by $H_{bm}^2(G,\underset{\sim}{U}(1)_I)$. If G is second countable locally compact, one can show that $H_{bm}^2(G,\underset{\sim}{U}(1)_I) = H_b^2(G,\underset{\sim}{U}(1)_I)$.

Following Mackey, we shall now try to associate with every BUμ-rep of a Polish group G on $\mathfrak{h}$ a continuous unitary representation (CU-rep) on $\mathfrak{h}$ of a topological group constructed by means of G and μ. Consider the set of all ordered pairs (ζ,g), where $\zeta \in \underset{\sim}{U}(1)$ and $g \in G$. Equipped with the multiplication

$$(\zeta,g)(\zeta',g') = (\zeta\zeta'\mu(g,g'),gg') ,$$

where $\mu \in Z_b^2(G,\underset{\sim}{U}(1))$, it is a group denoted by G^μ with neutral element $(1,e_G)$. We have the following

<u>Lemma 2.</u> [4] Let G be a Polish group and let $\mu \in Z_b^2(G,\underset{\sim}{U}(1)_I)$. There exists a unique topology on G^μ, compatible with the group

structure, such that the quotient group $G^{\mu}/\mathrm{Ker}\ pr_2$ is topologically isomorphic with G and that the mapping $g \mapsto (1,g)$ of G into G^{μ} is Borel. The group G^{μ}, equipped with this topology, is Polish. ∎

We remark that $\underline{U}(1)$ and Ker pr_2 are topologically isomorphic through the mapping $\zeta \mapsto (\zeta, e_G)$.

A CU-rep w on $\mathfrak{h}$ of the Polish group G^{μ} of Lemma 2 is said to be <u>$\underline{U}(1)$-split</u> if $w(\zeta, e_G) = \zeta \mathrm{Id}_{\mathfrak{h}}$ for all $\zeta \in \underline{U}(1)$. If u is a BU$\mu$-rep of G on $\mathfrak{h}$, then $w:(\zeta,g) \mapsto \zeta u(g)$ is a $\underline{U}(1)$-split CU-rep of G^{μ} on $\mathfrak{h}$. Conversely, if w is a $\underline{U}(1)$-split CU-rep of G^{μ} on $\mathfrak{h}$, then $u:g \mapsto w(1,g)$ is a BUμ-rep of G on $\mathfrak{h}$. Now let u' be a BUμ'-rep of G on $\mathfrak{h}$ similar to u. Then $w':(\zeta,g) \mapsto \zeta u'(g)$ is a $\underline{U}(1)$-split CU-rep of $G^{\mu'}$ on $\mathfrak{h}$ similar to w in the sense that there exists a mapping $\xi:\underline{U}(1) \times G \to \underline{U}(1)$ such that $w'(\zeta,g) = \xi(\zeta,g)w(\zeta,g)$ for all $(\zeta,g) \in \underline{U}(1)\times G$. It follows that a BU$\mu$-rep u and a BU$\mu'$-rep u' of a Polish group G on $\mathfrak{h}$ are similar if and only if $w:(\zeta,g) \mapsto \zeta u(g)$ and $w':(\zeta,g) \mapsto \zeta u'(g)$ are similar $\underline{U}(1)$-split CU-reps on $\mathfrak{h}$ of G^{μ} and $G^{\mu'}$ respectively. In this case, the Polish groups G^{μ} and $G^{\mu'}$ are topologically isomorphic.

We can summarize all the previous considerations into the following

<u>Theorem 1.</u> [4] Let G be a Polish group and choose a Borel mapping $\Sigma : \underline{PU}(\mathfrak{h}) \to \underline{U}(\mathfrak{h})$ as in Lemma 1. If $\tilde{u}$ is a CUP-rep of G on $\underline{P}(\mathfrak{h})$, then

(i) $u = \Sigma \circ \tilde{u}$ is a Borel lifting of $\tilde{u}$ with multiplier, say μ ;

(ii) $w:(\zeta,g) \mapsto \zeta u(g)$ is a $\underline{U}(1)$-split CU-rep on $\mathfrak{h}$ of the Polish group G^{μ} of Lemma 2.

Conversely, if $\mu \in Z^2_b(G, \underline{U}(1)_I)$ and if w is a $\underline{U}(1)$-split CU-rep of G^{μ} on $\mathfrak{h}$, then

(i') $u:g \mapsto w(1,g)$ is a BUμ-rep of G on $\mathfrak{h}$;

(ii') $\tilde{u} = \Omega \circ u$ is a CUP-rep of G on $\underset{\sim}{P}(\mathfrak{h})$.

Besides, if w' is a $\underset{\sim}{U}(1)$-split CU-rep of $G^{\mu'}$ on $\mathfrak{h}$ similar to w, then its associated CUP-rep by (i') and (ii') is $\tilde{u}$. ∎

In conclusion, we have the following program for the study of CUP-reps of a Polish group G:

(a) Determine the cohomology group $H^2_b(G,\underset{\sim}{U}(1)_I)$ and then its subgroup $H^2_{bm}(G,\underset{\sim}{U}(1)_I)$.

(b) From every element of $H^2_{bm}(G,\underset{\sim}{U}(1)_I)$ pick a representative μ, construct the Polish group G^{μ}, and study its $\underset{\sim}{U}(1)$-split CU-reps. From each one of these representations we get a BUμ-rep and then a CU[μ]--rep. If μ' also is a representative of an element of $H^2_{bm}(G,\underset{\sim}{U}(1)_I)$, we obtain the same CUP-reps if and only if $[\mu] = [\mu']$.

In Sec. 3 we shall work out step (a) for the BMS-group.

Remark. A theory paralleling the one just sketched can be developed for continuous unitary/antiunitary projective representations (CUAP-reps) of a Polish group G on $\underset{\sim}{P}(\mathfrak{h})$ [4], i.e., when one admits the possibility that some group elements are represented by antiunitary operators in $\mathfrak{h}$. It is well known that such a generalized theory is needed when G contains the time inversion. However, if G is connected, every CUAP-rep is a CUP-rep. This is, in particular, the case of the BMS-group.

3. Application to the BMS-group [9]

The BMS-group is defined as follows [3]. Let τ stand for the normalized rotation invariant measure on $\underset{\sim}{S}_2$ (the Euclidean two-dimensional unit sphere), and consider the real Hilbert space $L^2_{\underset{\sim}{R}}(\underset{\sim}{S}_2,\tau)$ of all equivalence classes of τ-square-integrable real-valued functions on $\underset{\sim}{S}_2$. If f is such a function, we shall denote by $\tilde{f}(\in L^2_{\underset{\sim}{R}}(\underset{\sim}{S}_2,\tau))$ its

τ-equivalence class. Moreover, we shall write L^2_R short for $L^2_R(S_2,\tau)$. There exists a linear operation Φ on L^2_R of the neutral component L_0 of the Lorentz group such that, for each $\Lambda \in L_0$ and each $\tilde{f} \in L^2_R$, $\tilde{f}_\Lambda = \Phi(\Lambda)\tilde{f}$ is defined by $f_\Lambda(x) = K_{\Lambda^{-1}}(x) f(\Lambda^{-1}.x)$. Here, x denotes a point of S_2, the dot stands for the usual conformal operation of L_0 on S_2, and $K_{\Lambda^{-1}}(x) = (\Lambda^{-1})^0{}_\mu n^\mu$, where n is the lightlike four-vector $(1,x)$. It can be shown that the operation Φ is topological, so that we can define the BMS-group to be the (external) topological semidirect product B of L_0 by the additive group of L^2_R (that we shall denote again by L^2_R) relative to Φ, i.e., $B = L^2_R \times_\Phi L_0$. This means that B consists of all ordered pairs $(\tilde{f},\Lambda)$, where $\tilde{f} \in L^2_R$ and $\Lambda \in L_0$, with the multiplication

$$(\tilde{f},\Lambda)(\tilde{f}',\Lambda') = (\tilde{f} + \Phi(\Lambda)\tilde{f}', \Lambda\Lambda'),$$

and that its topology is the product one. Since L^2_R is separable, B is a Polish group.

We shall determine $H^2_{bm}(B,U(1)_I)$; for this goal we need the

<u>Theorem 2.</u> [9] If $H^1_b(L_0, Z^1_b(L^2_R,U(1)_I)\hat{I}^1_Z) = \{1\}$, then

$$H^2_b(B,U(1)_I) \approx H^2_{b,s}(L^2_R,U(1)_I)' \times Z^2_{b,a}(L^2_R,U(1)_I)^{L_0} \times H^2_b(L_0,U(1)_I). \quad (3.1)$$

Let us explain the notation. To begin with, we show that the symbol $H^1_b(L_0, Z^1_b(L^2_R,U(1)_I)\hat{I}^1_Z)$ has a meaning. For since $Z^1_b(L^2_R,U(1)_I)$ is the group of all continuous unitary characters of the additive group of L^2_R (by the theorem of Banach already quoted), there exists a canonical group isomorphism, say γ, of L^2_R onto $Z^1_b(L^2_R,U(1)_I)$ ([10],23.32). We topologize $Z^1_b(L^2_R,U(1)_I)$ by transport of structure via γ. Then, the operation $\hat{I}^1_Z$ of L_0 on $Z^1_b(L^2_R,U(1)_I)$ such that $(\hat{I}^1_Z(\Lambda)\nu)(\tilde{f}) = \nu(\Phi(\Lambda^{-1})\tilde{f})$ for all $\Lambda \in L_0$, all $\nu \in Z^1_b(L^2_R,U(1)_I)$, and all $\tilde{f} \in L^2_R$, is topological.

Let $Z^2_{b,a}(L^2_{R},U(1)_I)$ (resp. $Z^2_{b,s}(L^2_{R},U(1)_I)$) stand for the subgroup of all antisymmetric (resp. symmetric) elements of $Z^2_b(L^2_{R},U(1)_I)$. We can define $H^2_{b,s}(L^2_{R},U(1)_I) = Z^2_{b,s}(L^2_{R},U(1)_I)/B^2_b(L^2_{R},U(1)_I)$, because $B^2_b(L^2_{R},U(1)_I) \subseteq Z^2_{b,s}(L^2_{R},U(1)_I)$. The symbol $H^2_{b,s}(L^2_{R},U(1)_I)'$ denotes the subgroup of all elements $[\mu_s]$ of $H^2_{b,s}(L^2_{R},U(1)_I)$ such that there exists a Borel mapping $\omega : L^2_{R} \times L_0 \to U(1)$ satisfying

$$\mu_s(\Phi(\Lambda)\tilde{f},\Phi(\Lambda)\tilde{f}') = \mu_s(\tilde{f},\tilde{f}')\omega(\tilde{f}+\tilde{f}',\Lambda)\omega(\tilde{f},\Lambda)^{-1}\omega(\tilde{f}',\Lambda)^{-1}$$

and

$$\omega(\tilde{f},\Lambda\Lambda') = \omega(\Phi(\Lambda')\tilde{f},\Lambda)\omega(\tilde{f},\Lambda')$$

for all Λ,Λ' in L_0 and all $\tilde{f},\tilde{f}'$ in L^2_{R}. Finally, $Z^2_{b,a}(L^2_{R},U(1)_I)^{L_0}$ stands for the subgroup of $Z^2_{b,a}(L^2_{R},U(1)_I)$ whose elements μ_a are L_0-invariant in the sense that

$$\mu_a(\Phi(\Lambda)\tilde{f},\Phi(\Lambda)\tilde{f}') = \mu_a(\tilde{f},\tilde{f}') \tag{3.2}$$

for all Λ,Λ' in L_0 and all $\tilde{f},\tilde{f}'$ in L^2_{R}.

Now we have the following results:

(1) $H^1_b(L_0,Z^1_b(L^2_{R},U(1)_I)_{\hat{I}^1_Z}) = \{1\}$. Let Φ' be the topological operation of L_0 on L^2_{R} such that, for each $\Lambda \in L_0$ and each $\tilde{f} \in L^2_{R}$, $\Phi'(\Lambda)\tilde{f} = \tilde{f}'_\Lambda$ is defined by $f'_\Lambda(x) = K^{-3}_{\Lambda^{-1}}(x)f(\Lambda^{-1}.x)$ (cf.[3]). The group isomorphism γ considered above is an L_0-module isomorphism, i.e., satisfies $\gamma\circ\Phi'(\Lambda) = \hat{I}^1_Z(\Lambda)\circ\gamma$ for all $\Lambda \in L_0$. Hence, we have to show that $H^2_b(L_0,(L^2_{R})_{\Phi'}) = \{0\}$. This follows from the arguments given in ([11], Theorem 14.1).

(2) $H^2_{b,s}(L^2_{R},U(1)_I)' = \{1\}$. We prove that, if V is any separable real Banach space, $H^2_{b,s}(V,R_I) = \{0\}$ and then $H^2_{b,s}(V,U(1)_I) = \{1\}$.

(3) $Z^2_{b,a}(L^2_{R},U(1)_I)^{L_0} = \{1\}$. It is sufficient to prove that $Z^2_{b,a}(L^2_{R},R_I)^{L_0} = \{0\}$, with the L_0-invariance defined by (3.2). We show that $Z^2_{b,a}(L^2_{R},R_I)$ is the additive group of all continuous antisymmetric bilinear forms on $L^2_{R} \times L^2_{R}$. The result follows from (3.2) and the fact that in the tensor product of the representation $\Phi'|SO(3)$ by

itself there are no antisymmetric one-dimensional subrepresentations.

(4) $H_b^2(\underset{\sim}{L}_0, \underset{\sim}{U}(1)_I) \approx C_2$ (a cyclic group of order 2).

It follows from Theorem 2 and (1)-(4) that $H_b^2(B, \underset{\sim}{U}(1)_I) \approx C_2$. On the other hand, it is obvious that $H_{bm}^2(B, \underset{\sim}{U}(1)_I) = H_b^2(B, \underset{\sim}{U}(1)_I)$; so we are done.

Let 1 and μ denote, respectively, representatives of the neutral and of the other element of $H_{bm}^2(B, \underset{\sim}{U}(1)_I)$. As shown in Sec.2, all CUP-reps of B are obtained from $\underset{\sim}{U}(1)$-split CU-reps of the Polish groups B^1 and B^μ. However, every $\underset{\sim}{U}(1)$-split CU-rep of B^1 on $\mathfrak{h}$ is identifiable in a trivial way with a CU-rep of B on $\mathfrak{h}$. If $\tilde{u}$ is a CUP-rep of B on $\underset{\sim}{P}(\mathfrak{h})$, there exists a CU-rep u of $\tilde{B}$ (the universal covering group of B) on $\mathfrak{h}$ such that $\tilde{u} \circ \rho_B = \Omega \bullet u$, where ρ_B is the covering mapping, and conversely. This follows [6] from $H_b^2(\tilde{B}, \underset{\sim}{U}(1)_I) = \{1\}$, which can be shown along the lines of the proof for B using $H_b^2(\underset{\sim}{SL}(2, \underset{\sim}{C}), \underset{\sim}{U}(1)_I) = \{1\}$.

References

[1] G.W. MACKEY, Acta Math. 99, 265 (1958).

[2] L. AUSLANDER and C.C. MOORE, Mem.Am.Math.Soc. No. 62 (1966).

[3] P.J. McCARTHY, "Projective Representations of the Asymptotic Symmetry Group of General Relativity" in Proceedings of the 2nd International Colloquium on Group Theoretical Methods in Physics, Nijmegen, June 25-29, 1973.

[4] U. CATTANEO, "On Unitary/Antiunitary Projective Representations of Groups". To appear in Rep.Math.Phys.

[5] E. WIGNER, Ann.Math. 40, 149 (1939).

[6] V. BARGMANN, Ann.Math. 59, 1 (1954).

[7] J. DIXMIER, Trans.Am.Math.Soc. 104, 278 (1962).

[8] S. MACLANE, Homology. Springer-Verlag:Heidelberg, 1963.

[9] U.CATTANEO, "Multipliers of BUM-reps of the Bondi-Metzner-Sachs group". To appear.

[10] E. HEWITT and K.A. ROSS, Abstract Harmonic Analysis I. Springer--Verlag: Heidelberg, 1963.

[11] K.R. PARTHASARATHY and K. SCHMIDT, Positive Definite Kernels, Continuous Tensor Products, and Central Limit Theorems of Probability Theory, Lecture Notes in Mathematics 272. Springer--Verlag: Heidelberg, 1972.

THE HILBERT SPACE $L^2(SU(2))$ AS A REPRESENTATION SPACE FOR THE GROUP $(SU(2) \times SU(2)) Ⓢ S_2$

R. Dirl [+]

Talk presented at the 4th International Colloquium on Group Theoretical Methods in Physics, June 1975, University of Nijmegen, The Netherlands

[+] 1.Institut für theoretische Physik, Technische Hochschule Vienna, Austria

Abstract: The Hilbert space $L^2(SU(2))$ is used as a representation space for a (unitary) representation of the semi-direct product group $(SU(2) \times SU(2)) Ⓢ S_2$ and the corresponding group algebra. Special operators are constructed which are closely related to the representation theory of the groups SU(2) and S_2 and are irreducible tensor operators with respect to $(SU(2) \times SU(2)) Ⓢ S_2$. These operators are then used to define complete sets of irreducible tensor operators, to derive correlations between such special operators and to calculate two classes of Clebsch-Gordan coefficients of $(SU(2) \times SU(2)) Ⓢ S_2$. The results obtained for SU(2) can be generalized in a systematic way for any finite or compact continuous group.

1. Introduction

The Hilbert space $L^2(SU(2))$ is one of the symmetric homogeneous spaces which are of considerable interest in mathematical physics [1-5]. It is used here as a representation space for the semi-direct product group $G = (SU(2) \times SU(2)) \circledS S_2$ with normal subgroup $H = SU(2) \times SU(2)$. We construct several irreducible tensor operators (ITs) with respect to these groups [6,7] and discuss the question whether it is possible to trace back an arbitrary IT (with respect to G or H) to special operators which are closely related to the representation theory of the groups SU(2) and S_2.

2. Unitary irreducible representations of $G = (SU(2) \times SU(2)) \circledS S_2$

Since we are interested to define (in $L^2(SU(2))$) ITs with respect to G and H we need the matrix elements of the unitary irreducible representations (unirreps) of these groups. First of all let us recall briefly the definition of G ($\omega_0 = (0,0,0)$):

$$G = (SU(2) \times SU(2)) \circledS S_2 = \left\{(\omega_1,\omega_2|r):\ \omega_i \in SU(2),\ r(=e,s) \in S_2\right\} \quad (2.1)$$

$$(\omega_1,\omega_2|r)(\omega_1',\omega_2'|r') = (\omega_1\omega_{r1}',\omega_2\omega_{r2}'|rr') \quad (2.2),$$

where r1, r2 is a permutation of 1, 2. The matrix elements of the unirreps of H are given by

$$(D^{jj'}(\omega_1,\omega_2))_{mk,m'k'} = (D^j(\omega_1) \otimes D^{j'}(\omega_2))_{mk,m'k'} = D^j_{mm'}(\omega_1)D^{j'}_{kk'}(\omega_2) \quad (2.3),$$

where the special functions $D^j_{mm'}(\omega)$ are the elements of the well-known unirreps of SU(2) [8]. There are two different types of unirreps of G whose matrix elements are given by [9]

$$D^{jj\varkappa}_{m_1m_2,m_1'm_2'}(\omega_1,\omega_2|r) = D^j_{m_1,m_{r1}'}(\omega_1)D^j_{m_2,m_{r2}'}(\omega_2)(-1)^{\varkappa r} \quad (2.4)$$

$$D^{j_1j_2}_{r'm_1m_2,r''m_1'm_2'}(\omega_1,\omega_2|r) = D^{j_1}_{m_1m_1'}(\omega_{r'1})D^{j_2}_{m_2m_2'}(\omega_{r'2})\,\delta_{r'r,r''} \quad (2.5),$$

where the symbols ri have the same meaning as before. Since the unirreps of SU(2) are all equivalent to their complex conjugates the same property holds for the unirreps of H and G. In order to state the Wigner-Eckart theorem for these groups we need also the corresponding Clebsch-Gordan coefficients (CG-coefficients). Those who refer to H are trivially composed from the CG-coefficients of SU(2). Those for G are less trivial. The first step which has to be done if we want to calculate them is to determine the multiplicities [10] occuring in the Kronecker products

$$D^{\beta}(\omega_1,\omega_2|r) \otimes D^{\beta'}(\omega_1,\omega_2|r) \simeq \sum_{\beta''} \oplus\, m_{\beta\beta'\beta''}\, D^{\beta''}(\omega_1,\omega_2|r) \tag{2.6}$$

For the sake of brevitiy we have denoted the labels $jj\varkappa$ or j_1j_2 by the symbols β,β',β''. By using the corresponding character formula we obtain the multiplicities

$$m_{jj\varkappa,j'j'\varkappa',j''j''\varkappa''} = \Delta(jj'j'')\,\delta_{\varkappa'',\varkappa+\varkappa'} \tag{2.7}$$

$$m_{jj\varkappa,j'j'\varkappa',j_1j_2} = \Delta(jj'j_1)\,\Delta(jj'j_2) \tag{2.8}$$

$$m_{jj\varkappa,j'_1j'_2,j''_1j''_2} = \Delta(jj'_1j''_1)\,\Delta(jj'_2j''_2) + \Delta(jj'_1j''_2)\,\Delta(jj'_2j''_1) \tag{2.9}$$

$$m_{j_1j_2,j'_1j'_2,j''_1j''_2} = \Delta(j_1j'_1j''_1)\,\Delta(j_2j'_2j''_2) + \Delta(j_1j'_1j''_2)\,\Delta(j_2j'_2j''_1) + \Delta(j_1j'_2j''_1)\,\Delta(j_2j'_1j''_2) + \Delta(j_1j'_2j''_2)\,\Delta(j_2j'_1j''_2) \tag{2.10}$$

where $\Delta(abc)$ denotes the usual triangle symbol. This shows that the group G is not simply reducible [10] and that only in the first two cases the corresponding CG-coefficients are uniquely determined up to a phase factor. The CG-coefficients which we denote by

$$\left\{ \begin{Bmatrix} \beta & \beta' & \beta'' \\ p & p' & p'' \end{Bmatrix}_w : \begin{array}{l} \beta(\beta',\beta'') = jj\ ;\quad p(p',p'') = m_1m_2 \\ \beta(\beta',\beta'') = j_1j_2;\quad p(p',p'') = r'\,m_1m_2 \\ w = 1,2,\,..\,,m_{\beta\beta'\beta''} \end{array} \right\} \tag{2.11}$$

must satisfy the usual orthogonality relations and must furthermore decompose the reducible representation (2.6) into the desired dirct sum of unirreps.

$$\sum_{\substack{p_1p'_1\\p_2p'_2}} \begin{Bmatrix} \beta & \beta' & \beta'' \\ p_1 & p'_1 & p''_1 \end{Bmatrix}_w^{*} D^{\beta}_{p_1p_2}(\omega_1,\omega_2|r)\, D^{\beta'}_{p'_1p'_2}(\omega_1,\omega_2|r) \begin{Bmatrix} \beta & \beta' & \beta'' \\ p_2 & p'_2 & p''_2 \end{Bmatrix}_{w'} = D^{\beta''}_{p''_1p''_2}(\omega_1,\omega_2|r)\,\delta_{ww'} \tag{2.12}$$

3. $L^2(SU(2))$ as a representation space for $H = SU(2)\times SU(2)$, $G = (SU(2)\times SU(2))\,\circledS\, S_2$ and the corresponding group algebras

The Hilbert space $L^2(SU(2))$ which we use as a representation for the above mentioned groups is the set of all complex valued square integrable functions where the scalar product is given in the usual way. An orthonormalized basis is given by the set

$$\left\{ Q^j_{mk} = (-1)^{j+k}\sqrt{2j+1}\, D^{j\,*}_{m,-k} :\ j = 0,1/2,1,\,..\quad ;\ -j \leq m,k \leq j \right\} \tag{3.1}$$

The definition of the left- and right-regular representation [11,12] allows to use

$L^2(SU(2))$ as a carrier space for a unitary representation U of H:

$$U: (\omega_1,\omega_2) \longrightarrow U(\omega_1,\omega_2)$$
$$\left[U(\omega_1,\omega_2)f\right](\omega) = f(\omega_1^{-1}\omega\omega_2) \qquad (3.2).$$

This representation U can be extended to a unitary representation V of G by

$$V: (\omega_1,\omega_2|r) \longrightarrow V(\omega_1,\omega_2|r) = U(\omega_1,\omega_2)W(r)$$
$$\left[V(\omega_o,\omega_o|s)f\right](\omega) = \left[W(s)f\right](\omega) = f(\omega^{-1}) \qquad (3.3).$$

It should be noted that V is isomorphic to SO(4,R) Ⓢ S_2 where the normal subgroup SO(4,R) is the homomorphic image of H.

$$W(s)U(\omega_1,\omega_2)W^+(s) = U(\omega_2,\omega_1) \qquad (3.4)$$

By means of the definitions (3.2,3) we obtain

$$U(\omega_1,\omega_2)Q^j_{mk} = \sum_{pq} D^j_{pm}(\omega_1)D^j_{qk}(\omega_2)Q^j_{pq} \qquad (3.5)$$
$$W(s)Q^j_{mk} = (-1)^{2j}\, Q^j_{km} \qquad (3.6)$$

This shows that the basis (3.1)decomposes the unitary representation V (U) into a direct sum of unirreps $D^{jj\varkappa(j)}(\omega_1,\omega_2|r)$ with $\varkappa(j) = 2j \bmod 2$ $(D^{jj}(\omega_1,\omega_2))$ each unirrep occuring only once [12].

In order to use them for the definition of some special ITs with respect to H we introduce the so-called 'units' of the group algebras $^{(i)}\mathcal{A}(SU(2))$ (i = left and right) [13] by

$$^{(i)}E^j_{pq} = (2j+1)\int d\mu(\omega_i)D^{j*}_{pq}(\omega_i)\begin{cases} U(\omega_1,\omega_o) & \text{for } i = \text{left} \\ U(\omega_o,\omega_2) & \text{for } i = \text{right} \end{cases} \qquad (3.7).$$

They represent bases of $^{(i)}\mathcal{A}(H)$ (i = left or right). These representations are characterized by the fact that only the operators

$$E^{jj}_{mk,m'k'} = {}^{(L)}E^j_{mm'}\;{}^{(R)}E^j_{kk'} \;;\; j = 0,1/2,1,\;..\;,\; -j\leqslant m,m',k,k'\leqslant j \qquad (3.8)$$

do not vanish identically. The representations of $^{(i)}\mathcal{A}(G)$ (i = left or right) are analoguously defined; for similar reasons as before only the operators

$$E^{jj\varkappa(j)}_{mk,m'k'} = \tfrac{1}{2}(E^{jj}_{mk,m'k'} + (-1)^{\varkappa(j)}E^{jj}_{mk,k'm'}W(s)) \qquad (3.9)$$

occur, where the operators $E^{jj}_{mk,m'k'}$ appearing in (3.9) are given by (3.8).

4. ITs with respect to G = (SU(2) x SU(2)) Ⓢ S_2 and H = SU(2) x SU(2)

The general definition of an IT with respect to a given group G' is given in terms of the transformation properties of its components [14]. In case G' is compact continuous, i.e. if G' is the semi-direct product of a connected compact continuous group and a finite group, then there is an equivalent definition which uses the commutation relations of the IT-components with the elements of the Lie algebra of the normal subgroup. Since G is of this type the 'global' definition is given by

$$V(\omega_1,\omega_2|r)\, T^{\beta}_{p}\, V^{+}(\omega_1,\omega_2|r) = \sum_{q} D^{\beta}_{qp}(\omega_1,\omega_2|r)\, T^{\beta}_{q} \qquad (4.1),$$

where (in accordance with (2.4,6)) β respectively p(q) means either $JJ\varkappa$ or J_1J_2 ($J_1 < J_2$) respectively p = MK or r' MK. The 'local' definition reads

$$\left[{}^{(L)}J_{\pm}\, ,\, T^{JJ\varkappa}_{MK}\right] = \sqrt{(J \mp M)(J \pm M + 1)}\; T^{JJ\varkappa}_{M\pm 1,K}$$

$$\left[{}^{(L)}J_{3}\, ,\, T^{JJ\varkappa}_{MK}\right] = M\, T^{JJ\varkappa}_{MK}$$

$$\left[{}^{(R)}J_{\pm}\, ,\, T^{JJ\varkappa}_{MK}\right] = \sqrt{(J \mp K)(J \pm K + 1)}\; T^{JJ\varkappa}_{M,K\pm 1}$$

$$\left[{}^{(R)}J_{3}\, ,\, T^{JJ\varkappa}_{MK}\right] = K\, T^{JJ\varkappa}_{MK} \qquad (4.2)$$

$$W(s)\, T^{JJ\varkappa}_{MK}\, W^{+}(s) = (-1)^{\varkappa}\, T^{JJ\varkappa}_{KM} \qquad (4.3)$$

$$\left[{}^{(L)}J_{\pm}\, ,\, T^{J_1J_2}_{eMK}\right] = \sqrt{(J_1 \mp M)(J_1 \pm M + 1)}\; T^{J_1J_2}_{e,M\pm 1,K}$$

$$\left[{}^{(L)}J_{3}\, ,\, T^{J_1J_2}_{eMK}\right] = M\, T^{J_1J_2}_{eMK}$$

$$\left[{}^{(R)}J_{\pm}\, ,\, T^{J_1J_2}_{eMK}\right] = \sqrt{(J_2 \mp K)(J_2 \pm K + 1)}\; T^{J_1J_2}_{e,M,K\pm 1}$$

$$\left[{}^{(R)}J_{3}\, ,\, T^{J_1J_2}_{eMK}\right] = K\, T^{J_1J_2}_{eMK} \qquad (4.4)$$

$$W(s)\, T^{J_1J_2}_{rMK}\, W^{+}(s) = T^{J_1J_2}_{sr,M,K} \qquad (4.5).$$

(The operators ${}^{(i)}J_{\pm}$, ${}^{(i)}J_3$, i = left and right are the elements of the corresponding Lie algebras). Because of

$$W(s)\, {}^{(L)}J_j = {}^{(R)}J_j\, W(s) \qquad (4.6)$$

which is a consequence of (3.4), it is clear that (4.4) and (4.6) with r = e suffice to define ITs. The 'local' definition is of course much more appropriate for the construction of ITs than the 'global' one. Since if we know ITs with respect to H (which are defined by (4.2,4) or the corresponding 'global' definition) we are able to 'induce' by means of (4.3,5) ITs with respect to G. This is just the 'inverse' of the subduction problem. If ITs with respect to G are given and we restrict this group to H then there are two possibilities: Whereas in the first case ($\beta = JJ\varkappa$) an IT remains an IT it decomposes always into two ITs in the second case ($J_1 < J_2$).

In order to decide which ranks $\beta = JJ\varkappa$ or J_1J_2 (JJ') of ITs can be realized in $L^2(SU(2))$ it suffices to remember that V (U) is isomorphic to SO(4,R) Ⓢ S_2 (SO(4,R)) or to investigate the corresponding Wigner-Eckart theorems. Because of (2.7,8) we are fortunately confronted with the multiplicity free cases and omit therefore the dummy index w. We use the Wigner-Eckart theorem for G and H in the form

$$\left\langle Q^{j}_{mk}\,,\, T^{\beta}_{p}\, Q^{j'}_{m'k'}\right\rangle = \left\{\begin{matrix}\beta & j'\,j'\varkappa(j') \\ p & m'\,k'\end{matrix}\,\middle|\,\begin{matrix} jj\varkappa(j) \\ mk\end{matrix}\right\}\{jj\varkappa(j)\,\|\,T^{\beta}\,\|\,j'\,j'\varkappa(j')\} \tag{4.7}$$

$$\left\langle Q^{j}_{mk}\,,\, T^{JJ'}_{MK}\, Q^{j'}_{m'k'}\right\rangle = (JM,j'\,m'|jm)(J'\,K,j'\,k'|jk)(j\,\|\,T^{JJ'}\,\|\,j') \tag{4.8}$$

In (4.7) we have to take for β (p) either $JJ\varkappa$ (p = MK) or J_1J_2 (p = r' MK). Therefore the following ranks with respect to G are allowed:

$$JJ\varkappa: \quad J = 0,1/2,1,\,.. \quad ; \varkappa = 0,1 \tag{4.9}$$

$$J_1J_2: \quad J_i = 0,1,2,\,... \quad \text{or} \quad J_i = 1/2,3/2,\,.. \tag{4.10}$$

(4.9,10) show also the allowed ranks for H.

An operator $\mathcal{O}$ whose domain of definition contains the basis (3.1) can be decomposed into IT-components with respect to G in the following way [9,15]:

$$\mathcal{O} = \sum_{\beta p} T^{\beta}_{pp}[\mathcal{O}] \tag{4.11}$$

$$T^{\beta}_{pq}[\mathcal{O}] = (n_\beta/2)\sum_{r}\iint d\mu(\omega_1)d\mu(\omega_2)D^{\beta *}_{pq}(\omega_1,\omega_2|r)V(\omega_1,\omega_2|r)\mathcal{O}V^{+}(\omega_1,\omega_2|r) \tag{4.12}$$

where n_β denotes the dimension of the unirrep $D^{\beta}(\omega_1,\omega_2|r)$. Analoguous equations hold for the decomposition into ITs with respect to H.

5. Special ITs with respect to H = SU(2) x SU(2)

The first type of ITs which we introduce consists of elements of the so-called tensor basis of $^{(i)}\mathcal{A}(SU(2))$. They are defined by the matrices U^j relating the unirreps of SU(2) to their complex conjugates and the CG-coefficients of SU(2) [9,16]

$$ {}^{(i)}T_M^{j;J} = \sum_k (-1)^{j+k} (j\ M+k, j\ -k|JM)\ {}^{(i)}E^j_{M+k,k} \tag{5.1} $$

For fixed j,J they form for i = left an IT of rank J0 and for i = right an IT of rank 0J with respect to H. For similar reasons as in (3.8) only the operators

$$ {}^{jj}T_{MK}^{JJ'} = {}^{(L)}T_M^{j;J}\ {}^{(R)}T_K^{j;J'} \tag{5.2} $$

being IT-components of rank JJ' do not vanish identically. The set of IT-components (5.2) represents a basis of ${}^{(i)}\mathcal{A}(H)$ equivalent to (3.8).

The second type of ITs which we define are closely related to the matrix elements of the unirreps of SU(2) [9,17].

$$ \left[Q^{RR}_{MK}\, f\right](\omega) = Q^R_{MK}(\omega) f(\omega) \tag{5.3} $$

They form for fixed R an IT of rank RR.

These special ITs allow to construct much more general ones. Thereby it can be shown that only the following three types of ITs are linearly independent.

$$ {}^jT_{ab}^{(J\underline{B})AB} = \sum_M (J\ a-M, BM|Aa)\ {}^{(L)}T^{j;J}_{a-M}\ Q^{BB}_{Mb} \tag{5.4} $$

$$ {}^jT_{ab}^{\{\underline{A}J\}AB} = \sum_K (A\ b-K, JK|Bb)\ Q^{AA}_{a,b-K}\ {}^{(R)}T^{j;J}_K \tag{5.5} $$

$$ {}^{jj'}T_{ab}^{(J\underline{R}J')AB} = \sum_{MK} (J\ a-M, RM|Aa)(R\ b-K, J'\ K|Bb)\ {}^{(L)}T^{j;J}_{a-M}\ Q^{RR}_{M,b-K}\ {}^{(R)}T^{j';J'}_K \tag{5.6} $$

The reason for this are correlations of the type

$$ {}^jT_{0b}^{(\underline{J}J)0J} = \sum_M (J\ -M, JM|00)\ Q^{JJ}_{-M,b}\ {}^{(L)}T_M^{j;J} = {}^jT_{0b}^{(J\underline{J})0J} = {}^{(R)}T_b^{j;J} \tag{5.7} $$.

The following set of operators

$$ \left\{ {}^{jj'}T_{ab}^{(2j,\underline{j+j'},2j')AB} \ :\ j,j' = 0,1/2,1,\ ..\ ;\ |j-j'| \leqslant A,B \leqslant j+j' \quad -A \leqslant a \leqslant A\ ;\ -B \leqslant b \leqslant B \right\} \tag{5.8} $$,

whose elements are given by (5.6) forms a complete set of ITs for the linear operator space $V_Q = \left\{\mathcal{O} : \mathcal{D}_{\mathcal{O}} \subset \{Q^j_{mk}\}\right\}$. Therefore the expansions are of the form (together with (4.11))

$$ T^{AB}_{ab,ab}\left[\mathcal{O}\right] = \sum_{jj'} c^{jj'(AB)}_{ab}\ {}^{jj'}T_{ab}^{(2j,\underline{j+j'},2j')AB} \tag{5.9} $$,

where the coefficients $c^{jj'(AB)}_{ab}$ are given by the quotient of the corresponding reduced matrix elements.

It is rather obvious that in constructing and correlating ITs the elements of the group algebras can be replaced by elements of the enveloping algebras. The operators

$$ {}^{(i)}J_{\pm 1} = \mp\frac{1}{\sqrt{2}}\,{}^{(i)}J_{\pm}\ ;\quad {}^{(i)}J_{o} = {}^{(i)}J_{3} \qquad i = \text{left and right} \tag{5.10} $$

forming an IT of rank 1,0 and 0,1 are correlated in the following way [18]:

$$ {}^{(R)}J_{K} = \sum_{M}(1\ {-M},1M|00)\ Q^{11}_{-M,K}\ {}^{(L)}J_{K} = \sum_{M}(1\ {-M},1M|00)\,{}^{(L)}J_{-M}\ Q^{11}_{MK} \tag{5.11} $$

(and vice versa). This is of some interest for practical calculations envolving elements of the left- and right-Lie algebras. Furthermore it is well-known that the operators (5.10) can be used for the definition of ITs of rank J0 and 0J which belong to the enveloping algebras. Denoting the IT-components of such an IT of rank J0 by

$$ {}^{(L)}Z^{J}_{M} = \left\{\text{power series in } {}^{(L)}J_{M'}\right\} \tag{5.12} $$

we obtain analoguously to (5.7) immediately the IT-components of an IT of rank 0J.

$$ {}^{(R)}Z^{J}_{K} = \sum_{M}(J\ {-M},JM|00)\ Q^{JJ}_{-M,K}\ {}^{(L)}Z^{J}_{M} \tag{5.13} $$

Much more ITs which are composed of the operators (5.3) and (5.12,13) can be constructed in the same way as it was done in (5.4-6).

6. Special ITs with respect to $G = (SU(2) \times SU(2)) \circledS S_2$

The first type of ITs which we introduce represents again parts of the tensor basis of the group algebras ${}^{(i)}\mathcal{A}(G)$. These operators can be defined by means of the matrices U^{β} relating the unirreps of G to their complex conjugates and the CG-coefficients of G. Likewise they can directly be constructed by menas of the already mentioned 'induction' (c.f.(4.3,5)) provided we know ITs with respect to H. Since this is the case we obtain with the aid of

$$ W(s)\ {}^{(L)}T^{j;J}_{M} = {}^{(R)}T^{j;J}_{M}\ W(s) \tag{6.1} $$

and (5.2) the following IT-components (where the notations for the IT-components are analoguous to (5.1))

$$ T^{jj\varkappa(j);JJ0}_{MK} = {}^{jj}T^{JJ}_{MK} \tag{6.2} $$

$$ T^{jj\varkappa(j);J_1J_2}_{eMK} = {}^{jj}T^{J_1J_2}_{MK}\ ;\quad T^{jj\varkappa(j);J_1J_2}_{sMK} = {}^{jj}T^{J_2J_1}_{KM} \tag{6.3} $$

These are the only ITs within the group algebra which do not vanish identically.

The ITs of the second type are just the operators (5.3). Because of

$$W(s)\; Q^{RR}_{MK}\; W^{+}(s) = (-1)^{2R}\; Q^{RR}_{KM} \tag{6.4}$$

these operators transform (for fixed R) according to the unirreps $D^{jj\varkappa(j)}(\omega_1,\omega_2|r)$.

The special ITs with respect to H introduced in Sec.5 can be extended by means of (4.3,5) to ITs with respect to G. In the case A = B we obtain e.g. the relations

$$W(s)\; {}^{jj'}T^{(J\underline{R}J')AA}_{ab}\; W^{+}(s) = (-1)^{2R}\; {}^{jj'}\hat{T}^{(J\underline{R}J')AA}_{ba} \tag{6.5}$$

with

$${}^{jj'}\hat{T}^{(J\underline{R}J')AA}_{ab} = \sum_{MK} (J\; b\text{-}K, RK|Ab)(R\; a\text{-}M, J'\, M|Aa)\; {}^{(R)}T^{j;J}_{b-K}\; Q^{RR}_{a-M,K}\; {}^{(L)}T^{j';J'}_{M} \tag{6.6}.$$

This means that the operators

$${}^{jj'}_{\pm}T^{(J\underline{R}J')AA\varkappa}_{ab} = \frac{1}{2}\left({}^{jj'}T^{(J\underline{R}J')AA}_{ab} \pm {}^{jj'}\hat{T}^{(J\underline{R}J')AA}_{ab}\right) \tag{6.7}$$

are already components of ITs of the rank $AA\varkappa$ with $\varkappa = 2R, 2R+1$. In the case $A < B$ we obtain by means of (4.5) the following ITs:

$$\left\{ {}^{jj'}T^{(J\underline{R}J')AB}_{rab} = W(r)\; {}^{jj'}T^{(J\underline{R}J')AB}_{ab}\; W^{+}(r):\;\; r = e,s\;;\; -A \leqslant a \leqslant A\;;\; -B \leqslant b \leqslant B \right\} \tag{6.8}.$$

It is now easy to show that the set of ITs (6.7,8) with $J = 2j$, $R = j+j'$, $J' = 2j'$ forms an other set of ITs which is complete with respect to V_Q.

Finally these special ITs allow to calculate for the two multiplicity free cases (2.7,8) the CG-coefficients for G. This can be done by comparing the Wigner-Eckart theorems (4.7) and (4.8) which give the desired CG-coefficients up to a factor which has to be determined by means of (2.12).

$$\left\{\begin{matrix} jj\varkappa & j'\,j'\,\varkappa' \\ mk & m'\,k' \end{matrix}\;\middle|\;\begin{matrix} j''j''\varkappa'' \\ m''k'' \end{matrix}\right\} = (jm, j'\, m'|j''m'')(jk, j'\, k'|j''k'')\;\delta_{\varkappa'', \varkappa+\varkappa'} \tag{6.9}$$

$$\left\{\begin{matrix} J_1J_2 & j'\,j'\,\varkappa' \\ rM_1M_2 & m'\,k' \end{matrix}\;\middle|\;\begin{matrix} j''j''\varkappa'' \\ m''k'' \end{matrix}\right\} = (J_{r1}M_{r1}, j'\, m'|j''m'')(J_{r2}M_{r2}, j'\, k'|j''k'')\frac{(-1)^{(\varkappa'+\varkappa'')r}}{\sqrt{2}} \tag{6.10}$$

$$\left\{\begin{matrix} jj\varkappa & j'\,j'\,\varkappa' \\ mk & m'\,k' \end{matrix}\;\middle|\;\begin{matrix} J_1J_2 \\ rM_1M_2 \end{matrix}\right\} = (jm, j'\, m'|J_{r1}M_{r1})(jk, j'\, k'|\, J_{r2}M_{r2})(-1)^{(\varkappa+\varkappa')r} \tag{6.11}$$

(Here the symbol ri indicates once again a permuted index).

7. Concluding remarks

The foregoing considerations have shown that we succeeded in defining complete sets of ITs (with respect to the groups G = (SU(2) x SU(2)) Ⓢ S_2 and H = SU(2) x SU(2)). They are composed of special ITs which are closely related to the representation theory of the groups SU(2) and S_2. Besides this we found correlations between such special ITs which are of some interest for practical applications, especially to simplify extremely the calculations of the left- and right-Lie algebras. Finally we were also able to calculate quite generally two different classes of CG-coefficients of G.

References

1 S Malin J.Math.Phys.16(1975)679
2 Dj Šijački J.Math.Phys.16(1975)298
3 B L Beers, R S Millman J.Math.Phys.16(1975)11
4 B L Hu J.Math.Phys.15(1974)1748
5 B L Hu Phys.Rev.D5(1973)1048
6 R Dirl 'The Hilbert space L^2(SU(2)) as a representation space for the group SU(2) x SU(2)' J.Phys.A(submitted for publication)
7 R Dirl 'The Hilbert space L^2(SU(2)) as a representation space for the group (SU(2) x SU(2)) Ⓢ S_2' J.Phys.A(submitted for publ-cation)
8 M E Rose 'Elementary Theory of Angular Momentum'(Wiley,New York,1957)
9 R Dirl, P Kasperkovitz 'Gruppentheorie, Anwendungen in der Atom- und Festkörperphysik'(Vieweg,Wiesbaden,1975 in press)
10 M Hamermesh 'Group Theory and Its Application to Physical Problems'(Addison-Wesley,Reading Mass.,1962)
11 I M Gelfand, M I Graev, N Ya Vilenkin 'Generalized Functions',vol.5(Academic, New York,1966)
12 A J Coleman 'Induced and Subduced Representations' in 'Group Theory and Its Applications' edited by M Loebl (Academic,New York,1968)
13 M A Naimark 'Normed Rings'(Noordhoff,Groningen,1964)
14 J S Lomont 'Applications of Finite Groups'(Academic,New York,1959)
15 R Dirl Nuovo Cim.23B(1974)417,441
16 P Kasperkovitz, R Dirl J.Math.Phys.15(1974)1203
17 B R Judd, E E Vogel Phys.Rev.B11(1975)2427
18 E R Marshalek Phys.Rev.C11(1975)1426

INDUCTION FROM A NORMAL NILPOTENT SUBGROUP

N. Giovannini

Institute for Theoretical Physics, University of Nijmegen ++

Abstract

The problem of the factors systems involved in the induction procedure (in the sense of Mackey) for the unitary representations of group extensions, is considered for the case where the normal subgroup is a nilpotent Lie group. A useful explicit expression is given for these factor systems which becomes especially simple for a large class of such extensions.

++ Postal address: Faculteit Wis- en Natuurkunde, Toernooiveld, Nijmegen, (the Netherlands)

Introduction

The theory of induction of representations for separable locally compact topological groups, as begun in the last years of the 19th century by Frobenius [1] for the finite groups, is now quite complete, especially since the well known work of Wigner [2] and Mackey [3] . In the simplest and commonest case the theory provides an explicit procedure for the construction of all irreducible unitary representations of a (regular) inessential extension $G \cong N \wedge_{\varphi} H$ with N normal abelian, from the representations of N and of certain subgroups of the factor group H. It has also been shown by Mackey [4] that whether the semi-directness or the abelian condition is left out, it becomes necessary to consider projective representations of these subgroups of H as well.

While the occuring factor systems are easy to find when N is still abelian, the problem becomes much more difficult in the general case. In this talk we consider this problem when N is a nilpotent Lie group, making profit of the work of Kirillov on the structure of the irreducible unitary representations for this case [5-6] . In the first part we remind briefly the theory of Kirillov and prove for a special kind of classes some additive, useful properties on the dual $\hat{N}$ of N. In the second part, we use Mackey's theory of induction and derive then an explicit expression of the factor sets involved. This expression is then made simpler for a large class of such **extensions**.

1. The dual of N

We first review briefly the structure of the set $\hat{N}$ of classes of irreducible unitary representations of a nilpotent Lie group N, as described by Kirillov [6] . For simplicity, we assume that N is simply connected, the generalization giving no difficulty, as is well known. Let thus $\underline{n}$ be the Lie algebra of N, $\underline{n}'$ the dual space of $\underline{n}$ and $\mathrm{coAd}_{\underline{n}'}(N)$ the co-adjoint representation of N on $\underline{n}'$, defined for $\nu \in \underline{n}'$ by

$$(\mathrm{coAd}_{\underline{n}'}(n_1)\nu)(x) \overset{\mathrm{def}}{=} \nu(\mathrm{Ad}_{\underline{n}}(n_1^{-1})x) \qquad (1.1)$$

with $n_1 \in N$, $x \in \underline{n}$. Let then $\mathcal{O}$ be the orbit of an element $\nu \in \underline{n}'$, i.e. the set of all images of ν under the action (1.1) of N. Since (1.1) defines a representation of N, the set of orbits is a partition of $\underline{n}'$. Consider then in each such orbit one arbitrary (but fixed) element ν and consider a subalgebra $\underline{\ell} \leq \underline{n}$ such that

$$[\underline{\ell}, \underline{\ell}] \leq \operatorname{Ker} \nu \tag{1.2}$$

This subalgebra $\underline{\ell}$ is called subordinate to ν. The following map T_ν on the Lie subgroup L of N generated by $\underline{\ell}$ ($L = \{\exp \underline{\ell}\}$)

$$T_\nu (\exp x) \stackrel{\text{def}}{=} \exp i\,\nu(x) \qquad , \; x \in \underline{\ell} \tag{1.3}$$

is then a one dimensional unitary representation of L.

Let us further induce this representation to a representation $V_{\nu,\underline{\ell}}$ of N by

$$(V_{\nu,\underline{\ell}})(n)\, f(\lambda) \stackrel{\text{def}}{=} T_\nu(\lambda \cdot n \cdot (\lambda')^{-1}) f(\lambda') \tag{1.4}$$

where λ, λ' are members of an (arbitrary but fixed) set of representatives of the (right) coset decomposition of N with respect to L, with λ' fixed by the condition $\lambda\; n\; (\lambda')^{-1} \in L$, and $f(\lambda)$ is a measurable and quadratic integrable function on the coset space with respect to the (quasi-) invariant (under N) measure μ and with values in the carrier space of the representation T_ν. By a slight abuse of language, and since they are in 1-to-1 Borel correspondance the parameter λ describes, as is usual, both the coset space N/L and a fixed choice of coset representatives.

The following then holds

Theorem 1.1 (Kirillov)

(i) Any irreducible unitary representation of N is obtained in this way, up to equivalence.

(ii) $V_{\nu,\underline{\ell}}$ is irreducible if and only if dim $\underline{\ell}$ is maximal

(iii) $V_{\nu,\underline{\ell}} \sim V_{\nu',\underline{\ell}'}$, both irreducible if and only if $\mathcal{O}_\nu = \mathcal{O}_{\nu'}$

(iv) $\hat{N}$, the dual of N, is always of type I

From now on $\underline{\ell}$ will always be assumed to be a subordinate subalgebra to ν of maximal dimension , i.e. equivalently, a real polarization at ν.
We now distinguish in $\hat{N}$ the following two types:

Def: the class $[\hat{n}]$ of $V_{\nu,\underline{\ell}}$ is said to be of type a if and only if it is possible to choose $\underline{\ell}$ ideal. Else it will be called of type b.

It is easy to verify that the implied condition does not depend on the choice of a particular element on the orbit $\mathcal{O}_\nu$ so that, together with theorem 1.1 this condition depends effectively only on the class $[\hat{n}]$ to which $V_{\nu,\underline{\ell}}$ belongs.

Clearly any $\hat{N}$ contains elements of type a. We shall call N to be of type a if the corresponding $\hat{N}$ contains only elements of type a. In order to show that an interestingly large number of nilpotent Lie groups is of type a, but not all, let us give some examples

(i) any nilpotent Lie group whose central series has length ≤ 2 is of type a.

(ii) any nilpotent Lie group of dimension less or equal to 5 is of type a (follows from direct check in the corresponding Lie algebras, as classified by Dixmier [7]).

(iii) any direct product of nilpotent Lie groups of type a is of type a.

(iv) not any nilpotent Lie group is of type a: consider for example the 10 dimensional Lie algebra of 5 x 5 matrices $\underline{\underline{A}}$ with $A_{\mu\nu} = 0$ if $\mu \geq \nu$.
It is a quite easy and useful exercise to show that it is not possible for the linear form $v(\underline{\underline{A}}) \overset{\text{def}}{=} A_{15}$ to find a real polarization $\underline{\ell}$ which is an ideal.
We do not know however at the moment of a simple immediate characterization of groups of type a.
Let us indicate without proof (details will be published elsewhere) some useful properties of nilpotent Lie groups of type a:

Proposition 1.2 let N be of type a, $\nu \in \mathcal{O}_\nu$ in $\underline{n}'$, $[\underline{\ell},\underline{n}] \subseteq \underline{\ell}$ a corresponding ideal subordinate to ν (and of maximal dimension) then

(i) $\underline{\ell}$ is his own centralisator with respect to ν , i.e. $\forall\, x \in \underline{n}$
$\nu([\underline{\ell},x]) = 0$ if and only if $x \in \underline{\ell}$

(ii) The coset space N/L is (Borel) isomorphic with the orbit of T_ν under N and is 1-to-1 characterized by the classes of elements of $\mathcal{O}_\nu$ which coincide with eachother when restricted to $\underline{\ell}$.

We now turn to the general induction procedure from a normal nilpotent subgroup N. First however, we shall present shortly the more general theory in a language which is not usual in this context but which is at our opinion the best adapted one to it.

2. General inducing from a nilpotent normal subgroup

Mackey's theory of induced representations is a well known procedure, when at least applied to a (regular) semidirect product $G = N \wedge_{\varphi} H$, with G separable locally compact and N abelian (φ denoting the canonical homomorphism from H to Aut (N)) [3] . Less known perhaps is the more general case where G, separable locally compact, is any (regular) extension of a group N, not necessarily abelian, by a group H, i.e. appears in the following exact sequence of groups

$$1 \longrightarrow N \longrightarrow G \longrightarrow H \longrightarrow 1 \quad , \quad m, \varphi \tag{2.1}$$

characterized by a factor set m: H x H $\longrightarrow$ N, with $m(h_1,h_2h_3) + \varphi(h_1)\, m(h_2,h_3) = m(h_1,h_2) + m(h_1h_2, h_3)$, and a map φ : H $\longrightarrow$ Aut (N) satisfying $\varphi(h_1)\cdot \varphi(h_2) = \mu(m(h_1,h_2))\, \varphi(h_1h_2)$, μ being the canonical epimorphism from N onto In(N), the group of inner automorphisms of N. The essential difference, in this more general case, is that, as shown by Mackey [4] , whether if N is no more abelian or the extension (2.1) does not split, no longer ordinary representations of the adequate subgroups of H have to be considered, but certain projective ones. It is the purpose of this section to calculate explicitly the factor sets which are then involved, for the case where N is a nilpotent Lie group.
Since the dependence of m (as in (2.1)) in these factor sets is easy to find, we assume first, in order to inlight the notation, that m = 0.

Let us indicate briefly, in a way convenient for our purposes, the essential steps of the explicit generalized Wigner-Mackey construction procedure of induced representations. Let therefore $[\hat{n}] \in \hat{N}$, the dual of N, with $\hat{n}$ a representant of this class of irreducible representations.

One defines from φ a map $\hat{\varphi}$ on H with

$$\hat{\varphi}(h) \;:\; \hat{N} \longrightarrow \hat{N} \quad , \quad \hat{\varphi}(h)\, [\hat{n}] \equiv [\hat{n}_h]$$

in the canonical way, i.e. by

$$\hat{n}_h \; (n) \; = \; \hat{n} \; (\varphi^{-1}(h) \; n) \tag{2.2}$$

The set of all classes $\{[\hat{n}_h]\}$ generated by $\hat{\varphi}$ from $[\hat{n}]$ is called the orbit of $[\hat{n}]$ under H and is denoted $\mathcal{O}_{[\hat{n}]} \subseteq \hat{N}$. The action (2.2) defines directly a subgroup $H_{\hat{n}}$ of H, which we shall call the homogeneous little group, by

$$h \in H_{\hat{n}} \iff [\hat{n}_h] = [\hat{n}]$$

i.e. $h \in H_{\hat{n}}$ if and only if there exists a unitary operator $S(h) \in \mathcal{U}(\mathcal{H}(\hat{n}))$, $\mathcal{H}(\hat{n})$ being the representation space of $\hat{n}$, such that

$$\hat{n}_h \; (n) \; = \; S \; (h)^{-1} \; \hat{n} \; (n) \; S \; (h) \quad , \quad \forall \, n \in N \tag{2.3}$$

The map $S: h \longrightarrow \mathcal{U}(\mathcal{H}(\hat{n}))$ is in general not a homomorphism, but, from (2.3), using that φ is now a homomorphism in (2.2) and that $\hat{n}$ is irreducible, it follows from Schur's Lemma that it can be choosen to be a projective one, satisfying thus

$$S \; (h) \; S \; (h') \; = \; \tau \; (h, h') \; S \; (h \cdot h') \tag{2.4}$$

$h, h' \in H_{\hat{n}}$ and $\tau(h, h')$ some factor set in $\mathcal{U}(1)$, the unit circle of the complex plane.

It is quite clear from the above formulas that it is in general not easy to find explicitely operators $S(h)$ satisfying (2.3) and hence the factor systems τ we will need in the sequel. It is just the purpose of our contribution to solve this problem for the more special cases we are interested in.

The isotropy group $G_{\hat{n}} \subseteq G$ (defined as the subgroup of G leaving $[\hat{n}]$ invariant under the action $g: \hat{n}(n) \longrightarrow \hat{n}(g^{-1} \cdot n \cdot g)$, N being identified with its image as subgroup of G) appears then as an extension (here trivial because so is G) of N by $H_{\hat{n}}$, as shown in the following commutative diagram of exact sequences

$$\begin{array}{ccccccccc} 1 & \longrightarrow & N & \xrightarrow{(\iota)} & G_{\hat{n}} & \xrightarrow{\pi} & H_{\hat{n}} & \longrightarrow & 1 \quad , 0 , \varphi \\ & & \| & & \downarrow (\iota) & & \downarrow (\iota) & & \\ 1 & \longrightarrow & N & \xrightarrow{(\iota)} & G & \xrightarrow{\pi} & H & \longrightarrow & 1 \quad , 0 , \varphi \end{array} \tag{2.5}$$

(ι) denoting the injection monomorphism. We construct then in a first step a (projective) representation of $G_{\hat{n}}$ as follows. Let L be a projective representation of $H_{\hat{n}}$ with factor set ω and carrier space $\mathcal{H}(L)$ then it is easy to see that by defining for all $(n, h_{\hat{n}}) \in G_{\hat{n}}$

$$(\hat{n}_s \cdot L)(n, h_{\hat{n}}) \overset{def}{=} \hat{n}(n)\, S(h_{\hat{n}}) \otimes L(h_{\hat{n}}) \tag{2.6}$$

on the Hilbert space $\mathcal{H}(\hat{n}) \otimes \mathcal{H}(L)$, $\otimes$ denoting the external Kronecker product, we get a (projective) representation of $G_{\hat{n}}$ with factor set σ , where

$$\sigma(g_1, g_2) = \tau(\pi g_1, \pi g_2)\, \omega(\pi g_1, \pi g_2) \tag{2.7}$$

so that, choosing $\omega = \tau^{-1}$ on $H_{\hat{n}}$, we get an ordinary representation of $G_{\hat{n}}$, with carrier space $\mathcal{H}(\hat{n}) \otimes \mathcal{H}(L)$. The last step is now the following: we decompose H in right cosets with respect to $H_{\hat{n}}$ with coset representatives $\{ h_i \mid i \in I$, I some index set (Borel) isomorphic to $H/H_{\hat{n}} \}$ and similarly G with respect to $G_{\hat{n}}$, choosing now as coset representatives the image of the $\{h_i\}$ under a fixed section r: $H \longrightarrow G$, the set of representatives being then given by $\{(0, h_i) \mid i \in I\}$.

Let now $\mu_{\hat{n}}$ be a quasi-invariant ergodic measure on $\hat{N}$ not identically zero but vanishing on all orbits outside $\mathcal{O}_{[\hat{n}]}$. We assume, again in order to simplify the notation, that this measure is right and left invariant, the generalization being straightforward. Since we have also assumed that the action of H on $\hat{N}$ was regular (in the sense of Mackey [4]), this measure is unique up to a class and also transitive, i.e. concentrated on the orbit [4] . The coset space $G/G_{\hat{n}} \cong H/H_{\hat{n}}$ can then be identified with the orbit $\mathcal{O}_{[\hat{n}]}$ by the 1-to-1 Borel isomorphism $[\hat{n}_{h_i}] \leftrightarrow h_i$, so that we may use the parametrization $\{h_i\}$ to describe both spaces. We now consider on this space a vector valued function f

$$f : h_i \longrightarrow \mathcal{H}(\hat{n}) \otimes \mathcal{H}(L)$$

satisfying the two conditions

(i) $(f(h_i), \Phi)$ is $\mu_{\hat{n}}$-measurable, $\forall \Phi \in \mathcal{H}(\hat{n}) \otimes \mathcal{H}(L)$

(ii) $$\|f\|^2 \overset{\text{def}}{=} \int_{\mathcal{O}_{[\hat{n}]}} \|f(h_i)\|^2 \, d\mu_{\hat{n}}(h_i) < \infty \tag{2.8}$$

where scalar product and norm under the integrals are taken in $\mathcal{H}(\hat{n}) \otimes \mathcal{H}(L)$. Identifying functions equal almost everywhere, the set of functions satisfying the above conditions can be shown [3] to form a separable Hilbert space, $\mathcal{H}$, with scalar product

$$(f,g) = \int_{\mathcal{O}_{[\hat{n}]}} d\mu_{\hat{n}}(h_i) \; (f(h_i), g(h_i))$$

The induced representation is then defined on $\mathcal{H}$ as follows:
let $(n,h) \in G$, then

$$(\hat{n} \uparrow G)^L (n,h) f(h_i) \overset{\text{def}}{=} (\hat{n}_s . L) \; ((0,h_i) \; (n,h) \; (0,h_j)^{-1}) \; f(h_j) \tag{2.9}$$

where h_j is the (unique) coset representative satisfying

$$h_i \, h \, h_j^{-1} \in H_{\hat{n}}$$

The following then holds, and follows from [4] for this special case:

<u>Theorem 2.1</u> (Mackey)

Consider an orbit $\mathcal{O}_{[\hat{n}]} \leq \hat{N}$ and a transitive ergodic measure $\mu_{\hat{n}}$, concentrated on this orbit as described above. Then the representation (2.9) is unitary and irreducible if and only if L is. Two such representations are equivalent only if they are based on the same orbit. Moreover all irreducible unitary representations of G are obtained, up to equivalence, once and only once, when one induces once per orbit, and for each orbit one considers all projective, inequivalent unitary irreducible, ω-representations L of the corresponding $H_{\hat{n}}$, with ω satisfying the equation (2.7), σ being taken equal to 1.

Up to now we have made no use of the fact that N is nilpotent and the results described above are valid for any regular split extension (2.1).

Let thus now $[\hat{n}] \in \hat{N}$, and N nilpotent. It follows from theorem 1.1 (i) that for each class $[\hat{n}]$, one can choose a representant $V_{\nu, \underline{\ell}}$ as defined in (1.4) with carrier space

$$\mathcal{H}(V) = \int_{N/L}^{\oplus} \mathcal{H}(T) \, d\mu(\lambda) \tag{2.10}$$

and with T and λ as in section 1. It is then possible, using the results of this first section and the theory just mentioned to construct explicitely the operators S (h), and hence to calculate the factor system ω of (2.7).
We omit here again the details and mention directly first the following result, valid for any regular extension (2.1) (not necessarily semidirect)

Proposition 2.2.

Let G , separable locally compact, be any regular extension of a nilpotent Lie group N by a group H. Then, given $[\hat{n}] \in N$, $V_{\nu,\underline{\ell}} \in [\hat{n}]$, the intertwinings operator $S(h) \equiv S((0,h))$ of (2.3) ((0,h) in the little group $G_{\hat{n}}$ and the factor system ω of (2.7) on $H_{\hat{n}}$ needed for the Wigner Mackey generalized induction procedure are respectively given by

$$S((0,h)) = V_{\nu,\underline{\ell}}(n(h))$$

with $n(h) \in N$ such that $\mathrm{coAd}(n(h))\,\nu = \mathrm{coAd}(h)\,\nu$, and

$$\omega(h,h')\cdot \mathbb{1} = \left[S((0,h))\, S((0,h'))\, S((0,h)\cdot(0,h'))^{-1} \right]^{-1}$$

$$= V_{\nu,\underline{\ell}}(m(h,h')\, V_{\nu,\underline{\ell}}(n(h\cdot h')\; V_{\nu,\underline{\ell}}^{-1}(n(h'))\; V_{\nu,\underline{\ell}}^{-1}(n(h)),$$

Note that coAd (h) is in this general case not a homomorphism of $H_{\hat{n}}$ on $\underline{n}'$ but satisfies

$$\mathrm{coAd}(h)\ \mathrm{coAd}(h') = \mathrm{coAd}(m(h,h'))\ \mathrm{coAd}(hh')$$

The result for ω simplifies greatly whenever N is of type a and the extension (2.1) central, i.e. with the factor set m in the centre of N.
Indeed, if $\underline{\ell}$ is ideal, one may construct the following exact sequence of groups

$$1 \longrightarrow L \longrightarrow N \longrightarrow N/L \longrightarrow 1, \qquad \rho, \psi \tag{2.11}$$

with factor set $\rho \in Z^2(N/L, L)$ and $\psi: N/L \longrightarrow \mathrm{Aut}\, L$ defined canonically.
The following then holds

Proposition 2.3.

Let G, separable locally compact, be a central extension of a nilpotent Lie group N of type a by a group H. Then given $[\hat{n}] \in \hat{N}$, $V_{\nu, \underline{\ell}} \in [\hat{n}]$ with $\underline{\ell}$ ideal, the factor set ω, as described above is given, up to equivalence, by

$$\omega(h, h') = T_{\nu, \underline{\ell}}(m(h,h')) \, . \, T_{\nu, \underline{\ell}}^{-1}(\rho(n(h), n(h'))) \qquad (2.12)$$

where ρ is the factor set of (2.11) and n (h) is the (unique up to L) element of N satisfying the condition

$$\mathrm{coAd}(n(h)) \, \nu(\underline{\ell}) = \mathrm{coAd}(h) \quad \nu(\underline{\ell})$$

Again we omit the details of the proof.
For N abelian, (2.12) gives the well known result back, ρ being trivial. Further it is clear that the expression remains true if N is not of type a, but then only for the representations of N which are of type a.
This problem has been developed in the frame of a specific physical problem [8]. We refer to this paper for an application of these results. On the other side, the details of the proofs of the results we have mentioned in this communication will also be published subsequently [9] .

References

(1) G. Frobenius, Sitz. Preus. Akad. Wiss. (1898), pp 501-515

(2) E.P. Wigner, Ann. of Math. 40, 149 (1939)

(3) for a review of the theory see for example
G.W. Mackey, Lecture notes, U.P. Oxford (1971)

(4) G.W. Mackey, Acta Math. 99, 265 (1958)

(5) A. Guichardet, Seminaire Bourbaki 1962/63, 249

(6) A. Kirillov, Uspekhi Math. Nauk. 17, 57 (1962)

(7) J. Dixmier, Can. J. Math. 10, 321 (1958)

(8) N. Giovannini, Elementary particles in external e.m. fields
Preprint June 1975,(to be published)

(9) N. Giovannini, Induction from a normal nilpotent subgroup,
Preprint September 1975, (to be published).

SPINOR REPRESENTATIONS

R.C. King
(Mathematics Department, The University, Southampton, England.)

In 1935 Brauer and Weyl[1] published a paper on the spin representations of the orthogonal groups which has withstood the test of time extraordinarily well. In this paper the spin representation Δ of $O(n)$ was defined and the Kronecker square of this spin representation was decomposed into its irreducible parts for both $n = 2k$ and $n = 2k + 1$. These reductions take the form

$$O(2k) \qquad \Delta \times \Delta = [1^k] + \sum_{t=1}^{k} \left([1^{k-t}] + [1^{k-t}]^* \right) , \tag{1}$$

$$O(2k+1) \qquad \Delta \times \Delta = \sum_{t=0}^{k} [1^{k-t}] (*)^{k-t} . \tag{2}$$

Here $[1^p]$ denotes the pth-rank antisymmetric tensor representation of $O(n)$, whilst $[1^p]^*$ denotes the associate representation equivalent to $[1^{n-p}]$, which differs from $[1^p]$ in its representation matrices by the factor $\det A$, where A is the group element of $O(n)$. Furthermore it was pointed out that

$$O(2k) \qquad \bar{\Delta} = \Delta = \Delta^* = \bar{\Delta}^* , \tag{3}$$

$$O(2k+1) \qquad \bar{\Delta} = \Delta , \qquad \bar{\Delta}^* = \Delta^* , \tag{4}$$

where $-$ signifies the contragredient of a representation.

More generally the inequivalent irreducible tensor and spinor representations may be labelled by means of partitions $(\lambda) = (\lambda_1, \lambda_2, \ldots, \lambda_p)$, with $\lambda_1 \geqslant \lambda_2 \geqslant \cdots \geqslant \lambda_p > 0$, which specify the symmetry of the tensor indices, and the symbol Δ signifying the presence of a spinor index. These representations are then

$$O(2k) \qquad [\lambda],\ [\lambda]^* \text{ with } p < k , \qquad [\lambda] \text{ with } p = k ,$$

$$[\Delta, \lambda] \text{ with } p \leqslant k ,$$

$$O(2k+1) \qquad [\lambda],\ [\lambda]^*,\ [\Delta; \lambda],\ [\Delta; \lambda]^* \text{ with } p \leqslant k .$$

All these representations are self-contragredient, and the representations of $O(2k)$: $[\lambda]$ with $p = k$ and $[\Delta; \lambda]$ with $p \leqslant k$, are self-associate.

Littlewood[2] has given in terms of S-functions the connection not only between tensor representations of O(n) and U(n) but also between tensor representations of Sp(n) and U(n). These take the form[3]:

$$U(n) \downarrow O(n) \qquad \{\lambda\} \downarrow \sum_{\delta} [\lambda/\delta] = [\lambda/D] \quad , \tag{5}$$

$$O(n) \uparrow_{r} U(n) \qquad [\lambda] \uparrow_{r} \sum_{\gamma} (-)^{c/2}\{\lambda/\gamma\} = \{\lambda/C\} \quad , \tag{6}$$

$$U(n) \downarrow Sp(n) \qquad \{\lambda\} \downarrow \sum_{\beta} \langle\lambda/\beta\rangle = \langle\lambda/B\rangle \quad , \tag{7}$$

$$Sp(n) \uparrow_{r} U(n) \qquad \langle\lambda\rangle \uparrow_{r} \sum_{\alpha} (-)^{a/2}\{\lambda/\alpha\} = \{\lambda/A\} \quad , \tag{8}$$

where the irreducible representations of Sp(n) with $n = 2k$ are given by $\langle\lambda\rangle$ with $p \leq k$, and A, B, C, D denote infinite series of particular S-functions satisfying the conditions $AB = CD = 1$.

The reduction of the Kronecker product of irreducible representations $\{\lambda\}$ and $\{\mu\}$, corresponding to the mutual symmetrisation of the two sets of tensor indices associated with these representations, is given by the Littlewood-Richardson rule for outer products of S-functions:

$$U(n) \qquad \{\lambda\} \times \{\mu\} = \{\lambda.\mu\} \quad . \tag{9}$$

It is of course this rule which enables S-function quotients $\{\lambda/\mu\}$ to be evaluated. It follows that for the orthogonal group:

$$O(n) \qquad [\lambda] \times [\mu] = [((\lambda/C).(\mu/C))/D] \quad . \tag{10}$$

However this rule is exactly equivalent to the much simpler rule also given by Littlewood[4]

$$O(n) \qquad [\lambda] \times [\mu] = \sum_{\zeta} [(\lambda/\zeta).(\mu/\zeta)] \tag{11}$$

which may be justified by noting that the reduction procedure corresponds to carrying out contractions, involving the metric tensor, between the indices associated with (λ) and (μ). Multiple contractions are such that the sets of contracted indices in each set must share the same symmetry signified by (ζ).

In what follows the aim is to generalise the Kronecker product rule (11) to include cases in which either one or both of the tensor representations $[\lambda]$ and $[\mu]$ are replaced by spinor representations. To do this it is helpful to return to the paper of Brauer and Weyl and generalise their method of obtaining the particular results (1) and (2).

For simplicity $O(n)$ is considered first of all with $n = 2k$. The group elements are the $n \times n$ unitary matrices A satisfying the condition $A^T A = I$ which takes the form:

$$g^{ij} A^k_i A^\ell_j = g^{k\ell} \tag{12}$$

when written in terms of the metric tensor g^{ij}.

The defining transformations are:

$$A : x_i \to x_i' = A^j_i x_j \qquad i, j = 1, 2, \ldots, 2k \tag{13}$$

and the basis states of the tensor representation $[\lambda]$ are such that

$$A : t^{[\lambda]}_{i_1 i_2 \ldots i_r} \to t'^{[\lambda]}_{i_1 i_2 \ldots i_r} = A^{j_1}_{i_1} A^{j_2}_{i_2} \ldots A^{j_r}_{i_r} t^{[\lambda]}_{j_1 j_2 \ldots j_r} \tag{14}$$

where the superscript $[\lambda]$ signifies that the symmetry of the tensor indices is specified by (λ) and that they are traceless in the sense that:

$$g^{i_a i_b} t^{[\lambda]}_{i_1 i_2 \ldots i_r} = 0 \qquad a, b, = 1, 2, \ldots r \tag{15}$$

The fundamental spin representation Δ arises through the existence of a 2^k—dimensional representation Γ, of a Clifford algebra whose generators in this representation Γ, are matrices γ_i with $i = 1, 2, \ldots, n = 2k$ satisfying the relations

$$\gamma_i \gamma_j + \gamma_j \gamma_i = 2g_{ij} \quad . \tag{16}$$

This representation serves to define a group G_k of order 2^{k+1} whose elements are the matrices

$$\gamma_0^{a_0} \gamma_1^{a_1} \ldots \gamma_n^{a_n} \quad \text{with } a_i = 0 \text{ or } 1 \quad \text{for } i = 0, 1, 2, \ldots, n$$

where it has been convenient to denote $-I$ by γ_0. This group has only one faithful irreducible representation namely this representation Γ of dimension 2^k.

The defining relations (16) are invariant under the replacement of γ_i by $A_i^j \gamma_j$ provided that (12) is satisfied. Therefore, these new matrices together with $\gamma_0 = -I$ furnish another representation of dimension 2^k which it may be shown is equivalent to Γ. This implies the existence of a matrix $S(A)$ for each A satisfying (12) such that

$$A_i^j \gamma_j = S(A)^{-1} \gamma_i S(A) \tag{17}$$

It is a straightforward matter to verify that the matrices $S(A)$ constitute a double-valued representation of $O(n)$. It is double-valued in the sense that corresponding to each group element A there exist two matrices $S(A)$ and $-S(A)$ satisfying (17).
This representation $A \to S(A)$ is the spin representation, Δ, of $O(2k)$. The basis states of this representation are the spinors ψ_α transforming in accordance with the rule:

$$A : \psi_\alpha \to \psi'_\alpha = S(A)_\alpha^\beta \psi_\beta \qquad \alpha, \beta = 1, 2, \ldots, 2^k \ . \tag{18}$$

Clearly the set of basis states $x_i \psi_\alpha$ correspond to the Kronecker product representations $[1] \times \Delta$. This representation is reducible as may be seen by noting that γ-matrices may be used to project out an invariant subspace just as the metric tensor is used to project out invariant subspaces in going from $U(n)$ to $O(n)$. Indeed A induces the transformation:

$$\begin{aligned} A : g^{ij}\gamma_i x_j \psi \to g^{ij}\gamma_i A_j^k S(A) x_k \psi \\ = S(A) g^{ij} A_i^\ell A_j^k \gamma_\ell x_k \psi \\ = S(A) g^{\ell k} \gamma_\ell x_k \psi \end{aligned} \tag{19}$$

where use has been made of (17) and (12). Hence $\phi = g^{ij}\gamma_i x_j \psi$ forms the basis of the same spin representation Δ. The remaining linearly independent basis states in the set $x_i \psi$ given by $\psi_i = \left(x_i \psi - \frac{1}{n} \gamma_i \phi\right)$ satisfying $g^{ij}\gamma_i \psi_j = 0$ form the basis of another irreducible representation which may conveniently be denoted by $[\Delta;\ 1]$. The symbols Δ and 1 indicate the presence of a spinor and a tensor index in the basis states.

More generally the operation of multiplication by products of γ-matrices and contractions with the metric tensor giving terms of the form:

$$g^{i_1 j_1}\, g^{i_2 j_2} \ldots g^{i_s j_s}\, \gamma_{i_1} \gamma_{i_2} \ldots \gamma_{i_s}\, t^{[\lambda]}_{j_1 j_2 \ldots j_s \ldots j_r}\, \psi$$

commutes with the transformations induced by A. It follows from (15) and (16) that the Kronecker product $\Delta \times [\lambda]$ reduces in accordance with the rule:

$$O(2k) \qquad \Delta \times [\lambda] = \sum_s\, [\Delta;\ \lambda/1^s] \tag{20}$$

where the basis states of the irreducible representation $[\Delta;\ \rho]$ are such that

$$g^{i_a i_b}\, \psi^{[\Delta;\ \rho]}_{i_1 i_2 \ldots i_r} = g^{j i_a}\, \gamma_j\, \psi^{[\Delta;\ \rho]}_{i_1 i_2 \ldots i_r} = 0 \qquad a,\ b = 1,\ 2,\ \ldots,\ r \tag{21}$$

This formula (20) was derived by Littlewood[5] using the arguments presented here and was derived independently by Murnaghan[6] using character theory. Littlewood[2] went on to write $\Delta \times \{\lambda\}$ in terms of the representations $[\Delta;\ \rho]$ and thereby obtained a formula which he could invert to write $[\Delta;\ \rho]$ as a product of Δ and a series of S-functions $\{\lambda\}$. This may be used, as has been exemplified by Butler and Wybourne[7], to evaluate Kronecker products of the form $[\Delta;\ \lambda] \times [\mu]$ and $[\Delta;\ \lambda] \times [\Delta;\ \mu]$. This procedure involves at every stage both positive and negative terms whose cancellation occurs at the final stage of the analysis. This may be avoided by deriving the inverse of the formula (20) more directly as follows.

The summation in (20) extends as far as $s = n$ since there exist $n = 2k$ independent γ-matrices which may be used to construct an anti-symmetric product. However provided that (λ) is a partition into p parts with $p \leqslant n$ this summation may be extended indefinitely. It follows that

$$O(2k) \qquad \Delta \times [\lambda] = [\Delta;\ \lambda/Q] \tag{22}$$

where Q denotes the infinite series of S-functions

$$Q = \sum_{s} \{1^s\} \quad .$$

This has the advantage of allowing an inversion procedure since $QP = 1$ where

$$P = \sum_{m} (-)^m \{m\} \quad .$$

It follows that

$$O(2k)\ [\Delta;\ \lambda] = \sum_{m} (-)^m\ \Delta \times [\lambda/m] = \Delta \times [\lambda/P] \quad . \tag{23}$$

This relation allows further Kronecker products to be evaluated very easily. Indeed from (23), (11), (22) and the use of the identity

$$((\sigma).(\tau))/Q = (\sigma/Q).(\tau/Q) \quad ,$$

it follows that

$$[\Delta;\ \lambda] \times [\mu] = \Delta \times [\lambda/P] \times [\mu]$$

$$= \sum_{\zeta} \Delta \times [(\lambda/P\zeta).(\mu/\zeta)] = \sum_{\zeta} [\Delta;\ ((\lambda/P\zeta).(\mu/\zeta))/Q]$$

$$= \sum_{\zeta} [\Delta;\ (\lambda/P\zeta Q).(\mu/\zeta Q)] = \sum_{\zeta} [\Delta;\ (\lambda/\zeta).(\mu/\zeta Q)]$$

which gives the rule:

$$O(2k)\ [\Delta;\ \lambda] \times [\mu] = \sum_{\zeta,s} [\Delta;\ (\lambda/\zeta).(\mu/\zeta.1^s)] \quad . \tag{24}$$

This result corresponds of course to carrying out contractions between the tensor indices associated with (λ) and (μ) and multiplying by γ-matrices whose indices are contracted only with those associated with (μ). Any contractions with those associated with (λ) are prohibited by the conditions (21).

Spin-spin Kronecker products may be analysed in a similar way. It is convenient to work in terms of the contragredient spin representation $\bar{\Delta}$ which is associated with spinors $\bar{\psi}^\alpha$ transforming in accordance with the rule:

$$A : \bar{\psi}^\alpha \rightarrow \bar{\psi}'^\alpha = \bar{\psi}^\beta\ S(A)^{-1\,\alpha}{}_{\beta} \quad . \tag{25}$$

Clearly $\bar{\psi}^{\alpha}\ \psi_{\alpha}$ is an invariant and other invariant subspaces associated with the set of basis states $\bar{\psi}^{\alpha}\ \psi_{\beta}$ have bases of the form $\bar{\psi}^{\alpha}(\gamma_{i_1}\ \gamma_{i_2}\ \cdots\ \gamma_{i_s})^{\beta}_{\alpha}\ \psi_{\beta}$ leading to the reduction

$$O(2k) \quad \bar{\Delta} \times \Delta = \sum_s \left[1^s\right] = Q \tag{26}$$

where, as in (20), the summation extends as far as $s = n$ but may be extended indefinitely since all s-fold antisymmetric tensors in an n-dimensional space are zero for $s > n$. To obtain the reduction (26) it is necessary to use the defining relation (16) together with the tracelessness condition on the resulting tensors. The formula (26) is identical with that of Brauer and Weyl given by (1) as may be seen by noting that $\left[1^s\right] = \left[1^{n-s}\right]^*$ and $\left[1^k\right] = \left[1^k\right]^*$ for $n = 2k$.

The generalisation of this result may be obtained by using (23) which gives

$$\left[\overline{\Delta;\ \lambda}\right]\left[\Delta;\ \mu\right] = \bar{\Delta} \times \left[\lambda/P\right] \times \Delta \times \left[\mu/P\right]$$

$$= \sum_{s,\zeta} \left[1^s\right] \times \left[(\lambda/P\zeta).(\mu/P\zeta)\right]$$

$$= \sum_{s,t,\zeta} \left[(1^s/1^t).((\lambda/P\zeta).(\mu/P\zeta))/1^t\right]$$

$$= \sum_{stu\zeta} \left[(1^{s-t}).(\lambda/P\zeta\ 1^u).(\mu/P\zeta\ 1^{t-u})\right] .$$

Since all the summations extend to infinity and $P = Q^{-1}$ it follows

$$O(2k) \quad \left[\overline{\Delta;\ \lambda}\right] \times \left[\Delta;\ \mu\right] = \sum_{s,\zeta} \left[(1^s).(\lambda/\zeta).(\mu/\zeta)\right] . \tag{27}$$

This remarkably simple result corresponds to carrying out contractions between the tensor indices associated with (λ) and (μ) and inserting products of γ-matrices between the spinors. The tensor indices on the γ-matrices are uncontracted but are antisymmetrised in order to satisfy the criterion of irreducibility (14).

The formula appropriate to $O(2k + 1)$ may be found in a very similar manner. The only essential difference being that the Clifford algebra is extended from that generated by γ_i with $i = 1, 2, \ldots 2k$ to include γ_n with $n = 2k + 1$ satisfying the relation $\gamma_n = i^k\ \gamma_1\ \gamma_2\ \cdots\ \gamma_{2k}$. This relation is only invariant under transformation induced by $SO(2k + 1)$. The net result is that contractions with pairs of γ-matrices are invariant. This leads to the formula

$$O(2k + 1)\ \Delta \times \Delta = \sum_{s=0}^{k} \left[1^{2s}\right] = \sum_{s=0}^{k} \left[1^{s}\right]^{(*)^{s}} \tag{28}$$

where the second form follows from the identity $\left[1^{2s}\right] = \left[1^{2k+1-2s}\right]^*$ and is just the result (2) given by Bauer and Weyl. Similarly it is found that

$$O(2k + 1)\ \Delta \times [\lambda] = \sum_{s} [\Delta;\ \lambda/1^{s}]^{(*)^{s}} \tag{29}$$

$$[\Delta;\ \lambda] = \sum_{m} (-)^{m}\ \Delta \times [\lambda/m]^{(*)^{m}} \tag{30}$$

and more generally:

$$O(2k + 1)\ [\Delta;\ \lambda] \times [\mu] = \sum_{s,\zeta} [\Delta;\ (\lambda/\zeta).(\mu/\zeta.1^{s})]^{(*)^{s}} \tag{31}$$

$$O(2k + 1)\ \overline{[\Delta;\ \lambda]} \times [\Delta;\ \mu] = \sum_{s,\zeta} [(1^{s}).(\lambda/\zeta).(\mu/\zeta)]^{(*)^{s}} \ . \tag{32}$$

These formulae (31), (32) for $O(2k + 1)$, (24) and (27) for $O(2k)$ together with (11) for $O(n)$ with $n = 2k + 1$ and $n = 2k$ enable all Kronecker products of representations of $O(n)$ to be evaluated.

Unfortunately, in making use of these formulae terms of the form $[\nu]$ and $[\Delta;\ \nu]$ will arise for which (ν) is a partition into q non-vanishing parts with $q > k$. The same is of course true in analysing the product (9) appropriate to $U(n)$. In this case all terms of the product defined by partitions (ν) for which $q > n$ may simply be ignored since the corresponding tensors vanish identically. In other words there is a modification rule for representations of $U(n)$ given by;

$$U(n)\ \{\nu\} = 0 \text{ if } q > n\ . \tag{33}$$

The corresponding modification rules for tensor and spinor representations of $O(n)$ have been given elsewhere[3] and take the form

$$O(n)\ [\nu] = (-)^{x-1}\ [\nu - h]^{*} \text{ with } h = 2q - n\ , \tag{34}$$

$$O(n)\ [\Delta;\ \nu] = (-)^{x}\ [\Delta;\ \nu - h]^{*} \text{ with } h = 2q - n - 1\ , \tag{35}$$

where $(\nu - h)$ is the partition obtained from (ν) by the removal from the Young diagram corresponding to (ν) of a continuous boundary hook of length h, starting from the foot of the first column and ending in the x-th column. The corresponding representations $[\nu - h]$ and $[\Delta; \nu - h]$ vanish identically unless a regular Young diagram is obtained by this hook removal procedure so that $(\nu - h)$ is indeed a partition.

These modification rules are directly applicable to the terms appearing on the right hand sides of the formulae (11), (24) and (31). Furthermore in both (24) and (31) the summation over s only extends as far as $s = k$. In the case of the formulae (27) and (32) an infinite summation over s can be avoided by modifying the factor $(\lambda/z).(\mu/z)$ using (35). The summation over s then extends only as far as $s = k$. The results are then:

$$O(2k)\ \overline{[\Delta;\lambda]} \times [\Delta;\mu] = \sum_{s=0,\zeta}^{k} [(1^s).(\lambda/\zeta).(\mu/\zeta)]^{+(*)} , \qquad (36)$$

$$O(2k+1)\ \overline{[\Delta;\lambda]} \times [\Delta;\mu] = \sum_{s=0,\zeta}^{k} [(1^s).(\lambda/\zeta).(\mu/\zeta)]^{(*)^s} , \qquad (37)$$

where $+(*)$ signifies that in the final expression the associate of all representations that are not self-associate must be included, that is for the group $O(2k)$ terms of the form $[\nu] + [\nu]^*$ should appear if (ν) is such that $q < k$, whilst for both $O(2k)$ and $O(2k + 1)$ terms of the form $[\nu]$ with $q > k$ should be ignored in accordance with the modification rule (33) with $n = k$.

References

1. R. Brauer and H. Weyl, Amer. J. Maths. 57, 425 (1935).
2. D.E. Littlewood, "The Theory of Group Characters", Oxford University Press, Oxford, (1940).
3. R.C. King, J. Phys. A. Math. Gen. 8, 429 (1975).
4. D.E. Littlewood, Can. J. Maths. 10, 17 (1958).
5. D.E. Littlewood, "A University Algebra", Heinemann, London, (1950).
6. F.D. Murnaghan, "The Theory of Group Representations", Johns Hopkins Press, Baltimore, (1938).
7. P.H. Butler and B.G. Wybourne, J. de Physique 30, 655 (1969).

WEIGHT MULTIPLICITIES FOR THE CLASSICAL GROUPS.

R.C. King.

(Mathematics Department, The University, Southampton, England.)

1. Introduction.

If G is a semi-simple compact Lie group of rank k, then the maximal toroidal subgroup, T_G, of G is isomorphic to the group $T_k = U(1)\times U(1)\times \ldots \times U(1)$, which consists of a direct product of k groups $U(1)$. A group element of T_k takes the form $(e^{i\phi_1}, e^{i\phi_2}, \ldots e^{i\phi_k})$ where ϕ_j, for $j=1,2 \ldots , k$, is a real parameter. An arbitrary irreducible representation of T_k, and thus of T_G, is specified by $\{w_1\}\times\{w_2\}\times \ldots\times\{w_k\}$, and this representation is defined by the mapping:

$$(e^{i\phi_1}, e^{i\phi_2}, \ldots e^{i\phi_k}) \rightarrow e^{i(w_1\phi_1 + w_2\phi_2 + \ldots + w_k\phi_k)} \quad . \tag{1.1}$$

If a representation λ_G of G decomposes on restriction of the group elements to those of the subgroup T_G in accordance with the branching rule:

$$G \downarrow T_G \qquad \lambda_G \downarrow \sum_{\underset{\sim}{w}} m^{\underset{\sim}{w}}_{\lambda_G} \quad \{w_1\}\times\{w_2\}\times\ldots\times\{w_k\} \ , \tag{1.2}$$

then $\underset{\sim}{w} = (w_1, w_2, \ldots ,w_k)$ is said to be a weight vector of the representation λ_G, and its multiplicity is the coefficient $m^{\underset{\sim}{w}}_{\lambda_G}$.

To determine the weight vectors and their multiplicities it is therefore only necessary to evaluate certain branching rules. It is shown that in the case of covariant tensor irreducible representations of the group $U(k)$ this leads in a natural way to the use of both Gelfand patterns and Young tableaux. The generalisation to mixed tensor representations of $U(k)$ is also made and the group $Sp(2k)$ is treated in detail. Some comments are made on the tensor and spinor representations of $O(2k)$ and $O(2k+1)$, and some concluding remarks on the results obtained are presented.

2. Covariant Tensor Representations of $U(k)$.

The irreducible covariant tensor representations of $U(k)$ are specified by $\{\lambda\}$ where $(\lambda) = (\lambda_1,\lambda_2, \ldots ,\lambda_a)$ is a partition of ℓ into a non-vanishing parts with $a\leq k$. The branching rule appropriate to the restriction to the subgroup $U(k-1)\times U(1)$ takes the form:[(1)]

$$U(k)\downarrow U(k-1)\times U(1) \qquad \{\lambda\} \downarrow \sum_{\sigma,w_k} \{\sigma\} \times \{w_k\} \ , \tag{2.1}$$

with $\lambda_i \geq \sigma_i \geq \lambda_{i+1}$ and $w_k=\ell - s$, where (σ) is a partition of s into c non-vanishing parts with $c\leq k -1$.

It follows from the rules appropriate to S-function division enunciated by Littlewood[2] that the branching rule (2.1) may also be written in the form[3]:

$$U(k) \downarrow U(k-1)\times U(1) \qquad \{\lambda\} \downarrow \sum_{w_k} \{\lambda/w_k\} \times \{w_k\} \quad . \tag{2.2}$$

The repeated application of this rule to the chain

$$U(k) \downarrow U(k-1)\times U(1) \downarrow U(k-2)\times U(1)\times U(1) \downarrow \ldots \downarrow T_k \ , \tag{2.3}$$

yields the branching rules

$$\{\lambda\} \downarrow \sum_{w_k} \{\lambda/w_k\}\times\{w_k\} \downarrow \sum_{w_{k-1},w_k} \{\lambda/w_{k-1}w_k\}\times\{w_{k-1}\}\times\{w_k\}$$

$$\ldots \downarrow \sum_{\underset{\sim}{w}} \{\lambda/w_1w_2\ldots w_k\} \ \{w_1\}\times\{w_2\}\times\ldots\times\{w_k\} \ , \tag{2.4}$$

where $\ell = \sum_{c=1}^{k} w_i$. Thus the weight multiplicities may be evaluated using the formula

$$m_{\{\tilde{\lambda}\}}^{\underset{\sim}{w}} = \{\lambda/w_1w_2\ldots w_k\} \quad . \tag{2.5}$$

The relationship between S-function quotients and outer products of S-functions is such that

$$\{w_1\}.\ \{w_2\}.\ldots.\ \{w_k\} = \sum_{\lambda} \ m_{\{\tilde{\lambda}\}}^{\underset{\sim}{w}} \ \{\lambda\} \tag{2.6}$$

It follows from the fact that S-function multiplication is commutative that the symmetry group of the weight diagrams is the symmetric group associated with the permutations of the components of the weight vectors $\underset{\sim}{w}$. Furthermore, since the coefficients in (2.6) are known to be independent of k, the weight multiplicities of the covariant tensor representations of U(k) are k-independent.

This method of determining weight multiplicities, involving as it does the step by step reduction of a representation of U(k) into a set of one dimensional irreducible representations of the Abelian group T_k, yields two equivalent labelling schemes for the basis states of such a representation $\{\lambda\}$ of U(k). The repeated application of (2.1) gives rise to Gelfand patterns[5] in accordance with the extension of labels defined by

$$\left\{\begin{matrix} \lambda_1 & \lambda_2 & & \lambda_k \\ \sigma_1 & \sigma_2\ldots\sigma_{k-1} & & \\ & \vdots & & \\ & & & \end{matrix}\right\} \Longrightarrow \left\{\begin{matrix} m_{1k} & m_{2k} & \cdots\cdots & m_{kk} \\ m_{1k-1} & m_{2k-1} & \cdots & m_{k-1k-1} \\ & \vdots & & \\ & m_{12} \ \ m_{22} & & \\ & m_{11} & & \end{matrix}\right\}$$

The constraints applying to (2.1) are such that m_{ij} is a non-negative integer, that $m_{ij+1} \geq m_{ij} \geq m_{i+1j+1}$ and that

$$w_j = \sum_{i=1}^{j} m_{ij} - \sum_{i=1}^{j-1} m_{ij-1} \qquad \text{for } j = 1,2,\ldots,k. \tag{2.7}$$

Similarly the repeated application of (2.2) to the Young diagrams specified by S-functions gives rise to Young tableaux[6]:

Row lengths

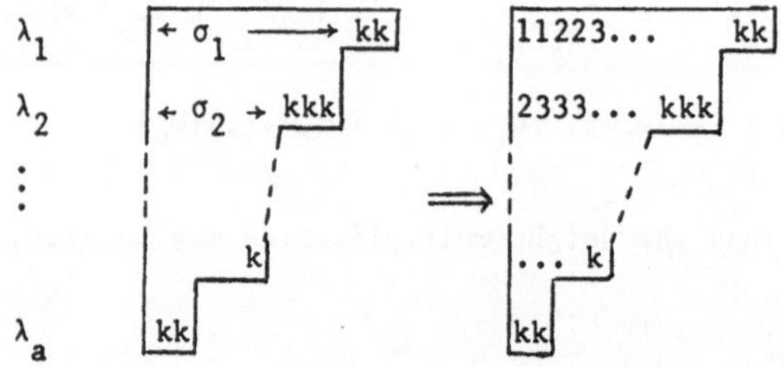

In this case the constraints are such that the numbers in the tableau are non-decreasing across each row from left to right and are strictly increasing down each column from top to bottom, and

$$w_j = \text{the number of } j\text{'s in the tableau.} \tag{2.8}$$

The multiplicity of each weight is the number of distinct Gelfand patterns, or equivalently the number of distinct Young tableaux, whose entries satisfy the given constraints.

For example in the case of the group U(5), for which $k = 5$, the multiplicity of the weight $\underset{\sim}{w} = (1,2,0,2,0)$ in the irreducible representation $\{\lambda\} = \{3,2\}$ is 2, corresponding to the existence of the two Young tableaux $\begin{smallmatrix}1&2&2\\4&4\end{smallmatrix}$ $\begin{smallmatrix}1&2&4\\2&4\end{smallmatrix}$ and the two

Gelfand patterns $\left\{\begin{matrix}3\;2\;0\;0\;0\\3\;2\;0\;0\\3\;0\;0\\3\;0\\1\end{matrix}\right\}$ $\left\{\begin{matrix}3\;2\;0\;0\;0\\3\;2\;0\;0\\2\;1\;0\\2\;1\\1\end{matrix}\right\}$

The symmetry of the weight diagram is exemplified by the fact that

$$m_{\{32\}}^{(22100)} = m_{\{32\}}^{(12020)} = m_{\{32\}}^{(20102)} = \ldots = 2 \ ,$$

and the k-independence of the weight multiplicities by the fact that

$$m_{\{32\}}^{(22100\ldots0)} = 2 \ .$$

3. Mixed Tensor Representations of U(k).

The irreducible mixed tensor representations of U(k) are specified[7] by $\{\bar{\mu};\lambda\}$ where (λ) and (μ) are partitions of ℓ and m respectively, into a and b non-vanishing parts such that $a + b \leq k$. The generalisation of the branching rule (2.1) takes the form:

$$U(k) \downarrow U(k-1)\times U(1) \qquad \{\bar{\mu};\lambda\} \downarrow \sum_{\tau,\sigma,p_k,q_k} \{\bar{\tau};\sigma\} \times \{p_k - q_k\} \tag{3.1}$$

with $\lambda_i \geq \sigma_i \geq \lambda_{i+1}$, $\mu_j \geq \tau_j \geq \mu_{j+1}$ and $w_k = p_k - q_k$ where (σ) and (τ) are partitions of s and t into c and d non-vanishing parts such that $p_k = \ell - s$, $q_k = m - t$ and $c + d \leq k-1$. This result corresponds to the fact that the appropriate generalisation of (2.2) is [3]

$$U(k) \downarrow U(k-1)\times U(1) \qquad \{\bar{\mu};\lambda\} \downarrow \sum_{p_k,q_k} \{\overline{\mu/q_k};\lambda/p_k\} \times \{p_k - q_k\} . \tag{3.2}$$

The repeated application of this rule to the chain (2.3) gives the result:

$$U(k) \downarrow T_k \quad \{\bar{\mu};\lambda\} \downarrow \sum_{\underset{\sim}{p},\underset{\sim}{q}} \{\overline{\mu/q_1q_2\cdots q_k};\lambda/p_1p_2\cdots p_k\}\{p_1-q_1\}\times\{p_2-q_2\}\times\ldots\times\{p_k-q_k\}. \tag{3.3}$$

It then follows from the definition, (1.2), of weights that

$$m^{\underset{\sim}{w}}_{\{\bar{\mu};\lambda\}} = \sum_{\underset{\sim}{p},\underset{\sim}{q}} \{\overline{\mu/q_1q_2\cdots q_k} ; \lambda/p_1p_2\cdots p_k\} \prod_{i=1}^{k} \delta^{w_i}_{p_i-q_i} \quad . \tag{3.4}$$

The corresponding generalisation of the Gelfand patterns arises as a result of the extension of labels defined by

$$\left\{\begin{matrix} \lambda_1\ \lambda_2\ldots\ \lambda_a 0\ldots 0 -\mu_b\ldots -\mu_2\ -\mu_1 \\ \sigma_1\ \sigma_2\ldots\sigma_c 0\ldots 0 -\tau_d\ldots -\tau_2 -\tau_1 \\ \vdots \\ \\ \end{matrix}\right\} \Longrightarrow \left\{\begin{matrix} m_{1k} \quad m_{2k} \quad \cdots\cdots \quad m_{kk} \\ m_{1k-1}\ m_{2k-1}\ \cdots m_{k-1k-1} \\ \vdots \\ m_{12}\ m_{22} \\ m_{11} \end{matrix}\right\}$$

The constraints applying to (3.1) are such that, once again, $m_{ij+1} \geq m_{ij} \geq m_{i+1j+1}$ and

$$w_j = \sum_{c=1}^{j} m_{ij} - \sum_{i=1}^{j-1} m_{ij-1} \qquad \text{for } j = 1,2,\ldots,k, \tag{3.5}$$

but now m_{ij} may be any integer: positive, negative or zero. Furthermore a generalisation of Young tableaux following immediately from (3.2) takes the form

Row lengths.

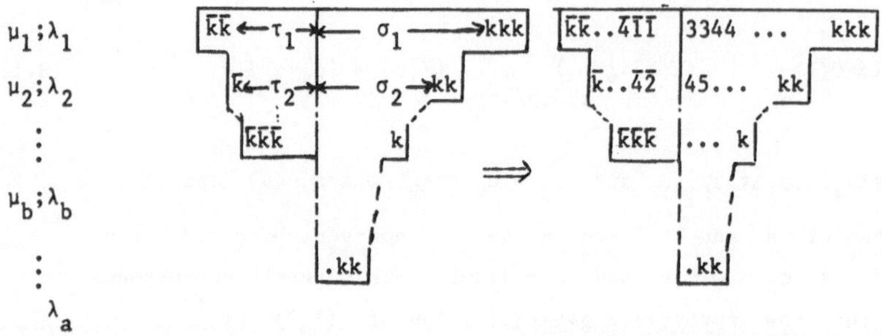

The numbers in the tableau are non-decreasing across each row from left to right and are strictly increasing in magnitude down each column from top to bottom where an entry $\bar{j}$ is to be interpreted as $-j$. In addition if the lowest rows in which j and $\bar{j}$ appear are the x-th and y-th then $x + y \leq j$. Finally:

$$w_j = \text{the number of } j\text{'s} - \text{the number of } \bar{j}\text{'s in the tableau.} \qquad (3.6)$$

Once more the multiplicity of each weight is the number of distinct patterns, or equivalently the number of distinct tableaux, whose entries satisfy the given constraints.

For example in the case of the group $U(5)$, for which $k=5$, the multiplicity of the weight $\underset{\sim}{w} = (0,1,\bar{1},1,\bar{1})$, where $\bar{1} = -1$, in the irreducible representation $\{\bar{\mu};\lambda\} = \{\bar{1}^3;21\}$ is 2, corresponding to the existence of the tableaux

$$\begin{array}{|c|cc|}\hline \bar{2} & 2 & 2 \\ \hline \bar{3} & 4 & \\ \hline \bar{5} & & \\ \hline\end{array} \qquad \begin{array}{|c|cc|}\hline \bar{3} & 2 & 4 \\ \hline \bar{4} & 4 & \\ \hline \bar{5} & & \\ \hline\end{array}$$

and the patterns

$$\left\{\begin{matrix} 2\;1\;\bar{1}\;\bar{1}\;\bar{1} \\ 2\;1\;\bar{1}\;\bar{1} \\ 2\;\bar{1}\;\bar{1} \\ 2\;\bar{1} \\ 0 \end{matrix}\right\} \qquad \left\{\begin{matrix} 2\;1\;\bar{1}\;\bar{1}\;\bar{1} \\ 2\;1\;\bar{1}\;\bar{1} \\ 1\;0\;\bar{1} \\ 1\;0 \\ 0 \end{matrix}\right\}$$

The symmetry group of the weight diagram is once again S_k since the multiplicities are invariant under permutations of the components of the weight vectors as illustrated by the fact that

$$m^{(01\bar{1}1\bar{1})}_{\{\bar{1}^3;21\}} = m^{(\bar{1}\bar{1}011)}_{\{\bar{1}^3;21\}} = m^{(11\bar{1}\bar{1}0)}_{\{\bar{1}^3;21\}} = \ldots = 2\ .$$

Now however, due to the cancellations that take place between p_j and q_j in defining w_j for $j = 1,2,\ldots,k$, the multiplicities are no longer k-independent. Indeed if the same example is considered for the group $U(k)$ the appropriate tableaux corresponding to the weight vector $(\bar{1}\bar{1}1100\ldots0)$

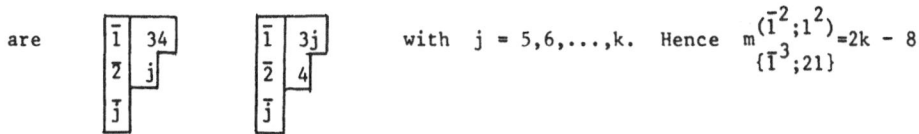

are  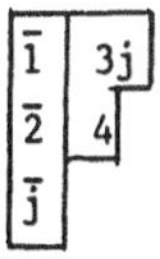 with $j = 5,6,\ldots,k$. Hence $m^{(\bar{1}^2;1^2)}_{\{\bar{1}^3;21\}} = 2k - 8$.

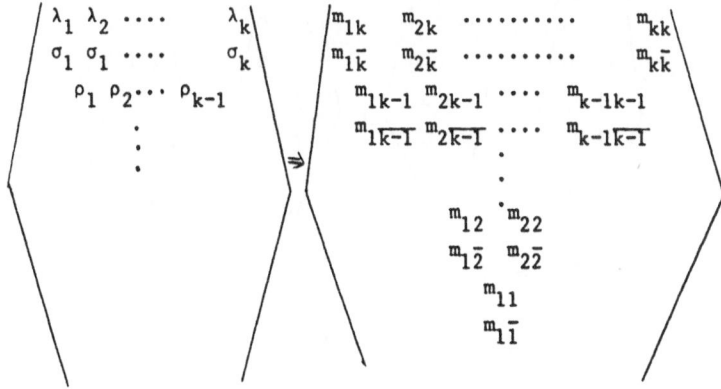

4. Representations of Sp(2k).

The irreducible representations of $Sp(2k)$ are specified by $\langle\lambda\rangle$ where (λ) is a partition of ℓ into a non-vanishing parts with $a \leq k$. Zhelobenko [8] has derived the branching rule

$$Sp(2k)\downarrow Sp(2k-2)\times U(1) \qquad \langle\lambda\rangle\downarrow \sum_{p_k,q_k} \langle\lambda/q_k p_k\rangle\times\{p_k-q_k\}, \tag{4.1}$$

with $\sigma_i \geq \rho_i \geq \sigma_{i+1}$, $\lambda_i \geq \sigma_i \geq \lambda_{i+1}$ and $w_k = \bar{p}_k - q_k$ where (σ) and (ρ) are partitions of s and r into c and e non-vanishing parts such that $p_k = s - r$, $q_k = \ell - s$ and $c \leq k$, $e \leq k-1$. In the notation appropriate to S-functions this takes the form

$$Sp(2k)\downarrow Sp(2k-2)\times U(1) \qquad \langle\lambda\rangle\downarrow \sum_{p_k,q_k} \langle\lambda/q_k p_k\rangle\times\{p_k-q_k\}. \tag{4.2}$$

The repeated application of this rule to the chain

$$Sp(2k)\downarrow Sp(2k-2)\times U(1)\downarrow Sp(2k-4)\times U(1)\times U(1)\downarrow\ldots\downarrow T_k, \tag{4.3}$$

gives

$$\langle\lambda\rangle\downarrow\ldots\downarrow\sum_{p,q} \langle\lambda/q_1p_1q_2p_2\ldots q_kp_k\rangle\ \{p_1-q_1\}\times\{p_2-q_2\}\times\ldots\times\{p_k-q_k\}, \tag{4.4}$$

so that the weight multiplicities are given by

$$m^{w}_{\langle\lambda\rangle} = \sum_{p,q} \langle\lambda/q_1p_1q_2p_2\ldots q_kp_k\rangle \prod_{i=1}^{k} \delta^{w_i}_{p_i-q_i}. \tag{4.5}$$

It follows from (4.1) that the Gelfand pattern generalisation takes the form:

$$\left\langle\begin{array}{cccc} \lambda_1 & \lambda_2 & \cdots & \lambda_k \\ \sigma_1 & \sigma_1 & \cdots & \sigma_k \\ & \rho_1 & \rho_2\cdots & \rho_{k-1} \\ & & \vdots & \end{array}\right\rangle \Rightarrow \left\langle\begin{array}{cccc} m_{1k} & m_{2k} & \cdots\cdots & m_{kk} \\ m_{1\bar{k}} & m_{2\bar{k}} & \cdots\cdots & m_{k\bar{k}} \\ m_{1k-1} & m_{2k-1} & \cdots & m_{k-1k-1} \\ m_{1\overline{k-1}} & m_{2\overline{k-1}} & \cdots & m_{k-1\overline{k-1}} \\ & \vdots & & \\ & m_{12} & m_{22} & \\ & m_{1\bar{2}} & m_{2\bar{2}} & \\ & & m_{11} & \\ & & m_{1\bar{1}} & \end{array}\right\rangle$$

The constraints appropriate to (4.1) are such that m_{ij} and $m_{i\bar{j}}$ are both non-negative integers, that $m_{ij} \geq m_{i\bar{j}} \geq m_{i+1j}$, $m_{i\bar{j}} \geq m_{ij-1} \geq m_{i+1\bar{j}}$ and

$$w_j = p_j - q_j \quad \text{where} \quad p_j = \sum_{i=1}^{j} m_{ij} - \sum_{i=1}^{j} m_{i\bar{j}} \quad \text{and} \quad q_j = \sum_{i=1}^{j} m_{i\bar{j}} - \sum_{i=1}^{j-1} m_{ij-1} \, .$$

In the same way it follows from (4.2) that the Young tableaux generalise to yield tableaux of the form:

Row lengths.

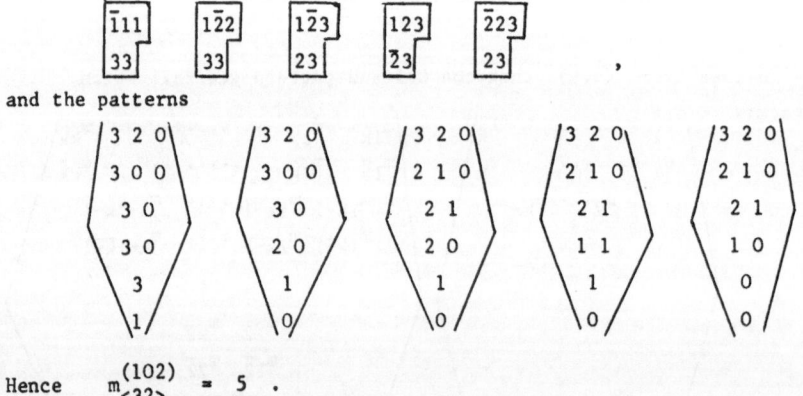

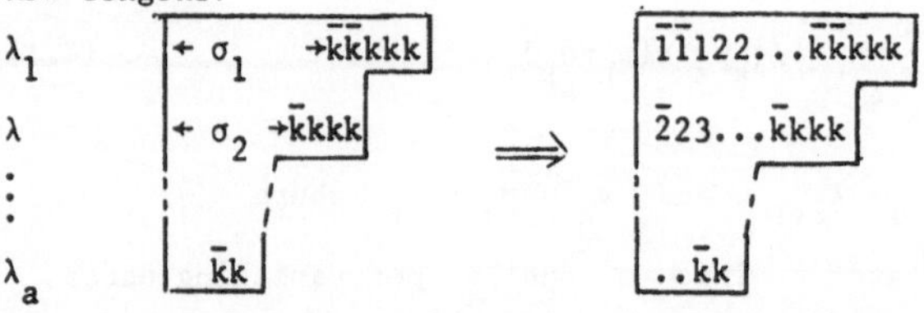

The constraints are such that with the ordering of the entries defined by $\bar{1}<1<\bar{2}<2<\ldots<\bar{k}<k$, the entries are non-decreasing across each row from left to right and are strictly increasing down each column from top to bottom. If the lowest rows in which j and $\bar{j}$ appear are the x-th and y-th then $x \leq j$ and $y \leq j$. Finally

$$w_j = \text{the number of } j\text{'s} - \text{the number of } \bar{j}\text{'s in the tableau.} \qquad (4.6)$$

As before the multiplicity of each weight is just the number of distinct patterns, or equivalently the number of distinct tableaux, whose entries satisfy the given constraints.

For example in the case of the group Sp(6) , for which $k=3$, the weight $\underset{\sim}{w} = (102)$ has multiplicity 5 in the irreducible representation $\langle\lambda\rangle = \langle 32\rangle$, corresponding to the existence of the tableaux

$$\begin{array}{|lll|}\hline \bar{1} & 1 & 1 \\ 3 & 3 & \\ \hline\end{array} \quad \begin{array}{|lll|}\hline 1 & \bar{2} & 2 \\ 3 & 3 & \\ \hline\end{array} \quad \begin{array}{|lll|}\hline 1 & \bar{2} & 3 \\ 2 & 3 & \\ \hline\end{array} \quad \begin{array}{|lll|}\hline 1 & 2 & 3 \\ \bar{2} & 3 & \\ \hline\end{array} \quad \begin{array}{|lll|}\hline \bar{2} & 2 & 3 \\ 2 & 3 & \\ \hline\end{array} \, ,$$

and the patterns

$$\begin{pmatrix} 3\;2\;0 \\ 3\;0\;0 \\ 3\;0 \\ 3\;0 \\ 3 \\ 1 \end{pmatrix} \quad \begin{pmatrix} 3\;2\;0 \\ 3\;0\;0 \\ 3\;0 \\ 2\;0 \\ 1 \\ 0 \end{pmatrix} \quad \begin{pmatrix} 3\;2\;0 \\ 2\;1\;0 \\ 2\;1 \\ 2\;0 \\ 1 \\ 0 \end{pmatrix} \quad \begin{pmatrix} 3\;2\;0 \\ 2\;1\;0 \\ 2\;1 \\ 1\;1 \\ 1 \\ 0 \end{pmatrix} \quad \begin{pmatrix} 3\;2\;0 \\ 2\;1\;0 \\ 2\;1 \\ 1\;0 \\ 0 \\ 0 \end{pmatrix}$$

Hence $m^{(102)}_{\langle 32\rangle} = 5$.

The symmetry group of the weight diagrams of $Sp(2k)$ is the hyperoctohedral group generated by the permutations of the components of the weight vectors and changes of the signs of these components. For example

$$m^{(102)}_{<32>} = m^{(210)}_{<32>} = m^{(\bar{1}02)}_{<32>} = m^{(\bar{2}\bar{1}0)}_{<32>} = \ldots = 5.$$

Once more the fact that $w_j = p_j - q_j$ for $j=1,2,\ldots,k$ leads to the weights being k-dependent. Extending the example to the case of the group $Sp(2k)$ yields for the weight $\underset{\sim}{w} = (210\ldots0)$ the tableaux

$$\begin{array}{|l|}\hline 112 \\ \bar{2}2 \\ \hline\end{array} \quad \begin{array}{|l|}\hline 11\bar{2} \\ 22 \\ \hline\end{array}$$

$$\begin{array}{|l|}\hline 112 \\ \bar{3}3 \\ \hline\end{array} \quad \begin{array}{|l|}\hline 112 \\ \bar{4}4 \\ \hline\end{array} \quad \ldots \quad \begin{array}{|l|}\hline 112 \\ \bar{k}k \\ \hline\end{array}$$

$$\begin{array}{|l|}\hline 11\bar{3} \\ 23 \\ \hline\end{array} \quad \begin{array}{|l|}\hline 11\bar{4} \\ 24 \\ \hline\end{array} \quad \ldots \quad \begin{array}{|l|}\hline 11\bar{k} \\ 2k \\ \hline\end{array}$$

$$\begin{array}{|l|}\hline 113 \\ 2\bar{3} \\ \hline\end{array} \quad \begin{array}{|l|}\hline 114 \\ 2\bar{4} \\ \hline\end{array} \quad \ldots \quad \begin{array}{|l|}\hline 11k \\ 2\bar{k} \\ \hline\end{array}$$

so that: $\quad m^{(21)}_{<32>} = 3k-4.$

5. Tensor and Spinor Representations of O(2k) and O(2k+1).

The irreducible tensor and spinor representations of $O(2k)$ and $O(2k+1)$ are specified by $[\lambda]$ and $[\Delta;\lambda]$ respectively, where (λ) is a partition of ℓ into a non-vanishing parts with $a \leq k$. The branching rules appropriate to these representations may be derived from those given elsewhere[3] and take the form:

$$O(2k)\downarrow O(2k-2)\times U(1) \qquad [\lambda] \downarrow \sum_{p_k q_k} [\lambda/q_k p_k] \times \{p_k - q_k\}\ , \tag{5.1}$$

$$[\Delta;\lambda] \downarrow \sum_{p_k q_k} [\Delta;\lambda/q_k p_k] \times (\{p_k - q_k + \tfrac{1}{2}\} + \{p_k - q_k - \tfrac{1}{2}\})\ . \tag{5.2}$$

The application of these results to the chain

$$O(2k)\downarrow O(2k-2)\times U(1)\downarrow O(2k-4)\times U(1)\times U(1)\downarrow \quad \ldots\ldots\downarrow T_k \tag{5.3}$$

leads to the formulae

$$O(2k) \qquad m^{\underset{\sim}{w}}_{[\lambda]} = \sum_{\underset{\sim}{p},\underset{\sim}{q}} [\lambda/q_1 p_1 q_2 p_2 \ldots q_k p_k] \prod_{i=1}^{k} \delta^{w_i}_{p_i - q_i}\ . \tag{5.4}$$

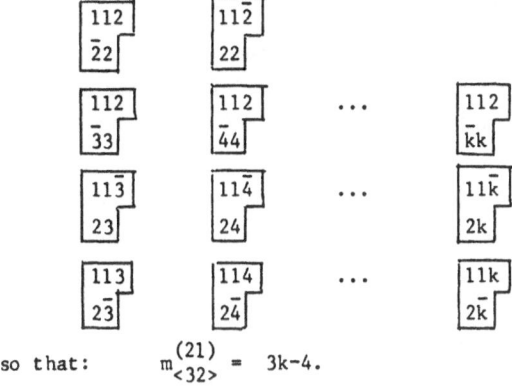

$$O(2k) \qquad m^{w}_{[\Delta;\lambda]} = \sum_{p,q,r} [\Delta; \lambda/q_1p_1q_2p_2 \cdots q_kp_k] \prod_{i=1}^{k} \delta^{w_i}_{p_i - q_i + r_i} \delta^{\frac{1}{2}}_{|r_i|} . \tag{5.5}$$

Corresponding formulae for $O(2k+1)$ may then be written down by noting the rules[3]

$$O(2k+1) \downarrow O(2k) \qquad [\lambda] \downarrow \sum_m [\lambda/m] , \tag{5.6}$$

$$[\Delta;\lambda] \downarrow \sum_m [\Delta;\lambda/m] . \tag{5.7}$$

It is then a straightforward task to draw up the appropriate generalisations of Gelfand patterns and Young tableaux. The former have of course been given Gelfand and Tseitlin[9]. In defining the latter care has to be taken in interpreting the S-function quotients. It is necessary to use the modification rules[3,10,11] for both tensor and spinor representations of the group $O(n)$. If this is done the resulting tableaux provide an easy way of deriving results such as:

$$O(2k) \qquad m^{(21)}_{[32]} = 3k-5 \qquad m^{(\Delta;11)}_{[\Delta;31]} = (3k^2 - 7k + 4)/2$$

$$O(2k+1) \qquad m^{(21)}_{[32]} = 3k-3 \qquad m^{(\Delta;11)}_{[\Delta;31]} = (3k^2 - k + 2)/2 .$$

6. Conclusions.

The branching rule associated with the subgroup chain leading from $U(k)$ to T_k was first used to calculate weight multiplicities by Delaney and Gruber[4], who exploited the connection with patterns and tableaux described here. The generalisation to the other classical groups was suggested by Gilmore[12,13] who did not however arrive at the generalised Young tableaux which are seen here to provide the most efficient means of calculating weight multiplicities. The great merit of using these tableaux is that results are readily obtained for groups of arbitrary rank. Considerable tables of results for the tensor and spinor representations of the groups $O(n)$ have been compiled by Plunkett[14]. It is hoped to publish these, and similar results for $U(k)$ and $Sp(2k)$, elsewhere.

References.

1. H. Weyl, "The Theory of Groups and Quantum Mechanics", Methuen, London, (1931).
2. D.E. Littlewood, "The Theory of Group Characters" Oxford University Press, Oxford, (1940).
3. R.C. King, J. Phys. A: Math. Gen. 8 429 (1975).
4. R.M. Delaney and B. Gruber, J. Math. Phys. 10 252 (1969).
5. I.M. Gelfand and M.L. Tseitlin, Dokl. Akad. Nauk. SSSR 71 825 (1950).
6. M. Hamermesh, "Group Theory", Addison Wesley, Reading, Mass. (1962).
7. Y.J. Abramsky and R.C.King, Nuovo Cimento A 67 153 (1970).
8. D.P. Zhelobenko , Russ. Math. Surveys 17 1 (1962).
9. I.M. Gelfand and M.L. Tseitlin, Dokl. Akad. Nauk. SSSR 71 1017 (1950).
10. R.C. King, J. Math. Phys. 12 1588 (1971).
11. M.J. Newell, Proc. Roy. Irish Acad. 54 153 (1951).
12. R. Gilmore, J. Math. Phys. 11 513 (1970).
13. R. Gilmore, J. Math. Phys. 11 1853 (1970).
14. S.P.O. Plunkett, M. Phil. Thesis, Southampton (1971).

CASIMIR OPERATORS OF SUBALGEBRAS OF THE POINCARE LIE ALGEBRA AND OF REAL LIE ALGEBRAS OF LOW DIMENSION†

J. PATERA, R.T. SHARP*, P. WINTERNITZ

Centre de Recherches Mathématiques
Université de Montréal
Montréal, Canada

and

H. ZASSENHAUS**

Department of Mathematics
Caltech, Pasadena
California

(Talk presented by J. Patera at the 4th International Colloquium on Group Theoretical Methods in Physics, Nijmegen, Netherlands, June 1975.)

1. Introduction

The Poincaré group generated by three space rotations L_1, L_2, L_3, three Lorentz boosts K_1, K_2, K_3, and four translations P_0, P_1, P_2, P_3 is the fundamental transformation group of special relativity. Many of its properties have been studied during the last 70 years by both mathematicians and physicists. Only recently, however, a list of its continuous subgroups has been obtained[1], thus providing an exhaustive description of all possible "subsymmetries" of continuous type.

For many applications in mathematical physics it is essential to know the (Casimir) operators which commute with all generators of each subalgebra S and are themselves elements of the enveloping algebra of S. Thus, for instance, these operators provide a convenient way of labelling representations of S and defining bases for the representations of the entire group, their eigenvalues are measurable physical quantities (quantum numbers), etc.[1].

The purpose of this contribution is to present the list of subalgebras of the Poincaré Lie algebra P (nonconjugate under the connected part of the Poincaré group) which have Casimir operators or other invariant operators, as explained below. A physical interpretation and further use of these operators is postponed to subsequent publications.

---

† Work supported in part by NATO.

* Permanent address: Department of Physics, McGill University, Montreal, Canada.

** Fairchild Distinguished Scholar; permanent address: Department of Mathematics, Ohio State University, Columbus, Ohio.

Since there are many subalgebras of dimensions 3 and 4 quite a few among them being isomorphic to each other, we found advantageous first to classify the subalgebras of dimension ≤ 4 into isomorphism classes, then to find the invariant operators for each isomorphic class, and then only to "interpret" the operators using generators of each nonconjugate subalgebra of P. For that we use the list of nonisomorphic real Lie algebras of low dimension as given by Mubarakzyanov[2]. We do not follow, however, the notations of Ref. 2 where algebras of different physical significance are often denoted by the same symbol.

The method for finding the invariant operators is briefly described in Sec. II together with the types of operators obtained. In Sec. III the Tab. I is described. It containes the real nonisomorphic Lie algebras of dimension ≤ 4, their Casimir and other invariant operators. In Sec. IV the isomorphisms between subalgebras of P are presented. Invariant operators of subalgebras of dimension 3 and 4 are given in Tab. II, and for subalgebras of dimension exceeding 4 they are in Tab. III.

II. A Method for Finding Invariants of Lie Algebras

The number of Casimir operators of a semisimple Lie algebra is equal to its rank. Therefore our problem is in the nonsemisimple Lie algebras for which no easy criterion is known to decide readily whether or not they possess a Casimir operator. (For other invariant operators see Ref. 3.)

Consider a Lie algebra $\mathcal{L}$ generated by $A_1,\ldots,A_n$ satisfying

$$[A_i,A_k] = f^{\ell}_{ik}A_{\ell}. \tag{1}$$

We represent the generators A_i as differential operators acting on a space of functions $F(a_1,\ldots,a_n)$ by putting

$$A_i = \sum_{k,\ell=1}^{n} f^{\ell}_{ik}a_{\ell}\frac{\partial}{\partial a_k} \qquad (i = 1,2,\ldots,n). \tag{2}$$

In order to find an operator valued function $P(A_1,\ldots,A_n)$ such that equation

$$[A_i,P(A_1,\ldots,A_n)] = 0 \quad \text{for} \quad i = 1,2,\ldots,n, \tag{3}$$

holds, we find first a numerical function $P(a_1,\ldots,a_n)$ annihilated by all operators A_i in (2). Thus we get a system of n linear homogeneous differential equations

$$\sum_{k,\ell=1}^{n} f^{\ell}_{ik}a_{\ell}\frac{\partial}{\partial a_k}P(a_1,\ldots,a_n) = 0, \qquad (i = 1,2,\ldots,n) \tag{4}$$

where some of the equations may be trivial, depending on actual values of the structure constants f^{ℓ}_{ik} of each algebra. Having found solutions $P(a_1,\ldots,a_n)$ of the system (4), we replace the variables a_i by A_i, symmetrize the solutions with respect to A_i's, and arrive thus at the invariant operators we were searching for.

Generally, three situations may arise:

(i) The system (4) does not have a solution different from zero. Then the algebra $\mathcal{L}$ does not have an invariant operator.

(ii) One or several solutions of (4) exist and are of polynomial form. Then we choose a convenient integrity basis for them (i.e. a minimal set of invariant operators such that any other invariant operator can be expressed through them as a polynomial) and call it the Casimir operators of the algebra (1).

(iii) Solutions of (4) are of nonpolynomial form. Then we have "generalized Casimir operators" which still commute with the algebra in the sense (4) and thus have fixed numerical values within each irreducible representation of $\mathcal{L}$. These invariants do not belong to the enveloping algebra of $\mathcal{L}$. A case of special interest is when these invariant operators are rational functions of the generators.

III. Isomorphic Classes of Real Lie Algebras of Dimension ≤ 4

All one-dimensional Lie algebras are isomorphic to each other. We denote them by A_1. The only generator is obviously also a Casimir operator of A_1.

An algebra of dimension 2 is isomorphic either to a direct sum A_1+A_1 which we denote by $2A_1$ (its two generators are again Casimir operators), or to a non-Abelian algebra generated by e_1, e_2:

$$[e_1 e_2] = e_1. \tag{5}$$

This algebra, denoted by A_2, does not have any invariant operator.

An algebra of dimension 3 which is a direct sum is isomorphic to one of the two:

$$3A_1 \equiv A_1+A_1+A_1 \quad \text{or } A_1+A_2. \tag{6}$$

The algebra $3A_1$ has all its generators as Casimir operators, for A_1+A_2 only the generator of A_1 is a Casimir operator. The remaining 3-dimensional algebras do not decompose into a direct sum of lower algebras. They are described by the entities $A_{3,1}, A_{3,2}, \ldots, A_{3,9}$ in Tab. I, which contains also their invariant operators.

Let us point out the $A^a_{3,5}$ and $A^a_{3,7}$ each represent a continuum of non-isomorphic algebras corresponding to different values of the parameter a; this fact is underlined by the appearance of the upper index a.

A 4-dimensional real Lie algebra is isomorphic either to one of the following direct sums

$$4A_1,\ 2A_1+A_2,\ 2A_2,\ A_1+A_{3,i} \qquad (i = 1,2,\dots,9) \qquad (7)$$

or to one of the algebras $A_{4,1}, A^a_{4,2}, \dots, A_{4,12}$ of Tab. I. The invariants of algebras (7) are obviously given, by the Casimir operator of each A_1 and the corresponding operators of $A_{3,i}$. For $A_{4,i}$ $(i = 1,\dots,12)$ the operators are listed in the Tab. I, whenever they exist.

The entities $A^a_{4,2}$, $A^{a,b}_{4,5}$, $A^{a,b}_{4,6}$, $A^b_{4,9}$, $A^a_{4,11}$ are again infinite families of non-isomorphic algebras for each value of the parameters whithin the specified range.

IV. Invariants of the Subalgebras of the Poincaré Lie Algebra

Representatives of conjugacy classes of subalgebras of the Poincaré Lie algebra P are denoted as in Tab. 3 and 4 of Ref. 1, i.e. either by $P_{i,k}$ or by $\tilde{P}_{m,n}$. Their generators are also found in those tables.

All 1-dimensional subalgebras of P are isomorphic to A_1 and their generators are also their Casimir operators. These algebras are:
$P_{11,6}$, $P_{12,10}$, $P_{13,9}$, $P_{14,9}$, $P_{15,8}$, $P_{15,9}$, $P_{15,10}$, $\tilde{P}_{12,23} \div \tilde{P}_{12,26}$, $\tilde{P}_{13,15}$, $\tilde{P}_{14,24}$, $\tilde{P}_{14,25}$, and $\tilde{P}_{14,26}$.

The subalgebras of dimension 2 which are not Abelian are all isomorphic to A_2 and they do not have an invariant operator. These algebras are: $P_{8,9}$, $P_{11,5}$, $P_{13,7}$, $\tilde{P}_{8,17}$, $\tilde{P}_{13,13}$.

The Abelian 2-dimensional subalgebras are all isomorphic to $2A_1$ and all their generators are also their Casimir operators. These algebras are:
$P_{9,6}$, $P_{10,5}$, $P_{12,7} \div P_{12,9}$, $P_{13,8}$, $P_{14,7}$, $P_{14,8}$, $P_{15,5} \div P_{15,7}$, $\tilde{P}_{10,6}$, $\tilde{P}_{12,19} \div \tilde{P}_{12,22}$, $P_{13,14}$, $\tilde{P}_{14,20} \div P_{14,23}$.

The subalgebras of P of dimension 3 and 4 are given in Tab. II together with their generators and invariants whenever these exist. Isomorphic classes are indicated.

The subalgebras of dimension ≥ 5 are given in Tab. III with their generators and invariants. None of these subalgebras are isomorphic to each other.

References

1. J. Patera, P. Winternitz, and H. Zassenhaus, J. Math. Phys. 16, 1597 (1975) and references therein.
2. G.M. Mubarakzyanov, Izv. Vyssh. Uch. Zav. 32, 114 (1963) (in Russian).
3. E.G. Beltrametti and A. Blasi, Phys. Lett. 20, 62 (1966).

TABLE I. Real Lie algebras of dimensions 3 and 4 which are not direct summs of Lie algebras of lower dimensions.

Name	Non-zero commutation relations	Invariants	Comments
$A_{3,1}$	$[e_2e_3] = e_1$	e_1	nilpotent (Algebra of Weyl group)
$A_{3,2}$	$[e_1e_3] = e_1$, $[e_2e_3] = e_1+e_2$	$e_1 \exp\left(-\frac{e_2}{e_1}\right)$	solvable
$A_{3,3}$	$[e_1e_3] = e_1$, $[e_2e_3] = e_2$	$\frac{e_2}{e_1}$	solvable, $D \cong T_2$
$A_{3,4}$	$[e_1e_3] = e_1$, $[e_2e_3] = -e_2$	e_1e_2	solvable, E(1,1)
$A^a_{3,5}$	$[e_1e_3] = e_1$, $[e_2e_3] = ae_2$ $(0<\|a\|<1)$	$e_2e_1^{-a}$	solvable
$A_{3,6}$	$[e_1e_3] = -e_2$, $[e_2e_3] = e_1$	$e_1^2+e_2^2$	solvable, E(2)
$A^a_{3,7}$	$[e_1e_3] = ae_1-e_2$, $[e_2e_3] = e_1+ae_2$ $(a>0)$	$(e_1^2+e_2^2)\left(\frac{e_1+ie_2}{e_1-ie_2}\right)^{ia}$	solvable,
$A_{3,8}$	$[e_1e_3] = -2e_2$, $[e_1e_2] = e_1$, $[e_2e_3] = e_3$	$2e_2^2+e_1e_3+e_3e_1$	semisimple, SU(1,1
$A_{3,9}$	$[e_1e_2] = e_3$, $[e_2e_3] = e_1$, $[e_3e_1] = e_2$	$e_1^2+e_2^2+e_3^2$	simisimple, SU(2)
$A_{4,1}$	$[e_2e_4] = e_1$, $[e_3e_4] = e_2$	e_1, $e_2^2-2e_1e_3$	nilpotent
$A^a_{4,2}$	$[e_1e_4] = ae_1$, $[e_2e_4] = e_2$, $[e_3e_4] = e_2+e_3$ $(a\neq 0)$	$\frac{e_2^a}{e_1}$, $e_2 \exp\left(-\frac{e_3}{e_2}\right)$	solvable, derived algebra $\sim 3A_1$
$A_{4,3}$	$[e_1e_4] = e_1$, $[e_3e_4] = e_2$	e_2, $e_1 \exp\left(-\frac{e_3}{e_2}\right)$	solvable, derived algebra $\sim 3A_1$

Name	Non-zero commutation relations	Invariants	Comments
$A_{4,4}$	$[e_1e_4] = e_1$, $[e_2e_4] = e_1+e_2$, $[e_3e_4] = e_2+e_3$	$e_1 \exp\left(-\frac{e_2}{e_1}\right)$, $\frac{2e_1e_3-e_2^2}{e_1^2}$	solvable, derived algebra ~ $3A_1$
$A^{ab}_{4,5}$	$[e_1e_4] = e_1$, $[e_2e_4] = ae_2$, $[e_3e_4] = be_3$ ($ab\neq 0$, $-1\leq b\leq a\leq 1$)	$\frac{e_1^a}{e_2}$, $\frac{e_1^b}{e_3}$	solvable, derived algebra ~ $3A_1$
$A^{ab}_{4,6}$	$[e_1e_4] = ae_1$, $[e_2e_4] = be_2-e_3$, $[e_3e_4] = e_2+be_3$ ($a\neq 0, b\geq 0$)	$\frac{e_1^{\frac{2b}{a}}}{e_2^2+e_3^2}$, $e_1^{\frac{i-b}{a}}(e_2-ie_3)+e_1^{\frac{-b-i}{a}}(e_2+ie_3)$	solvable, derived algebra ~ $3A_1$
$A_{4,7}$	$[e_2e_3] = e_1$, $[e_1e_4] = 2e_1$, $[e_2e_4] = e_2$, $[e_3e_4] = e_2+e_3$	none	solvable, derived algebra ~ $A_{3,1}$
$A_{4,8}$	$[e_2e_3] = e_1$, $[e_2e_4] = e_2$, $[e_3e_4] = -e_3$	e_1, $e_2e_3+e_3e_2-2e_1e_4$	solvable, derived algebra ~ $A_{3,1}$
$A^{b}_{4,9}$	$[e_2e_3] = e_1$, $[e_1e_4] = (1+b)e_1$, $[e_2e_4] = e_2$, $[e_3e_4] = be_3$ ($-1<b\leq 1$)	none	solvable, derived algebra ~ $A_{3,1}$
$A_{4,10}$	$[e_2e_3] = e_1$, $[e_2e_4] = -e_3$, $[e_3e_4] = e_2$	e_1, $2e_1e_4+e_2^2+e_3^2$	solvable, derived algebra ~ $A_{3,1}$
$A^{a}_{4,11}$	$[e_2e_3] = e_1$, $[e_1e_4] = 2ae_1$, $[e_2e_4] = ae_2-e_3$, $[e_3e_4] = e_2+ae_3$ ($a>0$)	none	solvable, derived algebra ~ $A_{3,1}$
$A_{4.12}$	$[e_1e_3] = e_1$, $[e_2e_3] = e_2$, $[e_1e_4] = -e_2$, $[e_2e_4] = e_1$	none	solvable, derived algebra ~ $2A_1$

TABLE II. Invariant operators of 3- and 4-dimensional subalgebras of the Poincaré Lie algebra

Class of isomorphism	Notation Ref. 1	Generators	Invariants	Comments
$3A_1$	$P_{10,4}$	$L_2+K_1,\ L_1-K_2,\ P_0-P_3$	all generators	Abelian
	$P_{12,5}$	$L_3,\ P_0,\ P_3$		
	$P_{13,6}$	$K_3,\ P_1,\ P_2$		
	$P_{14,4}$	$L_2+K_1,\ P_0-P_3,\ P_2$		
	$P_{15,2}$	$P_0-P_3,\ P_1,\ P_2$		
	$P_{15,3}$	$P_1,\ P_2,\ P_3$		
	$P_{15,4}$	$P_0,\ P_1,\ P_2$		
	$\tilde{P}_{10,11}$	$L_2+K_1,\ L_1-K_2+P_2,\ P_0-P_3$		
	$\tilde{P}_{14,13}$	$L_2+K_1-\frac{1}{2}(P_0+P_3),\ P_0-P_3,\ P_2$		
A_2+A_1	$P_{8,8}$	$L_2+K_1,\ K_3,\ P_2$	P_2	solvable
	$P_{9,5}$	$K_3,\ P_0-P_3,\ L_3$	L_3	
	$P_{13,5}$	$K_3,\ P_0-P_3,\ P_2$	P_2	
	$\tilde{P}_{13,5}$	$K_3+aP_2,\ P_0-P_3,\ P_1$ $(a>0)$	P_1	
$A_{3,1}$	$P_{14,5}$	$L_2+K_1,\ P_1,\ P_0-P_3$	P_0-P_3	nilpotent
	$P_{14,6}$	$L_2+K_1,\ P_2+bX_3,\ P_0-P_3$ $(b\neq0)$		
	$\tilde{P}_{10,12}$	$L_2+K_1-P_2,\ L_1-K_2-P_1+aP_2,\ P_0-P_3$ $(a>0)$		

TABLE II. (cont.)

Class of isomorphism	Notation Ref. 1	Generators	Invariants	Comments
$A_{3,1}$	$\tilde{P}_{10,13}$	$L_2+K_1+P_2,\ L_1-K_2+P_1+aP_2,\ P_0-P_3$ (a>0)	P_0-P_3	nilpotent
	$\tilde{P}_{10,14}$	$L_2+K_1-P_2,\ L_1-K_2-P_1,\ P_0-P_3$		
	$\tilde{P}_{10,15}$	$L_2+K_1+P_2,\ L_1-K_2+P_1,\ P_0-P_3$		
	$\tilde{P}_{14,14}$	$L_2+K_1-\frac{1}{2}(P_0+P_3),\ P_1,\ P_0-P_3$		
	$\tilde{P}_{14,15}$	$L_2+K_1-P_2,\ P_1,\ P_0-P_3$		
	$\tilde{P}_{14,16}$	$L_2+K_1+P_2,\ P_1,\ P_0-P_3$		
	$\tilde{P}_{14,17}$	$L_2+K_1-\frac{1}{2}(P_0+P_3),\ P_2+bP_1,\ P_0-P_3$ (b≠0)		
	$\tilde{P}_{14,18}$	$L_2+K_1-P_2,\ P_2+bP_1,\ P_0-P_3$ (b≠0)		
	$\tilde{P}_{14,19}$	$L_2+K_1+P_2,\ P_2+bP_1,\ P_0-P_3$ (b≠0)		
$A_{3,2}$	$\tilde{P}_{8,14}$	$L_2+K_1,\ K_3+aP_1,\ P_0-P_3$ (a>0)	$(P_0-P_3)\exp\left(-\frac{1}{a}\frac{L_2+K_1}{P_0-P_3}\right)$	solvable
	$\tilde{P}_{8,16}$	$L_2+K_1,\ K_3+bP_1+aP_2,\ P_0-P_3$ (a>0,b≠0)	$(P_0-P_3)\exp\left(-\frac{1}{b}\frac{L_2+K_1}{P_0-P_3}\right)$	
$A_{3,3}$	$P_{7,5}$	$K_3,\ L_2+K_1,\ L_1-K_2$	$\frac{L_1-K_2}{L_2+K_1}$	solvable, $D \cong T_2$
	$P_{8,7}$	$K_3,\ L_2+K_1,\ P_0-P_3$	$\frac{P_0-P_3}{L_2+K_1}$	

TABLE II. (cont.)

Class of isomorphism	Notation Ref. 1	Generators	Invariants	Comments
$A_{3,3}$	$\tilde{P}_{8,15}$	$K_3+aP_2,\ L_2+K_1,\ P_0-P_3\quad (a>0)$	$\frac{P_0-P_3}{L_2+K_1}$	solvable, $D \cong T_2$
$A_{3,4}$	$P_{11,4}$	$\cos c L_3+\sin c K_3,\ P_0,\ P_3\quad (0<c<\pi, c\neq\pi/2)$	$P_0^2-P_3^2$	solvable, E(1,1)
	$P_{13,4}$	$K_3,\ P_0,\ P_3$		
	$\tilde{P}_{13,12}$	$K_3+aP_2,\ P_0,\ P_3\quad (a>0)$		
$A_{3,6}$	$P_{6,4}$	$L_3,\ L_2+K_1,\ L_1-K_2$	$(L_2+K_1)^2+(L_1-K_2)^2$	solvable, E(2)
	$\tilde{P}_{6,9}$	$L_3+\frac{1}{4}(P_0-P_3),\ L_2+K_1,\ L_1-K_1$		
	$\tilde{P}_{6,10}$	$L_3-\frac{1}{4}(P_0-P_3),\ L_2+K_1,\ L_1-K_1$		
	$P_{11,3}$	$\cos c L_3+\sin c K_3,\ P_1,\ P_2\quad (0<c<\pi, c\neq\pi/2)$	$P_1^2+P_2^2$	
	$\tilde{P}_{12,15}$	$L_3+\frac{1}{4}(P_0+P_3),\ P_1,\ P_2$		
	$\tilde{P}_{12,16}$	$L_3-\frac{1}{4}(P_0+P_3),\ P_1,\ P_2$		
	$\tilde{P}_{12,17}$	$L_3+aP_0,\ P_1,\ P_2\quad (a>0)$		
	$\tilde{P}_{12,18}$	$L_3+bP_3,\ P_1,\ P_2\quad (b\neq 0)$		
$A_{3,7}$	$P_{5,4}$	$\cos c L_3+\sin c K_3,\ L_2+K_1,\ L_1-K_2\quad (0<c<\pi, c\neq\pi/2)$	$[(L_2+K_1)^2+(L_1-K_2)^2]\left\{\frac{L_2+K_1+i(L_1-K_2)}{L_2+K_1-i(L_1-K_2)}\right\}^{i\tan c}$	solvable, S(3)

TABLE II. (cont.)

Class of isomorphism	Notation Ref. 1	Generators	Invariants	Comments
$A_{3,8}$	$P_{4,4}$	L_3, K_1, K_2	$L_3^2-K_1^2-K_2^2$	semisimple, SU(1,1)
$A_{3,9}$	$P_{3,4}$	L_1, L_2, L_3	$L_1^2+L_2^2+L_3^2$	semisimple, SU(2)
$4A_1$	$P_{15,1}$	P_0, P_1, P_2, P_3	all generators	Abelian, T_4
A_2+2A_1	$P_{13,2}$	K_3, P_0-P_3, P_1, P_2	P_1, P_2	
$A_{3,1}+A_1$	$P_{10,3}$	$L_2+K_1, L_1-K_2, P_0-P_3, P_2$	L_2+K_1, P_0-P_3	
	$P_{14,2}$	$L_2+K_1, P_0-P_3, P_1, P_2$	P_0-P_3, P_2	
	$\tilde{P}_{10,9}$	$L_2+K_1, L_1-K_2-P_1, P_0-P_3, P_2$	L_2+K_1, P_0-P_3	
	$\tilde{P}_{10,10}$	$L_2+K_1, L_1-K_2+P_1, P_0-P_3, P_2$	L_2+K_1, P_0-P_3	
	$\tilde{P}_{14,10}$	$L_2+K_1-\frac{1}{2}(P_0+P_3), P_0-P_3, P_1, P_2$	P_0-P_3, P_2	
$A_{3,2}+A_1$	$\tilde{P}_{8,11}$	$K_3+aP_2, L_2+K_1, P_0-P_3, P_2$ (a>0)	$P_2, (P_0-P_3)\exp\left(\frac{1}{a}\frac{L_2+K_1}{P_0-P_3}\right)$	
$A_{3,3}+A_1$	$P_{8,4}$	$K_3, L_2+K_1, P_0-P_3, P_2$	$P_2, \frac{L_2+K_1}{P_0-P_3}$	
$A_{3,4}+A_1$	$P_{9,4}$	L_3, K_3, P_0, P_3	$L_3, P_0^2-P_3^2$	
	$P_{13,3}$	K_3, P_0, P_1, P_3	$P_1, P_0^2-P_3^2$	
	$\tilde{P}_{13,10}$	K_3+aP_2, P_0, P_1, P_3 (a>0)	$P_1, P_0^2-P_3^2$	

TABLE II. (cont.)

Class of isomorphism	Notation Ref. 1	Generators	Invariants	Comments
$A_{3,6}+A_1$	$P_{6,3}$	$L_3,\ L_2+K_1,\ L_1-K_2,\ P_0-P_3$	$P_0-P_3,\ (L_2+K_1)^2+(L_1-K_2)^2$	
	$P_{9,3}$	$L_3,\ K_3,\ P_1,\ P_2$	$K_3,\ P_1^2+P_2^2$	
	$P_{12,2}$	$L_3,\ P_0-P_3,\ P_1,\ P_2$	$P_0-P_3,\ P_1^2+P_2^2$	
	$P_{12,3}$	$L_3,\ P_1,\ P_2,\ P_3$	$P_3,\ P_1^2+P_2^2$	
	$P_{12,4}$	$L_3,\ P_0,\ P_1,\ P_2$	$P_0,\ P_1^2+P_2^2$	
	$\tilde{P}_{12,11}$	$L_3+\frac{1}{4}(P_0+P_3),\ P_0-P_3,\ P_1,\ P_2$	$P_0-P_3,\ P_1^2+P_2^2$	
	$\tilde{P}_{12,12}$	$L_3-\frac{1}{4}(P_0+P_3),\ P_0-P_3,\ P_1,\ P_2$	$P_0-P_3,\ P_1^2+P_2^2$	
	$\tilde{P}_{12,13}$	$L_3+aP_0,\ P_1,\ P_2,\ P_3$ $(a>0)$	$P_3,\ P_1^2+P_2^2$	
	$\tilde{P}_{12,14}$	$L_3+bP_3,\ P_0,\ P_1,\ P_2$ $(b\neq 0)$	$P_0,\ P_1^2+P_2^2$	
$A_{3,8}+A_1$	$P_{4,3}$	$L_3,\ K_1,\ K_2,\ P_3$	$P_3,\ L_3^2-K_1^2-K_2^2$	
$A_{3,9}+A_1$	$P_{3,3}$	$L_1,\ L_2,\ L_3,\ P_0$	$P_0,\ L_1^2+L_2^2+L_3^2$	
$A_{4,1}$	$P_{14,3}$	$L_2+K_1,\ P_0,\ P_1,\ P_3$	$P_0-P_3,\ P_0^2-P_1^2-P_3^2$	
	$\tilde{P}_{10,7}$	$L_2+K_1-\frac{1}{2}(P_0+P_3),\ L_1-K_2+\frac{b}{2}P_1,$ $P_0-P_3,\ P_2$ $(b\neq 0)$	$P_0-P_3,\ P_2^2+P_3^2-P_0^2+2(P_0-P_3)(L_2+K_1+bP_2)$	

TABLE II. (cont.)

Class of isomorphism	Notation Ref. 1	Generators	Invariants	Comments
$A_{4,1}$	$\tilde{P}_{10,8}$	$L_2+K_1-\frac{1}{2}(P_0+P_3)$, L_1-K_2, P_0-P_3, P_2	P_0-P_3, $P_2^2+P_3^2-P_0^2+2(P_0-P_3)(L_2+K_1)$	
	$\tilde{P}_{14,11}$	$L_2+K_1-\frac{1}{2}P_2$, P_0, P_1, P_3	P_0-P_3, $P_1^2+P_3^2-P_0^2$	
	$\tilde{P}_{14,12}$	$L_2+K_1+\frac{1}{2}P_2$, P_0, P_1, P_3	P_0-P_3, $P_1^2+P_3^2-P_0^2$	
$A_{4,2}^{1}$	$\tilde{P}_{7,7}$	K_3+aP_1, L_2+K_1, L_1-K_2, P_0-P_3 (a>0)	$\frac{L_1-K_2}{P_0-P_3}$, $(L_1-K_2)\exp\frac{B_3}{a(P_0-P_3)}$	
$A_{4,5}^{1,1}$	$P_{7,4}$	K_3, L_2+K_1, L_1-K_2, P_0-P_3	$\frac{L_2+K_1}{P_0-P_3}$, $\frac{L_1-K_2}{P_0-P_3}$	
$A_{4,6}^{\tan c,0}$	$P_{11,2}$	$\cos c L_3+\sin K_3$, P_0-P_3, P_1, P_2 $(0<c<\pi, c\neq\pi/2)$	$P_1^2+P_2^2$, $(P_0-P_3)^{i\cot c}(P_1-iP_2)+(P_0-P_3)^{-i\cot c}(P_1+iP_2)$	
$A_{4,6}^{a,a}$ (a ≡ tanc)	$P_{5,3}$	$\cos c L_3+\sin c K_3$, L_2+K_1, L_1-K_2, P_0-P_3 $(0<c<\pi, c\neq\pi/2)$	$[(L_2+K_1)^2+(L_1-K_2)^2](P_0-P_3)^{-2}$, $[L_1-K_2-i(L_2+K_1)](P_0-P_3)^{i\cot c-1}$ + $+[L_1-K_2+i(L_2+K_1)](P_0-P_3)^{-1-i\cot c}$	
$A_{4,10}$	$\tilde{P}_{6,7}$	L_3, $L_2+K_1-P_2$, $L_1-K_2-P_1$, P_0-P_3	P_0-P_3, $L_3(P_0-P_3)+\frac{1}{4}(L_2+K_1-P_2)^2+\frac{1}{4}(L_1-K_2-P_1)^2$	
	$\tilde{P}_{6,8}$	L_3, $L_2+K_1+P_2$, $L_1-K_2+P_1$, P_0-P_3	P_0-P_3, $L_3(P_0-P_3)+\frac{1}{4}(L_2+K_1+P_2)^2+\frac{1}{4}(L_1-K_2+P_1)^2$	
$A_{4,9}^{0}$	$P_{8,5}$	K_3, L_2+K_1, P_0-P_3, P_1	none	
	$P_{8,6}$	K_3, L_2+K_1, P_0-P_3, P_2+bP_1 (b≠0)	none	

TABLE II. (cont.)

Class of isomorphism	Notation Ref. 1	Generators	Invariants	Comments
$A^0_{4,9}$	$\tilde{P}_{8,12}$	K_3+aP_2, L_2+K_1, P_0-P_3, P_1 $(a>0)$	none	
	$\tilde{P}_{8,13}$	K_3+aP_2, L_2+K_1, P_0-P_3, P_2+bP_1 $(a>0, b\neq 0)$	none	
$A_{4,12}$	$P_{2,4}$	L_3, K_3; L_2+K_1, L_1-K_2	none	

TABLE III. Invariants of subalgebras of dim ≥ 5 of the Poincaré Lie algebra.

Subalgebra	Dim	Generators	Invariants	Comments
P_1	10	$L_1, L_2, L_3, K_k, K_2, K_3, P_0, P_1, P_2, P_3$	$m^2 \equiv P_0^2-\vec{P}^2$, S*)	Poincaré
$P_{2,1}$	8	$L_1-K_2, L_2+K_1, L_3, K_3, P_0, P_1, P_2; P_3$	$m^2, L_3 - \frac{P_2}{P_0-P_3}(L_2+K_1) - \frac{P_1}{P_0-P_3}(L_1-K_2)$	maximal solvable
$P_{2,2}$	7	$L_1-K_2, L_2+K_1, L_3, K_3, P_0-P_3, P_1, P_2$	$L_3 - \frac{P_2}{P_0-P_3}(L_2+K_1) - \frac{P_1}{P_0-P_3}(L_1-K_2)$	
$P_{3,1}$	7	$L_1, L_2, L_3, P_0, P_1, P_2, P_3$	$m^2, P_0, \vec{L}\cdot\vec{P}$	$E(3) \cong P_0$
$P_{4,1}$	7	$L_3, K_1, K_2, P_0, P_1, P_2, P_3$	$m^2, P_3, L_3P_0-K_1P_2+K_2P_1$	$E(2,1) \cong P_3$
$P_{5,1}$	7	$\cos cL_3+\sin cK_3, L_1-K_2, L_2+K_1, P_0, P_1, P_2, P_3$ $(0<c<\pi, c\neq\pi/2)$	m^2	$S(3) \cong T_4$
$P_{6,1}$	7	$L_3, L_1-K_2, L_2+K_1, P_0, P_1, P_2, P_3$	$m^2, P_0-P_3, \vec{L}\cdot\vec{P}+K_1P_2-K_2P_1-L_3P_0$	$E(2) \cong T_4$
$P_{7,1}$	7	$K_3, L_1-K_2, L_2+K_1, P_0, P_1, P_2, P_3$	m^2	
$P_{1,2}$	6	$L_1, L_2, L_3, K_1, K_2, K_3$	$\vec{L}\cdot\vec{K}, \vec{L}^2-\vec{K}^2$	$SL(2,C) \sim O(3,1)$
$P_{3,2}$	6	$L_1, L_2, L_3, P_1, P_2, P_3$	$\vec{P}^2, \vec{L}\cdot\vec{P}$	$E(3)$
$P_{4,2}$	6	$L_3, K_1, K_2, P_0, P_1, P_2$	$P_0^2-P_1^2-P_2^2, L_3P_0+K_2P_1-K_1P_2$	$E(2,1)$
$P_{5,2}$	6	$\cos cL_3+\sin cK_3, L_1-K_2, L_2+K_1, P_0-P_3, P_1, P_2$	none	
$P_{6,2}$	6	$L_3, L_1-K_2, L_2+K_1, P_0-P_3, P_1, P_2$	$P_0-P_3, (P_0-P_3)L_3-P_1(L_1-K_2)-P_2(L_2+K_1)$	

*) S denotes the square of the Pauli-Lubanski spin operator.

TABLE III. (cont.)

Subalgebra	Dim	Generators	Invariants	Comments
$P_{7,2}$	6	K_3, L_1-K_2, L_2+K_1, P_0-P_3, P_1, P_2	none	
$P_{8,1}$	6	K_3, L_2+K_1, P_0, P_1, P_2, P_3	m^2, P_2	
$P_{9,1}$	6	L_3, K_3, P_0, P_1, P_2, P_3	m^2, $P_0^2-P_3^2$	
$P_{10,1}$	6	L_1-K_2, L_2+K_1, P_0, P_1, P_2, P_3	m^2, P_0-P_3	
$\tilde{P}_{6,5}$ $(\varepsilon=1)$ $\tilde{P}_{6,6}$ $(\varepsilon=-1)$	6	$L_3+\frac{\varepsilon}{4}(P_0+P_3)$, L_1-K_2, L_2+K_1, P_1, P_2, P_0-P_3	$m^2+4\varepsilon(P_0L_3-P_2K_1+P_1K_2-\vec{L}\cdot\vec{P})$, P_0-P_3	
$P_{2,3}$	5	L_1-K_2, L_2+K_1, L_3, K_3, P_0-P_3	$\frac{(L_1-K_2)^2+(L_2+K_1)^2}{(P_0-P_3)^2}$	
$P_{7,3}$	5	K_3, L_1-K_2, L_2+K_1, P_0-P_3, P_2	$\frac{L_2+K_1}{P_0-P_3}$	
$P_{8,2}$	5	K_3, L_2+K_1, P_0-P_3, P_1, P_2	P_2	
$P_{8,3}$	5	K_3, L_2+K_1, P_0, P_1, P_3	$P_0^2-P_1^2-P_3^2$	
$P_{9,2}$	5	L_3, K_3, P_0-P_3, P_1, P_2	$P_1^2+P_2^2$	
$P_{10,2}$	5	L_1-K_2, L_2+K_1, P_0-P_3, P_1, P_2	P_0-P_3	
$P_{11,1}$	5	$\cos c L_3+\sin c K_3$, P_0, P_1, P_2, P_3 $(c\neq\pi/2, 0<c<\pi)$	m^2, $P_0^2-P_3^2$, $(P_0-P_3)^{2\cot c}\left(\frac{P_2+iP_1}{P_2-iP_1}\right)^i$	

TABLE III. (cont.)

Subalgebra	Dim	Generators	Invariants	Comments
$P_{12,1}$	5	L_3, P_0, P_1, P_2, P_3	m^2, P_0, P_3	
$P_{13,1}$	5	K_3, P_0, P_1, P_2, P_3	m^2, P_1, P_2	
$P_{14,1}$	5	$L_2+K_1, P_0, P_1, P_2, P_3$	m^2, P_0-P_3, P_2	
$\tilde{P}_{7,6}$	5	$K_3+aP_1, L_1-K_2, L_2+K_3, P_0-P_3, P_2$ $(a>0)$	$(P_0-P_3)^a \exp\left\{-2\frac{L_2+K_1}{P_0-P_3}\right\}$	
$\tilde{P}_{8,10}$	5	$K_3+aP_1, L_2+K_1, P_0, P_1, P_3$ $(a>0)$	$P_0^2-P_1^2-P_3^2$	
$\tilde{P}_{10,6}$	5	$L_3+K_1+\frac{1}{2}(P_0+P_3), L_1-K_2, P_0-P_3, P_1, P_2$	P_0-P_3	

THE MAXIMAL SOLVABLE SUBALGEBRAS OF THE REAL CLASSICAL LIE ALGEBRAS. II

Marcel PERROUD

Centre de Recherches Mathématiques,
Université de Montréal
Montréal, P.Q.
Canada

ABSTRACT

A list of all conjugacy classes of maximal solvable subalgebras of the Lie algebras $g\ell(n, \mathbb{R})$, $su^*(2r)$, $su(p,q)$, $so^*(2r)$, $sp(2r, \mathbb{R})$ and $usp(2p,2q)$ is proposed and a representative of each class is given in terms of a matrix algebra.

INTRODUCTION

The results presented in this communication have been the subject of 3 publications [1,2,3] and are part of the more general task of determining all the conjugacy classes of the subgroups of some Lie groups of physical interest[1,4,5] undertaken by Patera, Winternitz and Zassenhaus.

Except for the cases of $g\ell(n, \mathbb{R})$, $su^*(2r)$ and $so(p,q)$ where p and g are both odd numbers, there always exists a particular class of maximal solvable subalgebras: the class of the *compact Cartan subalgebras* which are abelian and of dimension equal to the rank of the Lie algebra.

The other conjugacy classes are determined by induction on the dimension of the algebra. This has the advantage of furnishing a explicit procedure for constructing representatives of each class. Three steps are to be considered. We adopt the standard matrix representations for the algebras under consideration. The first step consists of showing that every maximal solvable subalgebra S can be written in one of the two alternative forms

$$S = \begin{pmatrix} A_1 & A_2 & A_3 \\ 0 & A_5 & A_6 \\ 0 & 0 & A_9 \end{pmatrix} \quad \text{or } S = \begin{pmatrix} A_1 & A_2 \\ 0 & A_4 \end{pmatrix}$$

depending on whether the standard representation does or does not admit an invariant bilinear form. One then has to ensure that the blocks A_1 (and A_9) are maximal solvable subalgebras of $g\ell(\alpha, \mathbb{R})$ with $\alpha \leq 2$. In the second step it must be shown that the block A_4 and A_5 is a maximal sovable subalgebra of an algebra belonging to the same type as the starting Lie algebra, but of lower dimension. Finally, the third step consists in verifying that a necessary and sufficient condition for two maximal solvable subalgebras S' and S" to be conjugate (by an inner automorphism of the Lie algebra) is that $A_1' \sim A_1''$ and $A_4' \sim A_4''$ or $A_5' \sim A_5''$.

The number of conjugacy classes is often expressible in terms of the Fibonnaci numbers

$$F_n = \frac{1}{\sqrt{5}}\left[\left(\frac{1+\sqrt{5}}{2}\right)^n - \left(\frac{1-\sqrt{5}}{2}\right)^n\right].$$

(Recall that $F_{n+2} = F_{n+1}+F_n$; $F_0 = 0$, $F_1 = 1$.)

RESULTS

S denotes a representative of a conjugacy class of maximal solvable subalgebras; B stands for the matrix associated with an invariant bilinear form (if such exists) of the fundamental representation of the algebra and N (with some subscripts) denotes the number of conjugacy classes. Also we put

$$J = \begin{pmatrix} 0 & 1 \\ -1 & 0 \end{pmatrix}, \quad K = \begin{pmatrix} 0 & 1 \\ 1 & 0 \end{pmatrix}.$$

1. $g\ell(n, \mathbb{R})$

$$S = \begin{pmatrix} A_1 & & * \\ & \ddots & \\ 0 & & A_k \end{pmatrix} \qquad 1 \leq k \leq n$$

with

$$A_j = \begin{pmatrix} a_j & -b_j \\ b_j & a_j \end{pmatrix} \quad \text{or} \quad A_j = (c_j);\ a_j, b_j, c_j \in \mathbb{R}$$

$$N_n = F_n.$$

Each class is characterized by an ordered set of numbers (deg $A_1,\ldots,$deg A_k), with deg $A_j = 1$ or 2 and

$$\sum_{j=1}^{k} \deg A_j = n.$$

The elements in the upper triangle of S ((∇)) are any real numbers. The $s\ell(n, \mathbb{R})$ case is simply obtained by requiring trS = 0.

2. $su^*(2r)$

$$su^*(2r) = \left\{X \in s\ell(2r,\mathbb{C});\ T\bar{X} = XT\right\}$$

with

$$T = \text{diag}(\underbrace{J,\ldots,J}_{r})$$

$$(su^*(2) \simeq su(1,1))$$

$$S = \begin{pmatrix} A_{11} & \cdots & A_{1r} \\ & \ddots & \vdots \\ 0 & & A_{rr} \end{pmatrix}$$

with

$$A_{jj} = \begin{pmatrix} a_{jj} & 0 \\ 0 & a_{jj} \end{pmatrix} ; \quad A_{j\ell} = \begin{pmatrix} a_{j\ell} & b_{j\ell} \\ -\bar{b}_{j\ell} & \bar{a}_{j\ell} \end{pmatrix}, \qquad j < \ell;$$

$$a_{j\ell}, b_{j\ell} \in \mathbb{C} \quad \text{and} \quad \sum_{j=1}^{r} (a_{jj}+\bar{a}_{jj}) = 0$$

$$N_{2r} = 1$$

3. <u>su(p,q).</u> $(p \geq q)$.

$$su(p,q) = \left\{ X \in s\ell(p+q,\mathbb{C}), X^{\dagger}B+BX = 0 \right\}.$$

a) Compact cartan subalgebra:

$$S = \mathrm{diag}(ia_1,\ldots,ia_{p+q}); \quad B = \mathrm{diag}(\underbrace{1,\ldots,1}_{p},\underbrace{-1,\ldots,-1}_{q})$$

with

$$\sum_{j=1}^{p+q} a_j = 0, \qquad a_j \in \mathbb{R}.$$

b) others (possible if $q > 0$)

$$S = \begin{pmatrix} \lambda_1 & & & & & & * \\ & \ddots & & & & & \\ & & \lambda_s & & & & \\ & & & C & & & \\ & & & & -\bar{\lambda}_s & & \\ & & & & & \ddots & \\ 0 & & & & & & -\bar{\lambda}_1 \end{pmatrix}_{1 \leq s \leq q}, \quad B = \begin{pmatrix} 0 & & & & & & 1 \\ & & & & & \cdot^{\cdot^{\cdot}} & \\ & & & & 1 & & \\ & & & \tilde{B} & & & \\ & & 1 & & & & \\ & \cdot^{\cdot^{\cdot}} & & & & & \\ 1 & & & & & & 0 \end{pmatrix}$$

with $\lambda_j \in \mathbb{C}$ and C is either a compact Cartan subalgebra of $su(p-s,q-s)$ or $C = \emptyset$ (which implies $p=q$).

$$\tilde{B} = \mathrm{diag}(\underbrace{1,\ldots,1}_{p-s},\underbrace{-1,\ldots,-1}_{q-s}) \quad \text{or} \quad \tilde{B} = \emptyset$$

$$N_{p,q} = q+1.$$

4. <u>so(p,q).</u> $(p \geq q)$.

$$so(p,q) = \left\{ X \in s\ell(p+q, \mathbb{R}); X^{T}B+BX = 0 \right\}.$$

a) Compact Cartan subalgebra (if p and q are not both odd numbers).

$$S = \mathrm{diag}(A_1,\dots,A_t); \quad B = \mathrm{diag}(\underbrace{1,\dots,1}_{p},\underbrace{-1,\dots,-1}_{q})$$

with

$$A_k = (0), \quad k = \begin{cases} p & \text{if } p \text{ even} \\ p+1 & \text{if } p \text{ odd} \end{cases}, \quad t = \frac{k+q}{2}$$

$$A_j = \begin{pmatrix} 0 & -a_j \\ a_j & 0 \end{pmatrix}, \qquad j \neq k;\ a_j \in \mathbb{R}.$$

b) others (possible if $q \neq 0$).

$$\begin{pmatrix} A_1 & & & & & & * \\ & \ddots & & & & & \\ & & A_s & & & & \\ & & & C & & & \\ & & & & D_s & & \\ & & & & & \ddots & \\ 0 & & & & & & D_1 \end{pmatrix}_{1\le s\le q} ; \quad B = \begin{pmatrix} 0 & & & & & 1 \\ & & & & ⋰ & \\ & & & \tilde{B} & & \\ & & 1 & & & \\ & ⋰ & & & & \\ 1 & & & & & 0 \end{pmatrix}$$

with

$$A_j = (c_j) \ \text{ or } \ A_j = \begin{pmatrix} a_j & -b_j \\ b_j & a_j \end{pmatrix}, \quad D_j = -A_j;$$

$a_j, b_j, c_j \in \mathbb{R}$. C is either a compact Cartan subalgebra of $so(p-t,q-t)$,

$$t = \sum_{j=1}^{s} \deg A_j, \ \text{(if it is possible) or } C = \emptyset;$$

$$\tilde{B} = \mathrm{diag}(\underbrace{1,\dots,1}_{p-t},\underbrace{-1,\dots,-1}_{q-t}) \quad \text{or} \quad \tilde{B} = \emptyset.$$

$$N_{p,q} = \begin{cases} -1+3F_q+2F_{q-1} & \text{if } p+q \text{ is odd} \\ \frac{1}{2}((-1)^q-1)+2F_q+F_{q-1} & \text{if } p+q \text{ is even} \end{cases}$$

Each class is characterized by an ordered set of numbers $(\deg A_1,\dots,\deg A_s)$.

5. $so^*(2r)$

$$so^*(2r) = \left\{ X \in s\ell(2r,\mathbb{C}); X^T B + BX = 0; T\bar{X} = XT \right\}$$

$$\left(so^*(2) \simeq so(1,1)\right).$$

a) Compact Cartan subalgebra

$$S = \mathrm{diag}(A_1,\dots,A_r); \quad B = \mathrm{diag}(\underbrace{iK,\dots,iK}_{r}).$$

with

$$A_j = \begin{pmatrix} ia_j & 0 \\ 0 & -ia_j \end{pmatrix}; \quad a_j \in \mathbb{R}.$$

b) others

$$\begin{pmatrix} A_1 & & & & & & * \\ & \ddots & & & & & \\ & & A_s & & & & \\ & & & C & & & \\ & & & & D_s & & \\ & & & & & \ddots & \\ 0 & & & & & & D_1 \end{pmatrix}_{1\le s\le r} \qquad B = \begin{pmatrix} 0 & & & & i \\ & & & \cdot^{\cdot^{\cdot}} & \\ & & i\tilde{B}i & & \\ & \cdot^{\cdot^{\cdot}} & & & \\ i & & & & 0 \end{pmatrix}$$

with

$$A_j = \begin{pmatrix} \lambda_j & 0 \\ 0 & \bar{\lambda}_j \end{pmatrix}, \quad D_j = -\bar{A}_j : \lambda_j \in \mathbb{C}.$$

C is either a compact Cartan subalgebra of $so^*(2r-4s)$ or $C = \emptyset$ and

$$\tilde{B} = \mathrm{diag}(\underbrace{iK,\dots,iK}_{r-2s}) \text{ or } \tilde{B} = \emptyset$$

$$N_r = r+1.$$

6. <u>sp(2r, ℝ)</u>

$$sp(2r,\mathbb{R}) = \{X \in sl(2r,\mathbb{R});\ X^T B + BX = 0\}$$

$$(sp(2,\mathbb{R}) \simeq sl(2,\mathbb{R})).$$

a) Compact Cartan subalgebra

$$S = \mathrm{diag}(A_1,\dots,A_r); \quad B = \mathrm{diag}(\underbrace{J,\dots,J}_{r})$$

with

$$A_j = \begin{pmatrix} 0 & -a_j \\ a_j & 0 \end{pmatrix}; \quad a_j \in \mathbb{R}.$$

b) others

$$\begin{bmatrix} A_1 & & & & & * \\ & \ddots & & & & \\ & & A_s & C & & \\ & & & D_s & & \\ & & & & \ddots & \\ 0 & & & & & D_1 \end{bmatrix}_{1\le s\le r} \quad ; \quad B = \begin{bmatrix} 0 & & & & 1 \\ & & & \cdot^{\cdot^{\cdot}} & \\ & & {}_{-1}\widetilde{B}^{1} & & \\ & \cdot^{\cdot^{\cdot}} & & & \\ -1 & & & & 0 \end{bmatrix}$$

with

$$A_j = (c_j) \quad \text{or} \quad A_j = \begin{bmatrix} a_j & -b_j \\ b_j & a_j \end{bmatrix}, \quad D_j = -A_j;$$

$a_j, b_j, c_j \in \mathbb{R}$. C is either a compact Cartan subalgebra of sp $(2r-2t, \mathbb{R})$,

$$t = \sum_{j=1}^{s} \deg A_j, \text{ or } \quad C = \emptyset \text{ and}$$

$$\widetilde{B} = \operatorname{diag}\,(\underbrace{J,\ldots,J}_{r-t}) \text{ or } B = \emptyset$$

$$N_r = F_{r+2}-1.$$

Each class is characterized by an ordered set of numbers $(\deg A_1,\ldots, \deg A_s)$.

7. <u>usp(2p, 2q)</u>

$$\mathrm{usp}(2p,\ 2q) = \{X \in \mathrm{sl}(2p+2q,\ \mathbb{C}); X^T B + BX = 0; U\bar{X} = XU\}$$

with

$$U = \operatorname{diag}(\underbrace{J,\ldots,J}_{p},\underbrace{-J,\ldots,-J}_{q})$$

$$(\mathrm{usp}(2,0) \simeq \mathrm{su}(1,1)).$$

a) Compact Cartan subalgebra

$$S = \operatorname{diag}(A_1,\ldots,A_r)\ ;\ r = p+q;\ B = \operatorname{diag}(J,\ldots,J)$$

with

$$A_j = \begin{bmatrix} ia_j & 0 \\ 0 & -ia_j \end{bmatrix};\ a_j \in \mathbb{R}.$$

b) others

$$\begin{bmatrix} A_1 & & & & & \star \\ & \ddots & & & & \\ & & A_s & & & \\ & & & C & & \\ & & & & D_s & \\ & & & & & \ddots \\ 0 & & & & & D_1 \end{bmatrix}_{1\le s\le q} \quad ; \quad B = \begin{bmatrix} 0 & & & & 1 \\ & & & \iddots & \\ & & -1 & \tilde{B} & 1 \\ & \iddots & & & \\ -1 & & & & 0 \end{bmatrix}$$

with

$$A_j = \begin{pmatrix} \lambda_j & 0 \\ 0 & \bar{\lambda}_j \end{pmatrix}, \quad D_j = -\bar{A}_j; \quad \lambda_j \in \mathbb{C}.$$

C is either a compact Cartan subalgebra of usp (2p-2s, 2q-2s) or C = ∅ and

$$\tilde{B} = \text{diag}(\underbrace{J,\dots,J}_{r-2s}) \quad \text{or } \tilde{B} = \emptyset$$

$$N_{p,q} = q+1;$$

In these five latter cases, the elements in the upper triangle ((∇)) of S are completely determined by the defining conditions of the algebras (for example $S^TB+BS = 0$ and $U\bar{S} = SU$ for usp (2p,2q)).

REFERENCES

1. J. Patera, P. Winternitz and H. Zassenhaus, J. Math. Phys. 15, 1378 (1974).
2. J. Patera, P. Winternitz and H. Zassenhaus, J. Math. Phys. 15, 1932 (1974).
3. M. Perroud, The Maximal Solvable Subalgebras of the Real Classical Lie Algebras, CRM-521, Université de Montréal. (To be published.)
4. J. Patera, P. Winternitz and H. Zassenhaus, J. Math. Phys. 16, 1597 (1975).
5. J. Patera, P. Winternitz and H. Zassenhaus, J. Math. Phys. 16, 1615 (1975).

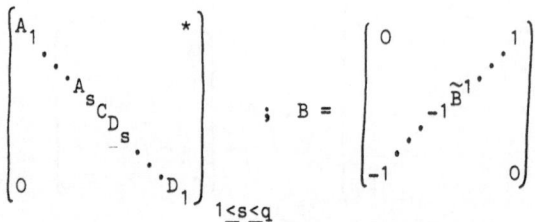

PHYSICS AND DEFORMATION THEORY OF FINITE AND INFINITE LIE ALGEBRAS

J.F. Pommaret - Centre de Physique Théorique
Ecole Polytechnique - Paris

A finite dimensional Lie algebra can be given as a set of structure constants satisfying the well known Jacobi relations. Such an algebra can also be considered as a point on a convenient algebraic variety and a deformation is an other point on the same variety. Two Lie algebras are called equivalent if one can be get from the other by a change of basis of their common underlying vector space. A Lie algebra is rigid if it is equivalent to any small deformation.

This deformation theory has been used in the study of dynamical groups and in general Relativity. For example, the Poincaré algebra can be deformed into the rigid, semi-simple de Sitter's algebras.

This general algebraic mechanism has been extended to infinite dimensional Lie algebras, but it is very difficult to apply it, even on easy examples.

On an other hand, physicists are used to deform structures defined by tensor fields or other geometric objects, by introducing a small parameter. This is the basic idea of General Relativity and other unitary theories, where the parameter is the inverse of the light speed.

However, the Lie algebras that can be introduced in such cases by looking at the infinitesimal transformations that let unchanged the later objects, are, in general, infinite dimensional ones.

A link between those two kinds of deformation theories, though it has been conjectured since a long time, has never been given.

Our aim is to produce it and to illustrate it with the example of contact structures, to be met in analytical mechanics. In particular, a constant is exhibited, similarly to the structure constants in the case of a Lie algebra, or to the curvature constant in the case of a Riemannian structure.

WHY DO ALL PHYSICAL LAWS PRESENT THE SAME FEATURES?

In order to describe a continous phenomenon, one has to introduce a tensor field, or, more generally, a geometric object ω, with local coordinates $\omega^{\tau}(x)$ over a convenient manifold X of finite dimension n (mainly R^4). The meaning is that, for any finite transformation of the base space X, $x' = \varphi(x)$, we know how the ω^{τ} are transformed:

$$\omega' = \varphi^*(\omega) :$$

$$\omega^{\tau}(x) = \Phi^{\tau}\left(\omega'(\varphi(x)), \partial_{\mu}\varphi^{k}\right) \qquad 1 \leq |\mu| = \mu_1 + \dots + \mu_n \leq q$$

To any geometric object, we can associate the Lie pseudogroup Γ of transformations of X that let ω unchanged, and the converse is also true.

It is a matter of fact that all pseudogroups to be met in physics are associated to geometric objects such that the Φ^{τ} involve the derivatives $\partial_{\mu}\varphi^{k}$ in a rational way.

Conversely, it can be shown that any transitive algebraic Lie pseudogroup Γ, defined by a formally integrable, involutive system ($\mathcal{E}$) of partial differential polynomial equations of order q, gives rise to such a kind of geometric objects. In particular $\Phi^{\tau}(\omega(y), \partial_{\mu}y^{k}) = U^{\tau}(y, \partial_{\mu}y^{k})$ are differential invariants of Γ and $\omega^{\tau}(x)$ their values at the identity $y^i = x^i$

Moreover, when ($\mathcal{E}$) is given the above properties - and this is always possible - the $\omega^{\tau}(x)$ must satisfy differential relations, called compatibility conditions, that can be brought into the following form:

$$A^{\nu i}_{\tau}(\omega)\frac{\partial\omega^{\tau}}{\partial x^{i}} + B^{\nu}(\omega) = 0$$

where A and B are polynomials in ω^{τ}. In particular, we may note that x is not appearing explicitely.

One will observe that physical laws, expressed by subsystems of the former ones, are thus systems of partial differential polynomials in the ω^{τ}.

At this time, as in physics, we do want to linearise the preceding system, setting: $\omega^{\tau}_{t}(x) = \omega^{\tau}(x) + t\,\Omega^{\tau}_{1}(x) + \dots$

In fact, in many cases, Γ contains the translations. Then the $\omega^{\tau}(x)$ can be taken as -1,0,+1 , and the linearised first order system of partial differential equations for Ω_1 has constant coefficients.

As a conclusion, we believe that the study of algebraic Lie pseudogroups must appear to be a useful tool in physics.

INTRODUCTION

Our aim will be to show why some physical arguments have led us to mix together differential and algebraic theories that, a priori, had nothing to do one with each other. The later theories and their relations, sketched on the following picture, will be presented in a self contained way.

Deformation theory of Lie algebras
⇓
Algebraic geometry (linear algebraic groups) ⟹ Deformation theory of structures ⟸ Formal theory of systems of partial differential equations
⇑
Lie pseudogroups

PHYSICAL BACKGROUND

The kind of situation to be met in physics is as follows: In order to describe a continous phenomenon, one has to introduce a tensor field, or, more generally, a geometric object ω, with local coordinates $\omega^\tau(x)$, over a convenient C^∞ paracompact manifold X of finite dimension n, mainly R^4. This fact agrees with a description of the universe by local observers, and the meaning is that, for any finite transformation $\bar{x} = \varphi(x)$ of the base space X or of its local coordinates, the ω^τ are transformed accordingly to the rules:

$$\bar{\omega} = \varphi(\omega) \quad : \quad \omega^\tau(x) = \phi^\tau\left(\bar{\omega}(\varphi(x)), \partial_\mu \varphi^k\right)$$

$$1 \leq |\mu| = \mu_1 + \cdots + \mu_n \leq q$$

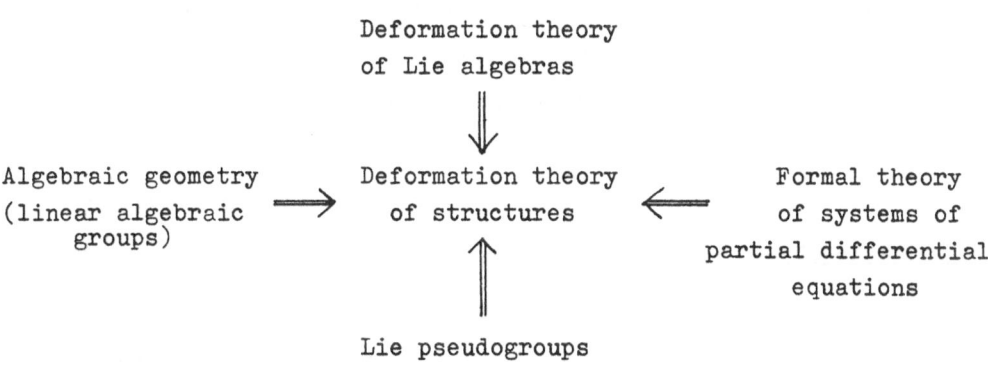

We then express the relation between field and matter by a system of p.d.e. : $I^{\nu}(\omega^{\tau}, \partial_{\mu}\omega^{\tau}) = M^{\nu}(x)$ or simply $I(\omega \frac{\partial \omega}{\partial x}) = M(x)$.

EXAMPLES: 1) Maxwell equations for the electromagnetic field.

2) Einstein equations for the gravitational field.

We make the following important remarks:

1) The coordinates x^i appear only in the right members.As for the left members I ,they are only functions of the ω^{τ} and their derivatives of different orders,(quasi)linear in the top order ones, and rational,that is to say quotients of polynomials,in the other derivatives and in the ω^{τ} .

Thus,in vacuum,the equations $I(\omega, \frac{\partial \omega}{\partial x}) = 0$ can be taken as differential polynomials with constant coefficients (2 d).

2) The $M(x)$ have in general to satisfy some compatibility conditions. For the two examples given above,they are the well known divergence relations.

3) In vacuum,that is to say when $M(x) = 0$,their may exist,at least locally,a potential,that is to say a way to know the $\omega^{\tau}(x)$ as functions of some $A(x)$ and their derivatives,such that $I\left(\omega(A, \frac{\partial A}{\partial x}), \frac{\partial \omega}{\partial x}(A, \frac{\partial A}{\partial x})\right) \equiv 0$. This is of course well known for Maxwell equations,but it is still an open problem for Einstein equations.

4) People do not like to manipulate non linear systems of p.d.e..

As Maxwell equations are already linear,there is nothing to do.

On the contrary,one has to linearise Einstein equations,and the mechanism to apply will be described.

• One has to choose a standard or special field,solution of the equations in vacuum:for our purpose it is the Minkowski metric.

•• Then we have to introduce a small parameter t (say 1/c in our case) and take a Taylor expansion: $\omega_t(x) = \omega(x) + t\,\underset{1}{\Omega}(x) + \cdots$

... Finally, we introduce partial differential operators $\mathcal{D}$, $\mathcal{D}_1$, $\mathcal{D}_2$ and rewrite 3) :

When $\mathcal{D}_1 . \underset{1}{\Omega} \equiv 0$ there may exist functions $\xi(x)$ such that $\underset{1}{\Omega} = \mathcal{D}.\xi$, and $\mathcal{D}_1 \circ \mathcal{D} \equiv 0$; moreover, in order that $\mathcal{D}_1 \underset{1}{\Omega} = \underset{1}{M}$, we must have $\mathcal{D}_2 . \underset{1}{M} \equiv 0$, and $\mathcal{D}_2 \circ \mathcal{D}_1 \equiv 0$.

For the reader not familiar with differential geometry, we recall some classical facts.

Let E be a vector bundle over X, with local coordinates (x^i, u^k) $i=1,..,n$, $k=1,..,\dim E$. We introduce the vector bundle $J_q(E)$ over X , with local coordinates (x^i, u^k, u^k_μ) $1 \leqslant |\mu| \leqslant q$, where the u^k_μ are transformed like the derivatives $\partial_\mu u^k$ under a change of coordinates.

<u>DEFINITION</u>: A linear partial differential operator $E \xrightarrow{\mathcal{D}} F$ of order q is given in local coordinates by the formulas:

$$u^k(x) \rightsquigarrow v^l(x) = \sum_{|\mu| \leqslant q} A^{l\mu}_k(x)\, \partial_\mu u^k(x)$$

where $l=1,..,\dim F$ and we call Ⓗ the set of sections of E , solutions of the system of p.d.e. (Σ): $\mathcal{D} u = 0$

We associate with $\mathcal{D}$ the unique morphism $\phi : J_q(E) \longrightarrow F$, such that $\mathcal{D} = \phi \circ j_q$ and given in local coordinates by the formulas:

$$j_q : u^k(x) \rightsquigarrow \partial_\mu u^k(x) \quad , \quad \phi : u^k_\mu(x) \rightsquigarrow \sum_{|\mu| \leqslant q} A^{l\mu}_k(x)\, u^k_\mu(x)$$

We define the r° prolongation of ϕ as a morphism:

$$p_r(\phi) \; : \; J_{q+r}(E) \longrightarrow J_r(F)$$

obtained by taking the derivatives of v^l up to order r , and we call $\sigma_r(\phi)$ the q+r-top order part of $p_r(\phi)$.

The kernel of $\sigma_r(\phi)$, called g_{q+r} , is defined by: $\sum_{|\mu|=q} A^{l\mu}_k(x)\, u^k_{\mu+\nu} = 0$ where $|\mu| = q$, $|\nu| = r$ and we easily check that g_{q+r} is uniquely determined by g_q .

<u>DEFINITION</u>:Let $E, F_0, F_1, \ldots, F_p, \ldots$ be vector bundles over X ,and let $\mathcal{D}, \mathcal{D}_1, \ldots, \mathcal{D}_p, \ldots$ be linear partial differential operators.We say that the sequence:

$$0 \longrightarrow \Theta \longrightarrow E \xrightarrow{\mathcal{D}} F_0 \xrightarrow{\mathcal{D}_1} F_1 \longrightarrow \cdots \longrightarrow F_{p-1} \xrightarrow{\mathcal{D}_p} F_p \longrightarrow \cdots$$

is a differential complex of finite length if $\mathcal{D}_p \circ \mathcal{D}_{p-1} \equiv 0$ and $F_p \equiv 0$ if p is big enough.

<u>EXAMPLE</u>:Let T (T^*) be the tangent (cotangent) bundle of X,and let $\Lambda^p T^*$ be the vector bundle over X ,the sections of which are exterior p-forms.We have the Poincare complex:

$$0 \longrightarrow \Theta \longrightarrow \Lambda^0 T^* \xrightarrow{d} \Lambda^1 T^* \xrightarrow{d} \cdots \xrightarrow{d} \Lambda^n T^* \longrightarrow 0$$

where d is just exterior differentiation and $\Lambda^0 T^*$ is written for $X \times R$.

Now the preceeding situation is just equivalent to exhibit a differential complex,and we are mainly interested in the initial part described by the following picture:

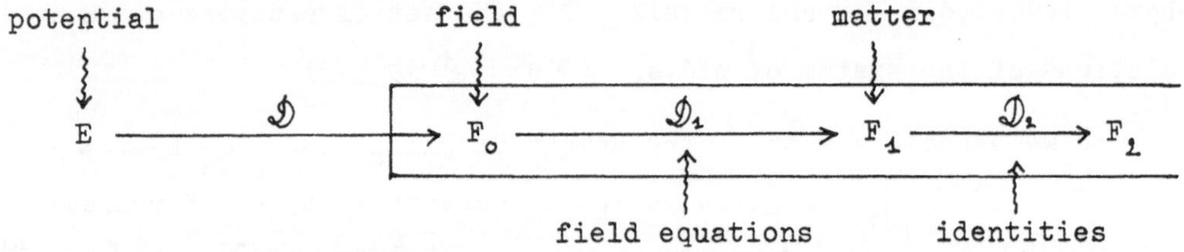

We are led to our first problem:

<u>PROBLEM I</u>: . Given any $\mathcal{D}$,does there exist such a complex and a reason to forget $\mathcal{D}$?

.. What kind of $\mathcal{D}$ must be taken in order to give a physical meaning to the <u>truncated complex</u> thus obtained ?

Now we have seen that the main system of p.d.e. to deal with is a non linear one, and we state our second problem:

PROBLEM II: Why are the field equations in vacuum given by a set of differential polynomials ?

Finally we will look for the different kinds of "deformations" that are used in physics.

Some methods of deformation have been introduced by Spencer (4,5) and others, since 1957, in order to describe structures on manifolds and their perturbations. For example, the deformation of a riemannian structure goes through the deformation of a metric tensor as in general relativity.

Now we show that it is possible to pass from the inhomogeneous Galilee group in two variables:

$$x' = x + vt + a_x \quad , \quad t' = t + a_t$$

to the inhomogeneous Lorentz group in two variables:

$$x' = x \operatorname{ch}\lambda + t \operatorname{sh}\lambda + b_x \quad , \quad t' = x \operatorname{sh}\lambda + t \operatorname{ch}\lambda + b_t$$

In fact we have just to introduce a parameter $1/c$ and consider the group with the infinitesimal generators:

$$I_x = \frac{\partial}{\partial x} \quad , \quad I_t = \frac{\partial}{\partial t} \quad , \quad I_r = t\frac{\partial}{\partial x} + \frac{1}{c}\, x \frac{\partial}{\partial t}$$

satisfying the following commutation relations:

$$[I_x , I_t] = 0 \quad , \quad [I_x , I_r] = \frac{1}{c}\, I_t \quad , \quad [I_t , I_r] = I_x$$

At this time, we only need to take first $c = \infty$, then $c=1$.

More generally (1), a finite dimensional Lie algebra $\mathfrak{g}$, with underlying vector space V , is determined by a set of structure constants c^i_{jk} , representing a map $V \wedge V \rightarrow V$, and satisfying the well known Jacobi relations:

$$\text{(J)} \qquad J(c) \equiv c^l_{ij} \cdot c^m_{kl} + c^l_{jk} \cdot c^m_{il} + c^l_{ki} \cdot c^m_{jl} = 0$$

Such an algebra can be considered as a point c on a convenient algebraic variety and a deformation c_t as an other point depending on a parameter t .

<u>DEFINITION</u>:Two Lie algebras with the same V are said <u>equivalent</u> if one can get from one to the other by a change of basis of V .

<u>DEFINITION</u>:A Lie algebra is called <u>rigid</u> if it is equivalent to any small deformation.

The main trick in the deformation theory of Lie algebras is to linearise the problem,setting: $c_t = c + t\, c_1 + \dots$
and looking at the infinitesimal (first order) deformation.

This theory has been developped since 1964 by Gerstenhaber and others.

We state now our third and last problem:

<u>PROBLEM III</u>: Is there a link between the deformation theory of geometric objects used for example in General Relativity and the later deformation theory of Lie algebras ?

We will now outline the solution of the three problems stated in the former pages.

MATHEMATICAL BACKGROUND

I / <u>FORMAL THEORY OF SYSTEMS OF P.D.E.</u> :

This powerful theory has been developped during the past ten years,mainly by Spencer,Quillen,Goldschmidt (4).In the linear case, the use of diagrams is a generalisation of the vector notations grad, div,rot and the exterior derivative d for exterior forms.

The meaning is that it allows one to look at p.d.e. systems in a deep and coherrent way,without any local writing,and to give intrinsic proofs that would be tedious by direct computations.

Of course, at the same time, it is also a new kind of abstraction to get used to.

REMARK: A main point is that we do not transform a given system of p.d.e. into an exterior system and thus we never lose any information on the base space X .

We now briefly review the principal results (2a,4).

. The key idea is that of involution. As we do not want to detail the definition, we will just say that it is a purely algebraic condition that can be easily verified on g_{q+r} for r big enough.

.. We may now suppose that g_q is involutive and that the first prolongation of (Σ) does not bring equations of order q that are not linear combinations of equations already in (Σ).

DEFINITION: We say that (Σ) (or $\mathcal{D}$) is formally integrable, involutive.

... We may also suppose that $\mathcal{D} = \phi \circ j_1$ with ϕ surjective, and get:

THEOREM: When $\mathcal{D}$ is such a formally integrable, involutive linear partial differential operator of order q, there exists a differential complex, called P-sequence:

$$P(\Theta) \qquad 0 \longrightarrow \Theta \longrightarrow E \xrightarrow{\mathcal{D}} F_0 \xrightarrow{\mathcal{D}_1} F_1 \xrightarrow{\mathcal{D}_2} \cdots \xrightarrow{\mathcal{D}_n} F_n \longrightarrow 0$$

of finite length n+1, with $\mathcal{D}_1, \mathcal{D}_2, \ldots$ first order, formally integrable, involutive linear partial differential operators (2a).

REMARK: In proving this theorem, we have, at one time, to forget $\mathcal{D}$ and consider only the truncated P-sequence:

$$P(\Omega) \qquad 0 \longrightarrow \Omega \longrightarrow F_0 \xrightarrow{\mathcal{D}_1} F_1 \xrightarrow{\mathcal{D}_2} \cdots \xrightarrow{\mathcal{D}_n} F_n \longrightarrow 0$$

made up only with first order operators.

The later remark gives an answer to the first part of problem I.

Now, what kind of $\mathcal{D}$ must be useful in physics ?

II / LIE PSEUDOGROUPS :

DEFINITION: A Lie pseudogroup Γ is a continous group of transformations $y = f(x)$ of R^n ,solutions of a system of p.d.e. :

$$(\mathcal{E}) \qquad H^\tau(x^i, y^k, \partial_\mu y^k) = 0 \qquad 1 \leq |\mu| \leq q$$

the finite equations of Γ . (2 b,c)

REMARK:The Lie pseudogroups were formerly known as infinite groups, in contrast to the finite groups,today known as Lie groups.

Because $y^i = x^i$ must be a solution,we may linearise ($\mathcal{E}$).

Setting $y^i = x^i + t\,\xi^i(x) + \ldots$ we get a linear system of p.d.e. :

$$(\Sigma) \qquad \mathcal{D}.\xi = 0$$

the infinitesimal equations of Γ . (ξ is a section of T)

The operator $\mathcal{D}$ has the property (just think about the Lie derivative of a tensor field !) that,if $\mathcal{D}.\xi_1 = 0, \mathcal{D}.\xi_2 = 0$ then $\mathcal{D}.[\xi_1, \xi_2] = 0$.

DEFINITION:Such an operator is called a Lie operator.

Let us now introduce new coordinates y^k_μ,called jet-coordinates,and transform them by a change of target $\bar{y} = \varphi(y)$ in the same way as the corresponding $\partial_\mu y^k$,while keeping the source x unchanged.

DEFINITION:A function $U(y, y^k_\mu)$ is called a differential invariant of Γ if $U(\bar{y}, \bar{y}^k_\mu) \equiv U(y, y^k_\mu)$ for any transformation $\bar{y} = \varphi(y) \in \Gamma$.

THEOREM:There exists a fundamental set of differential invariants such that ($\mathcal{E}$) can be written:

$$(\mathcal{E}) \qquad U^\tau(y, y^k_\mu) = \omega^\tau(x) \qquad \text{(Lie form)}$$

Moreover,under an arbitrary change of source $\bar{x} = \varphi(x)$ the ω^τ behave like the components of a geometric object:

$$\omega \rightarrow \bar{\omega} = \varphi(\omega) \quad : \quad \omega^\tau(x) = \phi^\tau(\bar{\omega}(\varphi(x)), \partial_\mu \varphi^k) \qquad 1 \leq |\mu| \leq q$$

and we have: $\omega = \Gamma(\omega) \quad : \quad U^\tau(y, y^k_\mu) \equiv \phi^\tau(\omega(y), y^k_\mu)$

Now we do want to effect a perturbation: $\omega_t(x) = \omega(x) + t \underset{1}{\Omega}(x) + \cdots$

At this time, (Σ) can be written:

$$(\Sigma) \qquad \underset{1}{\Omega}^{\tau} \equiv - L^{\tau \mu}_{k}(\omega(x))\, \partial_{\mu} \xi^{k} + \xi^{i} \frac{\partial \omega^{\tau}(x)}{\partial x^{i}} = 0$$

REMARK: If Γ contains the translations, as it does usually in physics, then the $\omega^{\tau}(x)$ must be constants that can be choosen as o or 1 .

THEOREM: In order to get a formally integrable involutive operator $\mathcal{D}$, the $\omega^{\tau}(x)$ must satisfy the following compatibility conditions:

$$(\mathrm{I}) \quad \begin{cases} I_{*}\left(\omega(x), \dfrac{\partial \omega(x)}{\partial x}\right) = 0 & (\text{ first kind }) \\ I_{**}\left(\omega(x), \dfrac{\partial \omega(x)}{\partial x}\right) = c & (\text{ second kind }) \end{cases}$$

DEFINITION: The constants c are called structure constants.

Linearising, we get $\mathcal{D}_1$ such that $\mathcal{D}_1 . \underset{1}{\Omega} \equiv 0$ when $\underset{1}{\Omega} = \mathcal{D}.\xi$.

THEOREM: In order to get a formally integrable involutive system (I), the structure constants must satisfy the following set of algebraic relations, called (generalised) Jacobi relations:

$$(\mathrm{J}) \qquad J(c) = 0$$

where the J are polynomials of order ≤ 2 .

DEFINITION: An algebraic Lie pseudogroup is a Lie pseudogroup defined by a system of polynomials in $\partial_{\mu} y^{k}$, $1 \leq |\mu| \leq q$ with coefficients C^{∞} in x, y.

THEOREM: The differential invariants of an algebraic Lie pseudogroup can be choosen as rational functions of the y^{k}_{μ}, $1 \leq |\mu| \leq q$ with coefficients C^{∞} in y. The compatibility conditions (I) are given by a set of differential polynomials with constant coefficients.

REMARK: The proof uses methods of algebraic geometry ($2d$) and we have to suppose that Γ is transitive, that is to say (ξ) and (Σ) do not contain equations of order 0 .

EXAMPLE: 1) All the classical examples, in particular those in which tensor fields are involved .

2)Let $\tilde{\Gamma}$ be the normaliser of Γ ,that is to say the largest group of transformations of R^n in which Γ is normal.Then $\tilde{\Gamma}$ is an algebraic pseudogroup $\forall\, \Gamma$.

Taking the field equations as subsystems of (I),called structured systems,we have answered to the second part of problem I and to problem II ;problem III only remains unsolved.

Let us now introduce a (small) parameter t and consider the new structure constants $c_t = c + t\,C + \dots$ satisfying $J(c_t)=0$, $\forall t$. Let $\omega_t(x)$ be solution of the system:

$$\begin{cases} I_*\left(\omega_t(x), \dfrac{\partial \omega_t(x)}{\partial x}\right) = 0 \\ I_{**}\left(\omega_t(x), \dfrac{\partial \omega_t(x)}{\partial x}\right) = c_t \end{cases}$$

For different choices of $\omega(x)$ we have a family of pseudogroups of transformations of X .In fact we have a fiber bundle $\mathcal{U} \overset{\omega}{\leftrightarrows} X$ and a structure over X is just a section ω of the later bundle.In particular,taking $\omega_t(x)$ as above,we have the family Γ_t .

DEFINITION:We say that two structures ω and $\bar{\omega}$ are equivalent if they give rise to the same pseudogroup Γ .

This definition is of course extended to the corresponding c and $\bar{c}$ and it is easy to show that it is a generalisation of the equivalence of two Lie algebras with the same underlying vector space.

The idea is to transfer the perturbation c_t of the structure constants c of the Lie algebra $\mathfrak{g}$ of a Lie group G to a perturbation G_t of G ,by means of a well known theorem of Lie.In fact,the $\omega^\tau(x)$ are just local coordinates for the left (or right) invariant Maurer-Cartan 1-forms defined on G,and the deformation theory can be developped as above.This is the last answer we needed.

We give examples, increasing the order of $(\mathcal{E})$ or (Σ), and giving $(\mathcal{E})$ in its Lie form.

1) Action of a Lie group : $G \times R \longrightarrow R$:

•) $\Gamma : y = x + a$, $(\mathcal{E})\ \frac{\partial y}{\partial x} = 1$, $(\Sigma)\ \frac{\partial \xi}{\partial x} = 0$

••) $\Gamma : y = ax + b$, $(\mathcal{E})\ \frac{\partial^2 y}{\partial x^2} / \frac{\partial y}{\partial x} = 1$, $(\Sigma)\ \frac{\partial^2 \xi}{\partial x^2} = 0$

•••) $\Gamma : y = \frac{ax+b}{cx+d}$, $(\mathcal{E})\ \frac{\frac{\partial^3 y}{\partial x^3}}{\frac{\partial y}{\partial x}} - \frac{3}{2}\left(\frac{\frac{\partial^2 y}{\partial x^2}}{\frac{\partial y}{\partial x}}\right)^2 = 0$, $(\Sigma)\ \frac{\partial^3 \xi}{\partial x^3} = 0$

2) Γ : transformations of R^n with jacobian = 1 .

$$(\mathcal{E})\quad \frac{\partial(y^1,\dots,y^n)}{\partial(x^1,\dots,x^n)} = 1 \qquad (\Sigma)\quad \sum_{i=1}^{n} \frac{\partial \xi^i}{\partial x^i} = 0 \quad (\operatorname{div} \xi = 0)$$

3) Γ : holomorphic transformations of the complex plane.

$(\mathcal{E})\ f(J) = J$, $(\Sigma)\ \mathcal{L}(\xi)J = 0$ where J is the mixed tensor $\begin{pmatrix} 0 & 1 \\ -1 & 0 \end{pmatrix}$

$(\mathcal{E})\ \frac{\partial y^2}{\partial x^2} - \frac{\partial y^1}{\partial x^1} = 0, \frac{\partial y^1}{\partial x^2} + \frac{\partial y^2}{\partial x^1} = 0$ (Cauchy-Riemann) , $(\Sigma)\ \frac{\partial \xi^2}{\partial x^2} - \frac{\partial \xi^1}{\partial x^1} = 0, \frac{\partial \xi^1}{\partial x^2} - \frac{\partial \xi^2}{\partial x^1} = 0$

4) $\Gamma :\quad y^1 = x^1 + a$, $y^2 = x^2 + g(x^1)$

$(\mathcal{E}) :\quad \frac{\partial y^1}{\partial x^1} = 1$, $\frac{\partial y^1}{\partial x^2} = 0$, $\frac{\partial(y^1,y^2)}{\partial(x^1,x^2)} = 1$

$(\Sigma)\quad \frac{\partial \xi^1}{\partial x^1} = 0$, $\frac{\partial \xi^1}{\partial x^2} = 0$, $\frac{\partial \xi^1}{\partial x^1} + \frac{\partial \xi^2}{\partial x^2} = 0$

$\Gamma = \Gamma_0$ can be deformed in order to get $\Gamma_1 \begin{cases} y^1 = f(x^1) \\ y^2 = x^2 f'(x^1) \end{cases}$ which is rigid.

CONCLUSION

Unfortunately, physicists are dealing with Lie pseudogroups, though they do not know what they are dealing with, because of the lack of a convenient mathematical treatment.

They just make out their own cooking for the cases they meet: actions of Lie groups, pseudogroups related to symplectic structures, riemannian structures, analytic structures, contact structures,...

The classical approach of Cartan, using Maurer-Cartan equations, has been generalised by Guillemin and Sternberg (Ref 6 in 5), in order to describe, when $\mathfrak{D}$ is a Lie operator, two differential complexes introduced by Spencer in the general case.More recently,Spencer, Goldschmidt and Malgrange have built up a new formalism to describe the same differential complexes (5),but it seems very difficult to use it properly in physics.

An algebraic attempt has also been made with infinite Lie algebras,but the methods are not easy to put into practice (3).

We have shown how to introduce an other differential complex, the P-sequence,that arises in a natural way from the study of any linear partial differential operator,and to construct its initial part when $\mathfrak{D}$ was a Lie operator.

We believe that those methods are to become a new powerful tool in mathematical physics.

BIBLIOGRAPHY

1) LEVY-NAHAS (M):Deformation and contraction of Lie algebras Jour. of math. Physics,vol 8,nº 6,1967,p1211-1222

2) POMMARET (J.F.):a)Etude interne des systèmes linéaires d'équations aux dérivées partielles:Ann.Inst.Henri Poincare,vol17,nº2,1972,p131
b)Théorie des déformations de structures: " ,vol18,nº4,1973,p285
c)Same title:Proc.3rd inter.coll.group methods in physics Marseille,C.N.R.S.,1974,p77-102
d)Pseudogroupes de Lie algebriques:C.R.Acad.Sc.,t280,1975,p1693

3) RIM (D.S.):Deformations of transitive filtred Lie algebras, Ann.of Math.,83,1966,p 339-357.

4) SPENCER (D.C.):Overdetermined systems of linear partial differential equations,Bull. A.M.S. ,1969,75,p 179-239.

5) SPENCER (D.C.) and KUMPERA (A.):Lie equations I ,Study nº 73 Princeton University Press 1972

WIGNER 3j-SYMBOLS AND THE LORENTZ GROUP

Ronald Shaw

Department of Applied Mathematics, University of Hull, Hull, England.

(Contribution to the 4th International Colloquium on Group Theoretical Methods in Physics, Nijmegen, June 1975)

Summary Use of a group $\mathcal{L}^{\uparrow}(C_2)$ which double covers $\mathcal{L}^{\uparrow}$, and of the spanning property of the spinor light cone, leads to a rapid derivation of the properties of Wigner's 3j-symbols. No obscure computations, choices of phase, or the like, are needed. The reality of the 3j-symbols follows from their invariance under the antilinear space reversal $\in \mathcal{L}^{\uparrow}_{-}(C_2)$. Some (possibly) new recursion relations are established. It is noted that classical invariant theory made use of the spinor light cone a century ago, and that the classical 3j-symbol takes integer values.

1. Introduction For some years now the author has been engaged in writing a unified coordinate-free account of what may be described briefly as "the mathematics of Minkowski space", or in a little more detail as "the linear, multilinear and antilinear algebra of Minkowski space M, the Lorentz group $\mathcal{L}$, and of associated spaces and groups". Hopefully, in the near future, the complete work will be published in book form. The present talk will describe a rather small subset of this work, namely that dealing with the 3j-symbols for the D^j representations of $\mathcal{L}$.

Traditional accounts[1,2] of the Wigner 3j-symbols and Clebsch-Gordan coefficients for the 3-dimensional rotation group can be criticized in that they make many of the important properties of the coefficients appear in a far from clear light. This lack of clarity is produced chiefly by (i) weakly-motivated choices of phase (ii) proofs involving computations of a somewhat complicated and murky nature (iii) treating the less symmetrical CG-coefficients before the more symmetrical 3j-symbols (iv) dealing with components of an object rather then the object itself. To correct these defects, the present account deals first with a certain trilinear invariant, then with its components (the 3j-symbols); next the trilinear invariant is used to introduce certain linear maps, and only then do we deal with the matrices of these maps (the CG-coefficients).

Important, but rather less traditional, accounts of the subject have been given by Schwinger[3], using certain operator methods, and by Bargmann[4], using function space methods. In contrast with these contributions the present account is mathematically simpler to the extent that it uses nothing more than the linear (and antilinear) algebra of finite-dimensional vector spaces. (However, at certain points, the present account would appear to be quite close to that of Bargmann.)

An essential ingredient of the present account is to treat the 3j-symbols as belonging to the representation theory of $SL(C_2)$ - which we view in its metrical guise of $Sp(C_2)$ - rather than that of $SU(C_2)$; moreover we use as well the antisymplectic transformations $ALSp(C_2)$, which adjoin to $Sp(C_2)$ to form a group $\mathcal{L}^{\uparrow}(C_2) = ALLSp(C_2)$ which (see Eq. (2.7)) double covers the orthochronous Lorentz group $\mathcal{L}^{\uparrow}$. Each choice of time-axis in Minkowski space defines a space inversion operator $\mathcal{P} \in ALSp(C_2)$, and picks out a corresponding $SU_{\mathcal{P}}(C_2)$ subgroup of $Sp(C_2)$, with inner product $(\ ,\ )_{\mathcal{P}}$ defined, as in Eq. (2.12), by

$$(\xi_1, \xi_2)_{\mathcal{P}} = [\mathcal{P}\xi_1, \xi_2], \quad \xi_i \in C_2 . \tag{1.1}$$

Starting from the one frame-independent bilinear form $[\ ,\ ]$ on C_2, we are of course at liberty at any stage to specialize our considerations to any one of the host of frame-dependent hermitian forms $(\ ,\ )_{\mathcal{P}}$.

Traditional accounts, which start out from SU(2), proceed in the opposite direction. They make the belated discovery of an antilinear operator $\mathcal{P}$ commuting with SU(2)-transformations, and so can introduce a bilinear form by $[\xi_1,\xi_2] = (\mathcal{P}^{-1}\xi_1,\xi_2)$. However, having started out from a particular hermitian form (,), they tend to concentrate upon it and to play down the role of the bilinear form $[\,,\,]$. But in fact it is the latter which is of paramount importance — even if in the end we specialize to SU(2) — since it is required in the definition of the fundamental trilinear invariant (in Section 4.1).

As a lead-in to the treatment of 3j-symbols in Section 4, several useful theorems concerning the space V^j of (2j+1)-component spinors will be stated (Theorems 3.2 - 3.6). The last of these is particularly noteworthy, in that it demonstrates that the crucial structure inherited by V^j from C_2 is not (in general) the metrical structure but is the <u>spinor light cone structure</u> N^j, defined in Section 3.1.

Most of the methods employed in this account are in essence far from modern, and many go back more than a century ago! In particular we do not scorn the use of classical bases, and draw the reader's attention to the fact that the <u>classical 3j-symbols take integer values</u>! Possibly more use could be made of this — compare, for example, the simplicity of the classical recursion relation (4.18) to its standard from (4.19).

Next, a brief word concerning notation. We use the logograms L, AL to denote maps which are, respectively, linear, antilinear; their combination ALL is used to denote "all" such maps, i.e. linear <u>and</u> antilinear ones.

Finally we point out that, due to lack of space, several proofs, including that of Theorem 3.6, have had to be omitted.

2. <u>The multiantilinear algebra of the Lorentz group</u>.

Only an abbreviated account of this topic will be given, tailored to the needs of the intended applications.

2.1 <u>The space C_2 of Lorentz 2-component spinors</u>.

Let C_2 denote a complex 2-dimensional vector space which is equipped with symplectic geometry by means of a (non-degenerate) skew-symmetric bilinear form $[\,,\,]$:

$$[\xi_1,\xi_2] = -[\xi_2,\xi_1] \in \mathbb{C}, \qquad \xi_i \in C_2 \tag{2.1}$$

Let $\mathcal{L}_+^\uparrow(C_2) \equiv Sp(C_2)$ and $\mathcal{L}_-^\uparrow(C_2) \equiv ALSp(C_2)$ denote the sets of isometries and anti-isometries of C_2. Together they form a subgroup $\mathcal{L}^\uparrow(C_2) \equiv ALLSp(C_2) \equiv Sp(C_2) \cup ALSp(C_2)$ of the group $GALL(C_2) \equiv GL(C_2) \cup GAL(C_2)$, the latter group consisting of all the linear and antilinear automorphisms of C_2, while the mappings belonging to the subgroup satisfy in addition the invariance property

$$[a\xi_1\,,\,a\xi_2] = [\xi_1,\xi_2]^a\,,\quad a \in \mathcal{L}^\uparrow(C_2). \tag{2.2}$$

Here λ^a, for $\lambda \in \mathbb{C}$, denotes λ or $\bar\lambda$ according as <u>a</u> is a linear or antilinear map.

2.2. <u>The space $V^{j,j'}$ of (j,j')-spinors</u>.

Let V^j denote the 2j th symmetrized tensorial power $\vee^{2j}C_2$ of C_2, and let $D^j(a)$, for $a \in GALL(C_2)$, denote the restriction to V^j of $\otimes^{2j}a$. Define also the spaces $V^{j,j'}$, and corresponding corepresentations $D^{j,j'}$ of $GALL(C_2)$, by

$$V^{j,j'} = V^j \otimes \overline{V^{j'}}\,,\quad D^{j,j'}(a) = D^j(a) \otimes \overline{D^{j'}}(a)\,,\quad a \in GALL(C_2), \tag{2.3}$$

where $\bar{V}$ denotes the antispace of V. In the case $j = j'$ we write $R^{j,j}$ for the <u>real</u> vector space consisting of those elements of $V^{j,j}$ which are real under the natural conjugation $\phi_1 \otimes \phi_2 \to \phi_2 \otimes \phi_1$, $\phi_i \in V^j$, and note that $D^{j,j}(a)$ can be thought of as a <u>real</u> operator upon $R^{j,j}$.

In particular we mention (i) the space C_3 of Lorentz complex 3-vectors (ii) real 4-dimensional Minkowski space M (iii) its complexification $M^{\mathbb{C}}$ (iv) the space of Weyl (5-component) spinors (v) the (real) space of trace-free Ricci tensors, given respectively by

$$
\begin{aligned}
&\text{(i)} \quad C_3 = V^1 = C_2 \vee C_2 \cong Sk(C_2,C_2) \cong L_0(C_2,C_2),\\
&\text{(ii)} \quad M = R^{\frac{1}{2},\frac{1}{2}} \cong ALSk(C_2,C_2)^5,\\
&\text{(iii)} \quad M^{\mathbb{C}} = V^{\frac{1}{2},\frac{1}{2}} = C_2 \otimes \bar{C}_2 \cong AL(C_2,C_2)^5, \qquad (2.4)\\
&\text{(iv)} \quad V^2 = \vee^4 C_2 \cong S_0(C_3,C_3) \cong (C_3 \vee C_3)_0 \cong \text{space of binary quartics}^{6,7},\\
&\text{(v)} \quad R^{1,1} \cong ALS(C_3,C_3) \cong S_0(M,M) \cong (M \vee M)_0^{7,12}.
\end{aligned}
$$

2.3 <u>Induced scalar products and isometries</u>. The space $V^{j,j'}$ inherits a non-degenerate scalar product $[\ ,\ ]$ from that on C_2. In particular that on V^j is defined to be the restriction to $\vee^{2j}C_2$ of the usual induced scalar product upon $\otimes^{2j}C_2$; thus if $\phi = \xi_1 \vee \xi_2 \vee \cdots \vee \xi_{2j}$ and $\psi = \eta_1 \vee \eta_2 \vee \cdots \vee \eta_{2j}$ are two (j,0)-spinors, their scalar product involves the permanent of the $(2j+1) \times (2j+1)$-matrix whose ik-element is the scalar product $[\xi_i, \eta_k]$:

$$(2j)!\,[\phi,\psi] = \text{permanent}\ \left([\xi_i,\eta_k]\right). \qquad (2.5)$$

The scalar product upon V^j clearly satisfies

$$[\phi,\psi] = (-)^{2j}[\psi,\phi]\,, \quad \phi,\psi \in V^j; \qquad (2.5a)$$

in particular the geometry on $C_3 = V^1$ is complex orthogonal. That[10] on $M = R^{\frac{1}{2},\frac{1}{2}}$ is real orthogonal, with signature (+ − − −) —as can be checked using the basis (2.10) below— so that M is indeed a Minkowski space.

If $a \in \mathcal{L}^{\uparrow}(C_2)$, then $D^{j,j'}(a)$ is clearly an isometry, or antiisometry, of $V^{j,j'}$. In particular

$$[D^j(a)\phi\,,\,D^j(a)\psi] = [\phi,\psi]^a\,, \quad a \in \mathcal{L}^{\uparrow}(C_2). \qquad (2.6)$$

However in general the group homomorphism $D^{j,j'}$ has image only some "small" subgroup of the isometry group of $V^{j,j'}$. As will be noted below, in Section 3.1 and footnote 16, the cases $(j,j') = (\frac{1}{2},\frac{1}{2})$, or $= (1,0)$, are <u>exceptional</u> in that the homomorphisms $D^{\frac{1}{2},\frac{1}{2}}: a \mapsto a \otimes \bar{a}$, and $D^1: a \mapsto a \vee a$, give rise to group <u>isomorphisms</u>

$$\mathcal{L}^{\uparrow}(C_2)/Z_2 \cong \mathcal{L}^{\uparrow}, \text{ and } \mathcal{L}^{\uparrow}(C_2)/Z_2 \cong ALIO_+(C_3). \qquad (2.7)$$

2.4 <u>Product bases</u>. Each <u>symplectic basis</u> $\{\xi,\eta\}$ for C_2, satisfying that is $[\xi,\eta] = 1$, gives rise to an associated product basis for $V^{j,j'}$. In particular the associated <u>standard basis</u> for V^j is $\{e^j_m\,;\ m = -j,\ldots,+j\}$, where e^j_m is defined by

$$e^j_m = \sqrt{\binom{2j}{m}}\ \xi^{j+m}\eta^{j-m}. \qquad (2.8)$$

(Here and below we use the abbreviation $\xi^{j+m}\eta^{j-m}$ to denote the symmetrized product $\xi \vee \xi \vee \cdots \vee \xi \vee \eta \vee \cdots \vee \eta$ of $j + m$ factors ξ and $j - m$ factors η.) The associated metric tensor in the space V^j is the <u>standard 1j-symbol</u>[11]:

$$\begin{pmatrix} j \\ m\ m' \end{pmatrix} \equiv [e^j_m, e^j_{m'}] = (-)^{j-m}\delta_{m,-m'}. \qquad (2.9)$$

(The relevant permanent — see Eq (2.5) — contains a $(j+m)\times(j+m)$ block of +1's and a $(j-m)\times(j-m)$ block of −1's).

The associated product basis $\{\xi\otimes\bar\xi,\ \eta\otimes\bar\eta,\ \xi\otimes\bar\eta,\ \eta\otimes\bar\xi\}$ in the space $M^{\mathbb{C}}$ is a null tetrad basis, from which we construct an associated (real) orthonormal basis[10] $\{x,y,z,t\}$ for M:

$$\sqrt{2}x = \xi\bar\eta + \eta\bar\xi,\quad \sqrt{2}y = -i(\xi\bar\eta - \eta\bar\xi),\quad \sqrt{2}z = \xi\bar\xi - \eta\bar\eta,\quad \sqrt{2}t = \xi\bar\xi + \eta\bar\eta, \tag{2.10}$$

whose metric tensor is diag(− − − +), in confirmation of the signature of M.

It is not difficult to find a set of canonical forms for $\mathcal{L}_-^{\uparrow}(C_2)$, as well as for $\mathcal{L}_+^{\uparrow}(C_2)$. One can then use the group isomorphisms[12] (2.7) to deduce a set of canonical forms for $\mathcal{L}_-^{\uparrow}$ and $ALO_+(C_3)$, as well as for $\mathcal{L}_+^{\uparrow}$ and $O_+(C_3)$. We content ourselves here with just two simple instances. Given the basis $\{\xi,\eta\}$, the simplest $\mathcal{L}_-^{\uparrow}(C_2)$-transformation is the conjugation[8] defined by $\xi\mapsto\xi,\ \eta\mapsto\eta$. It follows from Eq. (2.10) that at the M-level this transformation is <u>space reversal</u> with respect to the y-axis: $x\longmapsto x,\ y\mapsto -y,\ z\longmapsto z,\ t\longmapsto t$. At the V^j-level, this space reversal is the antilinear[8] map given simply by $e^j_m \longrightarrow e^j_m$, for each $m = -j, \ldots, j$.

Another simple $\mathcal{L}_-^{\uparrow}(C_2)$-transformation is that (antilinear[8]) map defined by its effect[15] $\xi\mapsto -\eta,\ \eta\mapsto\xi$ on the basis $\{\xi,\eta\}$. At the M-level it is <u>space inversion</u> $\{x,y,z,t\}\mapsto\{-x,-y,-z,t\}$. Noting that the basis $\{-\eta,\xi\}$ is left dual to the basis $\{\xi,\eta\}$, observe that at the V^j-level space inversion $\mathcal{P}$ is the antilinear[8] map given by[15] $\mathcal{P}e^j_m = e^m_j$, where $\{e^m_j\}$ is <u>left dual</u> to $\{e^j_m\}$ — i.e. e^m_j is defined by

$$\left[e^m_j,\ e^j_{m'}\right] = \delta^m_{m'}, \tag{2.11}$$

or equivalently by replacing $\{\xi,\eta\}$ by $\{-\eta,\xi\}$ in the definition of e^j_m in Eq. (2.8). We can use $\mathcal{P}$ to define a hermitian form $(\ ,\ )_{\mathcal{P}}$ on V^j by

$$(\phi_1,\phi_2)_{\mathcal{P}} = [\mathcal{P}\phi_1,\phi_2],\ \phi_i\in V^j, \tag{2.12}$$

w.r.t. which e^m_j is an orthonormal basis (in the <u>strict</u> sense). Thus each choice of space inversion and hence of frame (time-axis), in Minkowski space M results in a choice of <u>positive definite</u> unitary geometry for V^j via the inner product $(\ ,\ )_{\mathcal{P}}$.

2.5 <u>Classical bases and components</u>. While standard bases for V^j possess simple normalization properties, for many purposes — as was realized a century ago (bearing in mind Theorem 3.2 below) — it is better to avoid irrationalities and use instead <u>classical bases</u> of the type $\{E^j_\lambda;\ \lambda = 0,1,\ldots,2j\}$ where

$$E^j_\lambda = \binom{2j}{\lambda}\xi^{2j-\lambda}\eta^\lambda,\quad \lambda = 0,1,\ldots,2j. \tag{2.13}$$

The components of a general element $\phi\in V^j$ relative to the two types of basis will be denoted (ϕ^m) and (Φ^λ):

$$\phi = \sum_{m=-j}^{j}\phi^m e^j_m = \sum_{\lambda=0}^{2j}\Phi^\lambda E^j_\lambda. \tag{2.14}$$

The relation between the two sets of bases and components is thus

$$E^j_\lambda = \sqrt{\binom{2j}{\lambda}}\,e^j_{m'},\quad \phi^m = \sqrt{\binom{2j}{\lambda}},\quad \text{where } \lambda = j-m. \tag{2.15}$$

3. <u>The space $V^j = \vee^{2j}C_2$ of (j,0) - spinors</u>.

3.1 <u>The spinor light cone N^j</u>. An element $\phi\in V^j$ which is of the highly special form $\phi = \xi^{2j}$, for some non-zero $\xi\in C_2$, will be termed a <u>nil spinor</u>[16]. The <u>spinor light cone</u>[17,18] N^j of V^j is defined to consist of all the nil spinors of V^j. (These definitions can be generalized[16] in an obvious fashion to (j_1,j_2)-spinors.)

Clearly the image $T = D^j(a)$ of a "Lorentz transformation" $a \in \mathcal{L}^{\uparrow}(C_2)$ has the property of preserving the cone N^j, since $T\xi^{2j} = \eta^{2j}$, where $\eta = a\xi$. Conversely, if $T \in GALL(V^j)$ preserves N^j then Theorem 3.6 below implies that T is a scalar multiple of $D^j(a)$ for some $a \in \mathcal{L}^{\uparrow}(C_2)$. Consequently the crucial structural carried by the space V^j is the spinor light cone N^j, and not (at least when $j > 1$ ←see footnote 16) the metrical structure $[\ ,\ ]$.

3.2 Theorem Each choice of basis $\{\xi,\eta\}$ for C_2 gives rise to an isomorphism of V^j with the space of binary 2j-ics (i.e. the space of polynomials over $\mathbb{C}$ of homogeneous degree 2j in two indeterminates ξ,η).

3.3 Binomial theorem : $(\xi + z\eta)^{2j} = \sum_{\lambda=0}^{2j} z^{\lambda} E^j_{\lambda}$, $(z \in \mathbb{C},\ \xi,\eta \in C_2)$.

3.4 "Penrose's[19] theorem" (=[20] Fundamental theorem of algebra). Every element $\phi \in V^j$ is decomposable; that is there exist $\xi_1, \xi_2, \dots, \xi_{2j} \in C_2$ such that

$$\phi = \xi_1 \xi_2 \cdots \xi_{2j} \ (\equiv \xi_1 \vee \xi_2 \vee \cdots \vee \xi_{2j}).$$

Moreover, if $\phi \neq 0$, the factors $\xi_1, \dots, \xi_{2j}$ are subject to permutations and to rescalings of the type $\xi_i \mapsto \lambda_i \xi_i$, with $\lambda_1 \lambda_2 \cdots \lambda_{2j} = 1$, but are otherwise uniquely determined by ϕ.

3.5 Theorem. The spinor light cone N^j spans V^j.

3.6 Theorem. If $T \in GALL(V^j)$, then T preserves the spinor light cone N^j if and only if $T = D^j(a)$ for some $a \in GALL(C_2)$.

3.7 Remarks (a) Of the above five theorems, the odd one out is Theorem 3.4 is that it is peculiar to the dimension 2 of the base space C_2. The other four theorems generalize to dimension $n > 2$ (n-ary quantics, multinomial theorem, etc.); Theorem 3.2 can be paraphrased in the statement "symmetric algebra = coordinate-free polynomial algebra".

(b) Theorems 3.5 and 3.6 readily generalize to the case of (j_1, j_2)-spinors.

(c) On account of Theorem 3.5, a multilinear mapping $M : V^{j_1} \times V^{j_2} \times \dots \longrightarrow W$ is determined by its values $M(\xi_1^{2j_2}, \xi_2^{2j_2}, \dots)$ on nil spinors. The values $M(\phi_1, \phi_2, \dots)$ on general spinors can then be reconstituted by means of polarization, upon using Theorem 3.3. Of course, these are the methods familiar from Classical Invariant Theory (see Ch. 8A of Ref. 21). Before applying such methods to Wigner's 3j-symbols, let us give a very simple illustration of them.

3.8 Illustration: the bilinear invariant $[\ ,\ ] : V^j \times V^j \longrightarrow \mathbb{C}$.

A bilinear map $[\ ,\ ] : V^j \times V^j \to \mathbb{C}$ is determined by its values upon the nil spinors. The set of values defined by

$$[\xi_1^{2j}, \xi_2^{2j}] = [\xi_1, \xi_2]^{2j}, \quad \xi_i \in C_2, \tag{3.1}$$

is a possible one; for the degrees on either side tally, thereby guaranteeing the existence of the requisite polarized version of our specialized starting point. (We are supposing, for the sake of this illustration, that we do not already know this completely polarized version — namely that given in Eq. (2.5))

Upon choosing a symplectic basis $\{\xi,\eta\}$ for C_2, and writing $\xi_i = \xi + z_i\eta$, $(z_i \in \mathbb{C})$, so that $[\xi_1, \xi_2] = z_2 - z_1$, use of the binomial theorem in Eq. (3.1) yields the value of the classical 1j-symbol

$$\begin{bmatrix} & j & \\ \lambda_1 & & \lambda_2 \end{bmatrix} \equiv [E^j_{\lambda_1}, E^j_{\lambda_2}] \tag{3.2}$$

to be the coefficient of $z_1^{\lambda_1} z_2^{\lambda_2}$ in $(z_2 - z_1)^{2j}$. Thus

$$\begin{bmatrix} & j & \\ \lambda_1 & & \lambda_2 \end{bmatrix} = (-)^{\lambda_1} \binom{2j}{\lambda_1} \delta_{\lambda_1+\lambda_2, 2j} \ . \tag{3.3}$$

We thus obtain the well-known bilinear invariant

$$[\phi,\psi] = \sum_{\lambda=0}^{2j} (-)^{\lambda} \binom{2j}{\lambda} \bar{\phi}^{\lambda} \bar{\psi}^{2j-\lambda} \tag{3.4}$$

of two binary 2j-ics. (The invariance property (2.6) follows from Eq. (3.1) by virtue of the corresponding $\mathcal{L}^{\uparrow}(C_2)$-invariance property (2.2).)

In particular when j = 2, a general element $\phi = \sum \bar{\phi}^{\lambda} E^2_{\lambda}$ of V^2 is identified (under the isomorphism of Theorem 3.2) with the binary quartic

$$\phi = \bar{\phi}^0 \xi^4 + 4\bar{\phi}^1 \xi^3 \eta + 6\bar{\phi}^2 \xi^2 \eta^2 + 4\bar{\phi}^3 \xi \eta^3 + \bar{\phi}^4 \eta^4, \tag{3.5}$$

and we obtain the familiar quadratic invariant $\mathcal{J}$ of the binary quartic:

$$\mathcal{J} = \tfrac{1}{2}[\phi,\phi] = \bar{\phi}^0 \bar{\phi}^4 - 4\bar{\phi}^1 \bar{\phi}^3 + 3(\bar{\phi}^2)^2. \tag{3.6}$$

4. Trilinear invariants and the Wigner 3j-symbols

4.1 Trilinear invariant $[\ ,\ ,\ ]: V^{j_1} \times V^{j_2} \times V^{j_3} \to \mathbb{C}$.

The definition of this upon nil spinors by

$$\left[\xi_1^{2j_1}, \xi_2^{2j_2}, \xi_3^{2j_3}\right] = K[\xi_2,\xi_3]^{k_1} [\xi_3,\xi_1]^{k_2} [\xi_1,\xi_2]^{k_3}, \tag{4.1}$$

where K is a normalization constant, will succeed provided only that the "degrees tally": more precisely, non-negative integers[20] k_1, k_2, k_3 must exist such that

$$2j_1 = k_2 + k_3,\ 2j_2 = k_3 + k_1,\ 2j_3 = k_1 + k_2. \tag{4.2}$$

These equations can be solved, the solution being given uniquely by:

$$\begin{aligned} k_1 &= j_2 + j_3 - j_1, & k_2 &= j_3 + j_1 - j_2, & k_3 &= j_1 + j_2 - j_3, \\ &= J - 2j_1 & &= J - 2j_2 & &= J - 2j_3 \end{aligned} \tag{4.3}$$

where $J = j_1 + j_2 + j_3 = k_1 + k_2 + k_3$, provided only that j_1, j_2, j_3 from a triangle of integer perimeter:

$$j_2 + j_3 \leqslant j_1,\ j_3 + j_1 \leqslant j_2,\ j_1 + j_2 \leqslant j_3;\ J = \text{integer}. \tag{4.4}$$

Except for the arbitrariness in the choice of K, no other trilinear invariant exists (see Theorem 2.6A , or 6.1A, of reference 21 — or see Eq. (5.6) below) For a reason given later, we choose $K = K(j_1, j_2, j_3)$ to be

$$K(j_1,j_2,j_3) = \left\{[2j!]/[k!]\ (J+1)!\right\}^{\frac{1}{2}}, \tag{4.5}$$

where we have used the abbreviation $[p!] \equiv p_1! p_2! p_3!$.

4.2 Properties.

Eqs (2.5a), (4.1), (2.2) and immediately yield the invariance property

$$\left[D^{j_1}(a)\phi_1, D^{j_2}(a)\phi_2, D^{j_3}(a)\phi_3\right] = [\phi_1,\phi_2,\phi_3]^a, \quad a \in \mathcal{L}^{\uparrow}(C_2), \tag{4.6}$$

and also the permutational symmetry property

$$\left[\phi_{\sigma(1)}, \phi_{\sigma(2)}, \phi_{\sigma(3)}\right] = (\mathrm{sgn}\,\sigma)^J [\phi_1,\phi_2,\phi_3], \quad \phi_i \in V^{j_i}. \tag{4.7}$$

Since $K(0,j,j) = (2j+1)^{-\frac{1}{2}}$, we also have

$$[1,\phi,\psi] = [\phi,\psi]/(2j+1)^{\frac{1}{2}}, \quad \phi,\psi \in V^j. \tag{4.8}$$

4.3 Standard and classical 3j-symbols.

These are defined respectively by

$$\begin{pmatrix} j_1 & j_2 & j_3 \\ m_1 & m_2 & m_3 \end{pmatrix} = \left[e^{j_1}_{m_1}, e^{j_2}_{m_2}, e^{j_3}_{m_3}\right] \quad \text{and} \quad \begin{bmatrix} j_1 & j_2 & j_3 \\ \lambda_1 & \lambda_2 & \lambda_3 \end{bmatrix} = K^{-1}\left[E^{j_1}_{\lambda_1}, E^{j_2}_{\lambda_2}, E^{j_3}_{\lambda_3}\right]. \tag{4.9(a), (b)}$$

By Eqs (2.15), (4.5) they determine each other by means of

$$\begin{bmatrix} j_1 & j_2 & j_3 \\ \lambda_1 & \lambda_2 & \lambda_3 \end{bmatrix} = \left\{\frac{[k!]\,(J+1)!}{[\kappa!][\lambda!]}\right\}^{\frac{1}{2}} \begin{pmatrix} j_1 & j_2 & j_3 \\ m_1 & m_2 & m_3 \end{pmatrix}, \tag{4.10}$$

where $\lambda_i = j_i - m_i, \kappa_i = j_i + m_i$.

4.4 Properties of 3j-symbols. Various well known properties of the 3j-symbols follow immediately from the general invariance property (4.6). In particular, upon setting $\phi_i = e^{j_i}_{m_i}$ and choosing a $\in \hat{\mathcal{L}}(C_2)$ to be a suitable (a) screw (b) π-rotation (c) space reversal[21a] (d) space inversion[21a] – namely such that $D^j(a)$ maps e^j_m on to (a) $\alpha^{2m} e^j_m$ (b) $i^{2j} e^j_{-m}$ (c) e^j_m (d) e^m_j – we obtain the results:

(a) $\begin{pmatrix} j_1 & j_2 & j_3 \\ m_1 & m_2 & m_3 \end{pmatrix} = 0$ if $\sum m_i \neq 0$;

(b) $\begin{pmatrix} j_1 & j_2 & j_3 \\ m_1 & m_2 & m_3 \end{pmatrix} = (-)^J \begin{pmatrix} j_1 & j_2 & j_3 \\ -m_1 & -m_2 & -m_3 \end{pmatrix}$;

(c) the 3j-symbols are real;

(d) the fully covariant and fully contravariant[22] forms of the 3j-symbol are equal:

$$\begin{pmatrix} j_1 & j_2 & j_3 \\ m_1 & m_2 & m_3 \end{pmatrix} = \begin{pmatrix} m_1 & m_2 & m_3 \\ j_1 & j_2 & j_3 \end{pmatrix} \left(= \left[e^{m_1}_{j_1}, e^{m_2}_{j_2}, e^{m_3}_{j_3} \right] \right).$$

Similarly the permutational properties of the 3j-symbols follow from Eq. (4.7), while (4.8) yields[11]

$$\begin{pmatrix} 0 & j & j \\ 0 & m & n \end{pmatrix} = (2j+1)^{-\frac{1}{2}} \begin{pmatrix} j \\ m\,n \end{pmatrix}.$$

Orthogonality properties are considered in Section 5.

4.5 Computation Setting $\xi_i = \xi + z_i\eta$ in Eq. (4.1) and using the binomial theorem (as in Section 3.8), immediately yields the value of the classical 3j-symbol to be the coefficient of $z_1^{\lambda_1} z_2^{\lambda_2} z_3^{\lambda_3}$ in $(z_3 - z_2)^{k_1} (z_1 - z_3)^{k_2} (z_2 - z_1)^{k_3}$. Thus the classical 3j-symbol takes integer values:

$$\begin{bmatrix} j_1 & j_2 & j_3 \\ \lambda_1 & \lambda_2 & \lambda_3 \end{bmatrix} = \sum (-)^{q_1+q_2+q_3} \binom{k_1}{q_1}\binom{k_2}{q_2}\binom{k_3}{q_3}. \qquad (4.11)$$

Using Eq. (4.10) we deduce "Racah's formula[23]":

$$\begin{pmatrix} j_1 & j_2 & j_3 \\ m_1 & m_2 & m_3 \end{pmatrix} = \left\{ \frac{[k!]\,[\kappa!]\,[\lambda!]}{(J+1)!} \right\}^{\frac{1}{2}} \sum_{p,q} \frac{(-)^{q_1+q_2+q_3}}{[p!][q!]}, \qquad (4.12)$$

where the summation is over all non-negative integers p_i, q_i satisfying[24]

$$\begin{pmatrix} p_1 + q_1 & p_2 + q_2 & p_3 + q_3 \\ p_2 + q_3 & p_3 + q_1 & p_1 + q_2 \\ p_3 + q_2 & p_1 + q_3 & p_2 + q_1 \end{pmatrix} = \begin{pmatrix} k_1 & k_2 & k_3 \\ \lambda_1 & \lambda_2 & \lambda_3 \\ \kappa_1 & \kappa_2 & \kappa_3 \end{pmatrix}. \qquad (4.13)$$

(The summation in fact reduces to a single summation, the number of terms being $\mathrm{Min}(k_i, \lambda_i, \kappa_i) + 1$)

4.6 Illustration Consider the particular case $j_1 = j_2 = j_3 = 2$; upon adopting the abbreviation $[\lambda_1\lambda_2\lambda_3]$ for the corresponding classical 3j-symbol, we have from Section 4.5:

$$[\lambda_1\lambda_2\lambda_3] = \text{coef of } z_1^{\lambda_1} z_2^{\lambda_2} z_3^{\lambda_3} \text{ in } (z_2-z_3)^2 (z_3-z_1)^2 (z_1-z_2)^2. \qquad (4.14)$$

Thus $[024] = 1$, $[123] = 2$, $[033] = [411] = -2$, $[222] = -6$, and of course $[\lambda_1\lambda_2\lambda_3] = 0$ if $\sum\lambda_i \neq 6$. Thus the expression of the trilinear invariant $[\phi_1, \phi_2, \phi_3]$, $\phi_i \in V^2$, in terms of classical components involves (very low-lying) integers, and should be contrasted with the irrationalities of the standard expression. In particular, on setting $\phi_1 = \phi_2 = \phi_3 = \phi$, we derive the well-known expression[25] for the cubinvariant $\mathcal{J}$ of the binary quartic $\phi \in V^2$ of Eq. (3.5):

$$\mathcal{J} = [\phi,\phi,\phi]/6K = \frac{1}{6}\sum\begin{bmatrix}2 & 2 & 2\\ \lambda_1 & \lambda_2 & \lambda_3\end{bmatrix}\Phi^{\lambda_1}\Phi^{\lambda_2}\Phi^{\lambda_3}$$

$$= \Phi^0\Phi^2\Phi^4 + 2\Phi^1\Phi^2\Phi^3 - \Phi^0(\Phi^3)^2 - (\Phi^1)^2\Phi^4 - (\Phi^2)^3. \tag{4.15}$$

In terms of standard 3j-symbols and components:[26,27]

$$\mathcal{J} = \frac{(105)^{\frac{1}{2}}}{36}\sum\begin{pmatrix}2 & 2 & 2\\ m_1 & m_2 & m_3\end{pmatrix}\phi^{m_1}\phi^{m_2}\phi^{m_3} . \tag{4.16}$$

4.7 Recursion relations. To derive recursion relations for fixed values of j_1, j_2, j_3, use the infinitesimal version of Eq. (4.6). For varying j_1, j_2, j_3, adopt factorizations of the kind $[\xi_2,\xi_3]^{k_1} = [\xi_2,\xi_3]^{k_1'}[\xi_2,\xi_3]^{k_1''}$ in Eq. (4.1), set $\xi_i = \xi + z_i\eta$ and use the binomial theorem again to obtain the relations[32]

$$\begin{bmatrix}j_1 & j_2 & j_3\\ \lambda_1 & \lambda_2 & \lambda_3\end{bmatrix} = \sum_{\lambda_i'+\lambda_i''=\lambda_i}\begin{bmatrix}j_1' & j_2' & j_3'\\ \lambda_1' & \lambda_2' & \lambda_3'\end{bmatrix}\begin{bmatrix}j_1'' & j_2'' & j_3''\\ \lambda_1'' & \lambda_2'' & \lambda_3''\end{bmatrix} , \tag{4.17}$$

where $j_i' + j_i'' = j_i$. These are known in one or two special cases, at any rate when expressed in terms of standard symbols. For example, on setting $j_1' = 0$, $j_2' = j_3' = \frac{1}{2}$ we obtain

$$\begin{bmatrix}j_1 & j_2 & j_3\\ \lambda_1 & \lambda_2 & \lambda_3\end{bmatrix} = \begin{bmatrix}j_1 & j_2-\frac{1}{2} & j_3-\frac{1}{2}\\ \lambda_1 & \lambda_2 & \lambda_3-1\end{bmatrix} - \begin{bmatrix}j_1 & j_2-\frac{1}{2} & j_3-\frac{1}{2}\\ \lambda_1 & \lambda_2-1 & \lambda_3\end{bmatrix} \tag{4.18}$$

which in terms of standard symbols reads

$$\begin{pmatrix}j_1 & j_2 & j_3\\ m_1 & m_2 & m_3\end{pmatrix} = \left\{\frac{(j_2+m_2)(j_3-m_3)}{(J+1)(J-2j_1)}\right\}^{\frac{1}{2}}\begin{pmatrix}j_1 & j_2-\frac{1}{2} & j_3-\frac{1}{2}\\ m_1 & m_2-\frac{1}{2} & m_3+\frac{1}{2}\end{pmatrix} - \left\{\frac{(j_1-m_2)(j_3+m_3)}{(J+1)(J-2j_1)}\right\}^{\frac{1}{2}}\begin{pmatrix}j_1 & j_2-\frac{1}{2} & j_3-\frac{1}{2}\\ m_1 & m_2+\frac{1}{2} & m_3-\frac{1}{2}\end{pmatrix} \tag{4.19}$$

in agreement with Eq. 3.7.12 of reference 28. Perhaps[33] the general relations (4.17) are new?

5. The maps $\left\{\begin{matrix} & j_3 & \\ j_1 & & j_2\end{matrix}\right\}$ and Clebsch-Gordan coefficients.

We content ourselves here with a brief sketch, and will omit certain details[29].

First of all an invariant element[30] $h \in V^{j_1} \otimes V^{j_2} \otimes V^{j_3}$ is defined by

$$[\phi_1,\phi_2,\phi_3] = [h, \phi_1 \otimes \phi_2 \otimes \phi_3], \tag{5.1}$$

and so depends upon the choice of constant K in Eq. (4.1). If K is chosen to be real, then h is invariant under the natural action of $\mathcal{L}_-^{\uparrow}$ as well as $\mathcal{L}_+^{\uparrow}$; consequently $[h, h]$ is real and positive: $[h,h] = [\mathcal{P}h,h] = (h,h)_{\mathcal{D}} > 0$, since $(\ ,\ )_{\mathcal{D}}$ is positive definite. We may therefore fix K — and hence $[\ ,\ ,\ ]$ and h — by demanding $K > 0$ and $[h,h] = 1$. The actual value of K then turns out (after an apparently unavoidable computation) to be that given previously in Eq. (4.5).

Next a linear map $f : V^{j_3} \to V^{j_1} \otimes V^{j_2}$ is defined by

$$A[\phi_1 \otimes \phi_2 , f\phi_3] = [\phi_1, \phi_2, \phi_3] , \tag{5.2}$$

where $A = A(j_1,j_2,j_3)$ is a normalization constant. The invariance properties of $[\ ,\ ]$ and $[\ ,\ ,\ ]$ yield the intertwining property

$$\left[D^{j_1}(a) \otimes D^{j_2}(a)\right] \circ f = f \circ D^{j_3}(a), \quad a \in \mathcal{L}^{\uparrow}(C_2) . \tag{5.3}$$

Schur's lemma now tells us that f is injective. Let us denote Im f by $V^{j_3}_{j_1 j_2} \subset V^{j_1} \otimes V^{j_2}$; by restricting the target space of f we thus obtain a linear isomorphism

$$f_0 \equiv \left\{\begin{matrix} & j_3 & \\ j_1 & & j_2\end{matrix}\right\} : V^{j_3} \to V^{j_3}_{j_1 j_2} . \tag{5.4}$$

One next shows[29] that f_o is necessarily a scalar multiple of an isometry; consequently we will fix f_o, or equivalently the constant A, up to a sign by demanding that f_o equal an isometry. To compute $\pm$ A, note that upon identifying $L(V^{j_3}, V^{j_1}\otimes V^{j_2})$ with $V^{j_1}\otimes V^{j_2}\otimes V^{j_3}$ in the obvious way, we have $Af = (-)^{2j_3}h$. Hence

$$A^2 = [h,h]/[\tilde{f},f] = 1/\mathrm{tr}(\tilde{f}f) = (2j_3+1)^{-1},$$

(since $\tilde{f}f$ = identity operator $V^{j_3}\to V^{j_3}$). Hence

$$A = \varepsilon\,(2j_3+1)^{-\frac{1}{2}}, \text{ where } \varepsilon = \varepsilon(j_1,j_2,j_3) \text{ is a sign ambiguity.} \tag{5.5}$$

Next (using the inequivalence of D^j with $D^{j'}$ for $j \neq j'$) one shows[29] that the (non-singular) subspaces $V^{j_3}_{j_1 j_2}$ of $V^{j_1}\otimes V^{j_2}$ are mutually orthogonal, and one arrives at the decomposition

$$V^{j_1}\otimes V^{j_2} = \perp_{j_3}\, \delta(j_1,j_2,j_3)\; V^{j_3}_{j_1 j_2}, \tag{5.6}$$

after checking that the dimensions tally. Here $\delta(j_1,j_2,j_3)$ is defined to equal 1 if the triangle conditions (4.4) are satisfied, and to equal 0 otherwise. (Of course the fact the multiplicity δ is 0 or 1 confirms our previous assertions concerning the existence and uniqueness of a trilinear invariant $V^{j_1}\times V^{j_2}\times V^{j_3}\to \mathbb{C}$.) Upon introducing the map

$$\left\{{}^{\ *}_{j_1\ j_2}\right\} = \oplus_{j_3}\left\{{}^{\ j_3}_{j_1\ j_2}\right\}, \tag{5.7}$$

we obtain the intertwining property

$$\left[D^{j_1}(a)\otimes D^{j_2}(a)\right]\circ\left\{{}^{\ *}_{j_1\ j_2}\right\} = \left\{{}^{\ *}_{j_1\ j_2}\right\}\circ\left[\oplus_{j_3} D^{j_3}(a)\right], \tag{5.8}$$

it being understood that the summation over j_3 is restricted by the condition $\delta(j_1,j_2,j_3) = 1$.

Finally, Clebsch-Gordan coefficients $\langle j_1\, j_2\, m_1\, m_2 | \, j_3\, m_3\rangle$ and $\langle j_3\, m_3 | \, j_1\, j_2\, m_1\, m_2\rangle$ are defined to be the matrix elements of the linear isomorphism $\left\{{}^{\ *}_{j_1 j_2}\right\}$ and its inverse $\left\{{}^{j_1 j_2}_{\ *}\right\}$ with respect to standard bases in the relevant spaces. Thus

$$f^{j_3}_{m_3} = \sum_{m_1 m_2} e^{j_1 j_2}_{m_1 m_2}\langle j_1 j_2 m_1 m_2 | j_3 m_3\rangle,\quad e^{j_1 j_2}_{m_1 m_2} = \sum_{j_3 m_3} f^{j_3}_{m_3}\langle j_3 m_3 | j_1 j_2 m_1 m_2\rangle, \tag{5.9}$$

where $e^{j_1 j_2}_{m_1 m_2} \equiv e^{j_1}_{m_1}\otimes e^{j_2}_{m_2}$ and $f^{j_3}_{m_3} \equiv \left\{{}^{\ j_3}_{j_1 j_2}\right\} e^{j_3}_{m_3}$.

Eqivalently the CG-coefficients are given by[31]

$$\langle j_1 j_2 m_1 m_2 | j_3 m_3\rangle = \left[e^{m_1 m_2}_{j_1 j_2},\, f^{j_3}_{m_3}\right],\quad \langle j_3 m_3 | j_1 j_2 m_1 m_2\rangle = \left[f^{m_3}_{j_3},\, e^{j_1 j_2}_{m_1 m_2}\right]. \tag{5.10}$$

Setting $\phi_1 = e^{m_1}_{j_1}$, $\phi_2 = e^{m_2}_{j_2}$, $\phi_3 = e^{j_3}_{m_3}$ in Eq. (5.2) we obtain[22]

$$\frac{\varepsilon(j_1 j_2 j_3)\,\langle j_1 j_2 m_1 m_2 | j_3 m_3\rangle}{(2j_3+1)^{\frac{1}{2}}} = (-)^{2j_3}\begin{pmatrix} m_1 & m_2 & j_3\\ j_1 & j_2 & m_3\end{pmatrix} = (-)^{j_3-m_3}\begin{pmatrix} j_1 & j_2 & j_3\\ m_1 & m_2 & -m_3\end{pmatrix}. \tag{5.11}$$

Hence we may deduce properties of CG-coefficients from those in Section 4.4 of 3j-symbols. In particular the CG-coefficients are real; consequently upon using the (antilinear) space inversion map $e^j_m \mapsto e^m_j$ and $e^m_j \mapsto (-)^{2j} e^j_m$ in Eq. 5.10, we obtain $\langle j_1 j_2 m_1 m_2 | j_3 m_3\rangle = \langle j_3 m_3 | j_1 j_2 m_1 m_2\rangle$. In the case of orthogonality relations, however, it seems best to deduce those for the 3j-symbols from those for the CG-coefficients; for the latter are simply the expression of the fact that the matrix of the map $\left\{{}^{\ *}_{j_1 j_2}\right\}$ is the inverse of the matrix of the map $\left\{{}^{j_1 j_2}_{\ *}\right\} = \left\{{}^{\ *}_{j_1 j_2}\right\}^{-1}$.

The standard convention for fixing the sign $\varepsilon(j_1,j_2,j_3)$ is to demand that $\langle j_1 j_2 j_1 (j_3-j_1) | j_3 j_3\rangle$ be positive, which leads to

$$\varepsilon(j_1,j_2,j_3) = (-)^{k_2} = (-)^{j_3+j_1-j_2}. \tag{5.12}$$

Footnotes and References.

1. For some standard texts, see the bibliography to ref.2.
2. L.C. Biedenharn and H. van Dam (editors), The Quantum Theory of Angular Momentum, Academic Press 1965

3,4. See the articles by Schwinger and by Bargmann reprinted in ref.2.

5. The linear isomorphism $C_2 \otimes \bar{C}_2 \to AL(C_2, C_2)$ is given by $\xi \otimes \bar{\eta} \mapsto \xi\bar{\eta}$, where the latter denotes the antilinear dyad with effect $\zeta \longmapsto \xi\overline{[\eta,\zeta]}$. The isomorphism $R^{\frac{1}{2},\frac{1}{2}} \to ALSk(C_2, C_2)$ follows upon noting that the adjoint of the dyad $\xi\bar{\eta}$ is $-\eta\bar{\xi}$.
6. See Theorem 3.2 .
7. The linear isomorphisms $\vee^4 C_2 \to (C_3 \vee C_3)_0$ and $R^{1,1} \to (M \vee M)_0$ can be defined – see Theorem 3.5, footnote 16, and Remarks 3.7(b), (c) — by laying down that their effects upon nil spinors are respectively $\xi^4 \mapsto \xi^2 \otimes \xi^2$ and $\xi^2 \otimes \bar{\xi}^2 \longmapsto \xi\bar{\xi} \otimes \xi\bar{\xi}$.
8. Even though D^j represents space inversion antilinearly, the group $\mathcal{L}(C_2)$ is still useful in the construction of manifestly covariant corepresentations of the extended Poincare group P appropriate to the physically relevant UA-decomposition $P^{\uparrow} \cup P^{\downarrow}$; see Section 4.3 of Ref.9.
9. R. Shaw and J. Lever, Commun. Math. Phys. 38, 279 (1974).
10. The action $\Lambda(a) = a \otimes \bar{a}$ upon $R^{\frac{1}{2},\frac{1}{2}}$ is, for $a \in \mathcal{L}^{\uparrow}(C_2)$, isomorphic to the action $p \mapsto a \circ p \circ a^{-1}$ upon $ALSK(C_2, C_2)$ used in Ref.9; (however for $a \in \mathcal{L}^{\downarrow}(C_2)^9$, an extra minus sign has to be introduced in the first action if it is to correspond to the second). The scalar product $[\xi\bar{\xi}, \eta\bar{\eta}] = [\xi,\eta]\overline{[\xi,\eta]}$ upon $R^{\frac{1}{2},\frac{1}{2}}$ corresponds to the scalar product $[p,q] = -\,tr(p \circ q)$ upon $ALSk(C_2, C_2)$.
11. Our 1j-symbol is the transpose of that employed by Wigner in Ref.2.
12. The vector space isomorphisms (2.4) are also useful for solving canonical form problems. For example one can find canonical forms for an object $T \in ALS(C_3, C_3)$ i.e. for an antilinear map $T : C_3 \to C_3$ which is self-adjoint : $[T\phi, \psi] = \overline{[\phi, T\psi]}$, $\phi, \psi \in C_3$. (One way to do this is to use the "anti-Jordan" canonical form[13] for general antilinear operators.) Use of the isomorphism $ALS(C_3, C_3) \cong (M \vee M)_0$ then enables one to deduce a set of canonical forms for a trace-free Ricci tensor T, as given for example in Section 2 of Ref.14.

    Incidentally, since the square of an antilinear operator is a linear operator, $T \in ALS(C_3, C_3)$ implies $W \equiv T^2 \in S(C_3, C_3)$. Upon extracting the trace, we obtain the "Weyl square" $W_0 \in S_0(C_3, C_3)$ of the Ricci tensor T. The antilinear algebra way of introducing the Weyl square was in fact how the author first encountered it; for a possible use, see Section 5 of Ref.14.
13. R. Shaw (unpublished, 1969).
14. C. D. Collinson and R. Shaw, Intern. J. Theor. Phys., 6, 347 (1972).
15. Using the isomorphism $M \cong ALSk(C_2, C_2)$ again, $\mathcal{P}$ at the C_2-level is given by $\mathcal{P} = \sqrt{2}t\,(= \xi\bar{\xi} + \eta\bar{\eta}$, as in Eq. (2.10)), and so at the V^j-level it is $D^j(\sqrt{2}\,t)$, thus exhibiting clearly the dependence of $\mathcal{P}$ upon a particular time-axis.
16. A spinor $\phi \in V^{j_1,j_2}$ is said to be nil if it is of the form $\phi = \pm\xi^{2j_1} \otimes \bar{\xi}^{2j_2}$ for some non-zero $\xi \in C_2$ (the minus sign being needed only in the case $j_1 = j_2$). The set of nil spinors of V^{j_1,j_2} forms the spinor light cone N^{j_1,j_2}.

    The term "nil" is used rather than "null", so as to reserve the latter to refer (as in "null tetrad basis") to a non-zero spinor of zero length: $[\psi,\psi]=0$. Clearly every nil spinor is null. The cases when (j_1,j_2) equals $(\frac{1}{2},0)$, $(0,\frac{1}{2})$, $(1,0)$, $(0,1)$ or $(\frac{1}{2},\frac{1}{2})$ are exceptional in that every null spinor is nil in these cases (provided in the $(\frac{1}{2},\frac{1}{2})$ case we restrict our attention to the real space $M = R^{\frac{1}{2},\frac{1}{2}}$). In all the other cases there exist null spinors — for example the basic $(j,0)$-spinors c^j_m, with $m \neq 0$, $m \neq \pm j$ — which are null but not nil. Consequently in these other cases the images $D^{j_1,j_2}(a)$ of Lorentz transformations, which clearly preserve the cone N^{j_1,j_2}, can not be characterized entirely metrically.

17. I borrow this name from Dowker, J. S., - see Ref.18.
18. J. S. Dowker and M. Goldstone, Proc. Roy. Soc. A, 303, 381 (1968).
19. R. Penrose, Annals of Physics, 10, 171 (1960). 19a. Using Theorem 3.2.
20. In order that the requisite polarized version of the r.h. side should exist.
21. H. Weyl, The Classical Groups, Princeton University Press, 1946. 21a. As in § 2.4.
22. Caution : in dealing with mixed[30] forms of the 3j-symbols, note, by Eqs.(2.5a), (2.11), that the left dual of the basis $\{e^m_j\}$ is $\{(-)^{2j}e^j_m\}$.
23. This can be traced back, via Van der Waerden (1932) and Weitzenböck (1923) to Clebsch and Gordan (1872).
24. The notation is as in Bargmann's article (Rev.Mod. Phys 34, 829 (1962)), which is reprinted in Ref.2. At this point one can spot the Regge symmetries.
25. See any classical text on invariant theory. Since the corresponding trilinear invariant $\mathcal{J}(\ ,\ ,\ )$ is determined by its values upon nil spinors by $\mathcal{J}(\alpha^4,\beta^4,\gamma^4) = \frac{1}{6}[\beta,\gamma]^2[\gamma,\alpha]^2[\alpha,\beta]^2$, the latter, in the classical literature, is referred to as the "symbolic expression" of the cubinvariant $\mathcal{J} \equiv \mathcal{J}(\phi,\phi,\phi)$, and α,β,γ are said to be "equivalent symbols".
26. The expression agrees with that in Ref.27, after taking into account a factor $\sqrt{8}$ due to a different normalization.
27. J. A. Roche and J. S. Dowker, J. Phys. A1, 527 (1968).
28. A. R. Edmonds, Angular Momentum in Quantum Mechanics, Princeton University Press, 1960.
29. In particular we do not state certain useful general theorems concerning representations of groups by means of the isometries of a complex vector space equipped with orthogonal or symplectic geometry. For the most part these follow,by familiar methods, from Schur's Lemma.
30. The fully covariant standard 3j-symbols are the covariant components of h, i.e. with respect to the basis $\left\{e^{m_1}_{j_1} \otimes e^{m_2}_{j_2} \otimes e^{m_3}_{j_3}\right\}$; the mixed forms of the 3j-symbols are defined to be the corresponding mixed components of h. Consequently[22] take note of results such as
$$\begin{pmatrix} j_1 & j_2 & m_3 \\ m_1 & m_2 & j_3 \end{pmatrix} = (-)^{2j_3}\left[e^{j_1}_{m_1},\ e^{j_2}_{m_2},\ e^{m_3}_{j_3}\right].$$
31. Since $\left\{\begin{matrix} j \\ j_1 j_2 \end{matrix}\right\}$ is an isometry, $f^m_j \equiv \left\{\begin{matrix} j \\ j_1 j_2 \end{matrix}\right\} e^m_j$ is left dual to $f^j_m = \left\{\begin{matrix} j \\ j_1 j_2 \end{matrix}\right\} e^j_m$.
32. Similarly, using factorizations into n factors, we obtain corresponding relations involving products of n standard 3j-symbols, whose triples $j_1^{(s)} j_2^{(s)} j_3^{(s)}$ satisfy $\sum_{s=1}^{n} j_i^{(s)} = j_i$. (Eq. 4.11 is an $n=3$ instance!)
33. No! As witness to the success of the colloquium's poster sessions, I was informed by S. Ström of work by Bose and Patera - see Canad. J. Phys. 49, 947 (1971) - who in turn told me that my Eq(4.17) can be found in Vilenkin's book on Special Functions and Group Representations (but with no recognition of the integer-valued nature of the symbols).

CLASSICAL MECHANICS, QUANTUM MECHANICS, FIELD THEORY, STATISTICAL MECHANICS

DESCRIPTION OF SYMMETRIES IN INDEFINITE METRIC SPACES

L. BRACCI, G. MORCHIO and F. STROCCHI
Scuola Normale Superiore, Pisa, Italy

The characterization of group representations associated to symmetries in indefinite metric spaces is provided. The problem of describing a symmetry in an indefinite metric space is solved by proving the analogue of Wigner's theorem: ray transformations preserving the modulus of an indefinite metric scalar product can be implemented by linear or antilinear vector transformations which are generalized unitary or antiunitary operators with respect to the indefinite scalar product. Interesting features arise since such operators need not to be bounded.

The basic problem of identifying the physically relevant representations of symmetry groups in quantum mechanics has been solved by Wigner's theorem(1,2) on ray transformations in Hilbert spaces. In the proof of the theorem, a crucial rôle is played by the Hilbert space structure and, in particular, by the positivity of the scalar product. On the other hand it has become more and more evident that indefinite metric spaces may be more useful both for the discussion of physical problems as well as for more genuine mathematical questions. As far as the physical applications are concerned the growing evidence comes mainly from the theory of quantum fields for which the use of an indefinite metric has been advocated several times in the past(3,4) as a solution of the divergence problem in quantum field theory and as a method to obtain better regularity properties for theories previously regarded as untractable(5). For several physical interesting theories the use of an indefinite metric is not only a promising suggestion but an unavoidable feature if one wants to preserve some basic properties of the fields like relativistic covariance and locality(6).

From a mathematical point of view the interest of indefinite metric spaces has been pioneered by the Russian mathematicians and we refer to their papers for general motivations as well as for an exposition of the results obtained in that field[7]. It may be interesting to stress that indefinite metric spaces appear very useful also in solving stability problems in the classical theory of damped oscillations and in general as a powerful tool for solving system of differential equations. In particular canonical linear differential equations with a periodic Hamiltonian have been studied with success using indefinite metric spaces[8].

In the following by an indefinite metric space we mean a complex (separable) Hilbert space H equipped with a bounded, symmetric and not degenerate sesquilinear form $\langle \cdot , \cdot \rangle = (\cdot , \eta \cdot)$, where $(\cdot , \cdot)$ is the ordinary scalar product in H, such that η has a bounded inverse. Without loss of generality one can restrict to the case $\eta^2 = 1$, and this will be done in the sequel. In the applications mentioned above the basic quantity is the product $\langle \cdot , \cdot \rangle$ rather than $(\cdot , \cdot)$. For example it is in term of $\langle \cdot , \cdot \rangle$ that one computes all the physical quantities of the theory or one discusses the mathematical properties of a given differential operator.

It is therefore natural to ask the question analogue to the statement of Wigner's theorem: given a ray transformation between rays of the Hilbert space H, preserving the modulus of the product $\langle \cdot , \cdot \rangle$, is it possible to implement such a transformation by a linear or antilinear vector transformation? The answer to this question is crucial for the foundation of a quantum mechanical description of symmetries in indefinite metric spaces as well as for the existence of semigroups associated to time evolution, space translation etc.

<u>Definition 1</u>. A generalized unitary (antiunitary) operator V with $\bar{D}_V = H = \bar{\Delta}_V$ is such that

$$\langle Vx, Vy \rangle = \lambda \langle x, y \rangle \quad (= \lambda \langle y, x \rangle) \qquad \forall x, y \in D_V \tag{1}$$

$\lambda = \pm 1$, independent of x and y.

Remark. The occurrence of an indefinite metric operator in eq. (1) does not allow to conclude, as in the positive metric case, that V is bounded. This is a fundamental difference which makes generalized unitary operators much more difficult to treat than unitary operators. For example, the invariance of the Wightman functions under a given mapping V yields the validity of eq. (1) for the dense domain obtained by applying local fields to the vacuum. The extension of the mapping to every vector of H is often not possible.

Proposition 1[9]: A generalized unitary operator is a linear operator, it has an inverse which is a generalized unitary operator with the same λ and it is closable. A similar statement holds for generalized antiunitary operators.

Definition 2. Given a vector $x \in H$, the set of vectors of the form λx, with $|\lambda| = 1, \lambda \in \mathbb{C}$, is called the ray associated to x and it will be denoted by $\underline{x}$. The vector x is said to belong to $\underline{x}$: $x \in \underline{x}$.

For the rays of an indefinite metric space, one defines a semidefinite scalar product $\underline{x} \underset{\eta}{\cdot} \underline{y} = \underline{x} \cdot \eta \underline{y} \equiv |\langle x, y \rangle|$, $x \in \underline{x}$, $y \in \underline{y}$.

Definition 3. A symmetry operation T is an application defined on a set $\underline{D}$ of rays of an indefinite metric space H onto a set $\underline{D'}$ of rays of H such that

i) the set of vectors belonging to the rays of $\underline{D}$ is a linear manifold D, which is dense in H

ii) the set of vectors belonging to the rays of $\underline{D'}$ is a linear manifold D', which is dense in H

iii) T is one to one from $\underline{D}$ to $\underline{D'}$

iv) $\forall \underline{x}, \underline{y} \in \underline{D} \quad |\langle T\underline{x}, T\underline{y} \rangle| = |\langle \underline{x}, \underline{y} \rangle|$

It is obvious that generalized unitary or antiunitary operators induce symmetry operations. It is not trivial that all symmetry operation can be induced by generalized unitary or antiunitary operators.

Theorem. Let T be a symmetry operation, then

a) there exists an operator U, such that $\forall \underline{x} \in \underline{D}$, the vector Ux belongs to $T\underline{x}$

b) U is either a generalized unitary or antiunitary operator. Clearly the last two possibilities can occur only if the eigenvalues 1 and -1 of the metric operator η have the same multiplicity.

Proof. We will sketch the main lines of the proof. For a complete proof see ref. 9. One of the basic tools is the use of a set of vectors $\{e_i\}$, $e_i \in D$, $i = 1, -1, 2, -2, \ldots$ which have the following properties

A) $$(e_i, \eta e_j) = \langle e_i, e_j \rangle = \mathrm{sign}(i)\, \delta_{ij} \qquad (2)$$

B) $$\langle z, e_i \rangle = 0 \quad \forall e_i \Rightarrow z = 0 \qquad (3)$$

They are the analogue in D of the orthonormal complete set which plays a fundamental rôle in the usual version of Wigner's theorem ($\eta = 1$, $D = H$). The existence of the vectors $\{e_i\}$ is far from obvious in the present case, since in general the vectors of the subspaces $H^{\pm} = \frac{1}{2}(1 \pm \eta)H$ do not belong to D.

Moreover the vectors e_i can be found in such a way that also the set $\{e_i', e_i' \in T\underline{e_i}\}$ is a "complete orthonormal set with respect to η", i.e. the vectors e_i''s satisfy

A) $$\langle e_n', e_m' \rangle = \lambda\ \delta_{nm}$$

B) $$\langle z, e_n' \rangle = 0\ \forall e_n' \Rightarrow z = 0$$

Another basic step of the proof is that every ray associated to a finite linear combination of the e_i's is transformed into a ray associated to a finite linear combination of the corresponding vectors e_i'. The construction of the operator U is first given for special finite linear combinations of the e_i's and then extended to the generic vector of D, as in the standard proof of Wigner's theorem. Finally U is shown to be either a generalized unitary or antiunitary operator. A crucial and difficult point in the proof of the theorem is the extension of U from $D_0 = \{$ generated by finite linear combinations of the e_i's $\}$ to D, since U is unbounded. In general if $D_o \ni x_n \to x \in D$ as $n \to \infty$, Ux_n will not converge to a limit in H. To define U on any vector of D we first define U on a suitable sequence $x_n \in D_o$, n = 0, 1, -1, 2, -2, ... and show that at each step the definition of U on x_m is consistent with the definition of U on x_l, $l < |m|$.

One might think that the linearity of D' is an unnecessary assumption if one restricts the theorem to the proof of the existence of U on D_o. A counter example shows that this is not possible.

References

1) E.P. Wigner, Group Theory and Its Application to the Quantum Mechanics of Atomic Spectra, Academic Press, New York (1959).

2) V. Bargmann, Journ. Math. Phys. 5, 862 (1964).

3) P.A.M. Dirac, Proc. Roy. Soc. A180, 1 (1942).
W. Pauli, Rev. Mod. Phys. 15, 175 (1943).
W. Heisenberg, Introduction to the Unified Field Theory of Elementary Particles, Wiley and Sons, London (1966).

4) T.D. Lee and G.C. Wick, Nuclear Physics B9, 209 (1969); Nuclear Physics B10, 1 (1969); Phys. Rev. D2, 1033 (1970).

5) G. 't Hooft, Nuclear Physics B33, 173 (1971); Nuclear Physics B35, 167 (1971).

6) F. Strocchi and A.S. Wightman, Journ. Math. Phys. 15, 2198 (1974).

7) M.G. Krein, Second Mathematical Summer School Part I, Naukova Dumka Kiev (1965).

8) M.G. Krein and V.A. Jakubovic, Proc. Internat. Sympos. Non-Linear Vibrations (1961) vol. I Izdat Akad. Nauk. Ukrain. SSR Kiev, 1963.

9) L. Bracci, G. Morchio and F. Strocchi, Comm. Math. Phys. 41, 289 (1974).

PARTIAL DIAGONALIZATION OF BETHE-SALPETER TYPE EQUATIONS

C. CRONSTRÖM

Research Institute for Theoretical Physics
University of Helsinki
Helsinki, Finland

See: Ann. of Phys. (1975) 92, 262

We consider an equation of the Bethe-Salpeter type, with arbitrary potential and kernel, respectively, for space-like momentum transfer. The invariance group of the equation is then the Lorentz-group in three dimensions, the O(1,2) group. The standard procedure for the diagonalization of such equations (valid for square integrable solutions only) is generalized to include the case of power bounded solutions, by means of a generalized O(1,2) expansion formalism.
The result is a two-dimensional integral equation for the O(1,2) expansion coefficients. The right-most ℓ-plane singularities of these determine the asymptotic behaviour of the amplitudes as in ordinary Regge theory. The formalism can be applied to other dynamical equations possessing O(1,2) symmetry.

GROUP STRUCTURE FOR CLASSICAL LATTICE SYSTEMS OF ARBITRARY SPIN

A. Hintermann and C. Gruber
Laboratoire de Physique Théorique
Ecole Polytechnique Fédérale - Lausanne
Switzerland

Abstract

We equip lattice systems of arbitrary spin with group structures. Harmonic analysis is used to derive low and high temperature expansions of the partition function as well as duality relations among different models.
The Asano contraction is formulated without using the Griffiths transformation into an equivalent spin $\frac{1}{2}$ system. A necessary and sufficient condition is given to obtain the partition function as the Asano contraction of smaller systems. For a given system with spin $p > \frac{1}{2}$, the group structure is not unique. The consequences of this fact are discussed in the case of spin 1 models for which we give analyticity domains.

June 1975

Talk presented at the Fourth International Colloquium on Group Theoretical Methods in Physics, Nijmegen (June 1975).

0. INTRODUCTION

In the following, we would like to deduce some consequences of the group structures which can be defined on classical lattice systems of arbitrary spin.[1,2,3,4]. For spin $\frac{1}{2}$ systems these structures have been extensively discussed by C. Gruber and D. Merlini.[1].

In the usual <u>physical picture</u> of higher spin systems or multicomponent systems, there is no obvious group structure. However, generalizing the natural group structures defined for spin $\frac{1}{2}$ systems, we got an abstract picture, called <u>group picture</u>, of arbitrary spin systems in terms of abelian groups. Using harmonic analysis on these groups, the idea is to gain as much properties as possible of arbitrary spin lattice systems in terms of this group picture. e.g. high - and low temperature expansions, duality relations between different systems, equations for the correlation functions, Asano contractions for the partition function. Having these informations in the group picture, simple transformations immediately yield the analogue in the physical picture. This will be done explicitly in the spin 1 case.

I. GROUP STRUCTURES ASSOCIATED WITH SYSTEMS OF ARBITRARY SPIN

Let Λ be a finite subset of the lattice $\mathcal{L}$. We associate at each site $i \in \Lambda$ a compact abelian group $\mathcal{G}_i$ and we label the values of the spin variable at the site i with the elements of this group. Thus if $\mathcal{G}_i = \mathbb{Z}_q$, the integers modulo q, we will describe either a q component system, a system of spin $\frac{q-1}{2}$ or any other system having q local configurations. Therefore the configuration space of the system is identified with

$$\begin{array}{ccc} \mathcal{G}_\Lambda & = & \underset{i \in \Lambda}{\times} \mathcal{G}_i \\ \psi & & \psi \\ N & = & (n_1, \dots, n_{|\Lambda|}) \end{array}$$

The algebra of observables $\mathfrak{A}_\Lambda$ of the system is the set of continuous functions on $\mathcal{G}_\Lambda$. A natural basis of this algebra is the dual group $\hat{\mathcal{G}}_\Lambda = \underset{i \in \Lambda}{\times} \hat{\mathcal{G}}_i$. The Hamiltonian $-\beta H_\Lambda \in \mathfrak{A}_\Lambda$ for a given system admits the Fourier decomposition

$$-\beta H_\Lambda = \sum_{\chi \in \hat{\mathcal{G}}_\Lambda} J(\chi)\chi \qquad \left(= \int_{\hat{\mathcal{G}}_\Lambda} d\chi \, J(\chi)\chi \right) \quad , \quad J(\chi) \in \mathbb{C}$$

The potentials $J(\chi)$ are the Fourier coefficients of $-\beta H_\Lambda$ defining the <u>support of the interactions $\mathfrak{J}$</u>

$$\mathcal{J} = \{ \chi \; ; \; J(\chi) \neq 0 \}$$

The group $\bar{\mathcal{J}} \subset \hat{G}_\Lambda$ generated by the elements of $\mathcal{J}$ is called group of interactions. Moreover, for technical reasons, we define a set of generating bonds $\mathcal{B}$ as a set $\mathcal{B}$ together with a mapping $\pi : \mathcal{B} \longrightarrow \hat{G}_\Lambda$ such that

i) $\forall \chi \in \mathcal{J}$ $\exists$ at least one $b \in \mathcal{B}$ and r integer such that $\chi = (\pi(b))^r = \chi_b^r$

ii) $|\mathcal{B}|$ is minimal

Let us remark that

a) the choice of $\mathcal{B}$ is not unique;

b) dropping ii), we could choose $\mathcal{B} = \mathcal{J}$ which has been done by Greenberg [2], but this does not represent the best choice;

c) in the spin $\frac{1}{2}$ case there is no possible choice and $\mathcal{B} = \mathcal{J}$.

Using the bicharacter notation, the Hamiltonian writes

$$-\beta H_\Lambda(N) = \sum_{b \in \mathcal{B}} \sum_{r=1}^{d_b - 1} J_{b,r} \langle \chi_b^r ; N \rangle_\Lambda \qquad d_b \text{ order of } \chi_b$$

and
$$J_{b,r} = \begin{cases} J(\chi) & \text{if } \chi_b^r \text{ is associated with } \chi \\ 0 & \text{otherwise} \end{cases}$$

and we have for the partition function

$$Z_\Lambda\{\mathcal{J}\} = \sum_{N \in G_\Lambda} e^{-\beta H_\Lambda(N)} = \sum_{N \in G_\Lambda} \prod_{b \in \mathcal{B}} e^{\sum_{r=1}^{d_b-1} J_{b,r} \langle \chi_b^r ; N \rangle_\Lambda}$$

and for the correlation functions, we get with $\chi \in \hat{G}_\Lambda$

$$\langle \chi \rangle_\Lambda = (Z_\Lambda\{\mathcal{J}\})^{-1} \sum_{N \in G_\Lambda} \langle \chi ; N \rangle_\Lambda \, e^{-\beta H_\Lambda(N)}$$

Defining the internal symmetry group $\mathcal{S}$ by

$$\mathcal{S} = \{ S \; ; \; S \in G_\Lambda \; , \; H_\Lambda(SN) = H_\Lambda(N) \;\; \forall N \in G_\Lambda \}$$

we get

<u>Lemma</u> : $\mathcal{S} = \bar{\mathfrak{J}}^{\perp}$, $\mathcal{G}_{\Lambda}/\mathcal{S} \cong \bar{\mathfrak{J}}^{\wedge}$

From the set of bonds $\mathcal{B}$, we introduce the <u>group of graphs</u> $\mathcal{G}_{\mathcal{B}}$:

$$\mathcal{G}_{\mathcal{B}} = \times_{b\in\mathcal{B}} \mathcal{G}_b \quad , \quad \mathcal{G}_b \cong \mathbb{Z}_{d_b}$$

The fundamental groups in the spin systems are $\mathcal{G}_{\Lambda}$, $\mathcal{G}_{\mathcal{B}}$ and their dual groups $\hat{\mathcal{G}_{\Lambda}}$, $\hat{\mathcal{G}_{\mathcal{B}}}$, respectively. Let us define two homomorphisms between these groups, which are related to the high temperature (low temperature) expansions of the partition function :

$$\pi : \mathcal{G}_{\mathcal{B}} \longrightarrow \hat{\mathcal{G}_{\Lambda}} \qquad\qquad \gamma : \mathcal{G}_{\Lambda} \longrightarrow \hat{\mathcal{G}_{\mathcal{B}}}$$

$$L \rightsquigarrow \pi(L) = \prod_{b\in\mathcal{B}} \langle \chi_b^{\ell_b} ; \cdot\cdot \rangle_{\Lambda} \qquad\qquad N \rightsquigarrow \gamma(N)$$

The relation $\langle \gamma(N) ; L \rangle_{\mathcal{B}} = \langle \pi(L) ; N \rangle_{\Lambda}$ between bicharacters of $\mathcal{G}_{\mathcal{B}}$ and those of $\mathcal{G}_{\Lambda}$ defines the homomorphism γ . One easily verifies $\mathrm{Im}\,\pi = \bar{\mathfrak{J}}$, $\ker \gamma = \bar{\mathfrak{J}}^{\perp} = \mathcal{S}$. The kernel of π and the image of γ are by definition the high temperature group $\mathcal{K}$, the low temperature group Γ , respectively. Moreover, for these groups, we have the following isomorphisms

$$\mathcal{G}_{\mathcal{B}}/\mathcal{K} \cong \bar{\mathfrak{J}} \;,\quad \mathcal{G}_{\Lambda}/\mathcal{S} \cong \Gamma \;,\quad \mathcal{G}_{\Lambda}/\mathcal{S} \cong \bar{\mathfrak{J}}^{\perp} \quad \text{thus} \quad (\mathcal{G}_{\mathcal{B}}/\mathcal{K})^{\wedge} \cong \Gamma$$

and finally

<u>Lemma</u> : $\mathcal{K} = \Gamma^{\perp}$, $\Gamma = \mathcal{K}^{\perp}$

2. IMPLICATIONS OF THESE GROUP STRUCTURES

2.1. The high temperature (H.T.) expansion of the partition function is obtained from the Fourier decomposition of the Boltzmann factors

$$e^{\sum_{\ell=1}^{d_b-1} J_{b,\ell} \langle \chi_b^{\ell} ; N \rangle_{\Lambda}} = \sum_{\ell'=0}^{d_b-1} \langle \chi_b^{\ell'} ; N \rangle_{\Lambda} \, f_{b,\ell'}$$

with $f_{b,\ell'}$ the Fourier coefficients of the Boltzmann factors. Thus putting $t_{b,\ell} = \frac{f_{b,\ell}}{f_{b,0}}$ and using the group $\mathcal{G}_{\mathcal{B}}$ we get

$$Z_\Lambda\{J\} = \prod_{b\in\mathcal{B}} f_{b,0} \sum_{N\in\mathcal{G}_\Lambda} \sum_{L\in\mathcal{G}_\mathcal{B}} \prod_{b\in\mathcal{B}} t_{b,\ell_b} \langle \pi(L); N\rangle_\Lambda$$

and with the orthogonality relations, the H.T. expansion becomes

$$Z_\Lambda\{J\} = |\mathcal{G}_\Lambda| \prod_{b\in\mathcal{B}} f_{b,0} \sum_{K\in\mathcal{K}} \prod_{b\in\mathcal{B}} t_{b,k_b}$$

2.2. The low temperature (L.T.) expansion of the partition function is also obtained using the properties of the homomorphism π and γ. With any $b\in\mathcal{B}$, let $L_b\in\mathcal{G}_\mathcal{B}$ s.t. $(L_b)_{b'} = \delta_{bb'}$. Hence

$$\langle \chi_b^r ; N\rangle_\Lambda = \langle (\pi(L_b))^r ; N\rangle_\Lambda = \langle \pi(L_b^r); N\rangle_\Lambda = \langle \gamma(N) ; L_b^r\rangle_\mathcal{B}$$

Thus

$$Z_\Lambda\{J\} = \sum_{N\in\mathcal{G}_\Lambda} \prod_{b\in\mathcal{B}} e^{\sum_{r=1}^{d_b-1} J_{b,r} \langle \gamma(N); L_b^r\rangle_\mathcal{B}}$$

which implies

$$Z_\Lambda\{J\} = |\mathcal{S}| \prod_{b\in\mathcal{B}} e^{\sum_{r=1}^{d_b-1} J_{b,r}} \sum_{\gamma\in\Gamma} \prod_{b\in\mathcal{B}} e^{\sum_{r=1}^{d_b-1} J_{b,r}(\langle \gamma; L_b^r\rangle_\mathcal{B} - 1)}$$

and is called the L.T. expansion of the partition function.

2.3. The isomorphism $(\mathcal{G}_\mathcal{B}/\mathcal{K})^\wedge \cong \Gamma$ tells us that these two expansions are related by means of the Poisson formulae [3] .

2.4. Duality transformations among different models can be defined by identifying the high temperature group $\mathcal{K}$ with the configuration space $\mathcal{G}_{\Lambda^*}$ of the dual system. Choosing a minimal set of generators of $\mathcal{K}$ and associating to each generator a site i^*, the dual lattice is defined by $\Lambda^* = \{i^*\}$. Now we associate with the site i^* the group $\mathcal{G}_{i^*} = \mathbb{Z}_{q^*}$, q^* = order of the generator on the site i^*. The interactions in the dual system are computed taking into account that the H.T. expansion can also be viewed as the mean value of the function $\prod_{b\in\mathcal{B}} t_{b,k_b}$ over the group $\mathcal{G}_{\Lambda^*} = \times_{i^*\in\Lambda^*} \mathcal{G}_{i^*}$. The Fourier decomposition with respect to $\hat{\mathcal{G}}_{\Lambda^*}$ of the logarithm of this function directly gives the interactions for the dual model as well as the coupling constants. Since the choice of the generators of $\mathcal{K}$ is not unique, a given model has many duals. Taking generators with many b-components different from zero, the dual interactions will have complicated N-body interactions. Thus one generally chooses generators of $\mathbb{K}$ with the less b-components possible.

2.5. Equations for the correlation functions :

Let $\chi \in \mathcal{G}_\Lambda^\wedge$ and take $j \in \Lambda$ such that $\langle \chi ; N_j \rangle_\Lambda \neq 1$ for some $N_j \in \mathcal{G}_j$. Moreover for any $N \in \mathcal{G}_\Lambda$ we define $N = \bar{N} \cdot N_j$ with $N_j \in \mathcal{G}_j$. Then we can write with $\mathcal{G}_{\Lambda/j} = \underset{\substack{i \in \Lambda \\ i \neq j}}{\times} \mathcal{G}_i$

$$\langle \chi \rangle_\Lambda = (Z_\Lambda\{J\})^{-1} \int_{\mathcal{G}_{\Lambda/j}} d\mu_{\Lambda/j}(\bar{N}) \langle \chi ; \bar{N} \rangle_{\Lambda/j} \prod_{\substack{b \in \mathcal{B} \\ \langle \chi_b ; N_j \rangle_\Lambda = 1}} e^{\sum_{r=1}^{\alpha_b - 1} J_{b,r} \langle \chi_b^r ; \bar{N} \rangle_{\Lambda/j}} \int_{\mathcal{G}_j} d\mu_j(N_j) \langle \chi ; N_j \rangle_j \prod_{\substack{b \in \mathcal{B} \\ \langle \chi_b ; N_j \rangle_\Lambda \neq 1}} e^{\sum_{r=1}^{\alpha_b - 1} J_{b,r} \langle \chi_b^r ; \bar{N} N_j \rangle_\Lambda}$$

Restoring the trace over $\mathcal{G}_\Lambda$, we get with $\chi = \bar{\chi} \cdot \chi_j$, $\bar{\chi} \in \mathcal{G}_{\Lambda/j}^\wedge$, $\chi_j \in \mathcal{G}_j^\wedge$ the following set of equations

$$\langle \chi \rangle_\Lambda = \langle \bar{\chi} \cdot \frac{\int_{\mathcal{G}_j} d\mu_j(N_j) \langle \chi_j ; N_j \rangle_j \prod_{\substack{b \in \mathcal{B} \\ \langle \chi_b ; N_j \rangle_\Lambda \neq 1}} e^{\sum_{r=1}^{\alpha_b - 1} J_{b,r} \langle \chi_b^r ; N_j \rangle_j \bar{\chi}_b^r}}{\int_{\mathcal{G}_j} d\mu_j(N_j) \prod_{\substack{b \in \mathcal{B} \\ \langle \chi_b ; N_j \rangle_\Lambda \neq 1}} e^{\sum_{r=1}^{\alpha_b - 1} J_{b,r} \langle \chi_b^r ; N_j \rangle_j \bar{\chi}_b^r}} \rangle_\Lambda$$

The Fourier expansion of the quotient yields with $Y = \bigcup_{\substack{b \in \mathcal{B} \\ b_j \neq 0}} \bigcup_{\substack{i \in \Lambda/j \\ b_i \neq 0}} i$

$$\langle \chi \rangle_\Lambda = \sum_{\tilde{\chi} \in (\times_{i \in Y} \mathcal{G}_i)^\wedge} Th(\hat{\chi}) \langle \bar{\chi} \, \tilde{\chi} \rangle_\Lambda$$

$Th(\hat{\chi})$ being the Fourier coefficients of the quotient.

3. ASANO CONTRACTIONS

To define a polynomial, which is linear in each variable, associated with the L.T. expansion, let us introduce for any $b \in \mathcal{B}$ the α_b generalized activities

$$z_{b,m} = e^{\sum_{r=1}^{\alpha_b - 1} J_{b,r} (\langle m ; L_b^r \rangle_{\mathcal{B}} - 1)} \qquad m \in \mathbb{Z}_{\alpha_b}$$

(since $z_{b,0} = 1$ there are only $\alpha_b - 1$ independent ones). Thus this polynomial becomes

$$M(z_{\mathcal{B}}) = \sum_{M \in \mathcal{G}_{\mathcal{B}}} c_M z^M \quad , \quad z^M = \prod_{b \in \mathcal{B}} z_{b,m_b}$$

and $c_M = 1,(0)$ if $M \in \Gamma, (M \notin \Gamma)$, respectively. Let $\mathcal{B} = \bigcup_{i=1}^{n} \mathcal{B}_i$ be a finite covering of $\mathcal{B}$ and

$$P_i(z_{\mathcal{B}_i}) = \sum_{L_i \in \times_{b \in \mathcal{B}_i} \mathcal{G}_b} c_{i,L_i} \, z^{L_i} \quad , \quad z^{L_i} = \prod_{b \in \mathcal{B}_i} z_{b,\ell_{i_b}}$$

be a family of small polynomials

Definition : The polynomial $P(z_{\mathcal{B}}) = \sum_{L \in \mathcal{G}_{\mathcal{B}}} c_L z^L$ is the Asano contraction (A.C.) of the family of polynomials $\{P_i(z_{\mathcal{B}_i})\}$ if $c_L = \prod_{i=1}^{n} c_{i,L/\mathcal{B}_i}$ where $(L/\mathcal{B}_i)_b = \ell_b \, \delta_{b,\mathcal{B}_i}$, $\delta_{b,\mathcal{B}_i} = 1\ (0)$ if $b \in \mathcal{B}_i\ (b \notin \mathcal{B}_i)$.

This is a particular form of the usual A.C. which is defined in more general terms. Therefore we can apply Ruelle's theorem [5] without modification to relate the properties of zeros of the small polynomials to the ones of the contracted polynomial. The following theorem gives a necessary and sufficient condition that the polynomial associated to an expansion of the partition function is the A.C. of small polynomials :

Theorem Let $\mathcal{G} \subset \mathcal{G}_{\mathcal{B}}$ be any subgroup of $\mathcal{G}_{\mathcal{B}}$ and $\mathcal{B} = \bigcup_{i=1}^{n} \mathcal{B}_i$ be a finite covering of $\mathcal{B}$ then

$M(z_{\mathcal{B}}) = \sum_{R \in \mathcal{G}} z^R$ is the A.C. of $M(z_{\mathcal{B}_i}) = \sum_{R_i \in \mathcal{G}_i} z^{R_i}$, $\mathcal{G}_i \subset \times_{b \in \mathcal{B}_i} \mathcal{G}_b$ $i = 1, \dots, n$

if and only if $\mathcal{G}^{\perp}$ = group generated by $\bigcup_{i=1}^{n} \mathcal{G}_i^{\perp}$

where $\mathcal{G}_i = \{ L_i \in \times_{b \in \mathcal{B}_i} \mathcal{G}_b \, ; \, (L_i)_b = \ell_b \ \forall b \in \mathcal{B}_i$ for some $L \in \mathcal{G} \}$

This is a straightforward generalization of a theorem by Slawny [6] proved for the spin ½ case. In order to find the family of small polynomials, we proceed in the same way as in the spin ½ case[7] : First find $\mathcal{G}^{\perp}$, take $\{\mathcal{G}_i^{\perp}\}_{i=1,\dots,n}$ a finite family of subgroups of $\mathcal{G}^{\perp}$ which generates $\mathcal{G}^{\perp}$, then the covering is given by

$$\mathcal{B}_i = \bigcup_{L_i \in \mathcal{G}_i^{\perp}} \{ b \, ; \, (L_i)_b \neq 0 \}$$

Remark that due to the fact that for any bond $b \in \mathcal{B}$ we have $\alpha_b - 1$ variables, we will necessarily have analyticity properties in $\alpha_b - 1$ complex variables. e.g. for a spin 1 model we get analyticity domains in $e^{\beta h}$ and $e^{\beta\mu}$ simultaneously. Thus the above A.C. for higher spin systems is complementary to the ones introduced by Millard and Viswanathan using Griffiths transformation into an equivalent spin $\frac{1}{2}$ system [8] and their generalized A.C. given in [9] . Griffiths transformations and the resulting group structure have been discussed by Slawny [10] . They got analyticity domains in one complex variable, keeping the other interactions as real parameters.

4. APPLICATIONS TO SPIN 1 MODELS

Here, we limit ourselves to discuss the application to A.C., for other applications, we refer to [4] .

Let us consider the general spin 1 model $(\mathcal{G}_i = \mathbb{Z}_3 \;\; \forall i \in \Lambda)$ with field and two spin nearest neighbours (n.n.) interactions. Thus the set of bonds $\mathcal{B}$ satisfying i) and ii) is

$$\mathcal{B} = \{ b_r \; ; r \in \Lambda \} \cup \{ b^1_{rs} \; ; \; r,s \in \Lambda \quad r,s \text{ n.n.} \} \cup \{ b^2_{r,s} \; ; \; r,s \in \Lambda \, , \; r,s \text{ n.n.} \}$$

and the coupling constants we have are the two fields $J_{b_r,1}$, $J_{b_r,2}$ and the four n.n. coupling constants $J_{b^1_{rs},1}$, $J_{b^1_{rs},2}$, $J_{b^2_{rs},1}$, $J_{b^2_{rs},2}$.

Taking $\mathcal{G} = \Gamma$ we have $\mathcal{G}^{\perp} = K$. The simplest generators of K are defined on the n.n. sets $\Lambda_i = \{r,s\} \subset \Lambda$ and the covering sets $\mathcal{B}_i$ are

$$\mathcal{B}_i = \{ b_r , b_s , b^1_{rs} , b^2_{r,s} \}$$

Each bond of $\mathcal{B}_i$ defining two generalized activities. Thus computing $\mathcal{G}_i$ and putting $z_1 = z_{b_r,1}$, $z_2 = z_{b_r,2}$, $z_3 = z_{b_s,1}$, $z_4 = z_{b_s,2}$, $z_5 = z_{b^1,1}$, $z_6 = z_{b^1,2}$, $z_7 = z_{b^2,1}$, $z_8 = z_{b^2,2}$ the small polynomials become for i = 1,...,n

$$M(z_{\mathcal{B}_i}) = M(z_1,\dots,z_8) = 1 + z_1z_5z_7 + z_2z_6z_8 + z_3z_5z_8 + z_4z_6z_7 + z_1z_3z_6 + z_1z_4z_8 + z_2z_3z_7 + z_2z_4z_5$$

Again we use A.C. to discuss the domains free of zeros of this polynomial and we find that $M(z_1, \ldots, z_8)$ is the A.C. of 6 polynomials each depending on two variables only. The use of the theorem of Grace [11] combined with Ruelle's theorem [5] yields domains free of zeros of the small polynomials and finally of the polynomial associated with the L.T. expansion of the partition function.

5. THE PHYSICAL AND GROUP PICTURE OF SPIN 1 MODELS

In the usual physical language, the spin variables of a spin 1 model are $\sigma_i \in \{0, \pm 1\}$ $\forall i \in \Lambda$, and the most general Hamiltonian with fields and n.n. interactions writes

$$-\beta H_\Lambda\{\sigma\} = \sum_{\substack{ij \\ n.n.}} (\epsilon_{11}\sigma_i\sigma_j + \epsilon_{12}\sigma_i\sigma_j^2 + \epsilon_{21}\sigma_i^2\sigma_j + \epsilon_{22}\sigma_i^2\sigma_j^2) + h\sum_i \sigma_i + \mu\sum_i \sigma_i^2$$

The local transformation $\phi : \sigma_i \rightarrow n_i$ $\forall i \in \Lambda$ where $n_i \in \{0,1,2\} = \mathbb{Z}_3$ maps the physical picture of any spin 1 model on the corresponding group picture. Thus $-\beta H_\Lambda\{\sigma\}$ can be viewed as a function over G_Λ and Fourier transforms immediately yield the coupling constants $J_{b,r}$ in terms of the ϵ_{jk}, h, μ and conversely. There are 6 different local transformations ϕ to equip the phase space of the physical picture with a group structure, which are exhausted by the elements of the permutation group of 3 elements. (e.g. $\phi(-1) = 0$, $\phi(0) = 1$, $\phi(1) = 2$ etc.). Thus for the same spin 1 model, we get 6 different sets of generalized activities in terms of ϵ_{jk}, h, μ. Note that the small polynomials do not depend on ϕ but the domains free of zeros of the partition function in terms of the "physical" activities e^μ, e^h, $e^{\epsilon_{jk}}$ will depend on ϕ and in general we get for different maps different domains.

Taking as a typical spin 1 model the model of Lebowitz-Gallavotti [12] where $\epsilon_{11} = -\epsilon_{22} = \epsilon > 0$, $h, \mu \neq 0$ we find for fixed real values of ϵ the following analyticity regions (shaded) in <u>h and μ</u> of the partition function:

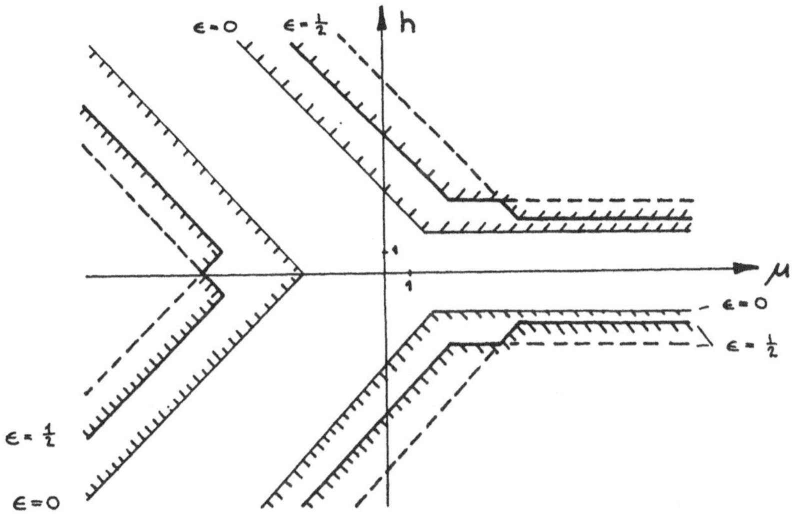

Similar analyticity domains are obtained for other spin 1 models. For the dilute Ising model, where $\epsilon_{11} > 0$, $\epsilon_{12} = \epsilon_{21} = \epsilon_{22} = 0$, $h, \mu \neq 0$ we improve bounds for the tricritical point given by Sarbach and Rys [13] .

References

[1] D. Merlini, C. Gruber, J.M.P. 13, (1972), 1814.

[2] W. Greenberg, Com. Math. Phys. 29, (1973), 163.

[3] C. Gruber, A. Hintermann, H.P.A. 47, (1974), 67.

[4] A. Hintermann, C. Gruber, in preparation.

[5] D. Ruelle, Phys. Rev. Lett., 26, (1971), 303.

[6] J. Slawny, Com. Math. Phys., 34, (1973), 271.

[7] C. Gruber, A. Hintermann and D. Merlini, Com. Math. Phys. 40, (1975), 83.

[8] K.Y. Millard, K.S. Viswanathan, Phys. Rev. B, 9, (1974), 2030.

[9] K.Y. Millard, K.S. Viswanathan, J.M.P. 15, (1974), 1821.

[10] J. Slawny, Ferromagnetic Spin Systems at Low Temperature, preprint.

[11] D. Ruelle, Com. Math. Phys. 31, (1973), 265.

[12] J.L. Lebowitz, G. Gallavotti, J.M.P. 12, (1971), 1129.

[13] S. Sarbach, F. Rys, Phys. Rev. B, 7, (1973), 3141.

Equivalent Lagrangians and quasicanonical transformations

by

G. MARMO

Istituto di Fisica Teorica dell'Università - Napoli

Istituto Nazionale di Fisica Nucleare - Sezione di Napoli

Introduction.

In this work, we analyze in terms of global differential geometry, the construction which leads in classical mechanics to Hamiltonian or to Lagrangian dynamics. In particular we are interested in the ambiguities which arise in the construction.

In classical mechanics usually position is assumed to be a fundamental observable. The set of all possible positions of a physical system S is assumed to have a differential structure (Levi-Civita e Amaldi: Lezioni di Meccanica Razionale[1]), i.e. has the structure of a differentiable manifold M, the dimension of M being the number of degrees of freedom of S. We assume therefore that the manifold M of positions of S is given, and that the experimental data consist of the set of all possible evolutions of S: i.e. a set of mappings $C(\mathbb{R}, M)$ from the time axis into M. If one tries to give a dynamical interpretation in term of ordinary differential equations (a "deterministic interpretation", see Ar'nold[2]) of these experimental data, one is compelled to enlarge the manifold M. This is because the system S can move in any direction starting at any point of M and so the experimental curves on M do not define directly any vector field on M.

The problem of writing a dynamics for S consists then in first enlarging M to M' by adding new variables to position and second in finding a vector field on M' whose flow projected on M gives the experimental curves on M. Two solutions to the problem are to use, instead of M, either TM, the tangent bundle to M (i.e. in a chart described by $q, \mathring{q}$), or T^*M, the cotangent space to M (i.e., in a chart described by q,p : the usual phase space) (For a general local treatment, see Synge [3]).
So we are led to deal with vector fields on TM or T^*M.
One problem is to analyze the ambiguities which arise in the "lifting", i.e. how many vector fields are there on TM or on T^*M whose flows projected on M give the required curves on M. We shall ignore this problem for now, but will treat it in a subsequent paper.

Here, the main problem we wish to analyze is the ambiguities in the Lagrangians or Hamiltonians which characterize a given vector field.

Equivalent Lagrangians.

On T^*M there is a canonical way to associate a vector field to a differential form and conversely a differential form to a vector field. Thus there exists an isomorphism (4)

$$\omega^\flat : \quad (T^*M) \rightleftarrows {}^*(TM)$$

Here $\omega^\flat$ comes from $\omega_o = d\theta_o$, θ_o being defined for all $m \in M, \alpha_m \in T^*_m M$ $w_m \in T_{\alpha m}(T^*M)$ and the projection $\tau^*_M : T^*M \longrightarrow M$, as $\theta_o(\alpha_m) \cdot W_m = -\alpha_m (T\tau^*_M \; W_m)$

So if $i_X \omega_o = dH$ and M is connected, H is determined up to a constant function on T^*M.

We shall here study the situation on TM.

On TM there is no "natural" way to associate vector fields and differential forms; but if we have a mapping $\phi \in C^\infty(TM, T^*M)$ it is possible to pull back ω_o onto TM, by the formula

$$(\phi_* \omega_o)(X, Y) = \omega_o \; (T\phi \; X, T\phi \; Y) \text{ for } \quad X, Y \in \mathfrak{X}(TM)$$

(There is a slight abuse of notations here as $T\phi$ acts on vectors, not on vector fields).

If we can characterize a vector field with a function, there is the possibility of getting a mapping ϕ from the function. Now for a given $L \in \mathcal{F}(TM)$ we can restrict L to the fiber through m, compute the differential and get an element in $T^*_m M$. Such operation is called fiber derivation [4)] and is denoted by FL. It has the property of making the following diagram commutative

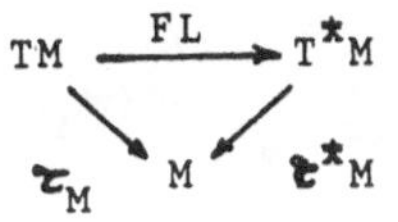

Pulling back ω_o using FL, we get a two form ω_L on TM. It is now possible to get a one form from X. If it is an exact one, we have

$$i_X \omega_L = dE$$

There is as yet no link between E and L, but if we impose that X defines the usual second order Euler-Lagrange's equations, we get $E(v) = FL(v) \cdot v - L(v)$. (Qualifications are necessary if ω_L turns out to be degenerate).

It is well known that different Lagrangians can lead to the same Euler-Lagrange's equation. We shall now try to analyze this point from a global point of view.

It is possible to get a partial order relation on the Lagrangians in the following way:

Def. Let $L, L' \in \mathcal{F}(TM)$, we say that L is subordinate to L' if for $\forall X$ such that $i_X\omega_{L'} = dE'$ we have $i_X\omega_L = dE$ where $E = A - L$, $E(v) = FL(v)\cdot v - L(v)$. We write $L \prec L'$. It is trivial that $L'' \prec L'$, $L' \prec L \Rightarrow L'' \prec L$.

Prop. For any function $L \in \mathcal{F}(TM)$, the set of functions subordinate to L is a real vector space.

This follows from linearity of $F : L \longmapsto FL$ (see Abraham [4] - prop. 17.6)

Def. L is equivalent to L', $L \equiv L'$, iff L is subordinate to L' and L' is subordinate to L.

Prop. $L' \prec L$ (L' subordinate to L) and $\operatorname{Ker} \omega_L^\flat = \operatorname{Ker} \omega_{L'}^\flat \Leftrightarrow L \equiv L'$.

Indeed $\operatorname{Ker} \omega_L^\flat$ is an ideal in $\mathfrak{X}(TM)$ [5]

so $$\mathfrak{X}(TM)/\operatorname{Ker}\omega_L^\flat \equiv \mathfrak{X}(TM)/\operatorname{Ker}\omega_{L'}^\flat$$

If $X \in [X]_L$ such that $i_X\omega_L = dE$ then

$$i_X\omega_{L'} = dE' \Rightarrow [X]_L \subset [X]_{L'} \Rightarrow [X]_L = [X]_{L'}$$

Corollary:

$L \equiv L'$, L regular $\Rightarrow L'$ regular

(this a generalization of the 1-dimensional case in Currie-Saletan [6]).

Prop. For $\forall f \in \mathcal{F}(M)$ such that $df \neq 0$, $(\tau_M)_* f$ is not subordinate to any Lagrangian

Indeed $\omega_{(\tau_M)_* f} = 0$, $dE_{(\tau_M)_* f} = (\tau_M)_* df$

$\Rightarrow$ there is no X such that

$$i_X\omega_{(\tau_M)_* f} = dE_{(\tau_M)_* f}$$

Starting with an $f \in \mathcal{F}(M)$, it is possible to define $\bar{f} : TM \longrightarrow \mathbb{R}$

$\bar{f}(m, v_m) = df(m)\cdot(v_m)$ (see Yano-Ishiara [7]). In a canonical chart

$$\bar{f}(q, \dot{q}) = \dot{q}_i \frac{\partial f}{\partial q_i}$$

Prop. For $\forall f \in \mathcal{F}(M)$, $\bar{f} \in \mathcal{F}(TM)$ is subordinate to every Lagrangian.

Indeed $(F\bar{f})_*\omega_0 = 0$, $A_{\bar{f}} = \bar{f}$, $E_{\bar{f}} = 0$ $\quad i_X \cdot 0 = 0$

Quasicanonical transformations.

Until now we have been associating differential forms to vector fields. If we want to do the converse, we need ω_L to be non degenerate (for ω degenerate see [5]). From now on, we will suppose our Lagrangians to be regular (i.e. ω_L non degenerate) or, if necessary, hyperregular (4).

If $L \equiv L'$, we have $i_X\omega_L = dE$, $i_X\omega_{L'} = dE'$.

If we pull back such equations to T^*M through FL^{-1}, we get

$$i_{X_L}\omega_o = d(E\circ FL^{-1}),\quad i_{X_L}(FL'\circ FL^{-1})_*\,\omega_o = d(E'\circ FL^{-1})$$

We thus learn that while going to T^*M starting from TM it is possible to get Hamiltonian equations with ω_o or with $(FL'\circ FL^{-1})\omega_o$, that is associating different differential forms with different symplectic structures with the same vector field. This fact suggests to use this machinary on (T^*M, ω_o) ignoring TM altogether.

Thus we have the definition:

Def. $\phi \in \mathrm{Diff}\,(T^*M)$ is a quasicanonical transformation for X_H, where $i_{X_H}\omega_o = dH$, if there exists a function $K \in \mathcal{F}(T^*M)$ such that $i_{X_H}\phi_*\omega_o = dK$.

As we are often interested in solving equations only in a local form, we can generalize this definition requiring only that

$$d(i_{X_H}\phi_*\omega_o) = 0.$$

In such a case the necessary and sufficient condition for ϕ being quasicanonical for X_H is $L_{X_H}\phi_*\omega_o = 0$.

In this way, we have introduced the notion of "homogeneous canonical transformations for a given vector field" globalyzing the one introduced by Curie and Saletan [6].

If one makes calculations in the simple case

$L \equiv L'$, $L' = KL + \bar{f}$ $\quad K \in \mathbb{R}$ then, one, finds locally

$$q_i \longmapsto q_i' = q_i$$
$$p_i \longmapsto p_i' = K p_i + \frac{\partial f}{\partial q_i}$$

We will give more details and results in a forthcoming paper.

References

1) Levi-Civita - Amaldi - Lezioni di Meccanica Razionale - Zanichelli, Bologna 1974

2) Ar'nold V.I. - Equations Différentielles ordinaires - Editions M I R - Moscow 1974

3) Synge J.L. - Handbuch der Physik, Band III/1 - Prinzipien der Klassischen Mechanik und Foldtheorie - Springer Verlag

4) Abraham R. - Marsden J. - Foundations of Classical Mechanics - Benjamin 1967

5) Marmo G. - Proceedings of the 2nd International Colloquium on Group Theoretical Methods - (Nijmegen 1973)

Lettere al Nuovo Cimento 13, 6 (1975)

6) Currie-Saletan - Nuovo Cimento 9 B, 143 (1972)
J. Math. Phys. 7, 967 (1966)

Gelman Y., Saletan E.J. - Nuovo Cimento 18 B, 53 (1973)

7) Yano K. - Ishihara S. - Tangent and Cotangent Bundles - M. Dekker, New York 1973.

GROUP THEORY OF MASSLESS BOSON FIELDS

U.H. Niederer

Institut für Theoretische Physik der Universität Zürich,

Schönberggasse 9, 8001 Zürich, Switzerland.

Abstract:

Free massless Boson fields are defined as manifestly covariant unitary representations of the Poincaré group for zero mass and integer spin s. The fields are tensors which, in the simplest case, belong to the representation $D(s,0) \oplus D(0,s)$ of the Lorentz group. They are characterized by wave equations of two types: (i) The symmetry conditions, which impose the requirement that the tensors indeed carry the representation $D(s,0) \oplus D(0,s)$, and (ii) the unitarity conditions, which turn out to be of the form

$$p_\lambda \psi_{\mu\nu\dots}(p) + p_\mu \psi_{\nu\lambda\dots}(p) + p_\nu \psi_{\lambda\mu\dots}(p) = 0.$$

In the case $s = 2$ the field is a 4th rank tensor, the symmetry conditions are the equations of the Riemann curvature tensor in the linearized vacuum theory of gravitation, and the unitarity conditions are the Bianchi identities.

1. Introduction

It has been shown recently [1] how wave equations for fields of given mass and spin can be obtained from manifestly covariant unitary representations of the Poincaré group, i.e. representations of the form

$$(U(a,\Lambda)\psi)(p) = e^{ia\cdot p} D(\Lambda)\psi(\Lambda^{-1}p) \quad , \quad (p^2 = m^2) \tag{1.1}$$

where D is a finite-dimensional representation of the Lorentz group. In the massive case the wave equations express the requirement that one spin only is present. In the massless case the situation is different. The little group E(2) is noncompact and the unitarity of the representation (1.1), automatic in the massive case due to the compactness of the little group SO(3), must be guaranteed by the requirement that the noncompact part of E(2) be trivially represented on the field ψ. It is this unitarity condition which appears as wave equation. It is shown in [1] that either of the following two equivalent conditions may be used as unitarity condition:

$$V^\lambda \psi(p) = i p^\lambda (M+N)\psi(p) \, , \tag{1.2}$$

$$W^\lambda \psi(p) = p^\lambda (M-N)\psi(p) \, , \tag{1.3}$$

where

$$V_\lambda = p^\mu M_{\mu\lambda} \quad , \quad W^\lambda = \tfrac{1}{2}\varepsilon^{\lambda\mu\rho\sigma} p_\mu M_{\rho\sigma} \, , \tag{1.4}$$

$M_{\mu\nu} = (\underset{\sim}{J}, \underset{\sim}{K})$ are the generators of D , $p^0 = |\underset{\sim}{p}|$, and (M,N) are operators which take the values (m,n) on the irreducible constituents D(m,n) of D . The sign in (1.2) is fixed relative to the convention that $\underset{\sim}{M} = \frac{1}{2}(\underset{\sim}{J} - i\underset{\sim}{K})$, $\underset{\sim}{N} = \frac{1}{2}(\underset{\sim}{J} + i\underset{\sim}{K})$ are the SU(2) $\otimes$ SU(2) generators of D(m,n).

The zero-component of (1.3) can be written as

$$h\,\psi(p) = (M-N)\,\psi(p)\,, \tag{1.5}$$

where $h = \underset{\sim}{p}\cdot\underset{\sim}{J} / |\underset{\sim}{p}|$ is the helicity. Thus, (1.5) implies the well-known fact that one helicity state only is presentin the irreducible representation D(m,n).

Applications of the unitarity condition to spin 1 and spin 2 are contained in [1] and [2]; they are also mentioned as special cases in section 4. In the present paper we want to apply the unitarity condition to the case of arbitrary integer spin s , or, more precisely, arbitrary helicities $\pm$ s. As can be seen from (1.5), we must choose representations with $m-n = \pm s$, and the simplest of these representations are D(s,0), D(0,s), D(s,0) $\oplus$ D(0,s) with helicities s, - s, $\pm$ s , respectively. It is to these representations that we restrict our attention. More general representations will be analyzed elsewhere.

The equations satisfied by the fields are of two types. One set of equations contains the conditions for the field to carry a specified representation D of the Lorentz group. Because these conditions have the form of symmetry properties for tensors, they will be called <u>symmetry conditions.</u> They are determined in section 2. The other set of equations contains the implications of the <u>unitarity condition</u>, which is evaluated in section 3. (A third type of equation, which is always present and will not be mentioned further, is the mass condition $p^2\,\psi(p) = 0$;

for s = 0 this is the only equation.) Finally, the results are discussed in section 4. We work in momentum space throughout.

2. Symmetry Conditions

In this section, we show that the representation $D(s,0) \oplus D(0,s)$, $s = 1,2,\ldots$, is carried by $2s$ - rank tensors $\psi_{\mu_1\nu_1\cdots\mu_s\nu_s}$ satisfying the following conditions:

$$\begin{aligned}
&(S1) \quad \psi_{\cdots\mu_i\nu_i\cdots} = -\psi_{\cdots\nu_i\mu_i\cdots}\,, \quad (i=1,\ldots,s) \\
&(S2) \quad \psi_{\cdot\cdot\mu_i\nu_i\cdot\cdot\mu_k\nu_k\cdot\cdot} = \psi_{\cdot\cdot\mu_k\nu_k\cdot\cdot\mu_i\nu_i\cdot\cdot}\,, \quad (i,k=1,\ldots,s\,,\ i\neq k) \\
&(S3) \quad \psi_{\cdot\cdot\mu_i\nu_i\cdot\cdot\mu_k\cdot\cdot} + \psi_{\cdot\cdot\mu_k\mu_i\cdot\cdot\nu_i\cdot\cdot} + \psi_{\cdot\cdot\nu_i\mu_k\cdot\cdot\mu_i\cdot\cdot} = 0\,, \quad (i,k=1,\ldots,s\,,\ i\neq k) \\
&(S4) \quad \psi_{\cdot\cdot\lambda\cdot\cdot}{}^{\lambda}{}_{\cdot\cdot} = 0\,.
\end{aligned} \tag{2.1}$$

If only one of the two representations D(s,0), D(0,s) is considered then the tensors have the additional property of being eigenstates of the duality operation, i.e.

$$(S5) \quad \varepsilon_{\mu_i\nu_i}{}^{\rho_i\sigma_i}\,\psi^{(\pm)}_{\cdot\cdot\rho_i\sigma_i\cdot\cdot} = \pm 2i\,\psi^{(\pm)}_{\cdot\cdot\mu_i\nu_i\cdot\cdot}\,, \quad \begin{cases} (+) \text{ for } D(s,0) \\ (-) \text{ for } D(0,s) \end{cases} \tag{2.2}$$

where $\varepsilon_{0123} = 1$, $i = 1,\ldots, s$.

To establish these conditions, we start from the well-known spinor representations D(s,0) and D(0,s). The representation D(s,0) is carried by the totally symmetric spinors $\phi_{A_1B_1\cdots A_sB_s}$. To define tensors in Minkowski space, we use the invariant spin-tensor

$$\omega_{\mu\nu}{}^{AB} = \frac{1}{2}\left(\sigma_{\mu}{}^{A}{}_{\dot{C}}\,\sigma_{\nu}{}^{B\dot{C}} - \sigma_{\nu}{}^{A}{}_{\dot{C}}\,\sigma_{\mu}{}^{B\dot{C}}\right), \tag{2.3}$$

where

$$\sigma_{\mu\,A\dot{B}} = (1, \underset{\sim}{\sigma})_{A\dot{B}}, \tag{2.4}$$

$\underset{\sim}{\sigma}$ are the Pauli matrices, and the raising/lowering of spinor indices is performed with $\varepsilon_{AB} = \varepsilon_{\dot{A}\dot{B}} = -\varepsilon^{AB} = -\varepsilon^{\dot{A}\dot{B}} = \begin{pmatrix} 0 & -1 \\ 1 & 0 \end{pmatrix}$.
The spin-tensor ω has the following important properties

$$\begin{aligned}
&\omega_{\mu\nu}{}^{AB} = \omega_{\mu\nu}{}^{BA} = -\omega_{\nu\mu}{}^{AB}, \\
&\omega_{\mu\nu\,AB}\,\omega^{\mu\nu\,CD} = 4\left(\delta_A^C\,\delta_B^D + \delta_A^D\,\delta_B^C\right), \\
&\omega_{\mu\nu\,AB}\,\omega_{\rho\sigma}{}^{AB} = 2\left(-i\varepsilon_{\mu\nu\rho\sigma} + g_{\mu\rho}g_{\nu\sigma} - g_{\mu\sigma}g_{\nu\rho}\right), \\
&\varepsilon_{\mu\nu}{}^{\rho\sigma}\,\omega_{\rho\sigma}{}^{AB} = 2i\,\omega_{\mu\nu}{}^{AB}.
\end{aligned} \tag{2.5}$$

We now define

$$\psi^{(+)}_{\mu_1\nu_1\cdots\mu_s\nu_s} = \omega_{\mu_1\nu_1}{}^{A_1B_1}\cdots\omega_{\mu_s\nu_s}{}^{A_sB_s}\,\phi_{A_1B_1\cdots A_sB_s}. \tag{2.6}$$

With frequent use of the properties (2.5), one eventually verifies the conditions (2.1) and (2.2) for the tensors $\psi^{(+)}$. One can also show that these conditions in turn imply the total symmetry of the spinors ϕ. It should be noted that the symmetry conditions (2.1), (2.2) are not necessarily independent of each other. The case of the representation D(0,s) and the corresponding tensors $\psi^{(-)}$ is similar except that all spinor indices are replaced by dotted indices and all equations are replaced by their complex conjugate. The conditions for the tensors ψ of the representation $D(s,0) \oplus D(0,s)$ are now obtained from the definition

$$\psi = \psi^{(+)} + \psi^{(-)}. \tag{2.7}$$

3. The Unitarity Condition

In this section we apply the unitarity condition (1.2) to the tensors $\psi_{\mu_1\nu_1 \cdots \mu_s\nu_s}(p)$ satisfying the symmetry conditions (S1-4) of section 2. We first state the result:

> The unitarity condition for the tensors $\psi_{\mu_1\nu_1 \cdots \mu_s\nu_s}(p)$ of the representation $D(s,0) \oplus D(0,s)$, $s = 1,2,\ldots$, is given by
>
> $$p_\lambda \psi_{\cdot\cdot\mu_i\nu_i\cdot\cdot}(p) + p_{\nu_i}\psi_{\cdot\cdot\lambda\mu_i\cdot\cdot}(p) + p_{\mu_i}\psi_{\cdot\cdot\nu_i\lambda\cdot\cdot}(p) = 0 \ ,(i=1,\ldots,s). \tag{3.1}$$
>
> For $s = 1$ there is the additional condition
>
> $$p_\lambda \psi^\lambda{}_\nu(p) = 0. \tag{3.2}$$

To prove this statement, we write the unitarity condition (1.2) as

$$\sum_{i=1}^{s} \left(V^{\lambda\mu_i}{}_\tau \psi^{\cdot\cdot\tau\nu_i\cdot\cdot} + V^{\lambda\nu_i}{}_\tau \psi^{\cdot\cdot\mu_i\tau\cdot\cdot} \right) = i s p^\lambda \psi^{\mu_1 \cdots \nu_s}, \tag{3.3}$$

where $V^{\lambda\mu}{}_\tau$ are the operators V^λ in the 4x4 Minkowski representation $D(\frac{1}{2},\frac{1}{2})$ and are given by

$$V^{\lambda\mu}{}_{\tau} = i\left(p^{\mu}\delta^{\lambda}{}_{\tau} - p_{\tau}g^{\lambda\mu}\right). \tag{3.4}$$

Inserting (3.4) into (3.3) and using the antisymmetry (S1) we have the condition

$$\sum_{i=1}^{s}\left(p^{(\lambda}\psi^{\cdot\cdot\mu_i\nu_i)\cdot\cdot} + p_{\tau}g^{\lambda\mu_i}\psi^{\cdot\cdot\tau\nu_i\cdot\cdot} + p_{\tau}g^{\lambda\nu_i}\psi^{\cdot\cdot\mu_i\tau\cdot\cdot}\right) = 0, \tag{3.5}$$

where (...) denotes the cyclic sum over the three indices inside the bracket. Contraction of (3.5) with $g_{\lambda\mu_1}$ and use of the symmetry conditions leads to the condition

$$p_{\lambda}\psi^{\lambda\nu_1\cdots\nu_s}(p) = 0. \tag{3.6}$$

Since the index λ in (3.6) can be shifted to any place with (S1,2), the condition (3.5) can be written as

$$\sum_{i=1}^{s} p^{(\lambda}\psi^{\cdot\cdot\mu_i\nu_i)\cdot\cdot} = 0. \tag{3.7}$$

Multiplying (3.7) with $\varepsilon_{\tau\lambda\mu_1\nu_1}$ we obtain

$$(s+2)\,\varepsilon_{\tau\lambda\mu_1\nu_1}p^{\lambda}\psi^{\mu_1\nu_1\cdots} = 0, \tag{3.8}$$

where we have used

$$\varepsilon_{\tau\lambda\mu_1\nu_1}\psi^{\mu_1\nu_1\cdots\lambda\cdots} = 0 \tag{3.9}$$

by (S3). Multiplication of (3.9) with $\varepsilon^{\tau\varrho\mu\nu}$ then proves (3.1).

It is easily seen that (3.1) and (3.6) in turn imply the original condition (3.5), thus together they are equivalent to (3.5). Furthermore, for $s \geqslant 2$, (3.6) follows from (3.1) because there exists an index, ρ say, which is not involved in the cyclic sum of (3.1) and which may be contracted with λ to yield

$$p_\lambda \psi_{\cdot\cdot\mu_i\nu_i\cdot\cdot}{}^{\lambda}{}_{\cdot\cdot} + p_{\nu_i} \psi_{\cdot\cdot\lambda\mu_i\cdot\cdot}{}^{\lambda}{}_{\cdot\cdot} + p_{\mu_i} \psi_{\cdot\cdot\nu_i\lambda\cdot\cdot}{}^{\lambda}{}_{\cdot\cdot} = 0 , \qquad (3.10)$$

where the second and third term vanish due to (S4). For s = 1 the condition (3.6) does not follow from (3.1) and must be imposed as an extra condition.

4. Conclusion

We have derived the wave equations for the simplest fields carrying a manifestly covariant unitary representation of the Poincaré group for zero mass and integer helicities $\pm s$. They are given by (2.1) and (3.1), (3.2). In particular, those equations which are differential equations stem from the condition that the representation is unitary. To show that the equations are quite familiar, we now specify for the cases s = 1,2

<u>s=1</u>: $$\psi_{\mu\nu} = -\psi_{\nu\mu} , \quad p_\lambda \psi_{\mu\nu} + p_\nu \psi_{\lambda\mu} + p_\mu \psi_{\nu\lambda} = 0 , \quad p_\lambda \psi^{\lambda}{}_{\nu} = 0 . \qquad (4.1)$$

(4.1) are the <u>Maxwell equations</u> for the electromagnetic field $\psi_{\mu\nu}(p)$.

<u>s=2</u>: $$\psi_{\mu\nu\rho\sigma} = -\psi_{\nu\mu\rho\sigma} = \psi_{\rho\sigma\mu\nu} , \qquad (4.2)$$

$$\psi_{\mu\nu\rho\sigma} + \psi_{\mu\sigma\nu\rho} + \psi_{\mu\rho\sigma\nu} = 0 , \qquad (4.3)$$

$$\psi^{\lambda}{}_{\nu\lambda\sigma} = 0 , \qquad (4.4)$$

$$p_\lambda \psi_{\mu\nu\rho\sigma} + p_\nu \psi_{\lambda\mu\rho\sigma} + p_\mu \psi_{\nu\lambda\rho\sigma} = 0 . \qquad (4.5)$$

We may identify $\psi_{\mu\nu\varrho\sigma}$ with the Riemann curvature tensor of general relativity. (4.2) and (4.3) are then the usual symmetry properties of this tensor, (4.4) tells us that the Ricci tensor $\psi_{\nu\sigma} = g^{\mu\varrho} \psi_{\mu\nu\varrho\sigma}$ vanishes, and (4.5) are the Bianchi identities in the linearized theory where covariant derivatives are replaced by ordinary derivatives. The field $\psi_{\mu\nu\varrho\sigma}(p)$ thus describes the <u>linearized theory of gravitation in vacuo.</u>

Up to now, we have restricted ourselves to the representations D(s,0) $\oplus$ D(0,s). A similar analysis can be made for the representations D(n+s,n) $\oplus$ D(n,n+s) , n = 0,$\frac{1}{2}$,1,..., s = 1,2,...; these are the most general representations compatible with helicities $\pm$ s and containing as few irreducible representations of the Lorentz group as possible. The representations are carried by tensors $\psi_{\lambda_1 \cdots \lambda_{2n} \mu_1 \nu_1 \cdots \mu_s \nu_s}$ which satisfy complicated symmetry conditions. However, it can be shown that the corresponding unitarity conditions then imply that these tensors can be written as derivatives of D(s,0) $\oplus$ D(0,s) tensors, viz.

$$\psi_{\lambda_1 \cdots \lambda_{2n} \mu_1 \nu_1 \cdots \mu_s \nu_s}(p) = p_{\lambda_1} \cdots p_{\lambda_{2n}} \psi_{\mu_1 \nu_1 \cdots \mu_s \nu_s}(p). \tag{4.6}$$

We therefore conclude that nothing is gained by this generalization of D(s,0) $\oplus$ D(0,s).

References

[1] U. Niederer, L.O'Raifeartaigh, Fortschritte der Physik 22, 131 (1974)

[2] U. Niederer, Group theory of the massless spin 2 field and gravitation, to appear in GRG-Journal.-

SOME CONSIDERATIONS ABOUT NELSON'S DERIVATION OF SCHROEDINGER EQUATION

E.Onofri

Istituto di Fisica dell'Università di Parma, Parma, I-43100
and
Istituto Nazionale di Fisica Nucleare, Sezione di Milano.

§ 0. Schroedinger equation for a particle moving in a potential V(x) in n-dimensions was rederived by Nelson by purely probabilistic considerations starting from Newtonian mechanics (see Nelson 1966, 1967). While there have been several attempts to generalize the theory to the relativistic domain and to the case of spinning particles (see DeLaPena-Auerbach 1969, 1971; Caubet 1975), the conceptual aspects of this approach have not encountered much consideration. I shall discuss the following points, in the following:

1) the contrast with the usual approach is entirely due to an improper use of the term "Markov process";

2) the measure on path space, defined by the stochastic process x(t) is, by construction, beyond any experimental verification;

3) from a technical point of view, the theory provides the simplest method to derive Feynman-Kac formula for the Green's function.

Points (1) and (2) are perhaps of "public domain", but I have not seen them stated explicitly. Point (3) is essentially due to Albeverio and Höegh-Krohn (1973) and can be found in Ezawa, Klauder and Shepp (1974).

§ 1 . The theory is given in terms of a stochastic process x(t) which is characterized by the following requirements: i) x(t) is a diffusion process with diffusion coefficient $\nu = \hbar/2m$, the drift coefficient b(x(t),t) being irrotational; ii) the process admits time-reversal invariance; iii) for a suitable definition of mean acceleration this quantity must be proportional to the classical force -grad V/m . Following Nelson's arguments, we end up with a process which is characterized by two functions R(x,t) and S(x,t) satisfying a system of two coupled non-linear differential equations of 1st order in time, which is equivalent to Schroedinger equation if we put $\psi(x,t) = \exp\{R(x,t)+iS(x,t)\}$. The point is that the drift coefficient $b(x,t) = \hbar/m \operatorname{grad}(R + S)$ is not a preassigned vectorfield, but is a function of the initial state of the particle. The conclusion is that the whole process $x(t)$ appears as some kind of envelope of a family of diffusion processes, but it is not a Markov process in itself. In fact, the knowledge of $x(t_0)$ (as a random variable) is not sufficient to determine the process; to do that, it is necessary to know also the expectation of $\frac{x(t_0) - x(t_0-h)}{h}$ (conditional on $x(t_0)$) for arbitrarily small h; this is characteristic of a non-Markov process. The essential feature of Markov processes, the transition function, can be defined, but it is dependent on the initial state of the particle; this is qualitatively different from the diffusion processes which arise in applications (heat equation, classical statistical mechanics, etc.) and there is no point to insist that they are the same.

§ 2. Let us suppose that a solution $\psi(x,t)$ of Schroedinger equation is known. Then we can write down the stochastic differential equation

$$dx(t) = b(x(t),t)\, dt + d\, w(t)$$

where $b(x,t) = \hbar/m \operatorname{grad}[\log|\psi| + \arg\psi]$ and $w(t)$ is the Wiener process with covariance $\hbar/m$. If we are able to solve this equation, the result is a stochastic process $x(t)$, that is a measure μ on path space. For example μ gives the probability that the trajectory of the particle belong to a certain cylinder set $C = \{x(t_1)\in\Gamma_1;\ x(t_2)\in\Gamma_2;\dots;\ x(t_n)\in\Gamma_n\}$. $\mu(C)$ is easily given in terms of the diffusion process $x(t)$, namely if $P(x,t;\Gamma,s)$ is the transition function, it holds:

$$\mu(C) = \int_{\Gamma_1}\cdots\int_{\Gamma_n} |\psi(x_1,t_1)|^2 dx_1\ P(x_1,t_1;dx_2,t_2)\dots P(x_{n-1},t_{n-1};dx_n,t_n)$$

The point that we want to stress is that this value of $\mu(C)$ is valid only if we do not try to check it! Actually a method to measure $\mu(C)$ consists in killing the particle at time $t_1, t_2,\dots,t_{n-1}$ if it is not in $\Gamma_1, \Gamma_2,\dots, \Gamma_{n-1}$ and counting the number of particles reaching Γ_n at time t_n. In this situation Quantum Mechanics gives the following expression for the probability of detecting the particle in Γ_n at time t_n:

$$P(x(t_n)\in\Gamma_n) =$$

$$\int_{\Gamma_n} dx_n \left| \int_{\Gamma_1}\cdots\int_{\Gamma_{n-1}} \psi(x_1,t_1)G(x_1,t_1;x_2,t_2)\ldots G(x_{n-1},t_{n-1};x_n,t_n)\, dx_1\ldots dx_{n-1}\right|^2$$

Of course this value cannot coincide with $\mu(C)$ since this latter is an additive functional of sets Γ's, while P is not. There is no contradiction, however; according to Nelson's theory, we must determine μ starting from the wave funtion, which is not the original one $\psi(x,t)$ but has been "projected" in $\Gamma_1, \Gamma_2, \ldots, \Gamma_{n-1}$ at times $t_1, t_2, \ldots, t_{n-1}$. We have to calculate this new wave function $\tilde{\psi}(x,t)$ and then to solve the stochastic differential equation for the new stochastic process $\tilde{x}(t)$, thus obtaining the new measure $\tilde{\mu}$ holding in this case. The conclusion is that the measure μ cannot give any additional information which is not explicitly contained in the wave function; consequently the tremendous task of solving the stochastic differential equation for a given $\psi(x,t)$ is unnecessary.

§ 3. The third point we wish to discuss is the technical value of Nelson's approach. Let us consider the ground state $\varphi_o(x)$ of the Hamiltonian $H = p^2/2m + V(x)$, and let E_o be the corresponding eigenvalue; $\varphi_o(x)$ can be taken to be real and nowhere vanishing. The Fokker-Planck equation for the process $x(t)$ is given by

$$\frac{\partial \rho}{\partial t} = - \operatorname{div}\left(\frac{\hbar}{m}\frac{\operatorname{grad}\varphi_o}{\varphi_o}\rho\right) + \frac{\hbar}{2m}\Delta\rho$$

Let $\rho = \varphi_o \psi$; it follows

$$-h\frac{\partial \psi}{\partial t} = -\frac{\hbar^2}{2m}\Delta\psi + (V(x)-E_o)\psi$$

which is formally Schroedinger equation with imaginary time (Albeverio and Hoegh-Krohn, 1973). It follows that the fundamental solution of the Fokker-Plank equation is given by

$$P(x,t;y,s) = \frac{\varphi_o(y)}{\varphi_o(x)}\, G(x,-it;y,-is)$$

This means that we can obtain the Green's function G through the following steps: calculate the ground state; calculate the transition probability density for the process associated to the ground state;

analitically continue to imaginary times.

A general formula is known for the transition probability density (see Gihman-Skorohod, 1972); the result, in our case, is just Feynman-Kac formula

$$\langle x \mid e^{-t(H-E_0)} \mid y \rangle = \int_{\substack{w(o)=x \\ w(t)=y}} \left[\exp - \frac{1}{\hbar} \int_0^t V\,(w(\tau))\, d\tau \right] d\mu_W$$

This connection between P(x,t;y,s) and the Green's function can be used to prove that E(x(t)x(s)) is just the analytic continuation of $\langle \varphi_0 | T(x_{op}(t)x_{op}(s)| \varphi_0 \rangle$ to imaginary times; in general, Feynman-Kac-Nelson formula can be obtained this way (Simon, 1974).

Acknowledgments

Stimulating discussions with F.Guerra, L.Accardi, A.Scotti, L.Galgani and M.Casartelli are gratefully acknowledged.

References

ALBEVERIO, S. and HOEGH-KROHN, R. (1973): "A remark on the connection between stochastic mechanics and the heat equation", Mat.Inst.Univ. Oslo, preprint No.27.

CAUBET, J.P. (1975): "Relativistic Brownian Motion", in "Probabilistic Methods in Differential Equations", Ed.by M.A.Pinsky, Springer-Verlag, Berlin.

DE LA PENA-AUERBACH, L. (1969): J.Math.Phys.10, 1620;
(1971): " " 12, 453.

EZAWA, H., KLAUDER, J.R. and SHEPP, L.A. (1974): Ann.of Phys. 88(2)588.

GIHMAN,I.I. and SKOROHOD, A.V. (1972): "Stochastic Differential Equations", Springer-Verlag, Berlin.

NELSON, E. (1966): Phys.Rev.150, 1079;
(1967): "Dynamical Theories of Brownian Motion", Princeton University Press, Princeton.

SIMON, B. (1974): "$P(\phi)_2$ Euclidean (Quantum) Field Theory", Princeton University Press, Princeton.

THE "GALILEAN" COMPONENTS OF A POSITION OPERATOR FOR THE PHOTON

M. J. PERRIN

Laboratoire de Physique Mathématique, Université de Dijon

Faculté des Sciences Mirande, 21000 - DIJON, France.

An acceptable solution to the old problem of the localizability of particles in relativistic quantum mechanics has been constructed by Newton and Wigner [1] within the framework of U.I.R. of the Poincare group.
However in this scheme no localization is possible for particles with zero mass and spin greater than 1/2. Recently there has been a renewed interest in this problem and a position operator for massless particles, with commuting components has been independently exhibited by J. Bertrand [2] and E. Angelopoulos et al. [3] by generalizing Wightman's reformulation [4] of the Newton-Wigner position operators.

In another context it is well known that the Poincare algebra contains the two-dimensional extended Galilean algebra. This allows to associate to any relativistic massless particle a Galilean one with a variable mass in a two-dimensional space which is orthogonal to its propagation direction : the "transverse plane" * . Then a "Galilean" position operator can be easily defined from the special Galilean

---

* In fact the orthogonal space to a light-like direction $\mathcal{T}^o(1)$ is a three-plane $\mathcal{T}^o(3)$ all vectors of which are space-like except the vectors which are collinear to the direction $\mathcal{T}^o(1)$. Then the "transverse plane" is a representative element of the class $\mathcal{T}^o(3)/\mathcal{T}^o(1)$ of the space-like two-planes orthogonal to the light-like direction $\mathcal{T}^o(1)$.

transformations generators in this plane. The so-obtained position operator is it related to the above mentioned position operators deduced from systems of imprimitivity ?

On the other hand the associated invariance algebra for massless particles is not only the Poincare algebra but the conformal one. In a previous paper [5] a non-relativistic decomposition of the conformal algebra involving the two-dimensional extended Schrödinger algebra, has been exhibited. Has this decomposition something to do with the definition of a massless particle position operator ?

We intend to answer both foregoing querries in what follows.

Let us suppose that the relativistic massless particle of quadrimomentum p^μ , $\mu = 0,1,2,3$ $(p^2 = 0)$, propagates along the third axis. Then we associate to it a Galilean particle of variable mass $M = p_0 + p^3$, the evolution of which with respect to the "time" $x_0 + x^3$ is described by the Hamiltonian $T = 1/2(p_0 - p^3)$. The generators of the two-dimensional extended Galilean algebra $\tilde{\mathcal{G}}_2$ are expressed in terms of the Poincare generators in the following manner * :

$$
\begin{aligned}
P_n &= p^n \qquad n = 1,2 \\
K_1 &= -(j_2 + k_1) \\
K_2 &= j_1 - k_2 \\
J &= j_3 \\
T &= \frac{1}{2}(p_0 - p^3) \\
M &= p_0 + p^3
\end{aligned}
\qquad (1)
$$

---

* We use again the notations of Ref. 5 : small letters denote the generators of the Poincare algebra and of the conformal one while capital letters denote the generators of the "non-relativistic" algebras i.e. the Galilean algebra and the Schrödinger algebra.

In Galilean quantum mechanics the position operator belongs to the field of quotients of the enveloping algebra of the extended Galilean algebra and is defined by :

$$Q_n = \frac{K_n}{M} \qquad (2)$$

Then from relations (1) the two components of the position operator in the "transverse plane" are given by :

$$Q_1 = -\frac{j_2 + k_1}{p_o + p^3} \quad , \quad Q_2 = \frac{j_1 - k_2}{p_o + p^3} \qquad (3)$$

Let us then consider the zeromass, discrete spin, positive energy I.U.R. of the Poincare group characterized by :

$$p^2 = 0 \quad , \quad p_o = |\underline{p}|$$

and the eigenvalues s of the helicity $\Sigma = \frac{\underline{p}\cdot\underline{j}}{|\underline{p}|}$ (2s ≷ 0,integer). Moreover for these representations the components of the Pauli-Lubanski vector :

$$W^\mu = -\frac{1}{2}\,\varepsilon_{\mu\nu\rho\sigma}\, M^{\nu\rho}\, p^\sigma$$

are such that $\quad W^\mu = s\,p^\mu \qquad (W^2 = 0)$

These representations are induced from the finite-dimensional unitary representations of $E(2) \otimes T_4 = (T_2 \,\square\, SO(2)) \otimes T_4$ where E(2) is the two-dimensional space Euclidean group which stabilizes a light-like vector.

Let us briefly recall the main points of the construction of these representations. Since the finite dimensional U.I.R. of E(2) are those in which the translational subgroup T_2 is trivially represented, the irreducible representations of the little group are one dimensional and are characterized by the eigenvalues s of the rotational generator in the chosen transverse plane. To any light-like vector ($p^2 = 0$) is associated a set of states $|[p],s\rangle$ such that :

$$\begin{aligned} p\,|[p],s\rangle &= p\,|[p],s\rangle \\ j(p)\,|[p],s\rangle &= s\,|[p],s\rangle \\ t_i(p)\,|[p],s\rangle &= 0 \qquad i = 1,2 \end{aligned} \qquad (4)$$

where $t_i(p)$ and $j(p)$ are the components of the restriction $W(p)$ of the Pauli-Lubanski vector in the tetrad $[p]$:

$$W(p) = \sum_{i=1,2} t_i(p) + j(p).p \qquad (5)$$

As usual $[p]$ denotes the set of vectors $\{p\ ,\ n_1(p),\ n_2(p),\ n_o(p)\}$ where $n_1(p)$ and $n_2(p)$ are two orthogonal space-like vectors, orthogonal to p, and $n_o(p)$ is a positive time-like vector.

Let $\mathcal{H}_p$ be the space spanned by the vectors $|[p],s\rangle$ with s staying fixed and for instance p varying on the future light-cone. In this space an element E of the little group is represented by :

$$e^{-i\alpha\, j(p)}\ |[p],s\rangle = e^{-is\alpha}|[p],s\rangle \qquad (6)$$

Then we can build an I.U.R. of the Lorentz group on $\mathcal{H}_p$. For this we use the representation of the little group of the vector $\overset{o}{p} = (1,0,0,-1)$ which is given by :

$$E(\overset{o}{p}) = \begin{bmatrix} 1+\frac{\alpha_1^2+\alpha_2^2}{2} & -(\alpha_1\cos\varphi+\alpha_2\sin\varphi) & \alpha_1\sin\varphi-\alpha_2\cos\varphi & \frac{\alpha_1^2+\alpha_2^2}{2} \\ \alpha_1 & -\cos\varphi & \sin\varphi & \alpha_1 \\ \alpha_2 & -\sin\varphi & -\cos\varphi & \alpha_2 \\ \frac{\alpha_1^2+\alpha_2^2}{2} & -(\alpha_1\cos\varphi+\alpha_2\sin\varphi) & \alpha_1\sin\varphi-\alpha_2\cos\varphi & \frac{\alpha_1^2+\alpha_2^2}{2}-1 \end{bmatrix}$$

and the associated reference tetrad which involves the vectors :

$$n_1(\overset{o}{p}) = (0,1,0,0) \quad , n_2(\overset{o}{p}) = (0,0,1,0) \quad , n_o(\overset{o}{p}) = (1,0,0,0)\ .$$

The Lorentz transformation which maps the tetrad $[\overset{o}{p}]$ on the tetrad $[p]$ is defined and usually denoted $[p]$; then by using the general induction method the operator $U^{a,\Lambda}$ which represents on $\mathcal{H}_p$ the Poincare transformation (a,Λ) is given by :

$$U^{a,\Lambda}|[p],s\rangle = e^{i(\Lambda p)a}\quad \mathcal{D}^{s}(\,[\Lambda p]^{-1}\,\Lambda\,[p]\,)\,|[p],s\rangle \qquad (8)$$

To construct the zeromass representations it is convenient to introduce the Lorentz transformation which possesses the following properties :

$$L(p)\,\overset{o}{p} = p \quad , \qquad L(p)\,u_j = u_j \tag{9}$$

where the vectors u_j , j = 1,2, are orthogonal to $\overset{o}{p}$ and p :

$$u_1 = \Big(\frac{p^1}{p_o+p^3}\,,\ 1\ ,\ 0\ ,\ -\frac{p^1}{p_o+p^3}\Big)$$
$$u_2 = \Big(\frac{p^2}{p_o+p^3}\,,\ 0\ ,\ 1\ ,\ -\frac{p^2}{p_o+p^3}\Big) \tag{10}$$

Then the vectors $\overset{o}{p}$, p, u_1 and u_2 form a complete (non orthonormed) basis in which L(p) can be written :

$$L(p) = \frac{1}{(\overset{o}{p}.p)}\ (p^o \otimes \overset{o}{p} + p \otimes p) - u_1 \otimes u_1 - u_2 \otimes u_2 \tag{11}$$

and it is easy to show that L(p) is a involutive automorphism.

From (8) and (11) it is then possible to exhibit a basis $\{\overset{o}{M}\}$ of operators M of $i\,dU^{a,\wedge}$ which represent the generators of the Lie algebra of the Lorentz group on $\mathcal{H}_p$ spanned by the vectors $|L(p),s\rangle$, namely $\{\overset{o}{M}\} = \{j,k\}$ is given by :

$$j_1 = -i\Big(p^2\frac{\partial}{\partial p^3} - p^3\frac{\partial}{\partial p^2}\Big) + s\,\frac{p^1}{p_o+p^3}$$
$$j_2 = -i\Big(p^3\frac{\partial}{\partial p^1} - p^1\frac{\partial}{\partial p^3}\Big) + s\,\frac{p^2}{p_o+p^3}$$
$$j_3 = -i\Big(p^1\frac{\partial}{\partial p^2} - p^2\frac{\partial}{\partial p^1}\Big) + s \tag{12}$$
$$k_1 = -i\,p_o\frac{\partial}{\partial p^1} - s\,\frac{p^2}{p_o+p^3}$$
$$k_2 = -i\,p_o\frac{\partial}{\partial p^2} + s\,\frac{p^1}{p_o+p^3}$$
$$k_3 = -i\,p_o\frac{\partial}{\partial p^3}$$

in which we recognize the operators used by J.S. Lomont and H.E. Moses in Ref. 6.* Then by inserting (12) in (3) we are led to the two following components of the position operator :

$$Q_1 = i\frac{\partial}{\partial p^1} - i\frac{p^1}{p_o+p^3}\frac{\partial}{\partial p^3}$$
$$Q_2 = i\frac{\partial}{\partial p^2} - i\frac{p^2}{p_o+p^3}\frac{\partial}{\partial p^3} \qquad (13)$$

These expressions are analogous to those of the two first components of the position operator obtained in Ref.2 and 3. Moreover, in these references a third component is also exhibited which describes the localization of the particle along its propagation direction, then we have looked for what could be such a component in our analogy with a non-relativistic problem.

The argument of W.Pauli [8] about the non-existence of time operator in quantum mechanics does not apply here because the associated Galilean particle is of variable mass and does not possess a discrete bound spectrum. Therefore we can look for a time position operator θ such that $[T,\theta] = i$. But such an operator commuting with Q_1 and Q_2 cannot be exhibited neither in the enveloping algebra of $\tilde{\mathcal{G}}_2$ nor in that of the conformal algebra, hence no satisfactory answer can be found from the Galilean analogy in what concerns the existence of a third component for the position operator of massless particles.

---

* The choice of another transformation [p] than L(p) is equivalent to make a change of basis. We show that :

$$M\,|[p],s\rangle = e^{i\varphi}\,\overset{\circ}{M}\,|L(p),s\rangle = \widehat{M}\,|[p],s\rangle$$

where $\widehat{M} = e^{i\varphi}\,\overset{\circ}{M}\,e^{-i\varphi}$ with $\varphi = \mathrm{Arc\ tg}\,\dfrac{(u_2.n_1(p))}{(u_1.n_1(p))} = -\,\mathrm{Arc\ tg}\,\dfrac{(u_1.n_2(p))}{(u_2.n_2(p))}$ (14)

It is worth noticing that the usual zeromass representation [7] is written with respect to a basis which corresponds to a traditional choice of [p] leading to

$$\varphi = \mathrm{Arc\ tg}\,\frac{p^1}{p^2}$$

Let us now come to the second point we want to discuss here.

Owing to the symmetric decomposition of the conformal algebra exhibited in Ref. 5 and 9, namely

$$\mathcal{SO}(4,2) \cong \begin{bmatrix} \mathcal{H}_2 \\ \mathcal{H}_2^\star \end{bmatrix} \boxtimes (\mathcal{SO}(2) \oplus \mathcal{SO}(2,1) \oplus \mathbb{R}^\wedge)$$

it is easy to see that the largest subalgebra of $\mathcal{SO}(4,2)$ which closes with the twO-dimensional position operator Q is $\widetilde{\mathcal{S}ch_2} \,\square\, \Lambda \cong \mathcal{H}_2 \,\square\, (\mathcal{SO}(2) \oplus \mathcal{SO}(2,1) \oplus \mathbb{R}^\wedge)$ which does not contain the Poincare algebra. Then it is interesting to build a representation of this algebra in the above defined Hilbert space $\mathcal{H}_p$. Such a representation can be constructed since $\widetilde{\mathcal{S}ch_2} \,\square\, \Lambda$ is a subalgebra of the conformal algebra and it is well known [10,11] that the zeromass, discrete spin U.I.R. of the Poincare group can be extended, in a unique way, to the most degenerate discrete series of I.U.R. of SU(2,2) which is locally isomorphic to the conformal group.

The following expressions of the dilatations and special conformal transformations are obtained :

$$\begin{aligned}
d &= -i(p.\frac{\partial}{\partial p} + i) \\
c_0 &= -p_0(\frac{\partial}{\partial p})^2 - 2i\frac{s}{p_0+p^3}(p^1\frac{\partial}{\partial p^2} - p^2\frac{\partial}{\partial p^1}) + 2\frac{s^2}{p_0+p^3} \\
c_1 &= -p^1(\frac{\partial}{\partial p})^2 + 2(p.\frac{\partial}{\partial p} + 1)\frac{\partial}{\partial p^1} + 2i\, s\frac{p^2}{p_0+p^3}\frac{\partial}{\partial p^3} - 2i\, s\frac{\partial}{\partial p^2} \qquad (15) \\
c_2 &= -p^2(\frac{\partial}{\partial p})^2 + 2(p.\frac{\partial}{\partial p} + 1)\frac{\partial}{\partial p^2} - 2i\, s\frac{p^1}{p_0+p^3}\frac{\partial}{\partial p^3} + 2i\, s\frac{\partial}{\partial p^1} \\
c_3 &= -p^3(\frac{\partial}{\partial p})^2 + 2i\frac{s}{p_0+p^3}(p^1\frac{\partial}{\partial p^2} - p^2\frac{\partial}{\partial p^1}) - 2\frac{s^2}{p_0+p^3} + 2(p.\frac{\partial}{\partial p} + 1)\frac{\partial}{\partial p^3}
\end{aligned}$$

The representation of the generators of $\widetilde{\mathcal{S}ch_2} \,\square\, \Lambda$ can be easily deduced through their expressions given in Ref. 5 and by using (12) and (15). We obtain :

$$P_n = p^n \qquad (n = 1,2)$$

$$K_1 = M\,Q_1 = -i(p_o+p^3)\frac{\partial}{\partial p^1} + i\,p^1\frac{\partial}{\partial p^3}$$

$$K_2 = M\,Q_2 = -i(p_o+p^3)\frac{\partial}{\partial p^2} - i\,p^2\frac{\partial}{\partial p^3}$$

$$M = p_o+p^3$$

$$J = j_3 = -\,i(p^1\frac{\partial}{\partial p^2} - p^2\frac{\partial}{\partial p^1}) + s. \qquad (16)$$

$$T = \frac{1}{2}(p_o-p^3)$$

$$C = \frac{1}{2}(c_o+c^3) = -\frac{1}{2}(p_o+p^3)(\frac{\partial}{\partial p})^2 + (p.\frac{\partial}{\partial p} + 1)\frac{\partial}{\partial p^3}$$

$$D = d - k_3 = -\,i(p.\frac{\partial}{\partial p} + 1 - p_o\frac{\partial}{\partial p^3})$$

$$\Lambda = -(d + k_3) = i\,(p.\frac{\partial}{\partial p} + 1 + p_o\frac{\partial}{\partial p^3})$$

It is worth noticing that the helicity s only appears in the rotation generator. This fact justifies the choice of the above used zeromass, discrete spin I.U.R. of the Poincare group since the Galilean structure lies in the transverse plane and since the helicity is by definition the spin projection onto the propagation direction.

To summarize we have shown that :

- the Galilean structure of the conformal algebra allows to construct a two component position operator for a massless particle which corresponds to the localizability of such a particle in a plane orthogonal to its propagation direction. But a third component, like the one proposed in Ref. 2 and 3 , cannot be exhibited from the "non-relativistic" analogy.
- the two-dimensional extended Schrödinger algebra and an extra dilatation which form a maximal subalgebra of the conformal one, is the largest subalgebra of $SO(4,2)$ which closes with the two "Galilean components" of the position operator.

REFERENCES :

1 T.D. NEWTON, E.P. WIGNER : Rev. Mod. Phys. 21 400 (1949).

2 J. BERTRAND : Nuov. Cim. 15 A 281 (1973).

3 E. ANGELOPOULOS, F. BAYEN, M. FLATO : Phys. Scripta 9 173 (1974).

4 A.S. WIGHTMAN : Rev. Mod.Phys. 34, 845 (1962).

5 G. BURDET, M. PERRIN, P. SORBA : Comm. Math. Phys. 34 85 (1973).

6 J.S. LOMONT and H.E. MOSES : Jour. Math. Phys. 3 405 (1962).

7 Iu. SHIROKOV : Sov. Phys. J.E.T.P. 6 919 (1958).

8 W. PAULI : Handbuch der Physik Vol. 24/1 143 .

9 M. PERRIN : Proc. 3rd Int. Coll. on Group. Th. Meth. Phys. 288 (1974).

10 D. STERNHEIMER : Jour. Math. Pures appl. 47 289 (1968).

G. MACK, I. TODOROV : Jour. Math. Phys. 10 2078 (1969).

11 J. MICKELSON, J. NIEDERLE : Jour. Math. Phys. 13 23 (1972).

Group Theoretic Aspects of Gibbs Space

A. Rieckers

Institut für Theoretische Physik

Universität Tübingen, Germany

1. Introduction

Since the days of Gibbs (Gibbs 1873) and Van der Waals (Van der Waals 1891, Korteweg 1891, Van der Waals and Ph. Kohnstamm 1912) geometrical and topological (in the sense of "analysis situs") considerations are characteristic for equilibrium thermodynamics. Later on Tisza (Tisza 1951, 1966) introduced new concepts and raised new questions concerning the geometrical interpretation of thermodynamic state spaces and coordinate transformations. He seems also to be the first to use the name "Gibbs space" for the set of extensive state (resp. density) variables. In the present investigation we deal with the following three of Tisza's problems:

(i) discussion of thermodynamic stability by means of separation of variables,

(ii) definition of "interaction" between thermodynamic state variables,

(iii) group theoretic properties of the separating coordinate transformations.

It turns out that the complete treatment of (i) requires curvilinear coordinate transformations in contradistinction to Tisza's linear (so-called restricted equiaffine) transformations, which are separating up to second order only. But many of the properties of the restricted equiaffine transformations can be generalized to the nonlinear case. More than that, the meaning of some features of the equiaffine matrices is understood more clearly from the general point of view. Thus, their triangular form can be related to the existence of the thermodynamic potentials in a distinguished order (here called condition L), and their principal minors being equal to unity can be connected to the transversality of certain curves in Gibbs space. Also the splitting of the equiaffine group into a descending series of subgroups can be explained in quite general terms. By studying not only the first but all finite order approximations to the nonlinear separating coordinate transformations one obtains a whole family of Lie groups which exhibit similar features as the restricted equiaffine group and which might be of interest for the discussion of higher order critical points.

If looked upon from the active point of view, the separating coordinate transformations describe in which way a locally ideal system develops into a system with interacting state variables and how the increasing interaction leads to instabilities and critical points. Or, to put into geometrical terms: a local piece of the state surface with separated curvatures (along the coordinate axes) is transformed into surface pieces of more and more mixed curvatures until there show up plaits with their plaitpoints of various kinds. Thus, the separating coordinate transformations may be useful for a general morphology of thermodynamic systems.

2. Model functions and thermodynamic stability

The equilibrium states of a thermodynamic system Σ can be described by means of a set of extensive variables $Z = (Z_1 \cdots Z_k, Z_{k+1})$ which satisfy conservation laws (Tisza 1961, Stumpf and Rieckers 1975). It is convenient to supplement Z by a set of formal (Tisza 1961) variables $\Omega = (\Omega_1 \cdots \Omega_\ell)$.

The Ω_i which count particles (or quasi particles or elementary cells) with special properties are related to Landau's order parameters (Landau 1937, Falk 1968). Their physical (reduced) values depend upon Z, but in a formal manner they are varied independently. Let $x := (Z_1 \cdots Z_k, \Omega_1 \cdots \Omega_\ell)/Z_{k+1} = (z_1 \cdots z_k, \omega_1 \cdots \omega_\ell) \in \mathbb{R}^n, n = k + \ell$, denote the densities, then the specific entropy function $s_\Sigma(x)$ contains all the thermodynamic information about Σ. Σ_1 and Σ_2 are of the same kind if x^1_i and x^2_i have the same dimension for all $1 \leq i \leq n$. The equivalence class of thermodynamic systems which are of the same kind as Σ is denoted by $\mathcal{C}(\Sigma)$. The set of ordered tuples of density variables, which can be assumed of the members of $\mathcal{C}(\Sigma)$ is called the Gibbs space $\mathcal{G}^{(n)}(\mathcal{C})$ of $\mathcal{C}$.

There is no single thermodynamic system $\Sigma \in \mathcal{C}$ whose state variables can assume all values $x \in \mathcal{G}^{(n)}(\mathcal{C})$. Nevertheless it is useful to extend the domain of definition of $s_\Sigma(x)$ to the whole of $\mathcal{G}^{(n)}(\mathcal{C})$. An extended fundamental function is called a model function. The states which can be assumed of Σ are named stable. Define

$$p_i(x) := \frac{\partial s(x)}{\partial x_i}, \quad s^*(x; x^\circ) := s(x) - \sum_{i=1}^{k} p_i(x^\circ) z_i . \tag{2.1}$$

Definition 2.1: A state x° is stable for Σ if

$$s^*(x^\circ; x^\circ) = \max_{x \in \mathcal{U}(x^\circ)} s^*(x; x^\circ) \tag{2.2}$$

where $\mathcal{U}(x^\circ)$ denotes a neighbourhood of x°.

The set of all ordered tuples of the form $(p_1 \cdots p_m, x_{m+1} \cdots x_n)$ which can be assumed of the members of $\mathcal{C}$ is denoted by $\mathcal{H}^{(m)}(\mathcal{C})$.

Definition 2.2: Let $s(x)$ be a twice differentiable model function. Define the mappings

$$\gamma^{(m)} : \mathcal{G}^{(n)} \to \mathcal{H}^{(m-1)} , \quad 1 \leq m \leq n ,$$

by
$$\gamma^{(m)}(x) = (p_1(x) \cdots p_{m-1}(X) , x_m \cdots x_n) , \tag{2.3}$$

where $\gamma^{(1)}$ is the identity map. We say that $s(x)$ satisfies the condition L in $\mathcal{U}(x^\circ)$, if $\gamma^{(m)-1} =: \varphi^{(m)}$ exists on $\mathcal{U}^{(m)}(x^\circ) := \gamma^{(m)} \mathcal{U}(x^\circ)$ for all $1 \leq m \leq n$.

The condition L guarantees the existence of the Legendre transformations of $s(x)$ (the so-called Massieu-Planck functions) in $\mathcal{U}^{(m)}(x^\circ)$ in a certain order.

Definition 2.3: Let $s(x)$ satisfy the condition L in $\mathcal{U}(x^\circ)$. We denote by $I_m \subset \mathbb{R}^1$ the open intervals, such that $(p_1^\circ \cdots p_{m-1}^\circ \, y_m \, x_{m+1}^\circ \cdots x_n^\circ) \in \mathcal{U}^{(m)}(x^\circ)$ for $y_m \in I_m$. We define n curves κ_m, $1 \leq m \leq n$, in $\mathcal{U}(x^\circ)$ as follows

$$\kappa_m := \{ x ; x = \varphi^{(m)}(p_1^\circ \cdots p_{m-1}^\circ \, y_m \, x_{m+1}^\circ \cdots x_n^\circ), \, y_m \in I_m \} . \tag{2.4}$$

Theorem 2.4 (Rieckers to be publ.): Let $s(x)$ satisfy the condition L in $\mathcal{U}(x^\circ)$. Then x° is stable, iff $s^*(x;x^\circ)$ is maximal in x° along the curves κ_m, $1 \leq m \leq n$.

Example 2.5: Consider the family of functions

$$s_a^*(x_1, x_2; 0) := (x_2^2 - x_1)(x_1 - x_2^2 - a e^{-1/x_2^2}) , \quad a \in \mathbb{R}^1 . \tag{2.5}$$

Then the s_a^* possess identical Taylor series at $x^\circ = (0,0)$ and they are all maximal in x° along all algebraic curves passing through the origin. Applying 2.4 one finds

$$s_a^*|_{\kappa_1} = - x_1^2$$
$$s_a^*|_{\kappa_2} = \frac{a^2}{4} e^{-2/x_2^2} , \tag{2.6}$$

which tells us that s_a^* has a minimum along the transcendental curve κ_2 if $a \neq 0$. Hence $x^\circ = (0,0)$ is stable only, if $a = 0$. This result cannot be obtained by studying the Taylor expansions of s_a^* .

3. General properties of the separating coordinate transformations

Definition 3.1: A mapping $\tau : \mathcal{U}(x^\circ) \to \mathcal{U}(x^\circ)$ is called a separating coordinate transformation if it is twice differentiable and satisfies

$$S: \qquad \kappa_m = \{ x;\ \tau(x_1^\circ \cdots x_{m-1}^\circ\ y_m\ x_{m+1}^\circ \cdots x_n^\circ),\ y_m \in I_m \}$$

and $\qquad 1 \leq m \leq n$

$$A_1: \qquad x_m = \tau_m(y_m, \cdots, y_n)\ .$$

The properties S and A_1 imply

$$A_2: \qquad y_m = \tau_m(y_m, x_{m+1}^\circ \cdots x_n^\circ)\ .$$

A_2 shows that τ is invertible. (This is another form to state that the κ_m are transversal to each other in $\mathcal{U}(x^\circ)$.) Consider the transformed fundamental function $\bar{s}(y) := s(x) \circ \tau(y)$. If s satisfies the condition L and τ has the property S, then the condition A_1 is necessary and sufficient that $\bar{s}$ fulfills the condition L too.

A separating coordinate transformation - which is not uniquely associated with a given model function by Def. 3.1 - transforms the coordinate axes of the new variables y_m into the curves κ_m (=subsets in the space of the original variables). In the new variables thermodynamic stability has to be investigated only along the coordinate axes. Therefore, one can say that in the new coordinates the mutual influence between the state variables is completely removed. And the deviation of the separating coordinate transformation τ from the identity map provides us with an analytical expression for the interactions between the original variables x_i . These concepts have to be tested by application to special models. But let us first state some general properties of τ .

One should clearly discriminate between those properties of τ which are related to a special thermodynamic model function and those which reflect a general mathematical structure.

Proposition 3.2: Let $\mathcal{A}_n$ denote the set of twice differentiable coordinate transformations $\alpha : \mathcal{U}(x^o) \to \mathcal{U}(x^o)$ which satisfy (the model independent) conditions A_1 and A_2 . Then $\mathcal{A}_n$ equipped with the o -composition is a group. ($\mathcal{A}_n$ can be made to a topological group.)

In general we do not know if the set of separating coordinate transformations $\{\tau_\Sigma , \Sigma \in \mathcal{C}\}$ constitutes a subgroup of $\mathcal{A}_n$.

The k-jet $j^k s$ of a k-times differentiable function (germ) s at x^o is defined as the equivalence class of all functions having the same Taylor expansions at x^o up to order k. For the transformation of $j^k s$ only the k-th Taylor polynomial $_k\alpha$ of a k-times differentiable coordinate transformation α has to be known. Define the k-jet composition as follows

$$_k\alpha' \overset{k}{o} {}_k\alpha'' := {}_k({}_k\alpha' o {}_k\alpha'') . \tag{3.1}$$

Proposition 3.3: Let $_k\mathcal{A}_n$ denote the set of all $_k\alpha$ with $\alpha \in \mathcal{A}_n$ and α k-times differentiable. Then $_k\mathcal{A}_n$ is a finite dimensional Lie group with respect to the $\overset{k}{o}$ -composition and is homomorphic to $\mathcal{A}_n$.

Proposition 3.4: All $_1\alpha \in {}_1\mathcal{A}_n$ have the form

$$_1\alpha(y) = x^o + \mu(y - x^o) , \tag{3.2}$$

where the linear transformation μ is given by a restricted equiaffine matrix $(\mu_{ij}), 1 \le i,j \le n$, i.e.

$$\mu_{ij} = 0, \; i > j \; ; \; \mu_{ii} = 1 \; ; \; \det(\mu_{ij})_{1 \le i,j \le m} = 1, \; 1 \le m \le n . \tag{3.3}$$

From Prop. 3.3 then follows that the set of all restricted equiaffine matrices constitutes a Lie group under matrix multiplication. The group $\mathcal{A}_n$ is thus a nonlinear generalization of this set of affine matrices. If the coordinate transformation is related via the separation condition S to a thermodynamic model function, it has the following special shape.

Proposition 3.5: The 1-jet approximation of a separating coordinate transformation is uniquely determined to be of the form

$$_1T(y) = x^0 + \mu(y - x^0)$$

with $$\mu_{ij} = \delta_{ij} + \sum_{\nu=1}^{n} \sum_{\mu=1}^{\nu-1} \delta_{i\mu} (\partial x_\mu / \partial x_\nu)_{K_\nu} \delta_{\nu j} \; . \tag{3.4}$$

Taking into account that along K_m the intensive variables $p_1, \dots, p_{m-1}$ are kept fix and the order of differentiation may be changed one immediately obtains the well known result

Proposition 3.6: Define $\bar{s}(y) := s(x) \circ {}_1T(y)$. Then

$$\frac{\partial^2 \bar{s}}{\partial y_i \partial y_j}(x^0) = \delta_{ij} \frac{\partial^2 s^{[1 \dots, i-1]}}{\partial x_i^2}(\gamma^{(i)}(x^0)) \; . \tag{3.5}$$

In general it is not assured that $\{ {}_1T_\Sigma \, , \, \Sigma \in \mathcal{C} \}$ is a subgroup of $_1\mathcal{A}_n$. But there are hints that this might be so.

Example 3.7: Let $\mathcal{C}$ be the class of simple fluid systems with $x_1 = u$ (energy per particle) and $x_2 = v$ (volume per particle). Then the 1-jet approximation of $T(y_1, y_2)$ reads as

$$\begin{pmatrix} u \\ v \end{pmatrix} = \begin{pmatrix} u^0 \\ v^0 \end{pmatrix} + \begin{pmatrix} 1 & (\partial u / \partial v)_T \\ 0 & 1 \end{pmatrix} \begin{pmatrix} y_1 - u^0 \\ y_2 - v^0 \end{pmatrix} \tag{3.6}$$

Since $(\partial u/\partial v)_T = -c_v J$, $J = (\partial T/\partial v)_u$ the Joule coefficient, which can take both signs as well as the value zero (globally for the ideal gas), the transformations (3.6) constitute the whole of ${}_1\mathcal{A}_2$.

4. Separating coordinate transformations for n = 2

For the construction of the full separating coordinate transformation we begin with the case n=2. From A_2 it follows that $\tau_2(y_2) = y_2$. For arbitrary $t_1(y_1, y_2)$ we may always write

$$\tau_1(y_1, y_2) = \gamma_1^{(2)}(t_1, y_2) \circ t_1(y_1, y_2) . \tag{4.1}$$

Condition S gives $t_1(x_1^\circ, y_2) = p_1^\circ$ and condition A_2 implies $t_1(y_1, x_2^\circ) = \gamma_1^{(2)}(y_1, x_2^\circ) =: p_1(y_1, x_2^\circ)$. The simplest function which is in accordance with these conditions is obviously $p_1(y_1, x_2^\circ)$ itself. Thus we take

$$\begin{aligned} x_1 &= \tau_1(y_1, y_2) = \gamma_1^{(2)}(p_1, y_2) \circ p_1(y_1, x_2^\circ) \\ x_2 &= \tau_2(y_2) = y_2 . \end{aligned} \tag{4.2}$$

Example 4.1: The full separating coordinate transformations for simple fluids (cf. Ex. 3.7) is given by

$$\begin{aligned} u &= \tau_1(y_1, y_2) = u(T, v) \circ T(y_1, v^\circ) \\ v &= \tau_2(y_2) = y_2 . \end{aligned} \tag{4.3}$$

The transformation (4.2) reduces to the identity map iff p_1 does not depend on y_2. For simple fluids this amounts to u(T, v) not depending upon v, a typical feature of the perfect gas. If $p_1 = p_1(y_1)$ then one has from the integrability condition that $p_2 = p_2(y_2)$. The intensities are functionally related only to their resp. conjugate densities. Geometrically $\tau \equiv 1$ characterizes a state surface for which the cross section with planes parallel to the (s, xi) plane, i=1, 2, are translations of one and the same curve. Surfaces of this type may show plaits, but no plaitpoints, i.e. no critical points. There is no mixture of curvature.

Proposition 4.2: If s(u, v) is a mean field model function of a simple fluid then (4.3) is globally separating, i.e.

$$s(x) \circ \tau(y) = \bar{s}_1(y_1) + \bar{s}_2(y_2), \tag{4.4}$$

where the functions $\bar{s}_i$ can be calculated from $\langle \varphi^{attr} \rangle$, Z^{rep} and $c_v(T)$. [Observe the much stronger separation property of τ in (4.4) in comparison with that of $_1\tau$ in (3.5). The strict separation of the attractive part φ^{attr} from the repulsive part φ^{rep} of the potnetial in a mean field model has its counterpart in the global separability of the entropy fundamental function.

At the present stage it is not clear if $\{\tau_\Sigma, \Sigma$ a simple fluid$\}$ is a subgroup of $\mathcal{A}_2$. It is remarkable that the set of Van der Waals gases constitutes the one parametric, additive subgroup

$$\left\{ \tau_a(y_1, y_2) = \begin{pmatrix} y_1 + a(1/v_0 - 1/y_2) \\ y_2 \end{pmatrix}, \ a \in \mathbb{R}^1 \right\}. \tag{4.5}$$

5. Separating coordinate transformations for n > 2

Similar arguments (including that of simplicity) to those which led us to (4.2) give for arbitrary n a binary transformation as follows.

Definition 5.1: For each pair (μ, ν) of natural numbers with $1 \leq \mu < \nu \leq n$ we define $\tau^{(\mu,\nu)}(y)$ by setting

$$\tau_m^{(\mu,\nu)}(y) := y_m, \quad m \neq \mu, \tag{5.1}$$

$$\tau_{\mu}^{(\mu,\nu)}(y) :=$$

$$\varphi_{\mu}^{(\nu)}(p_1^0 \cdots p_{\mu-1}^0\, p_\mu \cdots p_{\nu-1}\, y_\nu\, x_{\nu+1}^0 \cdots x_n^0) \circ \prod_{\lambda=\mu}^{\nu-1} P_\lambda(p_1^0 \cdots p_{\mu-1}^0\, y_\mu\, x_{\mu+1}^0 \cdots x_n^0), \tag{5.2}$$

where

$$P_\lambda(p_1 \cdots p_{\mu-1}\, x_\mu \cdots x_n) := P_\lambda(x_1 \cdots x_n) \circ \prod_{\lambda=1}^{\mu-1} \varphi_\lambda^{(\mu)}(p_1 \cdots p_{\mu-1}, x_\mu \cdots x_n). \tag{5.3}$$

Thus

$$\tau_{\mu}^{(\mu,\nu)}(y) = \tau_{\mu}^{(\mu,\nu)}(y_\mu, y_\nu). \tag{5.4}$$

The condition S is replaced by

$$S^{(\mu,\nu)}: \quad \tau_{\mu}^{(\mu,\nu)}(x_1^0 \cdots x_{\nu-1}^0\, y_\nu\, x_{\nu+1}^0 \cdots x_n^0) = \varphi_{\mu}^{(\nu)}(p_1^0 \cdots p_{\nu-1}^0\, y_\nu\, x_{\nu+1}^0 \cdots x_n^0). \tag{5.5}$$

The properties (5.1) and (5.4) define a subgroup $\mathcal{A}_n^{(\mu,\nu)} \subset \mathcal{A}_n$.

Proposition 5.2: Let x be in the range of $\tau^{(\mu,\nu)}$. Then

$$\tau_{m}^{(\mu,\nu)-1}(x) = x_m, \quad m \neq \mu, \tag{5.6}$$

$$\tau_{\mu}^{(\mu,\nu)-1}(x) =$$

$$\varphi_{\mu}^{(\nu)}(p_1^0 \cdots p_{\mu-1}^0\, p_\mu \cdots p_{\nu-1}\, x_\nu^0 \cdots x_n^0) \circ \prod_{\lambda=\mu}^{\nu-1} P_\lambda(p_1^0 \cdots p_{\mu-1}^0\, x_\mu\, x_{\mu+1}^0 \cdots x_{\nu-1}^0\, x_\nu\, x_{\nu+1}^0 \cdots x_n^0). \tag{5.7}$$

For $\mu \neq \mu'$ one has

$$\tau^{(\mu,\nu)} \circ \tau^{(\mu',\nu)} = \tau^{(\mu',\nu)} \circ \tau^{(\mu,\nu)}. \tag{5.8}$$

Definition 5.3:

$$\tau^{(\nu)} := \tau^{(\nu-1,\nu)} \circ \cdots \circ \tau^{(2,\nu)} \circ \tau^{(1,\nu)}. \tag{5.9}$$

One finds

$$\kappa_\nu = \{\tau^{(\nu)}(x_1^0 \cdots x_{\nu-1}^0\, y_\nu\, x_{\nu+1}^0 \cdots x_n^0),\ y_\nu \in I_\nu\} \tag{5.10}$$

$S^{(\nu)}$:

$$(x_1^0 \cdots x_{\mu-1}^0\, y_\mu\, x_{\mu+1}^0 \cdots x_n^0) = \tau^{(\nu)}(x_1^0 \cdots x_{\mu-1}^0\, y_\mu\, x_{\mu+1}^0 \cdots x_n^0), \quad \mu \neq \nu. \tag{5.11}$$

Furthermore

$$\tau_m^{(\nu)}(y) = \tau_m^{(\nu)}(y_m, y_\nu) \quad , \quad m < \nu \tag{5.12}$$

$$\tau_m^{(\nu)}(y) = y_m \quad , \quad m \geq \nu . \tag{5.13}$$

(5.11), (5.12) and (5.13) define a subgroup $\mathcal{A}_n^{(\nu)} \subset \mathcal{A}_n$.

Definition 5.4:

$$\tau^{(+\nu)} := \tau^{(\nu)} \circ \tau^{(\nu-1)} \circ \cdots \circ \tau^{(1)} . \tag{5.14}$$

One finds

$$K_\mu = \{ \tau^{(+\nu)}(x_1^o \cdots x_{\mu-1}^o y_\mu x_{\mu+1}^o \cdots x_n^o), y_\mu \in I_\mu \}, \mu \leq \nu , \tag{5.15}$$

$$(x_1^o \cdots x_{\mu-1}^o y_\mu x_{\mu+1}^o \cdots x_n^o) = \tau(x_1^o \cdots x_{\mu-1}^o y_\mu x_{\mu+1}^o \cdots x_n^o) \quad , \quad \mu > \nu . \tag{5.16}$$

Furthermore

$$\tau_m^{(+\nu)}(y) = \tau_m^{(+\nu)}(y_m, y_{m+1} \cdots y_\nu) \quad , \quad m < \nu \tag{5.17}$$

$$\tau_m^{(+\nu)}(y) = y_m \quad , \quad m \geq \nu . \tag{5.18}$$

(5.16), (5.17) and (5.18) define a subgroup $\mathcal{A}_n^{(+\nu)} \subset \mathcal{A}_n$.

From (5.15) and (5.16) one obtains the nontrivial result

Proposition 5.5: Define

$$\tau := \tau^{(+n)} = \prod_{1 \leq \mu < \nu \leq n}{}^{\!\circ >} \tau^{(\mu,\nu)} \tag{5.19}$$

where $\circ >$ denotes the ordered $\circ$-product, in which the larger ν stands left of the smaller one. Then τ is a separating coordinate transformation.

For a given fundamental function s(x), which satisfies condition L in $\mathcal{U}(x^o)$ there are many possibilities to construct separating coordinate transformations. τ from (5.19) is distinguished in that it allows for an easy calculation of its inverse by means of Prop. 5.2 and in that it reflects the structure of the 1-jet approximation as closely as possible. In fact, every restricted equiaffine matrix can be written in the form (5.19) if $\circ\!\!>$ is interpreted as the ordered matrix multiplication. The step by step construction of τ reveals also the following family of subgroups of $\mathcal{A}_n$:

$$\mathcal{A}_n \overset{n \geq m}{\supset} \mathcal{A}_n^{(+m)} \overset{m \geq \nu}{\supset} \mathcal{A}_n^{(\nu)} = \mathcal{A}_n^{(\nu-1,\nu)} \times \cdots \times \mathcal{A}_n^{(1,\nu)} . \tag{5.20}$$

To each subgroup there corresponds a set of density variables of Σ whose mutual interactions are tested by the separating transformations of this subgroup. The same inclusions as in (5.20) can be written for the k-jet approximations. The 1-jet approximations ${}_1\mathcal{A}_n^{x}$ are isomorphic to the following matrix groups

$${}_1\mathcal{A}_n^{(\mu,\nu)} \cong \{ (\mu_{ij});\ \mu_{ij} = \delta_{ij} + \delta_{i\mu}\, \mu_{\mu\nu}\, \delta_{\nu j} \} \tag{5.21}$$

$${}_1\mathcal{A}_n^{(\nu)} \cong \{ (\mu_{ij});\ \mu_{ij} = \delta_{ij} + \sum_{\lambda=1}^{\nu-1} \delta_{i\lambda}\, \mu_{\lambda\nu}\, \delta_{\nu j} \} \tag{5.22}$$

$${}_1\mathcal{A}_n^{(+\nu)} \cong \{ (\mu_{ij});\ \mu_{ij} = \delta_{ij} + \sum_{\varkappa=1}^{\nu} \sum_{\lambda=1}^{\varkappa-1} \delta_{i\lambda}\, \mu_{\lambda\varkappa}\, \delta_{\varkappa j} \} . \tag{5.23}$$

From the explicit construction of τ^{-1} one gets for the 1-jet approximation

$${}_1\tau^{-1}(x) = x^{\circ} + \mu^{-1}(x - x^{\circ})$$

$$\mu^{-1}_{ij} = \delta_{ij} - \sum_{\nu=1}^{n} \sum_{\mu=1}^{\nu-1} \delta_{i\mu} (\partial x_\mu / \partial x_\nu)\, p_1^{o} \cdots p_\mu^{o}\, \delta_{\nu j} . \tag{5.24}$$

References

Falk, G. (1968), "Thermodynamik", Springer, Berlin, Heidelberg, New York

Gibbs, J.W. (1873), In "Collected Works I", Zale Univ. Press, New Haven 1948

Korteweg, D.J. (1891), Arch. Néerlandaises 24, 57 and 295

Landau, L.D. (1937), Physik. Z. Sowjetunion 11, 26 and 545

Rieckers, A., to be published

Stumpf, H. and Rieckers, A. (1975), "Thermodynamik I", Vieweg, Braunschweig

Tisza, L. (1951), In "Phase Transformations in Solids", ed. R. Smoluchowski, J.E. Mayer and W.A. Weyl, J. Wiley and Sons, Inc. New York

Tisza, L. (1961), Ann. Phys. 13, 1

Tisza, L. (1966), "The Geometrical Interpretation of the Formalism of MTE" in "Generalized Thermodynamics", MIT Press, Cambridge (Mass.), London

Van der Waals, J.D. (1891), Arch. Néerlandaises 24, 1

Van der Waals, J.D. and Kohnstamm, Ph. (1912), "Lehrbuch der Thermodynamik", Teil II, Joh. Ambrosius Barth, Leipzig

APPROXIMATE SYMMETRY *

Joe Rosen

Department of Physics and Astronomy
Tel-Aviv University
Tel-Aviv 69978, Israel

A general, formal framework for approximate symmetry is proposed, based on the concept of metric relation in state spaces of systems. (This is a generalization of the general symmetry formalism, based on the concept of equivalence relation.) It is then quite straightforward to define: approximate symmetry transformation, approximate symmetry group, exact symmetry subgroup of this, goodness of approximation, and exact symmetry limit. For details see ref. 1.

---

* Work supported in part by the Israel Commission for Basic Research.

Reference:

1. J. Rosen, Tel-Aviv University report TAUP-438-74.

Cohomology of the action differential forms

P.B. Scheurer

Faculteit Wis-en Natuurkunde, Universiteit Nijmegen, Toernooiveld,
Nijmegen, Nederland.

Abstract: A general treatment of dynamics is given by using algebras of differential forms on suitable differentiable manifolds. Physics is introduced by giving these forms the physical dimension of an action on space-time or of an entropy on a thermodynamic manifold. Properties of differentiable manifolds are able to take account for various dynamical features, both classical and quantal.

From Newton's time on physicist's play with derivatives and differentials, but sometimes very ackwardly, like it can be seen in classical phenomenologic thermodynamics. Thus the use of such an appropriate tool as that one of the differentiable manifolds (DM) provides new insight how to treat more coherently various dynamical features, both classical and quantal, with an unique scheme of derivation in this theory. [1]

1.- As it will be shown here, physics of the various dynamics differs only from pure geometry of a DM in the fact that one has to endow the geometrical objects with only one extra physical dimension, giving thus a physical dimension to the unit. As well known, by using natural constants, it is possible to express every physical grandeur by the choice of an unique fundamental one, as length or mass. So in r=ct are length and time homogenous; in $E = h\nu$, time and energy; in $E = c^2 m$, mass and energy; in $E = kT$, temperature and energy; and the fine structure constant α accounts for the electric charge. When we realize moreover that it is only a historical prejudice which prevents us consider temperature as frequency, we can give to action and to entropy the same physical dimension, in the case that one of a number, because both provides us information on physical systems.

So on space-time we consider a) the dynamic 1-forms $\omega_1^1 = p_\mu(x)dx^\mu$ (with $p_\mu = mv_\mu(x)$ for a material point and $p_\mu(x) = m(x)\, v_\mu(x)$ for a fluid) and the electromagnetic one $\omega_1^2 = e/c\, A_\mu(x)\, dx^\mu$ and b) on a thermodynamic manifold we consider the entropy 1-form $dS(p) = \zeta_\alpha(p)dp^\alpha$

2.- A streaking feature in a DM is the duality between parametrized curves and numerical functions. Let us first recall that a DM is a topological space which locally looks like an Enclidean space, and where it is possible to differentiate and to integrate conventionally. Thus the local charts are pieced together by differentiable functions.

A parametrized curve c maps differentiably a segment I of the real line R into the manifold M (let say of dimension n and of element m)

$$c: \quad I\epsilon R \rightarrow M$$
$$t \mapsto m(t)$$

On the contrary, a numerical function f maps differentiably M into R

$$f: M \rightarrow R$$
$$m \mapsto f(m)$$

The natural coordinates on a chart $x^i : m \mapsto x^i(m)$ are such functions. One puts $f(x^i(t)) = \bar{f}(m(t))$

A convenient equivalence relation over de parametrized curves leads to the tangent vectors, which form a vector bundle T(M) over M. In natural coordinates such a vector is typically written $v = v^i\, \partial/\partial x^i$. It is a derivation on the numerical functions

1) $v(fg) = (vf)\, g + f\, (vg)$

Another convenient equivalence relation over the functions gives the differentials, which form a vector bundle $T^*(M)$ over M. In natural coordinates such a differential is typically written $df = \partial f/\partial x^i dx^i$. T (M) and T^* (M) are dual in the sense that the contraction

$$< \partial/\partial x^i,\ dx^j > = \delta_i^j \quad \text{(the Kronecker index)}$$

The geometrical nature of linear momentum as a tangent vector or position as a coordinate function provides immediately the Quantum Mechanical commutator between P et Q. For in 1) we have simply to choose v as $- i\hbar \dfrac{\partial}{\partial x}$, f(x) as x and g(x) as $\psi(x)$, the state function.

So the uncertainty relations reflect the duality between T(M) and $T^*(M)$.

3.- On a DM exists a graded algebra of differential p-forms, $0 \leq p \leq n$, with the external differentiation d as an antiderivation (recall $d^2 = dod = 0$.)

Applying this formalism to our 1-forms a) of action over space-time (DM of element $x=(\vec{x},t)$ and parameter the proper time τ) and b) of entropy over a DM of coordinates p^α, where one of this p^α is energy, we obtain immediately Hamilton, Euler, and Maxwell equations.

a^1) $\quad \omega_1^1 = p_\mu(x)\, dx^\mu + df(x) \qquad$ (f(x) is a mechanical gauge)

$$d\omega_1^1 = dp_\mu \wedge dx^\mu$$

As the induced 1-form over the parameter is closed

$$c^*\omega_1^1 = -E_0\, d\tau + \partial f/\partial\tau\, d\tau$$
$$dc^*\omega_1^1 = c^* d\omega_1^1 = 0,$$

from the closure of ω_1^1 we obtain Euler or Hamilton equation following the explicit (or not) dependence of p_μ on $\vec{x}$, and $dp_\mu(x(\tau))/d\tau$ can be interpreted as a force $F_\mu(\tau)$

a^2) $\quad \omega_1^2 = \dfrac{e}{c} A_\mu(x)\, dx^\mu + d\Lambda(x) \quad$ ($\Lambda(x)$ is a gauge)

As known, from this 1-form follows all classical electromagnetism

$$d\omega_1^2 = \frac{e}{c} \sum_{\mu<\nu} B_{\mu\nu}\, dx^\mu \wedge dx^\nu = \frac{e}{c} \sum_{i<k} B_{ik}\, dx^i \wedge dx^k + eE_k\, dx^k \wedge dt$$ with the usual

definitions $B_{ik} = \partial A_i/\partial x^k - \partial A_k/\partial x^i$

and $E_k = \partial A_k/c\partial t - \partial A_4 / \partial x^k$

We can put the Lorentz force in evidence

$$\frac{e}{c} B_{\mu\nu} w^\mu dx^\nu d\tau = (+ \frac{e}{c} B_{ik} v^k + eE\ ) dx^i \wedge dt$$

The Maxwell equations of the field are derived as

$$d^2 \omega_1^2 = 0 = \partial B_{\mu\nu}/\partial x^\rho\, dx^\rho \wedge dx^\mu \wedge dx^\nu$$

The corresponding equations of source are obtained by the star $*$-Hodge operation,

But if ω_1^2 is a differential $dA(x)$, where $A(x)$ is a phase as it is the case in a simply connected manifold (Poincaré's lemma: e.g.: a torus cut by a junction), the field $B_{\mu\nu}$ is null, but the A_μ can give rise to some effects as the Aharonov-Bohm effect.

b) On a thermodynamic manifold, from

$$dS(p) = \zeta_\alpha(p)\, dp^\alpha$$

we obtain immediately the thermodynamic Maxwell equations

$$d^2S = 0 = d\,\zeta_\alpha \wedge dp^\alpha$$

(as an example, try $dS = p/T\, dV + dE/T)$ [2]

4.- As the preceding equations $a^1)$ and $a^2)$ are obtained from closed 1-forms, where the gauges enter as cobounds $f(x)$ or $\Lambda\ (x)$, thus only the <u>classes of cohomology</u> of these forms are <u>physically relevant</u> (two closed forms are cohomologous if they differ only by a cobound)

5.- Consider the Laplacian operator Δ operating on the forms. A 1-form ω_1 is harmonic if $\Delta\omega_1 = 0$. Moreover consider a Killing vector X, a vector such that the Lie derivative of the metric $g_{\mu\nu}$ is null in the direction of X

$$\mathcal{L}_X\, g_{\mu\nu} = 0$$

Then a theorem of Borchner asserts that the contraction of a harmonic 1-form ω_1 by a Killing vector X gives a global invariant over the manifold:

$$< X, \omega_1 > = \text{const}$$

Quantum phenomena in coherent states are immediately obtained this way.

$a^1)$ $0 = < X^\mu(x)\, \partial_\mu\, ,\ p_\mu\ (x)\, dx^\mu > = p_i X^i - HT = nh - nh$

photon case $p = nh\nu/c$ and $H = nh\nu$

superfluid case : vortex quantification and

Anderson-Josephson effect

$a^2)$ $0 = < X^\mu\ (x)\, \partial_\mu ,\ 2\,\frac{e}{c} A_\mu\ (x)\, dx^\mu > =$

$$= 2\,\frac{e}{c}\,A_i\,X^i - 2\,eVT = nh\text{-}nh$$

Aharonov-Bohm effect ($A_i\,X^i$) or magnetic flux quantization ($B_{ik}\,S^{ik}$) and Josephson effect

b) $0 = \langle X^\alpha(p)\,\partial_\alpha\,,\,\zeta_\alpha(p)\,dp^\alpha\rangle = nk\text{-}nk$

(e.g. $0 = p/T\ V + E/T$ perfect gas case: $-pV = nkT$)

6.- A theorem asserts that the cohomology class of harmonic p-forms contains an unique element. So in space-time must exist a proportionality between $p_\mu\ (x)\ dx^\mu$ and $\frac{e}{c}\,A_\mu\ (x)\ dx^\mu$, i.e. a relation between h, e and c. By pure geometry are these natural constants interdependant. We could also choose $h = k$ by a convenient redefinition of the temperature (the degree Stueck, in symbole $\mu\!\!\backslash$)

7.- The essentially different geometrical nature of a coordinate and of a parameter permits to derive <u>both</u> the Newtonian and the Einsteinian dynamics <u>in the same dynamical structure.</u> For Newton the manifold is R^3 and the parameter the time t; for Einstein the manifold is R^4 and the parameter the proper time τ (for World-lines).

Instead of putting the Newton's axiom for the proportionality between velocity and linear momentum as $p^i(t) = mdx^i\ (t)\ /\ dt$, $i \varepsilon \{1, 2, 3\}$, it is better to introduce the mass into the parameter and to write is as

$$dx^i\ (t)\ /\ p^i(t) = dt/m.$$

We can extend this relation to a properly <u>Newton-Einstein's axiom:</u>

$$dx^i\ /p^i = cdt/cm = \text{def}\ dx^4\ /p^4 = d\tau/m_0$$

(where m_0 is the mass at rest corresponding to the new parameter). The extension is limited only by refering explicitly to electromagnetism in order to limit the manifold to four dimensions: the dual $*$-Hodge form of the electromagnetic 2-form of the field must be also the 2-form of the source (in vacuo). But by definition it is a $(n-2)$ -form ! So n=4.

From the definition $p^4 = cm$ one obtains immediately the Einstein's relation between mass and energy, by rewritting the relation between Lagrangian and Hamiltonian

$$Ldt = p_i dx^i - Hdt = \text{def}\ p_i dx^i + p_4\ dx^4 \qquad \text{So is } H = c^2 m$$

8.- Thus it is possible to account for the fundamental unicity of various dynamical phenomena by a unique dynamical structure given by differential forms on an appropriate manifold. In all cases the physics is obtained from geometry by adding a unique physical dimension, endowed with a unique natural constant. In this sense we can consider seriously the introduction of a new paradigm for dynamical phenomena, more powerful and more coherent than previous ones.

Bibliography

[1] P.B Scheurer, C.R. des Séances, SPHN Genève, NS 7, 1, 19-23 (1972); 7, 2-3, 89-96 (1972); and 8, 1-3, 32-37 (1973)

[2] E.C.G. Stueckelberg de Breidenbach et P.B. Scheurer, "Thermocinétique phénoménologique galiléenne", Birkhäuser Verlag, Basel, 1974.

CORRELATION INEQUALITIES IN A CLASS OF LATTICE SYSTEMS IN STATISTICAL MECHANICS

P.A. Vuillermot

Département de Physique, Université de Neuchâtel, Switzerland

H. Kunz

Laboratoire de Physique théorique EPFL, Lausanne, Switzerland

Ch.-Ed. Pfister

Zentrum für interdisziplinäre Forschung, Universität Bielefeld, Germany

Expectation and correlation inequalities are derived for a class of lattice systems whose configuration manifold is a compact connected real Lie group G. In particular, necessary and sufficient conditions are proved to ensure the positive definiteness of the corresponding Gibbs probability measures at every finite temperature, from which Griffith's first inequality emerges in a natural way for observables in the convex cone of positive definite, continuous functions on G. Necessary and sufficient conditions are also given to prove the non-negativity of mean order parameters for systems defined by G-invariant Hamiltonians. Higher-order correlation functions are also considered, and sufficient conditions are proved to derive Griffith's second inequality for observables defined from zonal spherical functions on G. This inequality holds at every finite temperature for some anisotropic vector models and for the isotropic Heisenberg-Stanley ferromagnet, and at least in the high-temperature, or weak-coupling limit for more complicated G-invariant vector models including lattice versions for liquid crystals.

EDUCATIONAL

WHAT IS SO "SPECIAL" ABOUT "RELATIVITY" ?

Jean-Marc Lévy-Leblond*
Physique Théorique
Université Paris VII.

Since many of the participants in this Colloquium hold teaching positions, I thought that we might try to consider teaching as a genuine scientific activity, deserving some explicit recognition. The real practice of physics cannot be reduced to the so-called "fundamental" research and we could perhaps fruitfully enlarge the scope of our discussions in meetings such as this one. I fully realize of course that most of our joys in the scientific life have little to do with teaching or popularizing science : publishing papers in top journals, giving talks at seminars, organizing colloquiums and summer schools, fighting over priorities, sending urgent preprints, exchanging invitations with colleagues across the borders, refereeing articles of competitors, consulting for private companies or official bodies, etc. How could you publish any work related to your teaching activity (except in some neglected - on this side of the ocean-journal such as the Am.J.Phys.), how could you hope to improve your academic position from such work, how could you use it to get funds from NATO or DGRST (that's a French agency)?

Yet I maintain that teaching practice (in a broad sense, including popularizing for instance) is a rich source of interesting problems. They derive from the necessary evolution in the conceptualization, formalízation and verbalization of physical theories all along their history. A theory can only emerge in a conceptual context deeply linked to the old ideas that the new ones are born to replace. As a result, it is a long and never-ending process to continually recast a physical theory on the basis of its contemporary collective practice. From that point of view, it must be said that "modern" physics is in a very bad shape and that, on the average, the gap does not seem to be narrowing between the way it is used in research activities on the one hand, and taught in classes on the other. It is as if we developed a kind of schizophrenic scientific personality, using for our teaching, approaches, concepts, terminologies, examples that our research activities at the same time prove to be unnecessayr, irrelevant or plainly wrong.

Obviously, there exist strong and constraining reasons for this state of affairs, deeply ingrained in the present social structure of science. Is it too radical a demand, however, to ask for the devoting of some interest to the very activities which, after all, many of us are largely paid for ? Or do we think that students only prevent us from fully dedicating ourselves to more noble tasks and that universities would be all the more better, would they be empty ? This might explain why our teaching often seems mainly aimed at driving the students out of the classes...To be frank, I must add that I also have more personal motivations for such a line of work : the growing competitivity and esoterism of contemporary "advanced" research make me rather uncomfortable. I find it at least as satisfying to leisurely develop my (along with others') understanding of "known" sectors of physics, than to try to catch up in a hurry with the latest theoretical gadget, only to see the fashion change at the very time when the truly interesting conceptual problems could be attacked.

Now, the impressing development of group-theoretical methods in physics during the past half-century, of which the existence of our Colloquium is a convincing proof, should not be without consequences on the recasting of established physical theories, hence on our teaching methods. Two years ago, I had sketched here the impact of these ideas upon ordinary ("non-relativistic") quantum theory[(1)]. I wish here to make some remarks in the same spirit about "relativity" theory. Indeed we now know that "relativity" is but one of the main applications of group-theoretical point of views in physics. An abundant literature exists and grows about the Lorentz and Poincaré groups, their properties, their representations, their generalizations, etc. Many communications at this Colloquium belong to this line of thought. Yet it is amazing to realize how little these ideas have permeated our teaching of relativity theory. Most of the textbooks and lectures still are based on the original approaches to the subject, which, justified as they were in their own historical context, should not be used uncritically today.

Of course, a first obstacle in the modernizing of our educational approaches lies in the terminology itself. It is clear that "the (special) theory of relativity" is badly misnamed from at least two points of view. First, it is a, and not the, theory of relativity. A perfectly consistent theory of relativity underlies the whole of classical mechanics. It is a specific, "Galilean" theory of relativity[(2)]-which then proved to be only approximate and had to be replaced by another one, the "Einsteinian" theory. One should carefully distinguish between the abstract principle of relativity, and the various concrete theories of relativity which may be set up. Second, the "relativist" terminology itself should be subject to caution, when we today put the emphasis on the invariants of the theory, that is, its "absolutist" aspects, (these remarks go back to A. Sommerfeld who wrote that the expression "theory of relativity" was a "widely misunderstood and not very fortunate name"). When one considers the far-out philosophical and ideological exploitations to which Einstein's work and name have been subjected, such an attitude appears all the more necessary. Of course, it cannot be hoped that a well-established and by now classical terminology could be changed at will. At least, let us explicitty discuss its historical sources[3], conceptual problems, and philosophical implications.

Now, coming closer to the specific issue that I wish to raise, I want to question the generalized use of the so-called "second postulate" in the introductory courses in special (Einsteinian) relativity, that is the hypothesis of frame-independance of the speed of light. True, this was the point of departure of Einstein's derivation of the Lorentz transformation formulas[4], through an analysis of space-time concepts in different reference frames. This analysis explicitly relies on the exchange of electromagnetic signals as means to synchronize clocks, measure lengths, etc. It thus dangerously lends itself to an ultra-positivistic (even though often implicit) interpretation, according to which the experimental techniques are the source of the theoretical structure. Also, it lends support to the idea that special relativity is in some ways intrinsically linked to electromagnetic phenomena. It may (and indeed often does) come as a surprise to the students that the same theory of space-time up to now seems to account for all physical phenomena, whether they depend upon electromagnetic, strong or weak interactions. Obviously, the logical ordering of the theoretical foundations here is inverse to the chronological order of discovery ; "relativity" nowadays is to be thought of as a general theory of space-time, which acts as a constraint, a "super-law"[5], on all specific physical phenomena taking place in this common arena. Electromagnetism thus has to be built upon special relativity rather than the reverse. Also, the frame independence of the speed of light now appears as a consequence of the dynamical behaviour allowed by Einsteinian relativity ; it is a specific property of objects with a zero invariant mass. As such, the speed-of-light is also the speed-of-neutrinos, to the extent of our present knowledge. Indeed, it could also be the speed of... nothing ; after all, we only know upper limits on the photon and neutrino masses[6]. It could very well be the case that for frequencies low enough, the speed of light appreciably departs from the... speed-of-light. This would in no way shake the foundations of special relativity, although it would invalidate most of its derivations.

An approach is possible, however, to the theory of relativity, which only relies on very general assumptions about the physical nature of space-time, without relying on any special property of any specific phenomenon. In such an approach, "c" appears, not as the speed-of-light (or of whatever else), but as a structural constant of space-time. Only a further dynamical analysis shows the possibility of objects with a null mass traveling with c as their (invariant) velocity. I will be content here with a general description of this approach the details of which may be found elsewhere[7]. The idea is to start from the general principle of relativity, as expressing the physical equivalence of frames of reference linked to one another by a family of transformations operating in space-time. A general argument first is used to show that, space and time translations apart, this set of transformation is necessarily a one-parameter family, if the resulting theory of relativity is to be neither empty nor trivial. The functional form of these transformations is then restricted by a series of successive general hypotheses, which can be conveniently ordered as follows :

1. homogeneity of space-time. This hypothesis is shown to imply linearity of the transformations, as well as uniformity of inertial motions. It also enables one to endow the parameter labeling the transformations with the physical meaning of a relative velocity. These transformations now are linear homogeneous transformations in space-time, the coefficients of which depend upon three unknown functions of the parameter.

2. isotropy of space-time, meaning, in the one-dimensional (in space) case, that both orientations of the space axis are physically equivalent. It is to be emphasized that such a requirement has nothing to do with the reflexion invariance of the physical laws for some specific class of phenomena ; parity is not conserved in weak interactions which nevertheless take place in a parity-invariant space.

3. group structure for the set of transformations between reference frames, under the composition law. This assumption translates the perhaps more intuitive idea of equivalence between these frames : reflexivity of the equivalence relation implies the presence of the identity transformation within the set, symmetry requires the existence of an inverse to any transformation and transitivity leads to associativity. The unknown functions of velocity which are the coefficients of the linear transformation, as a consequence of this group law obey functional equations with a very restricted set of solutions. Indeed, the only possible space-transformations consistent with the above hypotheses are shown to be of the form :

$$\begin{cases} x' = \dfrac{x - vt}{\sqrt{1-av^2}} \\ t' = \dfrac{t - avx}{\sqrt{1-av^2}} \end{cases} \qquad (1)$$

where a is an arbitrary real number. The simplest case, $a = 0$, now corresponds to the Galilean transformations, while $a > 0$ leads to the Lorentz ones. It is possible in the second case to write $a = c^{-2}$, where the constant c has the dimensions of a velocity, the numerical value of which is irrelevant since it only depends upon our choice of units. The only other possible case, namely $a < 0$, may be eliminated through a last assumption :

4. causality requires that at least some spatio-temporal intervals have the sign of their time component invariant under all transformations. This condition is met by time-like intervals for the Lorentz case, by all intervals for the Galilei one and by none in the case $a < 0$, where the transformation group is isomorphic to the rotation group in two dimensions.

It is seen that in this approach, Lorentz and Galilei transformations simultaneously emerge as possible candidates, leaving it to experiment to single out the Lorentz ones. It should be stressed, also, that the basic hypotheses of our construction, general as they may look, are by no means mild ones, and that "reasonable-looking" formulas may be written down, filling out any three of our four conditions and playing an all-important pedagogical function as counter-examples. For instance, in an expanding universe, time would not be homogeneous and the possible transformation formulas could be nonlinear, as exemplified by the De Sitter group. Indeed, a more general and abstract formulation of the same problem leads to a slightly larger set of admissible solutions(8). The work described here may be seen as a didactical adaptation, suited for introductory courses, of these considerations. Let me add here that the proposed approach also has the virtue of letting the students use a "group-theoretical method in physics" and appreciate the important role of a crucial mathematical tool. Mathematical concepts often are too long kept apart of their possible physical applications. Here is a true anecdote : we once gave to our students in an introductory relativity course, a problem sheet dealing with several generalizations of the Galilean or Lorentz formulas. The idea was to have them test these formulas for plausibility with respect to some general physical principles (precisely the ones used above). When asked for checking the group law, several students could not even understand the question. Rather appalled, we reminded them that already one year ago they had been exposed to a thorough mathematics course (French style) on basic algebra, with great emphasis on the group structure. "Oh, that's what you mean, a group as in mathematics" they replied, delighted and astonished with the idea of such a convergence...

The foregoing approach still shares with the traditional one the emphasis put on the "relativity" of space-time ; Lorentz (or Galilei) transformations indeed appear as generating a group acting on the space-time manifold. If a lesson is to be drawn, however, from the actual practice of relativistic physics, it is the primary importance of the energy and momentum concepts. The so-called relativistic "dynamics", much more than the corresponding "kinematics", lies at the basis of most experimental applications of the theory : particle accelerators, high-energy collisions, etc. In order to use this abundant material as a basis or as examples in our teaching of relativity theory, it would be most useful to have a direct approach to relativistic dynamics, and to derive the space-time properties from it. Indeed, teachers, and students even more, always suffer from the absence of concrete realistic demonstrations of the strange effects of Einsteinian relativity on the perspectives of space-time : "relativistic" rockets today have replaced the "relativistic" trains of the founding fathers, without being closer to actual physics. Is it not possible then to point out the inadequacies of Galilean mechanics and to look for a more correct theory by dealing first with the energy-momentum concepts ? A nice and neat evidence against the validity of Galilean concepts can be found (this is but a simple example), in bubble-chambers photographs of elastic collisions between two identical particles, one of which is at rest. The Galilean theory predicts that the trajectories of the outgoing particles always make a right angle. Any picture of a proton-proton collision, the initial energy of the moving one being higher than 500 MeV however unmistakably shows an acute angle. Starting from similar examples, one may make plausible the validity of energy-momentum conservation laws, provided that the expressions giving the energy and momentum of a particle as a function of its velocity be modified. The problem then is to look for suitable modifications. Two approaches at least could be followed, neither of which is fully developed yet, and will only be sketched here.

A first approach is due to Davidon[9]. It relies on the following three relationships :

$$\begin{cases} p = I \cdot v \\ dE = v \cdot dp \\ dI = a \cdot dE \end{cases} \qquad (2)$$

where a is a constant.

The first one may be considered as a definition of the "inertia" I of the object, possibly a function of its velocity v. The second one is the general expression linking the velocity to energy and momentum. Let me say at this point that I personally cannot claim to fully understand this basic relationship. I know it of course as Hamilton's first equation and as such, I know it to hold true for any physical system fitting into the canonical formalism. I have looked vainly, however, for an elementary argument, suitable for beginning students, which would do full justice to the meaning and generality of this relationship. Any such argument would be welcome. The third and really crucial relationship in (2) expresses the idea that any increase in the energy content of a body implies a proportional increase of its inertia. As emphasized by Davidon, this idea was already partially clear by the end of the 19th century and belongs explicitly to the prehistory of relativity theory. It might have been taken further already at that time. Any how, one may easily derive from (2) the following expressions :

$$\begin{cases} I = \dfrac{I_o}{\sqrt{1-av^2}} \\ p = \dfrac{I_o v}{\sqrt{1-av^2}} \\ E = \dfrac{I_o/a}{\sqrt{1-av^2}} \end{cases} \qquad (3)$$

A simple change in the notation, putting $I_o = m$ and $a = c^{-2}$ (see above - the speed-of-light here again appears as a general constant) gives the standard formulas. Of course, in the singular case $a = o$, the relationships (2) implie the Galilean formulas :

$$\begin{cases} I = I_o = C^{st} \\ p = I_o v \\ E = \frac{1}{2} I_o v^2 + U \end{cases} \qquad (4)$$

The interest of such a point of view is to bring into full light the relevant analogies and differences between the dynamical concepts in Galilean and Einsteinian relativities, as summarized by the following table (where $\mathcal{G}$ stands for Galilean mechanics, $\mathcal{E}$ for Einsteinian mechanics):

.../...

	invariant ? (from frame to frame)		conserved ? (in time)		related ? (up to a constant)
energy	$\mathcal{E}$: no	$\mathcal{G}$: no	$\mathcal{E}$: yes	$\mathcal{G}$: yes	energy ↔ inertia : $\mathcal{E}$
inertia	$\mathcal{E}$: no	$\mathcal{G}$: yes	$\mathcal{E}$: yes	$\mathcal{G}$: yes	inertia ↔ mass : $\mathcal{G}$
mass	$\mathcal{E}$: yes	$\mathcal{G}$: yes	$\mathcal{E}$: no	$\mathcal{G}$: yes	

The explicit introduction of the notion of "inertia" thus appears able to shed some light upon the conceptual interrelationships of the dynamical quantities. The analysis could be refined by introducing the internal energy U : an independent physical magnitude in Galilean dynamics, it is to be identified with the mass in the Einsteinian one. This fact accounts for a puzzling difference between the two cases : there is one more conservation law in Galilean dynamics (mass, beyond energy and momentum) ; but then, there is one more quantity, internal energy. The number of independent quantities thus is the same in both theories. The developement of these considerations should then be followed by a kinematical analysis. Lorentz transformations introduced from their energy-momentum realization should somehow be transferred to space-time, through a consideration of the velocity changes, probably.

A second possible dynamical approach relies on the assumption that a particle with velocity u in a certain reference frame has a velocity $u + \mathbf{V}$ in a second frame, the velocity of which is $\mathbf{V}$ with respect to the first one. That is, we postulate the conventional law of addition of velocities. "But we know it to be wrong !" you'll say - well, perhaps it just depends on the <u>words</u> we use ? Let us then suppose that in any physical process energy and momentum are conserved in the form :

$$\begin{cases} \sum E(u_k) - \sum E(u'_\ell) = 0 \\ \sum p(u_k) - \sum p(u'_\ell) = 0 \end{cases} \qquad (4)$$

where E(u) and p(u) are the functional forms of energy and momentum that we are looking for, and where the $\{u_k\}$ and $\{u'_\ell\}$ respectively are the velocities of the initial and final particles. For these conservation laws to hold in any other reference frame, we must have as well :

$$\begin{cases} \sum_k E(v_k+V) - \sum_l E(v'_l+V) = 0 \\ \sum_k p(v_k+V) - \sum_l p(v'_l+V) = 0 \end{cases} \tag{5}$$

for any value of the relative velocity V . According to Mac-Laurin then, one also gets conservation laws for all successive derivatives of the energy and momentum, hence an apparent infinity of conservation laws for E, $\frac{dE}{dv}$, $\frac{d^2E}{dv^2}$, etc., and p, $\frac{dp}{dv}$, $\frac{d^2p}{dv^2}$, etc. But only two independent such conservation laws are possible, so that the derivatives $\frac{dE}{dv}$, $\frac{dp}{dv}$, etc., in some way must be related to E and p themselves. The simplest possibility is a linear relationship, wich, according to the even and odd nature of E and p respectively, may only be of the form $\frac{dE}{dv} \propto p$ and $\frac{dp}{dv} \propto E$; the higher derivatives then are automatically taken care of. This entails the following expressions :

$$\begin{cases} E = E_0 \,\mathrm{ch}(Kv) \\ p = E_0 K \,\mathrm{sh}(Kv) \end{cases} \tag{6}$$

A change of notation $K = c^{-1}$, $E_0 = m\,c^2$ and, above all, $v = c\varphi$, brings us back to the standard formulas. Indeed, it clearly exhibits the fact that the additive parameter φ cannot in general be identified to the physical velocity $v = \frac{dE}{dp}$ that is, $v = c\,\mathrm{th}\varphi$, and is rather called the "rapidity" of the object. However a singular case where $\frac{dE}{dv} \propto p$ but $\frac{dp}{dv} = C^{st}$ (instead of giving back E) is possible, yielding the Galilean formulas. In that case of course, the two concepts "rapidity" and "velocity" merge into one. The virtue of this approach would lie into introducing the student right from the beginning to the concept of rapidity. Besides being ever more used in high-energy theory nowadays, this notion enables one to avoid some paradoxical features of the velocity in Einsteinian relativity. Namely, none of the "intuitive" features of the concept in Galilean relativity disappear, but they have to be looked for in two related but different concepts of Einsteinian relativity : the velocity v keeps its meaning as a time rate of spatial displacement for uniform motions as well as its general relationship to energy-momentum, $dE = v.dp$, while the rapidity φ obeys the additive composition law, with an infinite range (no upper limit). To be rigorous, <u>two</u> new names should have been invented in order to distinguish these two concepts from the unique Galilean velocity. Once more, abuse of language, if not avoidable should at least be explicitly mentioned. Let us note, finally, the deep group-theoretical basis of this approach - which after all is its reason for being exposed here. Indeed, the assumption of the existence of an additive parameter is related to the elementary result that any continuous one-parameter group is isomorphic to the real additive group. It is through this isomorphy that a canonical choice of parameter is commonly made. Even though such an abstract theorem need not be explicitly stated and used in an introductory relativity course, its existence nevertheless yields some strength and depth to these elementary considerations.

I do not doubt that other point of views yet exist which might be worth developing.

* Postal address : Laboratoire de Physique Théorique
Université Paris VII (tour 33-43 - 1er étage)
2, Place Jussieu
75221 - PARIS CEDEX 05.

REFERENCES

(1) J.M. Lévy-Leblond, Rivista del Nuovo. Cim. 4, 99 (1974).

(2) For a review, see J.M. Lévy-Leblond in "Group Theory and its Applications", Vol. 2, E. Loebl ed. Academic Press (New-York, 1971), p. 221 & seq.

(3) A fascinating account of the historical and social context of Einstein's work is to be found in L.S. Feuer, "Einstein and the Generations of Science" (Basic Books, N.Y. 1974) ; see especially pp. 59-60 a discussion of the ideological connotations of the "relativist" terminology.

(4) A. Einstein, Ann. Phys. 17, 891 (1905), reprinted in English translation in "The Principle of Relativity" (Dover, New-York).

(5) E.P. Wigner, Physics Today 17, 34 (1964).

(6) A review of our knowledge on the limits put upon the photon mass has been given by A.S. Goldhaber and M.M. Nieto, Revs. Mod. Phys. 43, 277 (1971).

(7) J.M. Lévy-Leblond, Am. J. Phys. (to be published) - preprint LPTHE 74/10. An independent, similar but less detailed, investigation has just been published by A.R. Lee and T.M. Kalotas, Am. J. Phys. 43, 434 (1975). These authors point out to the early roots of such criticisms of the "second postulate".

(8) H. Bacry and J.M. Lévy-Leblond, J. Math. Phys. 9, 1605 (1968).

(9) W.C. Davidon, Foundations of Physics 5, xxx (1975), (to be published).

ALPHABETICAL AUTHOR INDEX

Lecture Notes in Mathematics

Vol. 507: M. C. Reed, Abstract Non-Linear Wave Equations. VI, 128 pages. 1976.

Vol. 501: Spline Functions, Karlsruhe 1975. Proceedings. Edited by K. Böhmer, G. Meinardus, and W. Schempp. VI, 421 pages. 1976.

Vol. 495: A. Kerber, Representations of Permutation Groups II. V, 175 pages. 1975.

Vol. 490: The Geometry of Metric and Linear Spaces. Proceedings 1974. Edited by L. M. Kelly. X, 244 pages. 1975.

Vol. 489: J. Bair and R. Fourneau, Etude Géométrique des Espaces Vectoriels. Une Introduction. VII, 185 pages. 1975.

Vol. 485: J. Diestel, Geometry of Banach Spaces – Selected Topics. XI, 282 pages. 1975.

Vol. 484: Differential Topology and Geometry. Proceedings 1974. Edited by G. P. Joubert, R. P. Moussu, and R. H. Roussarie. IX, 287 pages. 1975.

Vol. 481: M. de Guzmán, Differentiation of Integrals in R^n. XII, 226 pages. 1975.

Vol. 480: X. M. Fernique, J. P. Conze et J. Gani, Ecole d'Eté de Probabilités de Saint-Flour IV–1974. Edité par P.-L. Hennequin. XI, 293 pages. 1975.

Vol. 477: Optimization and Optimal Control. Proceedings 1974. Edited by R. Bulirsch, W. Oettli, and J. Stoer. VII, 294 pages. 1975.

Vol. 474: Séminaire Pierre Lelong (Analyse) Année 1973/74. Edité par P. Lelong. VI, 182 pages. 1975.

Vol. 470: R. Bowen, Equilibrium States and the Ergodic Theory of Anosov Diffeomorphisms. III, 108 pages. 1975.

Vol. 468: Dynamical Systems – Warwick 1974. Proceedings 1973/74. Edited by A. Manning. X, 405 pages. 1975.

Vol. 464: C. Rockland, Hypoellipticity and Eigenvalue Asymptotics. III, 171 pages. 1975.

Vol. 463: H.-H. Kuo, Gaussian Measures in Banach Spaces. VI, 224 pages. 1975.

Vol. 461: Computational Mechanics. Proceedings 1974. Edited by J. T. Oden. VII, 328 pages. 1975.

Vol. 459: Fourier Integral Operators and Partial Differential Equations. Proceedings 1974. Edited by J. Chazarain. VI, 372 pages. 1975.

Vol. 458: P. Walters, Ergodic Theory – Introductory Lectures. VI, 198 pages. 1975.

Vol. 449: Hyperfunctions and Theoretical Physics. Proceedings 1973. Edited by F. Pham. IV, 218 pages. 1975.

Vol. 448: Spectral Theory and Differential Equations. Proceedings 1974. Edited by W. N. Everitt. XII, 321 pages. 1975.

Vol. 447: S. Toledo, Tableau Systems for First Order Number Theory and Certain Higher Order Theories. III, 339 pages. 1975.

Vol. 446: Partial Differential Equations and Related Topics. Proceedings 1974. Edited by J. A. Goldstein. IV, 389 pages. 1975.

Vol. 445: Model Theory and Topoi. Edited by F. W. Lawvere, C. Maurer, and G. C. Wraith. III, 354 pages. 1975.

Vol. 444: F. van Oystaeyen, Prime Spectra in Non-Commutative Algebra. V, 128 pages. 1975.

Vol. 443: M. Lazard, Commutative Formal Groups. II, 236 pages. 1975.

Vol. 442: C. H. Wilcox, Scattering Theory for the d'Alembert Equation in Exterior Domains. III, 184 pages. 1975.

Vol. 441: N. Jacobson, PI-Algebras. An Introduction. V, 115 pages. 1975.

Vol. 440: R. K. Getoor, Markov Processes: Ray Processes and Right Processes. V, 118 pages. 1975.

Vol. 439: K. Ueno, Classification Theory of Algebraic Varieties and Compact Complex Spaces. XIX, 278 pages. 1975.

Vol. 438: Geometric Topology. Proceedings 1974. Edited by L. C. Glaser and T. B. Rushing. X, 459 pages. 1975.

Vol. 437: D. W. Masser, Elliptic Functions and Transcendence. XIV, 143 pages. 1975.

Vol. 436: L. Auslander and R. Tolimieri, Abelian Harmonic Analysis, Theta Functions and Function Algebras on a Nilmanifold. V, 99 pages. 1975.

Vol. 435: C. F. Dunkl and D. E. Ramirez, Representations of Commutative Semitopological Semigroups. VI, 181 pages. 1975.

Vol. 434: P. Brenner, V. Thomée, and L. B. Wahlbin, Besov Spaces and Applications to Difference Methods for Initial Value Problems. II, 154 pages. 1975.

Vol. 433: W. G. Faris, Self-Adjoint Operators. VII, 115 pages. 1975.

Vol. 432: R. P. Pflug, Holomorphiegebiete, pseudokonvexe Gebiete und das Levi-Problem. VI, 210 Seiten. 1975.

Vol. 431: Séminaire Bourbaki – vol. 1973/74. Exposés 436–452. IV, 347 pages. 1975.

Vol. 430: Constructive and Computational Methods for Differential and Integral Equations. Proceedings 1974. Edited by D. L. Colton and R. P. Gilbert. VII, 476 pages. 1974.

Vol. 429: L. Cohn, Analytic Theory of the Harish-Chandra C-Function. III, 154 pages. 1974.

Vol. 428: Algebraic and Geometrical Methods in Topology, Proceedings 1973. Edited by L. F. McAuley. XI, 280 pages. 1974.

Vol. 427: H. Omori, Infinite Dimensional Lie Transformation Groups. XII, 149 pages. 1974.

Vol. 426: M. L. Silverstein, Symmetric Markov Processes. X, 287 pages. 1974.

Lecture Notes in Physics

Vol. 32: Particles, Quantum Fields and Statistical Mechanics. Proceedings 1973. Edited by M. Alexanian and A. Zepeda. V, 132 pages. 1975.

Vol. 33: Classical and Quantum Mechanical Aspects of Heavy Ion Collisions. Proceedings 1974. Edited by H. L. Harney, P. Braun-Munzinger, and C. K. Gelbke. VII, 311 pages. 1975.

Vol. 34: One-Dimensional Conductors GPS Summer School Proceedings, 1974. Edited by H. G. Schuster. VII, 371 pages. 1975.

Vol. 35: Proceedings of the Fourth International Conference on Numerical Methods in Fluid Dynamics, 1974. Edited by R. D. Richtmyer. V, 457 pages. 1975.

Vol. 36: R. Gatignol, Théorie Cinétique des Gaz à Répartition Discrète de Vitesses. II, 219 pages. 1975.

Vol. 37: Trends in Elementary Particle Theory. Proceedings 1974. Edited by H. Rollnik and K. Dietz. V, 472 pages. 1975.

Vol. 38: Dynamical Systems, Theory and Applications. Proceedings 1974. Edited by J. Moser. VI, 624 pages. 1975.

Vol. 39: International Symposium on Mathematical Problems in Theoretical Physics. Proceedings 1975. Edited by H. Araki. XII, 562 pages. 1975.

Vol. 40: Effective Interactions and Operators in Nuclei. Proceedings 1975. Edited by B. R. Barrett. XII, 339 pages. 1975.

Vol. 41: Progress in Numerical Fluid Dynamics. Proceedings 1974. Edited by H. J. Wirz. V, 471 pages. 1975.

Vol. 42: H II Regions and Related Topics. Proceedings 1975. Edited by D. Downes and T. L. Wilson. XII, 488 pages. 1975.

Vol. 43: Laser Spectroscopy. Proceedings 1975. Edited by S. Haroche, J. C. Pebay-Peyroula, T. W. Hänsch, and S. E. Harris. X, 466 pages. 1975.

Vol. 44: R. A. Breuer, Gravitational Perturbation Theory and Synchrotron Radiation. VI, 196 pages. 1975.

Vol. 45: Dynamical Concepts on Scaling Violation and the New Resonances in e^+e^- Annihilation. Edited by B. Humpert. VII, 248 pages. 1976.

Vol. 46: E. J. Flaherty, Hermitian and Kählerian Geometry in Relativity. VIII, 365 pages. 1976.

Vol. 47: Padé Approximants Method and Its Applications to Mechanics. Edited by H. Cabannes. XV, 267 pages. 1976.

Vol. 48: Interplanetary Dust and Zodiacal Light. Proceedings 1975. Edited by H. Elsässer and H. Fechtig. XII, 496 pages. 1976.

Vol. 49: W. G. Harter and C. W. Patterson, A Unitary Calculus for Electronic Orbitals. XII, 144 pages. 1976.

Vol. 50: Group Theoretical Methods in Physics. 4th International Colloquium. Nijmegen 1975. Edited by A. Janner, T. Janssen, and M. Boon. XIII. 629 pages. 1976.